MATHEMATICS FOR ENGINEERS AND SCIENTISTS

Volume 2

A SERIES OF PROGRAMMES ON MATHEMATICS
FOR
SCIENTISTS AND TECHNOLOGISTS

CONSULTANT EDITOR:

A.C. BAJPAI

PROFESSOR OF MATHEMATICAL EDUCATION AND DIRECTOR OF CAMET

Titles available in this series:

FORTRAN AND ALGOL	A.C. Bajpai, H.W. Pakes, R.J. Clarke, J.M. Doubleday and T.J. Stevens
MATHEMATICS FOR ENGINEERS AND SCIENTISTS - *Volume 1*	A.C. Bajpai, I.M. Calus, J.A. Fairley
MATHEMATICS FOR ENGINEERS AND SCIENTISTS - *Volume 2*	A.C. Bajpai, I.M. Calus, J.A. Fairley, D. Walker
NUMERICAL METHODS FOR ENGINEERS AND SCIENTISTS	A.C. Bajpai, I.M. Calus, J.A. Fairley
STATISTICAL METHODS FOR ENGINEERS AND SCIENTISTS	A.C. Bajpai, I.M. Calus, J.A. Fairley

CAMET

(CENTRE FOR ADVANCEMENT OF MATHEMATICAL EDUCATION IN TECHNOLOGY)

LOUGHBOROUGH UNIVERSITY OF TECHNOLOGY

MATHEMATICS FOR ENGINEERS AND SCIENTISTS

A STUDENTS' COURSE BOOK

Volume 2

A. C. BAJPAI

I. M. CALUS

J. A. FAIRLEY

D. WALKER

LOUGHBOROUGH UNIVERSITY OF TECHNOLOGY

JOHN WILEY & SONS

CHICHESTER · NEW YORK · BRISBANE · TORONTO

Library of Congress catalog card number
73–7377

ISBN 0 471 04374 5

Reprinted with corrections September, 1978
Reprinted with corrections May, 1981

Printed in Great Britain by J. W. Arrowsmith Ltd, Bristol

PREFACE

This is a volume of programmes on mathematics which is part of a course written for undergraduate science and engineering students in universities and colleges in all parts of the world. The emphasis is on the practical side of the subject and the more theoretical aspects have been omitted. Apart from numerical methods and statistics this book, together with Volume I, covers most of the mathematics required by such students at the undergraduate level. It comprises eight Units, numbered 6 to 13, thus following on from the five Units in Volume I. Programmes on allied topics are grouped together in each Unit. Although mathematical rigour has not always been emphasised in the programmes, they can serve as an introductory text for students taking honours mathematics courses, giving them some idea of how mathematics is used in other subjects.

The programmed method of presentation, used throughout, has many advantages. The development of the subject progresses in carefully sequenced steps, with the student proceeding at his own pace. At each stage he has an active part to play by answering a question or solving a problem, and thus learns by doing. By comparing his own answer with that given in the text, he obtains a continuous assessment of his understanding of the subject up to that point. Explanation of the material covered is given in greater detail than is usually to be found in conventional style textbooks, especially at points where, in their experience as teachers, the authors have found that students often have difficulty.

Where units have been used, the S.I. system has been adopted. Thus, for example, 5 A is the abbreviation for 5 amperes and t s for t seconds. However, the standard practice of using italic letters for quantities, e.g. *C* for capacitance, has not been followed, because italic lettering has been used for the answer frames. In this volume the notation ln is used for natural logarithms.

Where a reference is made in a programme to frames in the same programme only the frame numbers are given. When page numbers are also given these indicate that the reference is to frames in another programme.

In spite of careful checking by the authors, it is possible that the occasional error has crept through. They would appreciate receiving information about any such mistakes which might be discovered.

A debt of gratitude to the following is acknowledged with pleasure:

Loughborough University of Technology for supporting this venture.

Staff and students of the University and other institutions who have participated in the testing of these programmes.

Colleagues in other departments of the University for their advice on applications in their specialist fields.

Mr. C.J. Cook who, while he was a research student in CAMET, assisted in the production of a draft programme on Vector Analysis. Some material drawn from this has been used in Unit 11.

Mr. P.J. Wallace, former chief test plant engineer, D. Napier and Son (later Napier Aero Engines), for advice on vibration problems.

Mrs. June Russell for preparing the camera-ready copy from which the book has been printed.

John Wiley and Sons Ltd. for their help and cooperation.

Our thanks are also due to the *University of London* and the *Council of Engineering Institutions* for permission to use questions from their past examination papers. These are denoted by L.U. and C.E.I. respectively.

CONTENTS OF THIS VOLUME

Detailed contents of individual programmes will be found at the beginning of each Unit.

CONTENTS OF VOLUME I

This list is included to give you information about the previous volume.

UNIT 6

LINEAR DIFFERENTIAL EQUATIONS

- VARIATION OF PARAMETERS and SOLUTION IN SERIES

A.C. Bajpai
I.M. Calus
J.A. Fairley

Loughborough University of Technology

INSTRUCTIONS

This Unit comprises two programmes:

(a) Variation of Parameters
(b) Solution in Series

Each programme is divided up into a number of FRAMES which are to be worked *in the order given*. You will be required to participate in many of these frames and in such cases the answers are provided in ANSWER FRAMES, designated by the letter A following the frame number. Steps in the working are given where this is considered helpful. The answer frame is separated from the main frame by a line of asterisks: *********. Keep the answers covered until you have written your own response. If your answer is wrong, go back and try to see why. Do not proceed to the next frame until you have corrected any mistakes in your attempt and are satisfied that you understand the contents up to this point.

Before reading these programmes, it is necessary that you are familiar with the following

Prerequisites

For (a): First order differential equations, as covered in FRAMES 1-38 and 50-63, pages 5:1 to 5:19 and 5:26 to 5:35 of Vol I.

Second order differential equations, as covered in FRAMES 1-28, pages 5:51 to 5:64.

For (b): No special requirements for the main programme.

Partial differentiation, as covered in FRAMES 6-9, pages 1:211 to 1:214, for APPENDICES A and B.

CONTENTS

VARIATION OF PARAMETERS

FRAME 1

Introduction

In the differential equations

$$y' - 2y = e^{3x} \qquad (1.1)$$

$$y' - \frac{3}{x}y = x^2 \qquad (1.2)$$

$$y'' - 5y' + 4y = 2e^{2x} \qquad (1.3)$$

$$y'' + y = \operatorname{cosec} x \qquad (1.4)$$

$$x^2y'' + xy' - y = x^2e^x \qquad (1.5)$$

the dependent function y and its derivatives all appear to the first degree only, i.e. the d.e.'s are linear.

Equations (1.1) and (1.2) are of the form $\frac{dy}{dx} + Py = Q$, and can be solved by using an integrating factor (I.F.) $e^{\int Pdx}$, as shown in the programme "First Order Differential Equations" in this series. Equation (1.3) is of the form $ay'' + by' + cy = Q(x)$, i.e. a second order d.e. with constant coefficients, and various methods (trial solution, D-operator, Laplace Transform) for the solution of such equations have also been given in previous programmes. Although (1.4) is of the same type as (1.3), the function which appears on the R.H.S. is not one of the standard ones considered in the use of these methods. Equation (1.5) differs from (1.3) and (1.4) in that it does not have constant coefficients, and therefore cannot be solved by any of the methods so far considered.

This programme describes a technique which can be used in the solution of all the above equations, as well as in various other cases not considered here, such as problems involving systems of d.e.'s. Our aim is to give you a general appreciation of the concept involved in this technique by showing its application in a variety of situations, without necessarily implying that it should be used in preference to other methods where available.

FRAME 2

Reduced Equation

You will remember that the complementary function (C.F.) is that part of the general solution of

$$ay'' + by' + cy = Q(x) \qquad (2.1)$$

which is obtained by solving

$$ay'' + by' + cy = 0. \qquad (2.2)$$

The equation (2.2) is called the REDUCED EQUATION.

Note that:

(i) In (2.1) all terms involving y appear on the L.H.S.

(ii) In (2.2) the R.H.S. of (2.1), which consists of terms in x only, has been replaced by zero.

No doubt you can write down the reduced equations corresponding to equations (1.1) to (1.5).

****

2A

$y' - 2y = 0$ *is the reduced equation for (1.1)*

$y' - \frac{3}{x}y = 0$ " " " " " (1.2)

$y'' - 5y' + 4y = 0$ " " " " " (1.3)

$y'' + y = 0$ " " " " " (1.4)

$x^2y'' + xy' - y = 0$ " " " " " (1.5)

FRAME 3

First Order Linear Equations

Let us consider first the d.e. (1.1), i.e.

$$y' - 2y = e^{3x}$$

whose reduced equation is

$$y' - 2y = 0.$$

FRAME 3 continued

The general solution of this reduced equation, which is easily found by separating the variables, is $y = Ae^{2x}$.

To obtain the solution of the d.e. (1.1) we now set $y = ue^{2x}$, where u is a function of x, and then find what this function of x must be in order for (1.1) to be satisfied.

You will notice that in doing this, we are replacing the arbitrary constant A, i.e. the PARAMETER in the solution of the reduced equation, by a function u which varies with x. This procedure is therefore called the METHOD OF VARIATION OF PARAMETER.

FRAME 4

You may be wondering whether our assumption of a solution of the form $y = ue^{2x}$ is a justifiable one. In this example it is easy to show that it is, because you already know an alternative method for solving the equation. Multiplying throughout by the integrating factor e^{-2x}, the d.e. (1.1) can be rewritten

$$\frac{d}{dx}(e^{-2x}y) = e^{3x}.e^{-2x}$$

On integration, $e^{-2x}y = e^{x} + c$

$$\text{and so } y = e^{2x}(e^{x} + c) \quad \text{i.e. it is of the form } ue^{2x}.$$

In fact, it is easy to see that this form of solution can be assumed for any d.e. whose reduced equation is $y' - 2y = 0$.

Such a d.e. could be written $y' - 2y = Q(x)$.

Multiplying through by the I.F. e^{-2x} yields

$$\frac{d}{dx}(e^{-2x}y) = Q(x)e^{-2x}$$

$$e^{-2x}y = \int Q(x)e^{-2x}dx + c$$

$$y = e^{2x}\left[\int Q(x)e^{-2x}dx + c\right] \quad \text{which is of the form } ue^{2x}.$$

FRAME 5

Having, we hope, set your mind at rest on this point, let us return to the problem of finding the function u for which $y = ue^{2x}$ satisfies the d.e. $y' - 2y = e^{3x}$.

If $y = ue^{2x}$, $y' = 2ue^{2x} + u'e^{2x}$ so we must have

$$2ue^{2x} + u'e^{2x} - 2ue^{2x} = e^{3x} \quad (5.1)$$

$$u' = e^{x}$$

$$u = e^{x} + c$$

$$\therefore \quad y = e^{2x}(e^{x} + c)$$

FRAME 6

Let us now apply the method of variation of parameter to the d.e. (1.2)

$$\text{i.e } \frac{dy}{dx} - \frac{3}{x}y = x^2$$

whose reduced equation is $\frac{dy}{dx} - \frac{3}{x}y = 0$.

First, find the general solution of the reduced equation by separating the variables.

****

6A

$$\frac{dy}{y} = 3\frac{dx}{x}$$

$$y = cx^3$$

FRAME 7

Now set $y = ux^3$, and hence find the required solution.

****

7A

$$3ux^2 + u'x^3 - \frac{3}{x}ux^3 = x^2 \qquad (7A.1)$$

$$\text{Thus} \quad u' = \frac{1}{x} \quad \text{and hence} \quad u = \ln x + c$$

$$\therefore \; y = x^3(\ln x + c)$$

Notice that in equation (7A.1) the terms in u cancel out, as also happened in equation (5.1) in the previous example. Only u' remains, so that u can then be found by direct integration. This will, in fact, always happen when the trial solution of a linear first order d.e. is obtained by varying the parameter in the solution of the reduced equation.

FRAME 8

Solve the equation $L\frac{di}{dt} + Ri = E \sin \omega t$ by the method of variation of parameter. (This equation arises when an e.m.f. $E \sin \omega t$ is applied to a circuit consisting of an inductance L and a resistance R in series.)

****

8A

Reduced equation is $L\frac{di}{dt} + Ri = 0$

Solution of reduced equation is $i = Ae^{-Rt/L}$

Replacing A by a function of t gives

$$i = ue^{-Rt/L}$$

$$L\left(-\frac{R}{L}ue^{-Rt/L} + \frac{du}{dt}e^{-Rt/L}\right) + Rue^{-Rt/L} = E \sin \omega t$$

The terms in u cancel out, as they should do.

$$\frac{du}{dt} = \frac{E}{L}e^{Rt/L} \sin \omega t$$

$$u = \frac{E}{L}\frac{e^{Rt/L}}{\frac{R^2}{L^2} + \omega^2}\left(\frac{R}{L}\sin \omega t - \omega \cos \omega t\right) + c$$

$$= \frac{Ee^{Rt/L}}{R^2 + \omega^2L^2}(R \sin \omega t - \omega L \cos \omega t) + c$$

$$i = \frac{E}{R^2 + \omega^2L^2}(R \sin \omega t - \omega L \cos \omega t) + ce^{-Rt/L}$$

FRAME 9

Second Order Linear Equations

We shall now see how a similar technique can be applied to a second order d.e. such as (1.3), i.e.

$$y'' - 5y' + 4y = 2e^{2x}.$$

The reduced equation is $y'' - 5y' + 4y = 0$ and the solution of this (i.e. the C.F.) is $y = Ae^{x} + Be^{4x}$.

This has two parameters A and B. Replacing each of them by variables, u and v, say, we have

$$y = ue^{x} + ve^{4x}.$$

As with the first order equation considered in FRAMES 3 - 5, you already know another way of obtaining the solution. The C.F. has already been stated and the particular integral (P.I.) is easily found (by either trial solution or D-operator) to be $-e^{2x}$. Hence the general solution is

$$y = Ae^{x} + Be^{4x} - e^{2x}.$$

This can be written in a variety of ways, of which a few are

$$y = (A + e^{x})e^{x} + (B - 2e^{-2x})e^{4x}$$

$$y = Ae^{x} + (B - e^{-2x})e^{4x}$$

$$y = (A - e^{x})e^{x} + Be^{4x}$$

Each of these is of the form $y = ue^{x} + ve^{4x}$.

Again you can see that the assumed form of the solution is reasonable. You will, incidentally, notice that there are various possibilities for the functions u and v. Just three have been listed here; no doubt you can invent some more yourself. To reach the final solution it is sufficient to obtain just one of the various combinations of u and v.

FRAME 10

Let us now return to the solution of the d.e. (1.3) by the method of variation of parameters. (Parameter_s_ this time, as we have two of them.)

Having put $y = ue^x + ve^{4x}$
we then have $y' = ue^x + 4ve^{4x} + u'e^x + v'e^{4x}$.

It has already been shown that there are endless possibilities for u and v, any of which will give the required solution, so we choose those which can be obtained most easily.

As there are two functions u and v to be determined, two equations will be required. Substituting $ue^x + ve^{4x}$ for y in the original d.e. (1.3) will give one of these equations, but only one. The second equation is chosen to be

$$u'e^x + v'e^{4x} = 0 \tag{10.1}$$

because this leads to the simplest possible working, as will be seen when this example is being completed

Write down the expression for y", when the above condition is imposed.

****

10A

$$y'' = ue^x + 16ve^{4x} + u'e^x + 4v'e^{4x}$$

FRAME 11

We now have

$$y = ue^x + ve^{4x}$$

$$y' = ue^x + 4ve^{4x}$$

$$y'' = ue^x + 16ve^{4x} + u'e^x + 4v'e^{4x}$$

Substitution in $y'' - 5y' + 4y = 2e^{2x}$ leads to

$$u'e^x + 4v'e^{4x} = 2e^{2x} \tag{11.1}$$

You will notice that the u and v terms have cancelled out - this is because e^x and e^{4x} are solutions of the reduced equation.

FRAME 11 continued

We now have two equations (10.1) and (11.1) which have to be solved to obtain u and v. They involve u' and v' only, and can therefore be solved by simple algebra for u' and v', from which u and v can be found by integration. This straightforward process would not have been possible if any other equation had been chosen for (10.1).

Eliminating u' by subtraction gives

$$3v'e^{4x} = 2e^{2x}$$

$$v' = \frac{2}{3}e^{-2x}$$

$$v = c_1 - \frac{1}{3}e^{-2x}$$

Now find u, and hence write down the required solution.

****

11A

$$u'e^{x} = -\frac{2}{3}e^{-2x}e^{4x}$$

$$u' = -\frac{2}{3}e^{x}$$

$$u = c_2 - \frac{2}{3}e^{x}$$

$$y = (c_2 - \frac{2}{3}e^{x})e^{x} + (c_1 - \frac{1}{3}e^{-2x})e^{4x}$$

$$= c_2e^{x} + c_1e^{4x} - e^{2x}$$

You will notice that the terms involving the arbitrary constants constitute the C.F. and the other part is the P.I.

FRAME 12

As has already been pointed out, the d.e. which we have just solved using variation of parameters could have equally well been solved by any of the methods described in previous programmes. This is because the exponential function on the R.H.S. is one which was covered by each of these methods.

FRAME 12 continued

However, turning now to equation (1.4), i.e.

$$\frac{d^2y}{dx^2} + y = \text{cosec } x$$

you will find yourself at a loss if, for instance, you try to find the P.I. by trial solution or D-operator. This is where the method of variation of parameters comes into its own.

The reduced equation is $\frac{d^2y}{dx^2} + y = 0$ and the solution of this is $y = A\cos x + B\sin x$.

Replacing the parameters A and B by variables u and v, we have

$$y = u\cos x + v\sin x$$

$$y' = -u\sin x + v\cos x + \boxed{u'\cos x + v'\sin x}$$

As before, the expression involving u' and v' (shown in the dotted rectangle) is set equal to zero, i.e.

$$u'\cos x + v'\sin x = 0 \qquad (12.1)$$

Then $\quad y'' = -u\cos x - v\sin x - u'\sin x + v'\cos x$

Substitution in $y'' + y = \text{cosec } x$ leads to

$$-u'\sin x + v'\cos x = \text{cosec } x \qquad (12.2)$$

Notice that, as in the example in FRAMES 10 and 11, we have two equations in u' and v', without any u or v terms.

Now solve (12.1) and (12.2) for u' and v', find u and v, and hence write down the required solution.

****

12A

$$u' = -1$$

$$u = -x + c_1$$

$$v' = \frac{\cos x}{\sin x}$$

$$v = \ln \sin x + c_2$$

12A continued

Required solution is $y = (c_1 - x)\cos x + (c_2 + \ln \sin x)\sin x$

$$= c_1 \cos x + c_2 \sin x - x \cos x + \sin x \ln \sin x$$

FRAME 13

The d.e. $y'' - y = \dfrac{2}{1 + e^x}$, like the previous example, has a function on the R.H.S. which would present difficulties if you tried to use, for instance, trial solution or D-operator methods. You should now attempt its solution by the method of variation of parameters.

****

13A

Solution of reduced equation is $y = Ae^x + Be^{-x}$

Set $y = ue^x + ve^{-x}$

$$y' = ue^x - ve^{-x} + \boxed{u'e^x + v'e^{-x}}$$

Put $u'e^x + v'e^{-x} = 0$ (13A.1)

Then $y'' = ue^x + ve^{-x} + u'e^x - v'e^{-x}$

Substitution in the d.e. leads to

$$u'e^x - v'e^{-x} = \frac{2}{1 + e^x} \qquad (13A.2)$$

From (13A.1) and (13A.2), $2v'e^{-x} = -\dfrac{2}{1 + e^x}$

$$v' = -\frac{e^x}{1 + e^x}$$

$$v = -\ln(1 + e^x) + c_1$$

$$u' = \frac{1}{e^x(1 + e^x)}$$

$$= \frac{1}{e^x} - \frac{1}{e^x + 1} \quad \text{as} \quad \frac{1}{z(z + 1)} = \frac{1}{z} - \frac{1}{z + 1}$$

$$= e^{-x} - \frac{e^{-x}}{1 + e^{-x}}$$

13A continued

$$u = -e^{-x} + \ln(1 + e^{-x}) + c_2$$

Required solution is $y = e^{x}\left[c_2 - e^{-x} + \ln(1 + e^{-x})\right] + e^{-x}\left[c_1 - \ln(1 + e^{x})\right]$

FRAME 14

In the three second order differential equations which we have just solved the coefficients of y, $\frac{dy}{dx}$ and $\frac{d^2y}{dx^2}$ were constant, and we were therefore able to make use of the forms of the C.F. already known for such types of equation. These forms will not apply when the coefficients of y and its derivatives are not constant, as in equation (1.5). In such cases, therefore, the first problem is to find the solution of the reduced equation, and this may be difficult. Certain rules of guidance on suitable trial solutions can be given. However, this problem will not be dealt with here as the main concern of this programme is to show the application of the technique of variation of parameters. We shall therefore start from the point where the C.F. (or part of it) has been found.

FRAME 15

In the case of equation (1.5), i.e.

$$x^2y'' + xy' - y = x^2e^x$$

the solutions $y = Ax$ and $y = \frac{B}{x}$ of the reduced equation can be found, giving the C.F. as $Ax + \frac{B}{x}$.

Using the method of variation of parameters to solve (1.5) we now put $y = ux + \frac{v}{x}$ and proceed in the usual way.

By this time, you should have no difficulty in writing down y', and then the first equation in u' and v'.

****

15A

$$y' = u - \frac{v}{x^2} + \boxed{u'x + \frac{v'}{x}}$$

$$u'x + \frac{v'}{x} = 0 \qquad (15A.1)$$

FRAME 16

Now go ahead and write down y" and the second equation in u' and v'. Solve for u' and v', and hence find u and v so that you can write down the general solution of the d.e. (1.5).

****

16A

$$y'' = \frac{2v}{x^3} + u' - \frac{v'}{x^2}$$

Substitution in (1.5) leads to $x^2(u' - \frac{v'}{x^2}) = x^2 e^x$

$$u' - \frac{v'}{x^2} = e^x \qquad (16A.1)$$

Equations (15A.1) and (16A.1) give $u' = \frac{1}{2}e^x$

$$v' = -\tfrac{1}{2}x^2 e^x$$

Hence $u = \frac{1}{2}e^x + c_1$ *and* $v = -\frac{1}{2}x^2 e^x + \int xe^x dx$

$$= -\tfrac{1}{2}x^2 e^x + xe^x - e^x + c_2$$

Required solution is $y = (\frac{1}{2}e^x + c_1)x + (-\frac{1}{2}x^2 e^x + xe^x - e^x + c_2)\frac{1}{x}$

$$= c_1 x + \frac{c_2}{x} + (1 - \frac{1}{x})e^x$$

FRAME 17

Reference has already been made to the difficulties which may occur in trying to find the C.F. of a second order d.e. with variable coefficients. However even if only part of it can be spotted, it is still possible to use the method of variation of parameters to find the general solution of the d.e.

For instance, in the case of the equation just solved, i.e.

$$x^2 y'' + xy' - y = x^2 e^x$$

it is easily verified that $y = x$, and hence $y = Ax$, is a solution of the reduced equation. Suppose that the solution $y = \frac{1}{x}$, and therefore $y = \frac{B}{x}$, had escaped notice.

FRAME 17 continued

Proceeding on this basis, we vary the one parameter in the known part of the C.F. and put $y = ux$.

This gives $y' = u + u'x$

and $y'' = 2u' + u''x$.

What equation is obtained when these expressions are substituted in the d.e.?

****

17A

$$u''x + 3u' = e^x \qquad (17A.1)$$

FRAME 18

The substitution $p = u'$ in (17A.1) leads to a first order d.e. in p.

$$xp' + 3p = e^x$$

$$\text{i.e.} \quad p' + \frac{3}{x}p = \frac{1}{x}e^x$$

Multiply throughout by the appropriate integrating factor and hence solve for p.

****

18A

$$I.F. = e^{3\ln x} = x^3$$

$$\frac{d}{dx}(x^3p) = x^2e^x$$

$$x^3p = x^2e^x - \int 2xe^x dx$$

$$= x^2e^x - 2xe^x + 2e^x + c_1$$

$$p = \frac{1}{x}e^x - \frac{2}{x^2}e^x + \frac{2}{x^3}e^x + \frac{c_1}{x^3}$$

FRAME 19

We now have $u' = \frac{1}{x}e^x - \frac{2}{x^2}e^x + \frac{2}{x^3}e^x + \frac{c_1}{x^3}$

$$= \left(\frac{1}{x} - \frac{1}{x^2}\right)e^x + \left(-\frac{1}{x^2} + \frac{2}{x^3}\right)e^x + \frac{c_1}{x^3}$$

Noting that $\left(\frac{1}{x} - \frac{1}{x^2}\right)e^x + \left(-\frac{1}{x^2} + \frac{2}{x^3}\right)e^x = \frac{d}{dx}\left\{\left(\frac{1}{x} - \frac{1}{x^2}\right)e^x\right\}$, we have

$$u = \left(\frac{1}{x} - \frac{1}{x^2}\right)e^x - \frac{c_1}{2x^2} + c_2$$

$$y = ux = \left(1 - \frac{1}{x}\right)e^x - \frac{c_1}{2x} + c_2x$$

$$= c_2x + \frac{C}{x} + \left(1 - \frac{1}{x}\right)e^x \quad \text{where } C = -\frac{c_1}{2}$$

It is worth noting that, in general:

The d.e. in u (17A.1 in this case) is of the second order and does not contain u. It can therefore be reduced to a first order d.e. by the substitution $p = u'$.

FRAME 20

Miscellaneous Examples

In this frame, a collection of miscellaneous examples is given for you to try. Answers are supplied in FRAME 21 and hints have been provided in some cases.

Use the method of variation of parameters to solve the following equations.

1. $y' + y\cos x = \frac{1}{2}\sin 2x$

2. $y'' + n^2y = \sec nx$

3. $y'' - 2y' + y = e^x$

4. Given that $y = x$ and $y = e^x$ are solutions of the equation
$$(x - 1)y'' - xy' + y = 0$$
obtain the solution of
$$(x - 1)y'' - xy' + y = (x - 1)^2$$

FRAME 20 continued

5. It is easy to see that $y = x$ is a solution of

$$x^2y'' - 2xy' + 2y = 0$$

Use this result to solve the equation

$$x^2y'' - 2xy' + 2y = x^3$$

6. Standard trial solution techniques yield the solution $y = e^x$ for the equation $xy'' - (x + 1)y' + y = 0$. Use this to solve the equation $xy'' - (x + 1)y' + y = 2x^2e^x$.

FRAME 21

Answers to Miscellaneous Examples

1. Solution of reduced equation: $y = Ae^{-\sin x}$.

   Put $y = ue^{-\sin x}$

   $u = \int \sin x(e^{\sin x} \cos x)dx$, which can be integrated by parts.

   $y = c_1e^{-\sin x} + \sin x - 1$

2. C.F. is $A \cos nx + B \sin nx$.

   Put $y = u \cos nx + v \sin nx$.

   $$\left.\begin{aligned} u' \cos nx + v' \sin nx &= 0 \\ -u'n \sin nx + v'n \cos nx &= \sec nx \end{aligned}\right\} \text{give } u' = -\frac{1}{n}\tan nx \text{ and } v' = \frac{1}{n}$$

   $$y = (c_1 + \frac{1}{n^2}\log_e \cos nx)\cos nx + (c_2 + \frac{x}{n})\sin nx$$

3. C.F. is $Ae^x + Bxe^x$.

   Put $y = ue^x + vxe^x$.

   $$\left.\begin{aligned} u' + v'x &= 0 \\ u' + v'(x + 1) &= 1 \end{aligned}\right\} \text{give } u' = -x \text{ and } v' = 1$$

   $$y = (c_1 + c_2x + \frac{x^2}{2})e^x$$

   NOTE: This would be a "case of failure" if trial solution or D-operator methods were used.

FRAME 21 continued

4. Put $y = ux + ve^x$

$$\left.\begin{aligned} u'x + v'e^x &= 0 \\ u' + v'e^x &= x - 1 \end{aligned}\right\} \text{ give } u' = -1 \text{ and } v' = xe^{-x}$$

$$y = (c_1 - x)x + (c_2 - xe^{-x} - e^{-x})e^x$$

$$= c_1x + c_2e^x - x^2 - x - 1$$

5. Put $y = ux$

$u'' = 1$ (The d.e. in u is especially simple in this example because there is no term in u'.)

$$y = \left(\frac{x^2}{2} + c_1x + c_2\right)x$$

6. Put $y = ue^x$

$$p' + \left(1 - \frac{1}{x}\right)p = 2x \quad \text{where } p = u'$$

$$p = 2x + c_1xe^{-x}$$

$$y = c_2e^x - c_1(x + 1) + x^2e^x$$

SOLUTION IN SERIES

FRAME 1

Introduction

Previous programmes in this series have described methods for solving linear d.e.'s which are either (i) first order, or (ii) higher order with constant coefficients. Of these methods the only one that is applicable to linear d.e.'s of order greater than one, when the coefficients are not constant, is the method of variation of parameters, and even this is limited to cases where at least one solution of the reduced equation is known. A method which is more generally effective in the solution of linear d.e.'s with variable coefficients is based on the use of power series.

The idea of a function being expressed as a power series is no doubt familiar to you, and in fact it is sometimes this form that a digital computer uses when calculating the value of a function. Many of the solutions of d.e.'s which you have found in functional form could have been expressed in a series form. Thus, for example, the equation $\frac{d^2y}{dx^2} + n^2y = 0$ has the general solution

$$y = A \cos nx + B \sin nx$$

which could also be written as

$$y = A\left(1 - \frac{n^2x^2}{2!} + \frac{n^4x^4}{4!} - \ldots\right) + B\left(nx - \frac{n^3x^3}{3!} + \frac{n^5x^5}{5!} - \ldots\right)$$

Some differential equations which occur in science and engineering cannot be solved in terms of the standard functions known to us but yet one can find solutions for them in the form of an infinite series. In fact one can take the point of view that a differential equation is a way of defining a new function, values of which can be tabulated from the series for it. One such equation is Bessel's equation, which occurs in the theory of vibrations, heat flow and the propagation of electricity in conductors. The equation

$$x^2 \frac{d^2y}{dx^2} + x\frac{dy}{dx} + (x^2 - n^2)y = 0$$

is called Bessel's equation of order n and its solutions are called Bessel functions of order n.

FRAME 2

Method of Frobenius

Sometimes a series solution to a d.e. can be found by assuming a Maclaurin expansion, sometimes a trial solution

$$y = a_o + a_1x + a_2x^2 + \ldots\ldots + a_rx^r + \ldots\ldots$$

is successful. A more general method (the METHOD OF FROBENIUS) is to take as a trial solution

$$y = x^c(a_o + a_1x + a_2x^2 + \ldots + a_rx^r + \ldots) \qquad (2.1)$$

where a_o is the first non-zero coefficient in the expansion.

The existence of such a solution is dependent on certain properties of the coefficients in the d.e. Also, just as, you will recall, the expansion of a function in a Maclaurin series is only valid if the series converges, so also is the series solution of a d.e. dependent on the convergence of the series, if infinite, and it may therefore only be valid over a limited range of values of x. Furthermore, to be of any practical use the series should converge reasonably quickly. Problems of determining whether a series solution exists and, if it does, for what range of values it is convergent, will not be dealt with in this programme. The main purpose is to show how, in equations which have solutions of the form (2.1), these solutions can be found.

FRAME 3

The value of c and the coefficients a_r have to be found. This is done by substituting the trial solution in the given d.e. and equating the coefficients of like powers of x in the resulting identity. A few examples will show you how the method works.

Example 1 Solve the equation

$$4xy'' + 2y' + y = 0 \qquad (3.1)$$

FRAME 3 continued

The trial solution (2.1) gives

$$y = a_o x^c + a_1 x^{c+1} + a_2 x^{c+2} + \ldots$$

$$y' = a_o c x^{c-1} + a_1(c+1)x^c + a_2(c+2)x^{c+1} + \ldots\ldots\ldots\ldots$$

$$y'' = a_o c(c-1)x^{c-2} + a_1(c+1)cx^{c-1} + a_2(c+2)(c+1)x^c + \ldots\ldots\ldots\ldots\ldots\ldots$$

(These equations are basic to the method of Frobenius, and will be required in every example. You may find it useful to copy them out, to keep by you for easy reference.)

When these expressions are substituted in (3.1), an identity with zero on the R.H.S. is obtained. Thus the coefficient of each power of x on the L.H.S. must be zero. We shall start with the coefficient of the lowest power of x, so the first question to be answered is "What is the lowest power of x obtained when the trial solution is substituted in $4xy'' + 2y' + y$?" Write down your answer to this question. Don't bother about the coefficient itself at this stage.

****

3A

x^{c-1}, *coming from the* $4xy''$ *and* $2y'$ *terms.*

FRAME 4

The next step is to write down the coefficient of x^{c-1} and equate it to zero. The equation thus obtained is

$$4a_o c(c - 1) + 2a_o c = 0$$

$$a_o c(2c - 1) = 0$$

$a_o \neq 0$ (it was defined in FRAME 2 as the first non-zero coefficient)

$$\therefore c(2c - 1) = 0$$

This equation in c is called the INDICIAL EQUATION.
Here there are two possibilities for c, $c = 0$ and $c = \frac{1}{2}$.

FRAME 5

Continuing in this way, equating the coefficients of x^c, x^{c+1}, etc. to zero, the a's can be found.

x^c gives

$$4a_1(c + 1)c + 2a_1(c + 1) + a_o = 0$$

$$2a_1(c + 1)(2c + 1) + a_o = 0$$

$$a_1 = -\frac{a_o}{2(c + 1)(2c + 1)} \qquad (5.1)$$

We now have a_1 in terms of a_o. If you equate the coefficient of x^{c+1} to zero, you should obtain a_2 in terms of a_1.

****

5A

$$4a_2(c + 2)(c + 1) + 2a_2(c + 2) + a_1 = 0 \qquad (5A.1)$$

$$a_2 = -\frac{a_1}{2(c + 2)(2c + 3)} \qquad (5A.2)$$

FRAME 6

You could now continue this process, equating the coefficients of higher powers of x to zero, but by this time a pattern is emerging. If you can see it, you will be able to write down a_3 in terms of a_2, without reference to the series for y, y' and y" in FRAME 3. If you can't, proceed as before, this time equating the coefficient of x^{c+2} to zero and then see if you can write down a_4 in terms of a_3.

****

6A

$$a_3 = -\frac{a_2}{2(c + 3)(2c + 5)} \qquad (6A.1)$$

If you were not able to write this down straight away, you should have obtained the equation

$$4a_3(c + 3)(c + 2) + 2a_3(c + 3) + a_2 = 0$$

6A continued

Comparing this with 5A.1, you will see that (c + 3) replaces (c + 2) and (c + 2) replaces (c + 1), i.e. c has everywhere been replaced by (c + 1). When this is done in (5A.2), the relationship (6A.1) is obtained, from which in turn

$$a_4 = -\frac{a_3}{2(c + 4)(2c + 7)}$$

can be derived in a similar way.

FRAME 7

Having established a relationship between successive coefficients we now find what series will result from each of the values of c obtained from the indicial equation in FRAME 4.

Taking first $c = 0$, and substituting in (5.1), (5A.1) and (6A.1), we have

$$a_1 = -\frac{a_o}{2}$$

$$a_2 = -\frac{a_1}{4.3} = \frac{a_o}{4!}$$

$$a_3 = -\frac{a_2}{6.5} = -\frac{a_o}{6!}$$

You will notice that a_1, a_2, a_3 etc. can all be expressed in terms of a_o, which itself remains undetermined. Thus $c = 0$ yields the solution

$$a_o x^o(1 - \frac{x}{2!} + \frac{x^2}{4!} - \frac{x^3}{6!} + ...)$$

But $x^o = 1$, and as a_o can have any value, we can represent it by an arbitrary constant A. The solution becomes

$$A(1 - \frac{x}{2!} + \frac{x^2}{4!} - \frac{x^3}{6!} + ...)$$

Now, taking $c = \frac{1}{2}$, write down a_1, a_2 and a_3 in terms of a_o and hence obtain the solution

$$Bx^{\frac{1}{2}}(1 - \frac{x}{3!} + \frac{x^2}{5!} - \frac{x^3}{7!} + ...)$$

where B is another arbitrary constant.

****

7A

$$a_1 = -\frac{a_o}{3.2} = -\frac{a_o}{3!}$$

$$a_2 = -\frac{a_1}{5.4} = \frac{a_o}{5!}$$

$$a_3 = -\frac{a_2}{7.6} = -\frac{a_o}{7!}$$

Again, all coefficients are multiples of a_o, which is itself undetermined and can therefore be represented by an arbitrary constant B.

FRAME 8

We now have the two independent solutions

$$A\left(1 - \frac{x}{2!} + \frac{x^2}{4!} - \frac{x^3}{6!} + ...\right)$$

and $$Bx^{\frac{1}{2}}\left(1 - \frac{x}{3!} + \frac{x^2}{5!} - \frac{x^3}{7!} + ...\right)$$

and it can be shown that their sum is also a solution. (Your study of the equation $a\frac{d^2y}{dx^2} + b\frac{dy}{dx} + cy = 0$ will have familiarised you with this property of linear d.e.'s.) Furthermore, the sum of these two solutions will have the two arbitrary constants necessary to the general solution of a second order d.e.

$$\therefore y = A\left(1 - \frac{x}{2!} + \frac{x^2}{4!} - \frac{x^3}{6!} + \ldots\right) + Bx^{\frac{1}{2}}\left(1 - \frac{x}{3!} + \frac{x^2}{5!} - \frac{x^3}{7!} + \ldots\right)$$

is the general solution of the d.e. (3.1).

FRAME 9

Recurrence Relation

In Example 1 we equated to zero the coefficients of x^{c-1}, x^c, x^{c+1} etc. in the expression obtained when the trial solution was substituted in the L.H.S. of the d.e. We found that equating the coefficient of the lowest power of x (x^{c-1}) to zero gave the indicial equation from which values of the index c

FRAME 9 continued

could be found. From x^c onwards, equating the coefficients to zero gave relationships between successive coefficients i.e. a_1 and a_o, a_2 and a_1, and so on. These were:

$$a_1 = -\frac{a_o}{2(c+1)(2c+1)}$$

$$a_2 = -\frac{a_1}{2(c+2)(2c+3)}$$

$$a_3 = -\frac{a_2}{2(c+3)(2c+5)}$$

$$a_4 = -\frac{a_3}{2(c+4)(2c+7)}$$

The relationships are following a set pattern and it should therefore be possible to write down a general formula connecting any two successive coefficients.

Try to write down a relationship giving a_{r+1} in terms of a_r. Check that it works for the particular coefficients listed above and then turn to the answer frame to be quite sure that your answer is correct.

****

9A

$$a_{r+1} = -\frac{a_r}{2(c+r+1)(2c+2r+1)}$$

FRAME 10

This equation connecting a_{r+1} and a_r is an example of a RECURRENCE RELATION (sometimes called RECURSION FORMULA). In this particular case, it holds for all $r \geq 0$.

In fact, the formula for a_{r+1} could have been written down in the first place, without writing down the formulae for a_1, a_2 etc.

Equating the coefficient of x^c to zero gave a_1 in terms of a_o.

" " " " x^{c+1} " " " a_2 " " " a_1.

" " " " x^{c+2} " " " a_3 " " " a_2.

FRAME 10 continued

If instead we had taken the coefficient of the general term x^{c+r} and equated it to zero, this would have led to a relation between which two coefficients?

****

10A

a_{r+1} in terms of a_r.

FRAME 11

To write down the coefficient of x^{c+r} when the trial solution is substituted in $4xy'' + 2y' + y$, we shall require:

the coefficient of x^{c+r-1} in y''

" " " x^{c+r} " y'

" " " x^{c+r} " y.

These terms are shown below.

$$y = \quad a_o x^c + \ldots\ldots\ldots\ldots + \quad a_r x^{c+r} + a_{r+1} x^{c+r+1} + .$$

$$y' = \quad a_o c x^{c-1} + \ldots\ldots\ldots\ldots + a_{r+1}(c+r+1)x^{c+r} + \ldots\ldots$$

$$y'' = a_o c(c-1)x^{c-2} + \ldots\ldots + a_{r+1}(c+r+1)(c+r)x^{c+r-1} + \ldots\ldots\ldots\ldots$$

With the necessary information laid out in this way you should find it a simple matter to write down the coefficient of x^{c+r} in $4xy'' + 2y' + y$.

****

11A

$$4a_{r+1}(c + r + 1)(c + r) + 2a_{r+1}(c + r + 1) + a_r$$

simplifying to $2a_{r+1}(c + r + 1)(2c + 2r + 1) + a_r$

FRAME 12

Equating this coefficient to zero gives the recurrence relation

$$a_{r+1} = -\frac{a_r}{2(c + r + 1)(2c + 2r + 1)}$$

as previously arrived at in answer frame 9A. Although obtaining a recurrence relation from the coefficient of x^{c+r} is really the more efficient way of obtaining the a's for the series, you may feel it is easier to write down a_1, a_2, etc. as was done in Example 1. The important thing is that you understand what you are doing.

FRAME 13

The example which now follows is similar to Example 1 and should not present any difficulties when you are called upon to participate in solving it.

Example 2 Solve the equation

$$2x^2y'' - xy' + (1 - 2x)y = 0 \qquad (13.1)$$

You first have to decide what will be the lowest power of x when the trial solution $y = x^c(a_o + a_1x + a_2x^2 + ..)$ is substituted in the L.H.S. of (13.1). Having done that, equate its coefficient to zero, giving the indicial equation, and solve for c.

****

13A

Lowest power: x^c

$$2a_oc(c - 1) - a_oc + a_o = 0$$

$$(c - 1)(2c - 1) = 0$$

$$c = 1 \quad \text{or} \quad c = \tfrac{1}{2}$$

FRAME 14

Now find the series solution corresponding to each value of c (it will be sufficient to write down the first four terms) and hence the general solution of the d.e. (13.1).

****

14A

The coefficients of x^{c+1} *onwards all follow the same pattern. You could have equated the coefficient of* x^{c+1} *to zero, giving*

$$2a_1(c+1)c - a_1(c+1) + a_1 - 2a_0 = 0$$

$$\text{whence} \quad a_1 = \frac{2a_0}{c(2c+1)}$$

and it could then be seen that $a_2 = \dfrac{2a_1}{(c+1)(2c+3)}$, $a_3 = \dfrac{2a_2}{(c+2)(2c+5)}$

etc. Alternatively, you could have equated the coefficient of x^{c+r} *to zero, giving* $2a_r(c+r)(c+r-1) - a_r(c+r) + a_r - 2a_{r-1} = 0$

$$a_r = \frac{2a_{r-1}}{(c+r-1)(2c+2r-1)} \qquad \text{for} \qquad r \geq 1$$

This time the coefficient of x^{c+r} *gives* a_r *in terms of* a_{r-1}, *instead of* a_{r+1} *in terms of* a_r, *but you will realise that this does not matter - the essential feature of the recurrence relation in these examples is that it gives one 'a' in terms of a previous one.*

$c = 1$ *gives* $a_1 = \dfrac{2a_0}{1.3}$ $\quad a_2 = \dfrac{2a_1}{2.5}$ $\quad a_3 = \dfrac{2a_2}{3.7}$

Denoting a_0 *by an arbitrary constant A, this gives the solution*

$$Ax\left(1 + \frac{2}{1!3}x + \frac{2^2}{2!3.5}x^2 + \frac{2^3}{3!3.5.7}x^3 + \ldots\right) \qquad (14A.1)$$

$c = \frac{1}{2}$ *gives* $a_1 = \dfrac{2a_0}{1.1}$ $\quad a_2 = \dfrac{2a_1}{3.2}$ $\quad a_3 = \dfrac{2a_2}{5.3}$

Denoting a_0 *by an arbitrary constant B, this solution is*

$$Bx^{\frac{1}{2}}\left(1 + 2x + \frac{2^2}{2!3}x^2 + \frac{2^3}{3!3.5}x^3 + \ldots\ldots\right) \qquad (14A.2)$$

The general solution is the sum of (14A.1) and (14A.2).

FRAME 15

In both Examples 1 and 2, the recurrence relation was between a coefficient and the one before it, i.e. between a_{r+1} and a_r or a_r and a_{r-1}. This is not necessarily always the case, and in the next example you will find that alternate coefficients are related, i.e. a_r and a_{r-2}.

Example 3 Solve the equation

$$x^2y'' + xy' + (x^2 - \frac{1}{9})y = 0$$

Taking the usual trial solution $y = x^c(a_o + a_1x + a_2x^2 + \ldots)$ obtain the indicial equation and solve for c.

(If you have found difficulty in writing down the coefficients of the various powers of x using only the series for y, y' and y", as displayed in FRAMES 3 and 11, you may find the lay-out shown on Page 6:30 helpful.)

****

15A

$$a_oc(c - 1) + a_oc - \frac{1}{9}a_o = 0$$

$$c^2 - \frac{1}{9} = 0$$

$$c = \pm\frac{1}{3}$$

FRAME 16

Your experience of the series solution method up to this point will have shown you that, when the trial solution is substituted in the L.H.S. of the d.e., the only 'a' involved in the coefficient of the lowest power of x is a_o. In Examples 1 and 2 the coefficients of all powers of x after that satisfied the recurrence relation involving two 'a's.

In the present example, the indicial equation was obtained from the coefficient of x^c. Now write down the coefficient of the next higher power of x, i.e. x^{c+1}, and see how the situation this time differs from that in the previous examples.

****

Suggested lay-out (see FRAME 15) for obtaining the coefficients of powers of x required in the solution of

the equation $x^2y'' + xy' + (x^2 - \frac{1}{9})y = 0$

The following are written down using the expressions for y, y' and y" given in FRAME 3:

$$x^2y'' = a_oc(c-1)x^c + a_1(c+1)cx^{c+1} + a_2(c+2)(c+1)x^{c+2} + \ldots\ldots + a_r(c+r)(c+r-1)x^{c+r} + \ldots\ldots$$

$$xy' = a_ocx^c + a_1(c+1)x^{c+1} + a_2(c+2)x^{c+2} + \ldots\ldots + a_r(c+r)x^{c+r} + \ldots\ldots$$

$$x^2y = a_ox^{c+2} + \ldots\ldots + a_{r-2}x^{c+r} + \ldots\ldots$$

$$-\frac{1}{9}y = -\frac{1}{9}a_ox^c - \frac{1}{9}a_1x^{c+1} - \frac{1}{9}a_2x^{c+2} - \ldots\ldots - \frac{1}{9}a_rx^{c+r} - \ldots\ldots$$

Adding vertically:

Coefficient of $x^c = a_oc(c-1) + a_oc - \frac{1}{9}a_o$

" " $x^{c+1} = a_1(c+1)c + a_1(c+1) - \frac{1}{9}a_1$

" " $x^{c+2} = a_2(c+2)(c+1) + a_2(c+2) + a_o - \frac{1}{9}a_2$

and so on.

16A

$$a_1(c + 1)c \; + \; a_1(c + 1) \; - \; \frac{1}{9}a_1$$

This involves only a_1, *not* a_1 *and* a_o.

FRAME 17

Equating the coefficient of x^{c+1} to zero gives

$$a_1\{(c + 1)^2 \; - \; \frac{1}{9}\} \; = \; 0 \qquad (17.1)$$

Neither of the values of c obtained from the indicial equation will make the expression in the curly brackets zero, so we must have $a_1 = 0$.

If you now write down the coefficient of x^{c+2} and equate it to zero you will find you have a relation between a_2 and a_o. Try to see what difference between the coefficient of x^{c+2} and those of x^c and x^{c+1} has caused two 'a's to be involved instead of only one.

****

17A

$$a_2(c + 2)(c + 1) + a_2(c + 2) + a_o - \frac{1}{9}a_2 \; = \; 0 \qquad (17A.1)$$

All terms in $x^2y'' + xy' + (x^2 - \frac{1}{9})y \; = \; 0$ *are now contributing. The* x^2y *term did not make a contribution before.*

(17A.1) may be written

$$a_2\{(c + 2)^2 - \frac{1}{9}\} + a_o \; = \; 0$$

giving $$a_2 \; = \; - \frac{9a_o}{(3c + 5)(3c + 7)}$$

FRAME 18

Equating the coefficient of x^{c+r} to zero will give a_r in terms of a_{r-2}. Find this recurrence relation and say for what values of r it is valid.

****

18A

$$a_r\{(c+r)^2 - \tfrac{1}{9}\} + a_{r-2} = 0$$

$$a_r = -\frac{9a_{r-2}}{(3c+3r-1)(3c+3r+1)}, \quad \textit{Valid for } r \geq 2$$

FRAME 19

What can you now say about a_3, a_5, a_7 etc. when c is either $+\frac{1}{3}$ or $-\frac{1}{3}$?

****

19A

They are all zero, because $a_1 = 0$ *and* a_3, a_5, a_7 *etc. are all multiples of* a_1.

FRAME 20

Now put $c = \frac{1}{3}$ in the recurrence relation given in 18A. Use the result to write down a_2, a_4 and a_6, and hence the first four terms in the series solution corresponding to this value of c.

****

20A

$$a_r = -\frac{3a_{r-2}}{r(3r+2)}$$

$$a_2 = -\frac{3}{2.8}a_o$$

$$a_4 = -\frac{3}{4.14}a_2 = \frac{3^2}{2.4.8.14}a_o$$

$$a_6 = -\frac{3}{6.20}a_4 = -\frac{3^3}{2.4.6.8.14.20}a_o$$

$$Ax^{1/3}\left(1 - \frac{3}{2.8}x^2 + \frac{3^2}{2.4.8.14}x^4 - \frac{3^3}{2.4.6.8.14.20}x^6 \;\ldots\ldots\right)$$

FRAME 21

You can now complete the solution to this problem by finding the series corresponding to $c = -\frac{1}{3}$.

****

21A

$$a_r = -\frac{3a_{r-2}}{r(3r-2)}$$

$$Bx^{-1/3}\left(1 - \frac{3}{2.4}x^2 + \frac{3^2}{2.4.4.10}x^4 - \frac{3^3}{2.4.6.4.10.16}x^6 + \ldots\ldots\right)$$

The general solution of the d.e. is, of course, then obtained by adding this series to the one in 20A.

FRAME 22

In Example 1 the values of c were 0 and $\frac{1}{2}$, in Example 2 they were 1 and $\frac{1}{2}$ and in Example 3, $+\frac{1}{3}$ and $-\frac{1}{3}$. In none of these cases did the values of c differ by an integer. When this happens, the method of solution is affected in one of two ways, as will now be demonstrated by suitable examples.

Example 4   Solve the equation

$$y'' + xy' + y = 0$$

You can begin this in the usual way by obtaining the indicial equation and solving it.

****

22A

$$a_o c(c-1) = 0$$

$$c = 0 \text{ or } 1$$

FRAME 23

You will notice that the roots of the indicial equation differ by an integer, 1.

We shall now proceed in the usual way and see what happens.

The indicial equation was obtained by equating the coefficient of x^{c-2} to zero, so we turn our attention next to the coefficient of x^{c-1}, which you should now write down.

****

23A

$$a_1(c + 1)c$$

FRAME 24

Equating the coefficient of x^{c-1} to zero gives

$$a_1(c + 1)c = 0 \qquad (24.1)$$

You will notice that this is similar to equation (17.1) in Example 3, in that only a_1 occurs, not a_1 and a_0. However, it differs in that we cannot say that $(c + 1)c$ is not zero for either value of c, so it cannot be said that $a_1 = 0$ for both values of c.

When $c = 1$ we have $2a_1 = 0$ and therefore $a_1 = 0$

But when $c = 0$, equation (24.1) is satisfied whatever the value of a_1 so that the value of a_1 is left undetermined.

FRAME 25

Only the y" term contributed to the coefficients of x^{c-2} and x^{c-1}, but from x^c onwards contributions are made by all three terms on the L.H.S. of the d.e., i.e. by y", xy' and y. Therefore the coefficients of x^c, x^{c+1}, x^{c+2} etc. will all follow the same pattern and all 'a's from a_2 onwards will be given by a recurrence relation.

FRAME 25 continued

Now obtain this recurrence relation by equating to zero the coefficient of x^{c+r} in $y'' + xy' + y$.

****

25A

$$a_{r+2} = -\frac{a_r}{c+r+2} \quad for \quad r \geq 0$$

FRAME 26

Taking first $c = 0$, we have already seen that in this case the value of a_1 remains undetermined. The recurrence relation now becomes

$$a_{r+2} = -\frac{a_r}{r+2}$$

Write down the results obtained by substituting $r = 0,1,2,3$ in this relation.

****

26A

$r = 0 \qquad a_2 = -\frac{a_o}{2}$

$r = 1 \qquad a_3 = -\frac{a_1}{3}$

$r = 2 \qquad a_4 = -\frac{a_2}{4}$

$r = 3 \qquad a_5 = -\frac{a_3}{5}$

FRAME 27

The series deriving from $c = 0$ is therefore

$$x^o\left(a_o + a_1x - \frac{a_o}{2}x^2 - \frac{a_1}{3}x^3 + \frac{a_o}{2.4}x^4 + \frac{a_1}{3.5}x^5 - \ldots\ldots\right)$$

i.e. $a_o\left(1 - \frac{x^2}{2} + \frac{x^4}{2.4} - \ldots\right) + a_1\left(x - \frac{x^3}{3} + \frac{x^5}{3.5} \quad \ldots\ldots\right)$

FRAME 27 continued

As a_o and a_1 are undetermined, we can represent them by arbitrary constants A and B, so we have the solution

$$A\left(1 - \frac{x^2}{2} + \frac{x^4}{2.4} - \ldots\right) + B\left(x - \frac{x^3}{3} + \frac{x^5}{3.5} \ldots\right)$$

The present situation is rather puzzling, because this solution already has the two arbitrary constants required in the general solution. It looks as if we shall have too many arbitrary constants when we add the solution corresponding to $c = 1$.

Well, let's see what happens when $c = 1$. It was concluded in FRAME 24 that in this case $a_1 = 0$. The next step is to write down the recurrence relation and deduce expressions for a_2, a_3, a_4, a_5, and we suggest you do this.

****

27A

$$a_{r+2} = -\frac{a_r}{r+3}$$

$$a_2 = -\frac{a_o}{3}$$

$$a_3 = 0 \quad \text{as} \quad a_1 = 0$$

$$a_4 = -\frac{a_2}{5}$$

$$a_5 = 0$$

FRAME 28

The series corresponding to $c = 1$ is therefore

$$x\left(a_o - \frac{a_o}{3}x^2 + \frac{a_o}{3.5}x^4 - \ldots\right)$$

$$\text{i.e. } a_o\left(x - \frac{x^3}{3} + \frac{x^5}{3.5} - \ldots\right)$$

But this series already forms part of the solution obtained in FRAME 27 from $c = 0$, so it has nothing new to contribute. We see, therefore, that the solution given by the lower value of c was, in fact, the general solution

FRAME 28 continued

$$y = A\left(1 - \frac{x^2}{2} + \frac{x^4}{2.4} - \ldots\right) + B\left(x - \frac{x^3}{3} + \frac{x^5}{3.5} - \ldots\right)$$

FRAME 29

In Example 4, the method of obtaining the general solution of the d.e. was broadly the same as it was in the previous examples. However, this is not always the case when the roots of the indicial equation differ by an integer.

Bessel's equation of order n was referred to in the introduction, and in Example 3 it was solved for $n = \frac{1}{3}$. We shall now see what happens when n is an integer.

The equation is $x^2y'' + xy' + (x^2 - n^2)y = 0$

Starting with the usual trial solution, obtain the indicial equation.

****

29A

Coefficient of $x^c = (c^2 - n^2)a_0$

Indicial equation: $c^2 - n^2 = 0$

FRAME 30

The roots of the indicial equation are $c = \pm n$, and if n is an integer the difference between the roots will also be an integer.

Now equate the coefficient of x^{c+1} to zero and deduce the value of a_1.

****

30A

$a_1(c + 1)c + a_1(c + 1) - n^2a_1 = 0$

i.e. $\{(c + 1)^2 - n^2\}a_1 = 0$ *(30A.1)*

As neither value of c makes $\{(c + 1)^2 - n^2\}$ *zero, it can be concluded that*

$$a_1 = 0$$

FRAME 31

The term x^2y on the L.H.S. of the d.e. has not contributed to the coefficients of x^c and x^{c+1}, but from x^{c+2} onwards this will cease to be the case. Hence equating the coefficient of x^{c+r} to zero will give a recurrence relation valid for $r \geqq 2$. Find this recurrence relation (giving a_r in terms of a_{r-2}), and deduce the values of a_3, a_5, a_7 etc. for both values of c.

****

31A

$$a_r = -\frac{a_{r-2}}{(c+r)^2 - n^2} \qquad (31A.1)$$

$a_3 = a_5 = a_7 = \ldots = 0$ *for either value of c.*

FRAME 32

Assuming, for the sake of argument, that n is a positive integer, find the first four terms in the solution corresponding to $c = n$.

****

32A

$$a_r = -\frac{a_{r-2}}{r(2n+r)}$$

$$\therefore \quad a_2 = -\frac{a_o}{2(2n+2)} = -\frac{a_o}{2^2.1(n+1)}$$

$$a_4 = -\frac{a_2}{4(2n+4)} = -\frac{a_2}{2^2.2(n+2)}$$

$$a_6 = -\frac{a_4}{6(2n+6)} = -\frac{a_4}{2^2.3(n+3)}$$

Denoting a_o *by an arbitrary constant A, the solution is*

$$A\{x^n - \frac{x^{n+2}}{2^2 1!(n+1)} + \frac{x^{n+4}}{2^4 2!(n+2)(n+1)} - \frac{x^{n+6}}{2^6 3!(n+3)(n+2)(n+1)} + \ldots\}$$

If $A = \frac{1}{2^n n!}$ *the result is "Bessel's function of order n of the first kind" and is denoted by* $J_n(x)$.

FRAME 33

Turning next to $c = -n$, the recurrence relation then becomes

$$a_r = -\frac{a_{r-2}}{r(r-2n)}$$

so that we have

$$a_2 = -\frac{a_o}{2(2-2n)} = -\frac{a_o}{2^2.1(1-n)}$$

$$a_4 = -\frac{a_2}{4(4-2n)} = -\frac{a_2}{2^2.2(2-n)}$$

$$a_6 = -\frac{a_4}{6(6-2n)} = -\frac{a_4}{2^2.3(3-n)}$$

By considering in turn $n = 1$, $n = 2$, $n = 3$, can you see why $c = -n$ will not give a solution?

****

33A

If $n = 1$, a_2 *becomes infinite.*

" $n = 2$, a_4 " "

" $n = 3$, a_6 " "

By this time it should be obvious that n being a positive integer makes an 'a' infinite, so a series solution is not obtained by putting $c = -n$.

FRAME 34

In this type of situation, then, the usual procedure only gives one series solution (the one in 32A) which we know cannot be the general solution, having only one arbitrary constant. It is possible to find a second series solution, with the necessary second arbitrary constant, but the explanation of the method and the working involved are rather lengthy and we feel that most technologists will not find a detailed study of this particular problem sufficiently rewarding to justify the time spent. For those who are interested, a worked example is given in APPENDIX B.

FRAME 35

Mention must also be made of the case when the roots of the indicial equation are equal. This would happen, for instance, with Bessel's equation of order zero, where the indicial equation is $c^2 = 0$. Having only one value of c, we shall only get one series solution, with one arbitrary constant, using the method described in this programme. The remarks made in the previous frame apply equally to this case, and you can see a worked example in APPENDIX A, if you wish.

FRAME 36

Legendre's Equation

As has already been mentioned, Bessel's equation occurs in certain physical problems. It can arise after separation of the variables in a particular partial differential equation (Laplace's Equation). When a different system of coordinates is used, we get instead a d.e. of the form

$$(1 - x^2)y'' - 2xy' + n(n + 1)y = 0$$

This is known as Legendre's equation of order n.

Let us now solve Legendre's equation of order 1, i.e.

$$(1 - x^2)y'' - 2xy' + 2y = 0$$

Make a start by writing down and solving the indicial equation.

****

36A

Coefficient of $x^{c-2} = a_0c(c - 1)$
Equating this to zero gives $c = 0$ *or* 1.

FRAME 37

Now equate the coefficient of x^{c-1} to zero and state what conclusions can be drawn about the value of a_1.

****

37A

$$a_1(c + 1)c = 0$$

If $c = 0$, a_1 *can take any value i.e. it is undetermined.*

If $c = 1$, $a_1 = 0$.

FRAME 38

Now go ahead and find the values of a_2, a_3, a_4, a_5, a_6 for the case when $c = 0$, and hence write down the series solution corresponding to this value of c.

****

38A

$$a_2 = -a_o$$

$$a_3 = 0 \quad \text{and} \quad \therefore \quad a_5 = 0$$

$$a_4 = \frac{1}{3} a_2$$

$$a_6 = \frac{3}{5} a_4$$

Series solution is $x^o(a_o + a_1x - a_ox^2 - \frac{1}{3} a_ox^4 - \frac{1}{3} \cdot \frac{3}{5} a_ox^6 - \ldots)$

Denoting a_o *and* a_1 *by arbitrary constants A and B respectively this becomes*

$$A\left(1 - x^2 - \frac{1}{3} x^4 - \frac{1}{5} x^6 - \ldots\ldots\right) + Bx$$

FRAME 39

You will notice that this is similar to Example 4, in that the solution given by the smaller value of c has the two arbitrary constants necessary for the general solution. You should now check that the solution given by $c = 1$ is already included in the solution just found.

****

39A

If $c = 1$, $a_1 = 0$

$a_2 = 0$

$\therefore \quad a_3 = a_4 = a_5 = a_6 = \ldots\ldots = 0$

This leaves only a_0x, *which is already covered by the* Bx *in the solution obtained from* $c = 0$.

FRAME 40

The complete solution, then, of Legendre's equation of order 1 is

$$y = A\left(1 - x^2 - \frac{1}{3}x^4 - \frac{1}{5}x^6 - \ldots\ldots\right) + Bx \qquad (40.1)$$

You have seen that, as in Example 4 where the roots differed by an integer, the complete solution is given by the lower value of c.

The solution (40.1) differs from those obtained in all the previous examples in this programme in that it does not consist of two infinite series, but of one infinite series and a series which terminates after a limited number of terms (in this example, only one). This always happens with Legendre's equation when n is a positive integer. The terminating series is then a polynomial of degree n, and, if the arbitrary constant in each case is chosen so that the value of the polynomial is 1 when $x = 1$, the functions known as Legendre's polynomials are obtained. The notation $P_n(x)$ is used. Thus $P_1(x) = x$.

FRAME 41

$\sum$ notation

You are probably aware of the $\sum$ notation which can be used for the sum of a number of terms all of the same kind. Its advantage is that it is more compact. For example, instead of the trial solution being written as

$$y = x^c(a_0 + a_1x + a_2x^2 + \ldots + a_rx^r + \ldots)$$

FRAME 41 continued

it could be expressed as

$$y = x^c \sum_{r=o}^{\infty} a_r x^r \quad \text{or} \quad \sum_{r=o}^{\infty} a_r x^{c+r}.$$

This notation can be used throughout the whole of the working in solution in series problems and the rather cumbersome lay-outs incurred in writing out y, y', y" at length are then avoided. However, the non-mathematician is not usually sufficiently familiar with working with $\sum$ notation to be really happy with it, so it has not been used in this programme. For those who are interested, an example demonstrating its use is given in APPENDIX C.

FRAME 42

Miscellaneous Examples

In this frame some miscellaneous examples are given for you to try. Answers are supplied in FRAME 43 and hints have been provided in some cases.

Use a trial solution of the form (2.1) to solve the d.e.'s 1 - 4.

1. $3xy'' + (1 - 3x)y' - 3y = 0$

2. $2x^2y'' - xy' + (1 - x^2)y = 0$

3. $xy'' + 2y' + xy = 0$

4. $(x - x^2)y'' - 3y' + 2y = 0$

5. In a tubular gas preheater hot air is obtained by drawing cool air through a heated cylindrical tube. In a particular case, to find the temperature of the emergent hot air, it was necessary to solve the d.e.

$$\frac{d^2t}{dx^2} - 7000\frac{dt}{dx} - 3750x^{-\frac{1}{2}}t = 0$$

t being the difference between the temperature of the tube wall and the air inside it at a distance x from the inlet end.
By putting $x = y^2$ show that this equation becomes

$$y\frac{d^2t}{dy^2} - \left(1 + 14\,000\,y^2\right)\frac{dt}{dy} - 15\,000y^2t = 0$$

and then find the series solution for t as far as the term in y^5.

Answers to Miscellaneous Examples

FRAME 43

1. $y = A\left(1 + 3x + \frac{3^2x^2}{4} + \frac{3^3x^3}{4.7} + \ldots\right) + Bx^{2/3}\left(1 + x + \frac{x^2}{2!} + \frac{x^3}{3!} + \ldots\right)$

2. $y = Ax^{1/2}\left(1 + \frac{x^2}{3.2} + \frac{x^4}{3.7.2.4} + \frac{x^6}{3.7.11.2.4.6} + \ldots\right)$

$\quad + Bx\left(1 + \frac{x^2}{2.5} + \frac{x^4}{2.4.5.9} + \frac{x^6}{2.4.6.5.9.13} + \ldots\right)$

3. $c = -1$ or 0

$c = -1$ gives the solution

$$y = Ax^{-1}\left(1 - \frac{x^2}{2!} + \frac{x^4}{4!} - \ldots\right) + B\left(1 - \frac{x^2}{3!} + \frac{x^4}{5!} - \ldots\right)$$

$c = 0$ repeats the second series in this solution.

4. $c = 0$ or 4

$c = 0$ gives the solution

$$y = A\left(1 + \frac{2x}{3} + \frac{x^2}{3}\right) + Bx^4\left(1 + 2x + 3x^2 + 4x^3 + \ldots\right)$$

You may have thought that, because $a_3 = 0$, $a_4 = a_5 = a_6 = \ldots = 0$ too, but $a_4 = \frac{c+1}{c} a_3$, so if $a_3 = 0$ and $c = 0$, a_4 is indeterminate.

Or, think of it another way: $ca_4 = (c + 1)a_3$

If $c = 0$ and $a_3 = 0$, this relationship is satisfied whatever the value of a_4. The second arbitrary constant B is assigned to a_4.

$c = 4$ repeats the infinite series.

5. $\frac{dt}{dx} = \frac{1}{2y}\frac{dt}{dy}$ and $\frac{d^2t}{dx^2} = \frac{1}{4y^3}\left(y\frac{d^2t}{dy^2} - \frac{dt}{dy}\right)$

New d.e. then obtained by substitution.

$c = 0$ or 2

$c = 0$ gives $A(1 + 5000x^3 + 14\,000\,000x^5 + \ldots) + Bx^2(1 + 3500x^2 + 1000x^3 + \ldots)$

$c = 2$ repeats the second series in this solution.

The following point may be of interest:

To find the temperature of the emergent hot air it is necessary to insert the appropriate value of x into this solution. Unfortunately the series

FRAME 43 continued

is unsuitable for computational purposes as it converges very slowly (remember the remark in FRAME 2 about this) and consequently a very large number of terms must be calculated. The trouble arises because the coefficients of $\frac{dt}{dx}$ and $x^{-\frac{1}{2}}t$ are large in comparison with that of $\frac{d^2t}{dx^2}$.

A reasonable approximation can be obtained by neglecting the second derivative term which is due to the thermal conductivity of the air. The equation then becomes

$$\frac{dt}{dx} + \frac{15}{28}x^{-\frac{1}{2}}t = 0$$

which can easily be solved by separating the variables.

APPENDIX A

Roots of Indicial Equation equal

The method of solution in this case will be shown by considering Bessel's equation of order zero, i.e.

$$xy'' + y' + xy = 0$$

In FRAME 29 you were asked to write down the indicial equation for Bessel's equation of order n. Putting $n = 0$ in the result gives

$$c^2 = 0$$

whose roots are both equal to zero.

Putting $n = 0$ in the other results obtained for Bessel's equation of order n, the equation (30A.1) becomes $(c + 1)^2 a_1 = 0$

giving $a_1 = 0$ as for other values of n,

and the recurrence relation (31A.1) becomes

$$a_r = -\frac{a_{r-2}}{(c + r)^2}.$$

Hence $a_3 = a_5 = a_7 = \dots = 0$

and $a_2 = -\dfrac{a_o}{(c + 2)^2}$, $a_4 = -\dfrac{a_2}{(c + 4)^2}$, $a_6 = -\dfrac{a_4}{(c + 6)^2}$ etc.

Let us write

$$z = a_o x^c \{1 - \frac{1}{(c + 2)^2} x^2 + \frac{1}{(c + 2)^2 (c + 4)^2} x^4 - \frac{1}{(c + 2)^2 (c + 4)^2 (c + 6)^2} x^6 \dots\dots \}$$

This is the series we get when we substitute the expressions obtained above for a_1, a_2, a_3 etc. and it will be a solution of the d.e. when $c = 0$.

If the series denoted by z is substituted for y in the L.H.S. of the d.e., all the terms cancel out except one, $a_o c^2 x^{c-1}$. (If you wish to verify this for yourself you are advised to use the kind of lay-out shown on Page 6:30.) Putting this another way,

$$xz'' + z' + xz = a_o c^2 x^{c-1} \qquad \text{(A.1)}$$

and, of course, you will see that $c = 0$ makes the R.H.S. zero and hence makes z in that case a solution of the original d.e.

APPENDIX A continued

In what follows next, $\frac{\partial}{\partial m}(x^m)$ will be required and this will now be obtained in case you are not familiar with the result.

$$\text{If} \quad f = x^m$$

$$\ln f = m \ln x$$

$$\frac{1}{f}\frac{\partial f}{\partial m} = \ln x$$

$$\frac{\partial f}{\partial m} = f \ln x$$

$$\text{i.e.} \quad \frac{\partial}{\partial m}(x^m) = x^m \ln x$$

If both sides of equation (A.1) are differentiated partially with respect to c, we get

$$x\frac{\partial}{\partial c}z'' + \frac{\partial}{\partial c}z' + x\frac{\partial}{\partial c}z = a_o(2cx^{c-1} + c^2x^{c-1}\ln x) \qquad (A.2)$$

Now, as you know, differentiating z' with respect to c gives the same result as differentiating $\frac{\partial z}{\partial c}$ with respect to x. That is to say, it makes no difference whether we differentiate with respect to x before differentiating with respect to c, or vice versa.

∴ Equation (A.2) is equivalent to

$$x\left(\frac{\partial z}{\partial c}\right)'' + \left(\frac{\partial z}{\partial c}\right)' + x\frac{\partial z}{\partial c} = a_oc(2x^{c-1} + cx^{c-1}\ln x)$$

Now, putting $c = 0$ will make the R.H.S. zero, just as it did in equation (A.1). Hence, when $c = 0$, $\frac{\partial z}{\partial c}$, as well as z, is a solution of $xy'' + y' + xy = 0$ and this gives us the second solution which we require.

The next step is to find $\frac{\partial z}{\partial c}$, and if you refer back to the series which represents z, you may think the differentiation appears rather formidable.

It will help to make the working clearer if we write

$$z = a_ox^cV \quad \text{where} \quad V = 1 - \frac{1}{(c+2)^2}x^2 + \frac{1}{(c+2)^2(c+4)^2}x^4 - \frac{1}{(c+2)^2(c+4)^2(c+6)^2}x^6 + \ldots\ldots$$

APPENDIX A continued

Then $\frac{\partial z}{\partial c} = a_o(Vx^c \ln x + x^c \frac{\partial V}{\partial c})$

To find $\frac{\partial V}{\partial c}$, it will be necessary to differentiate terms like

$\frac{1}{(c+2)^2(c+4)^2}$ and $\frac{1}{(c+2)^2(c+4)^2(c+6)^2}$ with respect to c. If we can find out what happens in the case $\frac{1}{(c+2)^2(c+4)^2}$, perhaps we shall be able to deduce the corresponding results for the more difficult terms.

$$\text{Let} \quad w = \frac{1}{(c+2)^2(c+4)^2}$$

$$\ln w = -2\ln(c+2) - 2\ln(c+4)$$

$$\frac{1}{w}\frac{dw}{dc} = -\frac{2}{c+2} - \frac{2}{c+4}$$

$$\frac{dw}{dc} = -\frac{2}{(c+2)^2(c+4)^2}\left(\frac{1}{c+2} + \frac{1}{c+4}\right)$$

From this it can be seen that, similarly, differentiating

$\frac{1}{(c+2)^2(c+4)^2(c+6)^2}$ with respect to c will yield

$$-\frac{2}{(c+2)^2(c+4)^2(c+6)^2}\left(\frac{1}{c+2} + \frac{1}{c+4} + \frac{1}{c+6}\right).$$

Returning to $\frac{\partial z}{\partial c}$, we have

$$\frac{\partial z}{\partial c} = a_o x^c V \ln x + a_o x^c \frac{\partial V}{\partial c}$$

$$= z \ln x + a_o x^c \left\{\frac{2}{(c+2)^3}x^2 - 2\frac{\frac{1}{c+2} + \frac{1}{c+4}}{(c+2)^2(c+4)^2}x^4 + 2\frac{\frac{1}{c+2} + \frac{1}{c+4} + \frac{1}{c+6}}{(c+2)^2(c+4)^2(c+6)^2}x^6 \ldots\ldots\right\}$$

We now have the two series, z and $\frac{\partial z}{\partial c}$, which are solutions of the d.e. when c = 0.

APPENDIX A continued

Putting $c = 0$ and denoting a_o by an arbitrary constant A, z becomes

$$A\left(1 - \frac{1}{2^2}x^2 + \frac{1}{2^2.4^2}x^4 - \frac{1}{2^2.4^2.6^2}x^6 + \cdots\right)$$

Putting $c = 0$ and denoting a_o by an arbitrary constant B, $\frac{\partial z}{\partial c}$ becomes

$$B\left(1 - \frac{1}{2^2}x^2 + \frac{1}{2^2.4^2}x^4 - \frac{1}{2^2.4^2.6^2}x^6 + \cdots\right)\ln x$$

$$+ B\left(\frac{1}{2^2}x^2 - \frac{1+\frac{1}{2}}{2^2.4^2}x^4 + \frac{1+\frac{1}{2}+\frac{1}{3}}{2^2.4^2.6^2}x^6 - \cdots\right)$$

The general solution is

$$y = (A + B\ln x)\left(1 - \frac{1}{2^2}x^2 + \frac{1}{2^2.4^2}x^4 - \frac{1}{2^2.4^2.6^2}x^6 + \cdots\right)$$

$$+ B\left(\frac{1}{2^2}x^2 - \frac{1+\frac{1}{2}}{2^2.4^2}x^4 + \frac{1+\frac{1}{2}+\frac{1}{3}}{2^2.4^2.6^2}x^6 - \cdots\right)$$

A practical situation which leads to a d.e. whose indicial equation has equal roots is that of a transverse cooling fin on a pipe. Under certain conditions, taking the heat balance leads to

$$x(R - x)\frac{d^2t}{dx^2} + (R - 2x)\frac{dt}{dx} - K(R - x)t = 0$$

where, at a distance x from the rim of the fin, t is the excess temperature over that of the surrounding air. R is the radius of the fin and K is a constant.

The indicial equation is $c^2 = 0$ and so the method of solution is that described in this appendix. However as it is necessary for t to remain finite when $x = 0$ (i.e. on the rim) the coefficient of ln x must be zero and so only the simple part of the solution is required.

APPENDIX B

If you have not read APPENDIX A, you should do so before proceeding.

Roots of Indicial Equation differing by an integer and making an 'a' infinite.

The method of solution in this case will be shown by considering Bessel's equation of order one, i.e.

$$x^2y'' + xy' + (x^2 - 1)y = 0$$

In FRAME 29 you were asked to write down the indicial equation for Bessel's equation of order n. Putting $n = 1$ in the result gives

$$c^2 - 1 = 0$$

whose roots are -1 and +1, i.e. differing by 2.

Putting $n = 1$ in the other results obtained for Bessel's equation of order n, the equation (30A.1) becomes $\{(c + 1)^2 - 1\}a_1 = 0$

giving $a_1 = 0$ as for other values of n,

and the recurrence relation (31A.1) becomes $a_r = -\dfrac{a_{r-2}}{(c + r)^2 - 1}$.

Hence $a_3 = a_5 = a_7 = \ldots\ldots = 0$

and $a_2 = -\dfrac{a_o}{(c + 1)(c + 3)}$, $a_4 = -\dfrac{a_2}{(c + 3)(c + 5)}$, $a_6 = -\dfrac{a_4}{(c + 5)(c + 7)}$ etc.

Let us write

$$z = a_o x^c\left\{1 - \frac{1}{(c + 1)(c + 3)}x^2 + \frac{1}{(c + 1)(c + 3)^2(c + 5)}x^4 - \frac{1}{(c + 1)(c + 3)^2(c + 5)^2(c + 7)}x^6 + \ldots\ldots\right\}$$

This is the series we get when we substitute the expressions obtained above for a_1, a_2, a_3 etc. When $c = 1$, the solution

$$Ax\left(1 - \frac{1}{2.4}x^2 + \frac{1}{2.4^2.6}x^4 - \frac{1}{2.4^2.6^2.8}x^6 + \ldots\ldots\right)$$

is obtained, but if we substitute $c = -1$, the other root of the indicial equation, the coefficients become infinite because of the factor $(c + 1)$ in the denominators.

APPENDIX B continued

This difficulty can be overcome by replacing a_o by $(c + 1)k$, and replacing the previous condition $a_o \neq 0$ by the condition $k \neq 0$. We then have

$$z = kx^c\{(c + 1) - \frac{1}{c + 3}x^2 + \frac{1}{(c + 3)^2(c + 5)}x^4 - \frac{1}{(c + 3)^2(c + 5)^2(c + 7)}x^6 + ...\} \qquad (B.1)$$

If this series is substituted for y in the L.H.S. of the d.e., all the terms cancel out except one, $kx^c(c + 1)^2(c - 1)$. As in APPENDIX A, the working for this is not set out here - it is left to you to verify the result if you wish, and again the lay-out as shown on Page 6:30 is recommended.

We have, then,

$$x^2z'' + xz' + (x^2 - 1)z = kx^c(c + 1)^2(c - 1)$$

The situation is now similar to that arrived at when equation (A.1) was written down in APPENDIX A. There was then a factor c^2 on the R.H.S. and, because of that, $\frac{\partial z}{\partial c}$, as well as z, was a solution of the d.e. when $c = 0$. In the present case, the squared factor $(c + 1)^2$ on the R.H.S. makes $\frac{\partial z}{\partial c}$, as well as z, a solution of the d.e. when $c = -1$.

$\frac{\partial z}{\partial c}$ is found in much the same way as in APPENDIX A - this time we shall just state the result.

$$\frac{\partial z}{\partial c} = z \ln x + kx^c\left\{1 + \frac{1}{(c + 3)^2}x^2 - \frac{\frac{2}{c + 3} + \frac{1}{c + 5}}{(c + 3)^2(c + 5)}x^4 + \frac{\frac{2}{c + 3} + \frac{2}{c + 5} + \frac{1}{c + 7}}{(c + 3)^2(c + 5)^2(c + 7)}x^6 - .....\right\}$$

We have already found one solution, corresponding to $c = 1$, so if we are going to get two further solutions from $c = -1$, it would appear that we are going to have three solutions altogether, giving a general solution with three arbitrary constants whereas for a second order d.e. there should be only two. However, substituting $c = -1$ in (B.1) gives the solution

APPENDIX B continued

$$Bx^{-1}\left(-\frac{1}{2}x^2 + \frac{1}{2^2.4}x^4 - \frac{1}{2^2.4^2.6}x^6 + \ldots\ldots\right)$$

and this can be written

$$-\frac{B}{2}x\left(1 - \frac{1}{2.4}x^2 + \frac{1}{2.4^2.6}x^4 - \ldots\ldots\right)$$

which is essentially the same as the solution obtained using $c = 1$. So there are only two linearly independent solutions after all. The second one, obtained by putting $c = -1$ in $\frac{\partial z}{\partial c}$, is

$$-\tfrac{1}{2}Bx\log x\left(1 - \frac{1}{2.4}x^2 + \frac{1}{2.4^2.6}x^4 - \frac{1}{2.4^2.6^2.8}x^6 + \ldots\ldots\right)$$

$$+ Bx^{-1}\left(1 + \frac{1}{2^2}x^2 - \frac{1+\frac{1}{4}}{2^2.4}x^4 + \frac{1+\frac{1}{2}+\frac{1}{6}}{2^2.4^2.6}x^6 - \ldots\ldots\right)$$

The general solution is

$$y = (A - \tfrac{1}{2}B\log x)x\left(1 - \frac{1}{2.4}x^2 + \frac{1}{2.4^2.6}x^4 - \frac{1}{2.4^2.6^2.8}x^6 + \ldots\ldots\right)$$

$$+ Bx^{-1}\left(1 + \frac{1}{2^2}x^2 - \frac{1+\frac{1}{4}}{2^2.4}x^4 + \frac{1+\frac{1}{2}+\frac{1}{6}}{2^2.4^2.6}x^6 - \ldots\ldots\right)$$

APPENDIX C

An example using Σ notation

The equation which has been chosen to show the use of Σ notation is

$$2x^2y'' - xy' + (1 - 2x)y = 0$$

This was solved in Example 2 in FRAMES 13 - 14.

The trial solution is $\quad y = \sum_{r=o}^{\infty} a_r x^{c+r}$

$$\text{so} \quad y' = \sum_{r=o}^{\infty} a_r(c + r)x^{c+r-1}$$

$$\text{and} \quad y'' = \sum_{r=o}^{\infty} a_r(c + r)(c + r - 1)x^{c+r-2}$$

Substituting these expressions in $2x^2y'' - xy' + (1 - 2x)y$ leads to

$$2\sum_{r=o}^{\infty} a_r(c + r)(c + r - 1)x^{c+r} - \sum_{r=o}^{\infty} a_r(c + r)x^{c+r} + \sum_{r=o}^{\infty} a_r x^{c+r} - 2\sum_{r=o}^{\infty} a_r x^{c+r+1}$$

On combining the first three $\sum$'s, which all involve x^{c+r}, this becomes

$$\sum_{r=o}^{\infty} \{2(c + r)(c + r - 1) - (c + r - 1)\}a_r x^{c+r} - 2\sum_{r=o}^{\infty} a_r x^{c+r+1}$$

$$\text{i.e.} \quad \sum_{r=o}^{\infty} (2c + 2r - 1)(c + r - 1)a_r x^{c+r} - 2\sum_{r=o}^{\infty} a_r x^{c+r+1}$$

Now x^{c+r+1} when $r = 0$ is x^{c+r} when $r = 1$, so $\sum_{r=o}^{\infty} a_r x^{c+r+1}$ can be rewritten as $\sum_{r=1}^{\infty} a_{r-1} x^{c+r}$. This is done to bring all powers of x in the $\sum$'s down to the lowest one present (x^{c+r} in this case).

The next step is to equate to zero the various powers of x in

$$\sum_{r=o}^{\infty} (2c + 2r - 1)(c + r - 1)a_r x^{c+r} - 2\sum_{r=1}^{\infty} a_{r-1} x^{c+r}$$

When $r = 0$, only the first $\sum$ makes a contribution and gives the term

APPENDIX C continued

$(2c - 1)(c - 1)a_o x^c$. Equating this coefficient to zero, we get

$$(2c - 1)(c - 1) = 0 \quad \text{as} \quad a_o \neq 0.$$

This is, of course, the indicial equation, as obtained in answer frame 13A.

From $r = 1$ onwards, both $\sum$'s make a contribution.

$\therefore$ for $r \geq 1$, coefficient of $x^{c+r} = (2c + 2r - 1)(c + r - 1)a_r - 2a_{r-1}$

Equating this to zero gives the recurrence relation

$$a_r = \frac{2a_{r-1}}{(2c + 2r - 1)(c + r - 1)}$$

as obtained in answer frame 14A.

The roots of the indicial equation are $\frac{1}{2}$ and 1.

Taking first $c = 1$, the recurrence relation becomes

$$a_r = \frac{2a_{r-1}}{(2r + 1)r}$$

Repeated application of this formula leads to

$$a_r = \frac{2^r}{r!3.5.7.....(2r + 1)} a_o = \frac{2^r \, 2.4.6.....2r}{r!(2r + 1)!} a_o$$

$$= \frac{2^r.2^r \, r!}{r!(2r + 1)!} a_o = \frac{2^{2r}}{(2r + 1)!} a_o$$

The solution given by $c = 1$ can therefore be written in the form

$$Ax\{1 + \sum_{r=1}^{\infty} \frac{2^{2r}}{(2r + 1)!} x^r\}$$

where A is the arbitrary constant assigned to a_o.

(This is solution (14A.1) written in $\sum$ form.)

When $c = \frac{1}{2}$, the recurrence relation becomes

$$a_r = \frac{2a_{r-1}}{2r(r - \frac{1}{2})}$$

$$\text{i.e.} \quad a_r = \frac{2a_{r-1}}{r(2r - 1)}$$

APPENDIX C continued

By applying this formula repeatedly, and simplifying the result, we obtain

$$a_r = \frac{2^{2r-1}}{r(2r-1)!} a_o$$

Denoting a_o by an arbitrary constant B, we then have the solution

$$Bx^{\frac{1}{2}}\{1 + \sum_{r=1}^{\infty} \frac{2^{2r-1}}{r(2r-1)!} x^r\}$$

which is, of course, (14A.2) written in $\sum$ form.

The complete solution is

$$y = Ax\{1 + \sum_{r=1}^{\infty} \frac{2^{2r}}{(2r+1)!} x^r\} + Bx^{\frac{1}{2}}\{1 + \sum_{r=1}^{\infty} \frac{2^{2r-1}}{r(2r-1)!} x^r\}$$

UNIT 7

FURTHER PARTIAL DIFFERENTIATION

A.C. Bajpai
I.M. Calus
J.A. Fairley

Loughborough University of Technology

INSTRUCTIONS

This programme is divided up into a number of FRAMES which are to be worked *in the order given*. You will be required to participate in many of these frames and in such cases the answers are provided in ANSWER FRAMES, designated by the letter A following the frame number. Steps in the working are given where this is considered helpful. The answer frame is separated from the main frame by a line of asterisks: *********. Keep the answers covered until you have written your own response. If your answer is wrong, go back and try to see why. Do not proceed to the next frame until you have corrected any mistakes in your attempt and are satisfied that you understand the contents up to this point.

Before reading this programme, it is necessary that you are familiar with the following

Prerequisites

The contents of the Partial Differentiation programme in Unit 1, Vol I.

Determinants, as covered in FRAMES 1-14 and 34-37, pages 3:1 to 3:9 and 3:19 to 3:22.

Simple integration, as covered in FRAMES 17-24, pages 2:11 to 2:17.

Maxima and minima of functions of one variable, as covered in FRAMES 59-72, pages 1:136 to 1:147.

Taylor's series, as covered in FRAMES 37-49, pages 1:187 to 1:196.

CONTENTS

FRAME 1

Change of Independent Variables

In the earlier programme on Partial Differentiation, in Unit 1 of Vol I, it was seen that if V is a function of the two variables r and h then

$$\delta V \simeq \frac{\partial V}{\partial r}\delta r + \frac{\partial V}{\partial h}\delta h \qquad (1.1)$$

It was also seen that, if r and h both depended on the temperature T,

$$\frac{dV}{dT} = \frac{\partial V}{\partial r}\frac{dr}{dT} + \frac{\partial V}{\partial h}\frac{dh}{dT} \qquad (1.2)$$

Now suppose V is the volume of a sausage balloon such as can be bought at the Loughborough Fair held annually in the Market Place. Its shape can be regarded as that of a cylinder with hemispherical ends and then V is a function of the radius r and height h of the cylinder, so (1.1) applies. When you take the balloon home into the warm house, the temperature T will change and so also will r and h, as these depend on T. If the temperature were the only factor affecting r and h, (1.2) would apply. But as the volume increases so also does the pressure due to the skin of the balloon and consequently the increases in r and h are less than they would otherwise be. That is, r and h, and so also V, are now dependent on both T and the pressure p, and thus their derivatives with respect to T are partial instead of ordinary. (1.2) now takes the form

$$\frac{\partial V}{\partial T} = \frac{\partial V}{\partial r}\frac{\partial r}{\partial T} + \frac{\partial V}{\partial h}\frac{\partial h}{\partial T} \qquad (1.3)$$

Similarly, if the change w.r.t. p is considered,

$$\frac{\partial V}{\partial p} = \frac{\partial V}{\partial r}\frac{\partial r}{\partial p} + \frac{\partial V}{\partial h}\frac{\partial h}{\partial p} \qquad (1.4)$$

The fact that V is a function of r and h, which in turn are functions of p and T, can be illustrated diagrammatically as follows:

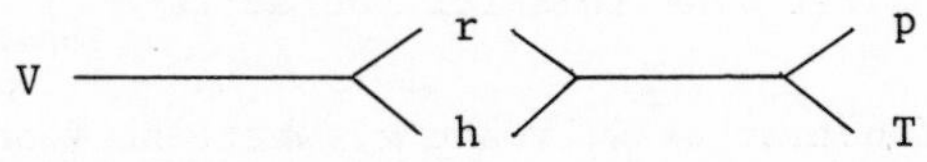

V —— < r, h > —— < p, T

Similarly, if z is a function of x and y, these being functions of r and θ, the chain diagram is

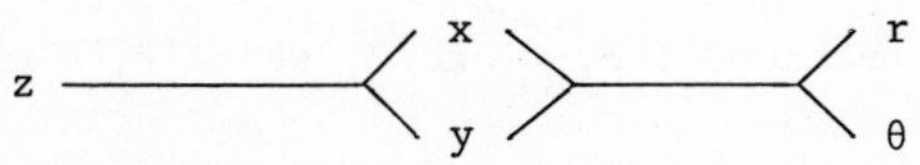

z —— < x, y > —— < r, θ

Now write down formulae similar to (1.3) and (1.4) for $\frac{\partial z}{\partial r}$ and $\frac{\partial z}{\partial \theta}$.

****

1A

$$\frac{\partial z}{\partial r} = \frac{\partial z}{\partial x}\frac{\partial x}{\partial r} + \frac{\partial z}{\partial y}\frac{\partial y}{\partial r}$$

$$\frac{\partial z}{\partial \theta} = \frac{\partial z}{\partial x}\frac{\partial x}{\partial \theta} + \frac{\partial z}{\partial y}\frac{\partial y}{\partial \theta}$$

FRAME 2

The equations you have just written down can be used when changing from Cartesian to polar coordinates. For example, the temperature at a point (x,y) in a certain plate is given by $T = (2/\pi)\tan^{-1}(y/x)$. Use formulae corresponding to those in 1A to find $\frac{\partial T}{\partial r}$ and $\frac{\partial T}{\partial \theta}$.

Then express T as a function of r and θ, and so verify your results.

****

2A

$$\frac{\partial T}{\partial r} = \frac{2}{\pi}\frac{-y}{x^2+y^2}\cos\theta + \frac{2}{\pi}\frac{x}{x^2+y^2}\sin\theta = 0$$

$$\frac{\partial T}{\partial \theta} = \frac{2}{\pi}\frac{-y}{x^2+y^2}(-r\sin\theta) + \frac{2}{\pi}\frac{x}{x^2+y^2}r\cos\theta = \frac{2}{\pi}$$

$$T = 2\theta/\pi$$

FRAME 3

The extension to functions of more than two variables should be obvious to you. To test whether this is so, see if you can write down formulae for the following partial derivatives. In (i) the chain diagram is given but in (ii) and (iii) you should first draw this for yourself.

(i) $\frac{\partial T}{\partial \theta}$ if T is a function of x, y and z, where x, y and z are functions of r, θ and φ, i.e. the chain diagram is T —— < x, y, z > —— < r, θ, φ

(ii) $\frac{\partial S}{\partial x}$ if S is a function of P, Q and R, where P, Q and R are functions of x and t.

(iii) $\frac{\partial u}{\partial t}$ if u is a function of v, where v is a function of r, s and t.

****

3A

(i) $$\frac{\partial T}{\partial \theta} = \frac{\partial T}{\partial x}\frac{\partial x}{\partial \theta} + \frac{\partial T}{\partial y}\frac{\partial y}{\partial \theta} + \frac{\partial T}{\partial z}\frac{\partial z}{\partial \theta}$$

(ii) S —< P, Q, R >— < x, t

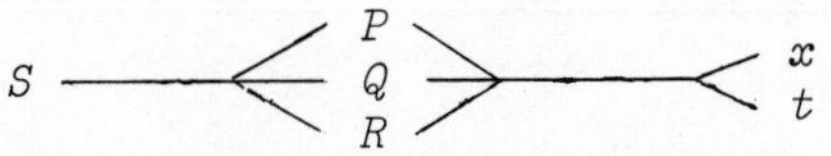

$$\frac{\partial S}{\partial x} = \frac{\partial S}{\partial P}\frac{\partial P}{\partial x} + \frac{\partial S}{\partial Q}\frac{\partial Q}{\partial x} + \frac{\partial S}{\partial R}\frac{\partial R}{\partial x}$$

(iii) u — v —< r, s, t

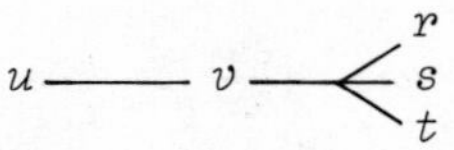

$$\frac{\partial u}{\partial t} = \frac{du}{dv}\frac{\partial v}{\partial t}$$

Note $\frac{du}{dv}$ *here as u is a function of only one variable.*

FRAME 4

To illustrate further the use of change of variable formulae, we shall find $\frac{\partial f}{\partial x}$ in terms of $\frac{\partial f}{\partial u}$ and $\frac{\partial f}{\partial v}$, f being a function of x and y where $x = u^2 + v^2$ and $y = 3uv$.

Now the relationship between the variables, in the form just given, may be described by the chain diagram

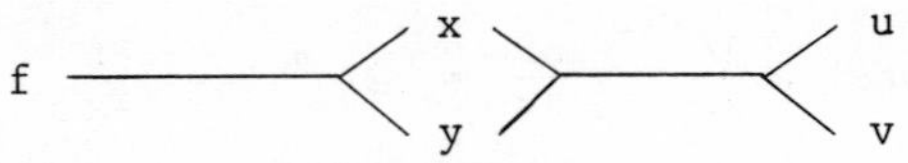

f —< x, y >— < u, v

and so

$$\frac{\partial f}{\partial u} = \frac{\partial f}{\partial x}\frac{\partial x}{\partial u} + \frac{\partial f}{\partial y}\frac{\partial y}{\partial u} \qquad (4.1)$$

$$\frac{\partial f}{\partial v} = \frac{\partial f}{\partial x}\frac{\partial x}{\partial v} + \frac{\partial f}{\partial y}\frac{\partial y}{\partial v} \qquad (4.2)$$

But if x and y are functions of u and v, it is also true to say that u and v are functions of x and y. This means that, alternatively, f can be regarded as a function of u and v, these being functions of x and y. The chain diagram in this case is

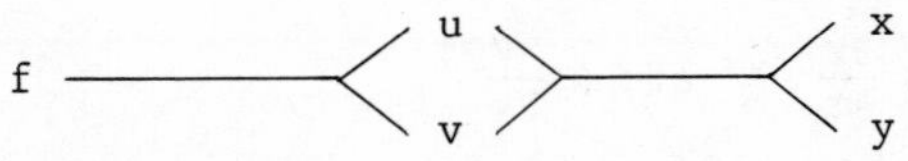

f —< u, v >— < x, y

and so

$$\frac{\partial f}{\partial x} = \frac{\partial f}{\partial u}\frac{\partial u}{\partial x} + \frac{\partial f}{\partial v}\frac{\partial v}{\partial x} \qquad (4.3)$$

$$\frac{\partial f}{\partial y} = \frac{\partial f}{\partial u}\frac{\partial u}{\partial y} + \frac{\partial f}{\partial v}\frac{\partial v}{\partial y} \qquad (4.4)$$

To find $\frac{\partial f}{\partial x}$ in terms of $\frac{\partial f}{\partial u}$ and $\frac{\partial f}{\partial v}$, it might seem that the obvious equation to use is (4.3). Can you see what difficulty will arise if you try to use this equation?

****

4A

$\frac{\partial u}{\partial x}$ here means $\left(\frac{\partial u}{\partial x}\right)_y$. The difficulty arises that you have not actually got u as a function of x and y, nor is this easy to obtain.

Similar remarks apply to $\frac{\partial v}{\partial x}$.

FRAME 5

To overcome this difficulty, we regard (4.1) and (4.2) as simultaneous equations in $\frac{\partial f}{\partial x}$ and $\frac{\partial f}{\partial y}$. However, as $x = u^2 + v^2$ and $y = 3uv$, they can be rewritten as

$$\frac{\partial f}{\partial u} = 2u\frac{\partial f}{\partial x} + 3v\frac{\partial f}{\partial y}$$

$$\frac{\partial f}{\partial v} = 2v\frac{\partial f}{\partial x} + 3u\frac{\partial f}{\partial y}$$

Eliminating $\frac{\partial f}{\partial y}$ gives $u\frac{\partial f}{\partial u} - v\frac{\partial f}{\partial v} = 2(u^2 - v^2)\frac{\partial f}{\partial x}$

whence $\frac{\partial f}{\partial x} = \frac{1}{2(u^2 - v^2)}\left(u\frac{\partial f}{\partial u} - v\frac{\partial f}{\partial v}\right)$

Now if V is a function of x and y where $x = r\cos\theta$, $y = r\sin\theta$, use the above technique to find $\frac{\partial V}{\partial x}$ and $\frac{\partial V}{\partial y}$ in terms of $\frac{\partial V}{\partial r}$ and $\frac{\partial V}{\partial \theta}$.

****

5A

$$\frac{\partial V}{\partial r} = \cos\theta\frac{\partial V}{\partial x} + \sin\theta\frac{\partial V}{\partial y}$$

$$\frac{\partial V}{\partial \theta} = -r\sin\theta\frac{\partial V}{\partial x} + r\cos\theta\frac{\partial V}{\partial y}$$

Solving,

$$\frac{\partial V}{\partial x} = \cos\theta\frac{\partial V}{\partial r} - \frac{\sin\theta}{r}\frac{\partial V}{\partial \theta} \qquad (5A.1)$$

$$\frac{\partial V}{\partial y} = \sin\theta\frac{\partial V}{\partial r} + \frac{\cos\theta}{r}\frac{\partial V}{\partial \theta} \qquad (5A.2)$$

FRAME 6

An equation which occurs in the study of fluid flow and other types of continuous movement, such as heat flow and magnetic or electric flux, is

$$\frac{\partial^2 V}{\partial x^2} + \frac{\partial^2 V}{\partial y^2} = 0 \qquad (6.1)$$

This is Laplace's equation in two dimensions and you will meet it several times in later Units in this book. V is a function of position in a plane and (6.1) is appropriate when Cartesian coordinates are used. Sometimes, however, it is more convenient to use polar coordinates and the equation then takes on a different form which we shall now proceed to find. This involves extending (5A.1) and (5A.2) a step further to give $\frac{\partial^2 V}{\partial x^2}$ and $\frac{\partial^2 V}{\partial y^2}$.

FRAME 7

First of all, we notice that (5A.1) can be written

$$\frac{\partial}{\partial x} V = \left(\cos\theta \frac{\partial}{\partial r} - \frac{\sin\theta}{r}\frac{\partial}{\partial\theta}\right) V$$

showing that the operators $\frac{\partial}{\partial x}$ and $\cos\theta \frac{\partial}{\partial r} - \frac{\sin\theta}{r}\frac{\partial}{\partial\theta}$ are equivalent,

$$\text{i.e.} \quad \frac{\partial}{\partial x} \equiv \cos\theta \frac{\partial}{\partial r} - \frac{\sin\theta}{r}\frac{\partial}{\partial\theta}$$

Thus if U is a function of x and y

$$\frac{\partial}{\partial x} U = \left(\cos\theta \frac{\partial}{\partial r} - \frac{\sin\theta}{r}\frac{\partial}{\partial\theta}\right) U$$

$$\text{i.e.} \quad \frac{\partial U}{\partial x} = \cos\theta \frac{\partial U}{\partial r} - \frac{\sin\theta}{r}\frac{\partial U}{\partial\theta}$$

$$\text{Similarly} \quad \frac{\partial}{\partial x}\left(\frac{\partial V}{\partial x}\right) = \left(\cos\theta \frac{\partial}{\partial r} - \frac{\sin\theta}{r}\frac{\partial}{\partial\theta}\right)\frac{\partial V}{\partial x}$$

$$\text{i.e.} \quad \frac{\partial^2 V}{\partial x^2} = \left(\cos\theta \frac{\partial}{\partial r} - \frac{\sin\theta}{r}\frac{\partial}{\partial\theta}\right)\left(\cos\theta \frac{\partial V}{\partial r} - \frac{\sin\theta}{r}\frac{\partial V}{\partial\theta}\right)$$

$$= \cos\theta \frac{\partial}{\partial r}\left(\cos\theta \frac{\partial V}{\partial r} - \frac{\sin\theta}{r}\frac{\partial V}{\partial\theta}\right) - \frac{\sin\theta}{r}\frac{\partial}{\partial\theta}\left(\cos\theta \frac{\partial V}{\partial r} - \frac{\sin\theta}{r}\frac{\partial V}{\partial\theta}\right)$$

$$\text{Now} \quad \frac{\partial}{\partial r}\left(\cos\theta \frac{\partial V}{\partial r} - \frac{\sin\theta}{r}\frac{\partial V}{\partial\theta}\right) = \cos\theta \frac{\partial^2 V}{\partial r^2} + \frac{\sin\theta}{r^2}\frac{\partial V}{\partial\theta} - \frac{\sin\theta}{r}\frac{\partial^2 V}{\partial r\partial\theta}$$

What is $\frac{\partial}{\partial\theta}\left(\cos\theta \frac{\partial V}{\partial r} - \frac{\sin\theta}{r}\frac{\partial V}{\partial\theta}\right)$?

****

7A

$$- \sin\theta \frac{\partial V}{\partial r} + \cos\theta \frac{\partial^2 V}{\partial\theta\partial r} - \frac{\cos\theta}{r}\frac{\partial V}{\partial\theta} - \frac{\sin\theta}{r}\frac{\partial^2 V}{\partial\theta^2}$$

FRAME 8

We now have

$$\frac{\partial^2 V}{\partial x^2} = \cos\theta\left(\cos\theta \frac{\partial^2 V}{\partial r^2} + \frac{\sin\theta}{r^2}\frac{\partial V}{\partial\theta} - \frac{\sin\theta}{r}\frac{\partial^2 V}{\partial r\partial\theta}\right)$$

$$- \frac{\sin\theta}{r}\left(- \sin\theta \frac{\partial V}{\partial r} + \cos\theta \frac{\partial^2 V}{\partial\theta\partial r} - \frac{\cos\theta}{r}\frac{\partial V}{\partial\theta} - \frac{\sin\theta}{r}\frac{\partial^2 V}{\partial\theta^2}\right)$$

$$= \cos^2\theta \frac{\partial^2 V}{\partial r^2} - \frac{2\sin\theta\cos\theta}{r}\frac{\partial^2 V}{\partial r\partial\theta} + \frac{\sin^2\theta}{r^2}\frac{\partial^2 V}{\partial\theta^2} + \frac{\sin^2\theta}{r}\frac{\partial V}{\partial r} + \frac{2\sin\theta\cos\theta}{r^2}\frac{\partial V}{\partial\theta}$$

By a similar process, see if you can find $\frac{\partial^2 V}{\partial y^2}$.

****

8A

$$\frac{\partial}{\partial y} \equiv \sin\theta \frac{\partial}{\partial r} + \frac{\cos\theta}{r}\frac{\partial}{\partial\theta}$$

$$\frac{\partial^2 V}{\partial y^2} = \sin^2\theta \frac{\partial^2 V}{\partial r^2} + \frac{2\sin\theta\cos\theta}{r}\frac{\partial^2 V}{\partial r\partial\theta} + \frac{\cos^2\theta}{r^2}\frac{\partial^2 V}{\partial\theta^2} + \frac{\cos^2\theta}{r}\frac{\partial V}{\partial r} - \frac{2\sin\theta\cos\theta}{r^2}\frac{\partial V}{\partial\theta}$$

FRAME 9

Substituting the expressions obtained for $\frac{\partial^2 V}{\partial x^2}$ and $\frac{\partial^2 V}{\partial y^2}$, (6.1) becomes

$$\frac{\partial^2 V}{\partial r^2} + \frac{1}{r}\frac{\partial V}{\partial r} + \frac{1}{r^2}\frac{\partial^2 V}{\partial\theta^2} = 0$$

FRAME 10

Returning now to the example in FRAME 4, you saw that a difficulty arose if you were given the equations $x = u^2 + v^2$ and $y = 3uv$ and you wanted to find $\left(\frac{\partial u}{\partial x}\right)_y$. In that particular problem, we were able to avoid having to work this out but in case you are wondering how it's done, the secret will now be revealed.

FRAME 10 continued

First, $$dx = \frac{\partial x}{\partial u} du + \frac{\partial x}{\partial v} dv \qquad (10.1)$$

and $$dy = \frac{\partial y}{\partial u} du + \frac{\partial y}{\partial v} dv \qquad (10.2)$$

giving $$dx = 2u\,du + 2v\,dv$$

and $$dy = 3v\,du + 3u\,dv$$

Regarding these as simultaneous equations in du and dv leads to

$$du = \frac{\begin{vmatrix} dx & 2v \\ dy & 3u \end{vmatrix}}{\begin{vmatrix} 2u & 2v \\ 3v & 3u \end{vmatrix}} \quad \text{and} \quad dv = \frac{\begin{vmatrix} 2u & dx \\ 3v & dy \end{vmatrix}}{\begin{vmatrix} 2u & 2v \\ 3v & 3u \end{vmatrix}}$$

i.e. $$du = \frac{3u\,dx - 2v\,dy}{6(u^2 - v^2)} = \frac{u}{2(u^2 - v^2)}dx - \frac{v}{3(u^2 - v^2)} dy$$

and $$dv = \frac{2u\,dy - 3v\,dx}{6(u^2 - v^2)} = -\frac{v}{2(u^2 - v^2)} dx + \frac{u}{3(u^2 - v^2)} dy$$

Now, if u and v are regarded as functions of x and y,

$$du = \frac{\partial u}{\partial x} dx + \frac{\partial u}{\partial y} dy \quad \text{and} \quad dv = \frac{\partial v}{\partial x} dx + \frac{\partial v}{\partial y} dy$$

Comparing the two expressions for du, it follows that

$$\frac{\partial u}{\partial x} = \frac{u}{2(u^2 - v^2)}$$

Also, although it wasn't asked for, from the same two expressions you could get

$$\frac{\partial u}{\partial y} = -\frac{v}{3(u^2 - v^2)}$$

Can you now write down $\frac{\partial v}{\partial x}$ and $\frac{\partial v}{\partial y}$ in terms of u and v?

****

10A

$$\frac{\partial v}{\partial x} = -\frac{v}{2(u^2 - v^2)} \qquad \frac{\partial v}{\partial y} = \frac{u}{3(u^2 - v^2)}$$

FRAME 11

Jacobians

You know that, when simultaneous equations are solved by determinants, the denominator in the expression for each unknown is the same. This was, of course, true of the expressions for du and dv in the last frame. If (10.1) and (10.2) had been solved for du and dv, what would the denominator determinant have been?

****

11A

$$\begin{vmatrix} \frac{\partial x}{\partial u} & \frac{\partial x}{\partial v} \\ \frac{\partial y}{\partial u} & \frac{\partial y}{\partial v} \end{vmatrix}$$

FRAME 12

This determinant is called the JACOBIAN of x,y with respect to u,v. Where no ambiguity can arise, it is often denoted by J. Otherwise the variables involved are indicated by writing it as

$$J\left(\frac{x,y}{u,v}\right) \quad \text{or} \quad \frac{\partial(x,y)}{\partial(u,v)}$$

What determinants would you understand the notations $J\left(\frac{x,y}{r,\theta}\right)$ and $\frac{\partial(u,v)}{\partial(x,y)}$ to mean?

****

12A

$$\begin{vmatrix} \frac{\partial x}{\partial r} & \frac{\partial x}{\partial \theta} \\ \frac{\partial y}{\partial r} & \frac{\partial y}{\partial \theta} \end{vmatrix} \quad \text{and} \quad \begin{vmatrix} \frac{\partial u}{\partial x} & \frac{\partial u}{\partial y} \\ \frac{\partial v}{\partial x} & \frac{\partial v}{\partial y} \end{vmatrix}$$

FRAME 13

In a similar way, if x,y,z are functions of u,v,w the Jacobian is

$$\begin{vmatrix} \frac{\partial x}{\partial u} & \frac{\partial x}{\partial v} & \frac{\partial x}{\partial w} \\ \frac{\partial y}{\partial u} & \frac{\partial y}{\partial v} & \frac{\partial y}{\partial w} \\ \frac{\partial z}{\partial u} & \frac{\partial z}{\partial v} & \frac{\partial z}{\partial w} \end{vmatrix}$$

The notation for this is $J\left(\frac{x,y,z}{u,v,w}\right)$ or $\frac{\partial(x,y,z)}{\partial(u,v,w)}$, or, if there is no possible ambiguity, just J again.

Write down and simplify

(i) $J\left(\frac{x,y}{r,\theta}\right)$ if $x = r\cos\theta$, $y = r\sin\theta$

(ii) $J\left(\frac{x,y,z}{r,\theta,\phi}\right)$ if $x = r\sin\theta\cos\phi$, $y = r\sin\theta\sin\phi$, $z = r\cos\theta$.

****

13A

(i)
$$\begin{vmatrix} \cos\theta & -r\sin\theta \\ \sin\theta & r\cos\theta \end{vmatrix} = r$$

(ii)
$$\begin{vmatrix} \sin\theta\cos\phi & r\cos\theta\cos\phi & -r\sin\theta\sin\phi \\ \sin\theta\sin\phi & r\cos\theta\sin\phi & r\sin\theta\cos\phi \\ \cos\theta & -r\sin\theta & 0 \end{vmatrix}$$

$$= r^2\sin\theta$$

FRAME 14

Differentiation under the Integral Sign

The evaluation of $\int_1^2 (4x^3 - 2tx + t^3)dx$, where t does not vary with x, is very simple.

It immediately gives $\left[x^4 - tx^2 + t^3x\right]_1^2$

$$= 15 - 3t + t^3$$

FRAME 14 continued

You will notice that this is a function of t only, and so can be differentiated w.r.t. t to give $-3 + 3t^2$,

$$\text{i.e.} \quad \frac{d}{dt}\int_1^2 (4x^3 - 2tx + t^3)dx = -3 + 3t^2 \qquad (14.1)$$

Now the integrand, being a function of x and t, can be differentiated partially w.r.t. t giving $-2x + 3t^2$. What is $\int_1^2 (-2x + 3t^2)dx$? Do you notice anything about the result?

****

14A

$-3 + 3t^2$, which is the same as (14.1).

FRAME 15

It has just been shown that

$$\frac{d}{dt}\int_1^2 (4x^3 - 2tx + t^3)dx = \int_1^2 \frac{\partial}{\partial t}\left(4x^3 - 2tx + t^3\right)dx$$

Now, find out whether $\frac{d}{dt}\int_2^4 (2x - t)^{3/2}\,dx = \int_2^4 \frac{\partial}{\partial t}\left(2x - t\right)^{3/2}dx$ by evaluating both sides of the equation.

****

15A

Each side gives $\frac{1}{2}\{(4 - t)^{3/2} - (8 - t)^{3/2}\}$

FRAME 16

The question now arises: Are the two results just obtained coincidental or are $\frac{d}{dt}\int_a^b f(x,t)dx$ and $\int_a^b \frac{\partial}{\partial t} f(x,t)dx$ always the same?

From what has been done already, you will realise that $\int_a^b f(x,t)dx$ is a function of t and so can be denoted by F(t). Increasing t by a small amount δt leads to

FRAME 16 continued

$$F(t + \delta t) = \int_a^b f(x, t + \delta t)dx$$

and so $$F(t + \delta t) - F(t) = \int_a^b f(x, t + \delta t)dx - \int_a^b f(x, t)dx$$

$$= \int_a^b \{f(x, t + \delta t) - f(x, t)\}dx$$

$$\therefore \quad \frac{F(t + \delta t) - F(t)}{\delta t} = \int_a^b \frac{f(x, t + \delta t) - f(x, t)}{\delta t}\,dx$$

Proceeding to the limit, as $\delta t \to 0$, gives

$$\frac{d}{dt}F(t) = \int_a^b \frac{\partial}{\partial t} f(x, t)dx$$

i.e. $$\frac{d}{dt}\int_a^b f(x, t)dx = \int_a^b \frac{\partial}{\partial t} f(x, t)dx$$

The process of replacing $\frac{d}{dt}\int_a^b f(x, t)dx$ by $\int_a^b \frac{\partial}{\partial t} f(x, t)dx$ is known as DIFFERENTIATION UNDER THE INTEGRAL SIGN.

FRAME 17

Once certain integrals have been found, differentiation under the integral sign can sometimes be used to deduce the values of various other integrals.

Begin by finding, by direct integration, $\int_0^2 e^{xt}dx$.

****

17A

$(e^{2t} - 1)/t$

FRAME 18

We now have $\int_0^2 e^{xt}dx = \frac{e^{2t} - 1}{t}$

What result is obtained if you now differentiate both sides of this equation w.r.t. t?

****

18A

Using $\frac{d}{dt}\int_0^2 e^{xt}dx = \int_0^2 \frac{\partial}{\partial t}(e^{xt})dx$ *leads to*

$$\int_0^2 xe^{xt}dx = \frac{(2t - 1)e^{2t} + 1}{t^2} \qquad (18A.1)$$

FRAME 19

Differentiating again w.r.t. t would now give the value of $\int_0^2 x^2e^{xt}dx$.

However, a more useful integral to look at is $\int_0^\infty e^{-st}dt$, where $s > 0$.

Verify, by direct integration, that its value is $\frac{1}{s}$ and then, by successive differentiation, obtain the values of $\int_0^\infty te^{-st}dt$ and $\int_0^\infty t^2e^{-st}dt$.

****

19A

$$\int_0^\infty e^{-st}dt = -\frac{1}{s}\Big[e^{-st}\Big]_0^\infty$$

$$= \frac{1}{s}, \quad \textit{as} \quad e^{-st} \to 0 \quad \textit{when} \quad t \to \infty \quad \textit{for} \quad s > 0.$$

Differentiating once w.r.t. s gives

$\int_0^\infty te^{-st}dt = \frac{1}{s^2}$, *and twice gives* $\int_0^\infty t^2e^{-st}dt = \frac{2}{s^3}$

Those of you who are familiar with the Laplace Transform will recognise these three integrals as $\mathcal{L}\{1\}$, $\mathcal{L}\{t\}$ *and* $\mathcal{L}\{t^2\}$. *The technique adopted here for finding these transforms is an alternative to that used in Unit 5 in Vol I, and you will meet it again if you read Unit 13.*

FRAME 20

As a final example on this topic we shall consider the following:

A uniform beam of length ℓ and weight w per unit length is clamped horizontally at one end and freely supported at the other. The strain energy U is given by

$U = \frac{1}{2EI}\int_0^\ell (\frac{1}{2}wx^2 - Rx)^2 dx$ where R is the reaction at the free end and is such that U is a minimum. Find the value of R which satisfies this condition.

****

20A

$$\frac{dU}{dR} = \frac{-1}{EI}\int_0^\ell x\left(\frac{1}{2}wx^2 - Rx\right)dx = -\frac{1}{EI}\left(\frac{w\ell^4}{8} - \frac{R\ell^3}{3}\right)$$

$\frac{dU}{dR} = 0$ *when* $R = \frac{3w\ell}{8}$ *and this obviously gives minimum U.*

FRAME 21

Taylor's Series in Two Dimensions

You are already familiar with Taylor's series in one dimension in the form

$$f(x + h) = f(x) + hf'(x) + \frac{h^2}{2!}f''(x) + \frac{h^3}{3!}f'''(x) + \dots$$

This relates the value of the function at the point Q(x + h) with

x x+h
O P Q

that at P(x).

Now, as you already know, we often have to consider functions which depend on more than one variable. Taylor's series can be extended to include such functions. Here we shall limit the discussion to the case where there are two independent variables, i.e. the connection between f(x+h, y+k) and f(x,y) will be found. In other words, the value of the function at the point Q(x+h, y+k) will be related with that at P(x,y).

FRAME 21 continued

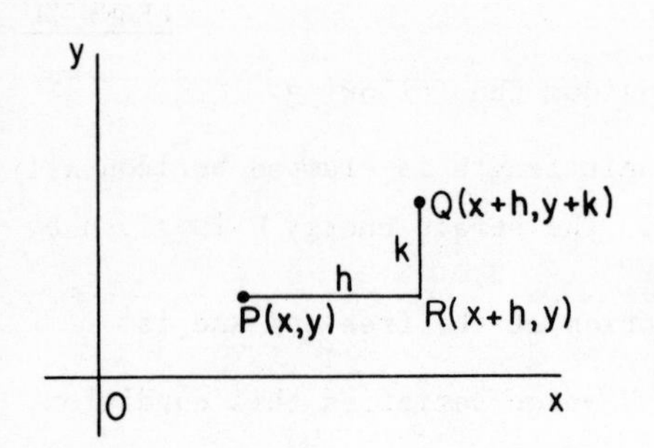

We start by relating the value of the function at R with that at P. Between P and R only x changes and so Taylor's series in one dimension will apply.

Thus $f(x+h,y) = f(x,y) + h\,f'_x(x,y) + \frac{h^2}{2!} f''_{xx}(x,y) + \ldots$ (21.1)

where $f'_x(x,y)$ means $\frac{\partial}{\partial x} f(x,y)$, and so on. (Very often the dashes are omitted in this notation.)

We next relate the value of the function at Q with that at R. Between R and Q only y changes and so once again Taylor's series in one dimension applies. See if you can write down the next two terms on the R.H.S. in

$$f(x+h,y+k) = f(x+h,y) +$$

****

21A

$k\,f'_y(x+h,y) + \frac{k^2}{2!} f''_{yy}(x+h,y)$

FRAME 22

We now have

$$f(x+h,y+k) = f(x+h,y) + k\,f'_y(x+h,y) + \frac{k^2}{2!} f''_{yy}(x+h,y) + \ldots \quad (22.1)$$

To express f(x+h,y+k) in terms of f(x,y) and its derivatives, (21.1) has to be used. This gives f(x+h,y) immediately, and the other terms on the R.H.S. of (22.1) can be found by differentiation.

Thus, going as far as second derivative terms,

$$f'_y(x+h,y) = f'_y(x,y) + h\,f''_{xy}(x,y) + \ldots$$

and $f''_{yy}(x+h,y) = f''_{yy}(x,y) + \ldots$

FRAME 22 continued

Substituting in (22.1) gives

$$f(x+h,y+k) = f(x,y) + h\,f'_x(x,y) + \frac{h^2}{2!}f''_{xx}(x,y) + \ldots$$
$$+ k\{f'_y(x,y) + h\,f''_{xy}(x,y) + \ldots\}$$
$$+ \frac{k^2}{2!}\{f''_{yy}(x,y) + \ldots\}$$

$$= f(x,y) + h\,f'_x(x,y) + k\,f'_y(x,y)$$
$$+ \frac{h^2}{2!}f''_{xx}(x,y) + hk\,f''_{xy}(x,y) + \frac{k^2}{2!}f''_{yy}(x,y) + \ldots \qquad (22.2)$$

This is TAYLOR'S SERIES in TWO DIMENSIONS. It can alternatively be expressed as

$$f(x+h,y+k) = f(x,y) + \left(h\frac{\partial}{\partial x} + k\frac{\partial}{\partial y}\right)f(x,y)$$
$$+ \frac{1}{2!}\left(h^2\frac{\partial^2}{\partial x^2} + 2hk\frac{\partial^2}{\partial x\partial y} + k^2\frac{\partial^2}{\partial y^2}\right)f(x,y) + \ldots$$

The R.H.S. can be written in the form

$$f(x,y) + \left(h\frac{\partial}{\partial x} + k\frac{\partial}{\partial y}\right)f(x,y) + \frac{1}{2!}\left(h\frac{\partial}{\partial x} + k\frac{\partial}{\partial y}\right)^2 f(x,y) + \ldots \qquad (22.3)$$

and this gives us a clue as to how the series continues. Can you guess what the next term in (22.3) will be?

****

22A

$$\frac{1}{3!}\left(h\frac{\partial}{\partial x} + k\frac{\partial}{\partial y}\right)^3 f(x,y)$$

FRAME 23

In the earlier programme on Partial Differentiation in Unit 1 it was suggested, though not actually proved, that if z is a function of two independent variables x and y then

$$\delta z \simeq \frac{\partial z}{\partial x}\delta x + \frac{\partial z}{\partial y}\delta y$$

The proof of this follows easily from Taylor's Series. From (22.2)

$$f(x+h,y+k) \simeq f(x,y) + h\frac{\partial}{\partial x}f(x,y) + k\frac{\partial}{\partial y}f(x,y)$$

FRAME 23 continued

Now, if $f(x,y) = z$, $h = \delta x$ and $k = \delta y$, so that $f(x+h,y+k) = z + \delta z$,

$$z + \delta z \simeq z + \frac{\partial z}{\partial x}\delta x + \frac{\partial z}{\partial y}\delta y$$

$$\text{i.e.} \quad \delta z \simeq \frac{\partial z}{\partial x}\delta x + \frac{\partial z}{\partial y}\delta y \qquad (23.1)$$

FRAME 24

Again, in the earlier programme the extension of this formula to a function of three independent variables x, y and t was mentioned as being

$$\delta z \simeq \frac{\partial z}{\partial x}\delta x + \frac{\partial z}{\partial y}\delta y + \frac{\partial z}{\partial t}\delta t$$

The proof of this follows from Taylor's Series in three dimensions just as (23.1) follows from (22.2). The extension of Taylor's Series to more than two dimensions is easily seen from the form (22.3). Thus, for example,

$$f(x+h,y+k,t+\ell) = f(x,y,t) + \left(h\frac{\partial}{\partial x} + k\frac{\partial}{\partial y} + \ell\frac{\partial}{\partial t}\right)f(x,y,t)$$
$$+ \frac{1}{2!}\left(h\frac{\partial}{\partial x} + k\frac{\partial}{\partial y} + \ell\frac{\partial}{\partial t}\right)^2 f(x,y,t) + \ldots$$

FRAME 25

Maxima and Minima

If you have a function of a single variable, say $z = f(x)$, you know how to find any maximum or minimum values of z. The problem can be interpreted as one of finding the turning points on the graph of $z = f(x)$. If z is a function of two independent variables x and y it is possible that, for certain values of these variables, z attains a maximum or minimum value. Now $z = f(x,y)$ can be interpreted as representing a surface.

z gives the height PQ of the point Q on the surface immediately above the point P in the plane Oxy. The problem of finding where maximum or minimum values, if any, occur can be regarded as finding the "tops of the hills" and the "greatest depths of the lakes". Some hills may be lower than others but their summits still represent maximum values in the mathematical sense of the word. Thus, in

FRAME 25 continued

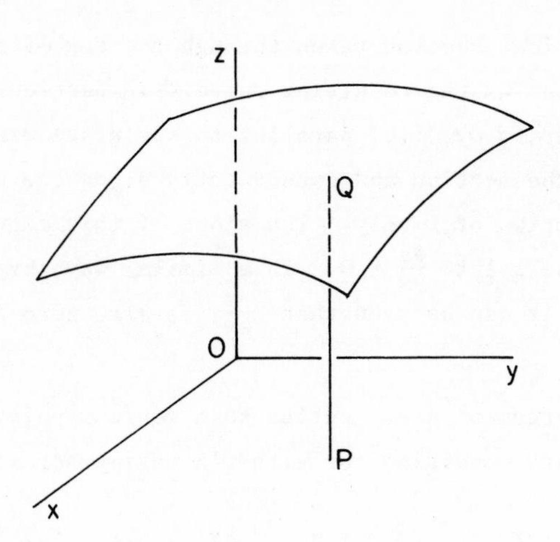

the contour map shown, there are maxima at A and B where, in each case, the height above sea level is greater than at any point in the immediate neighbourhood.

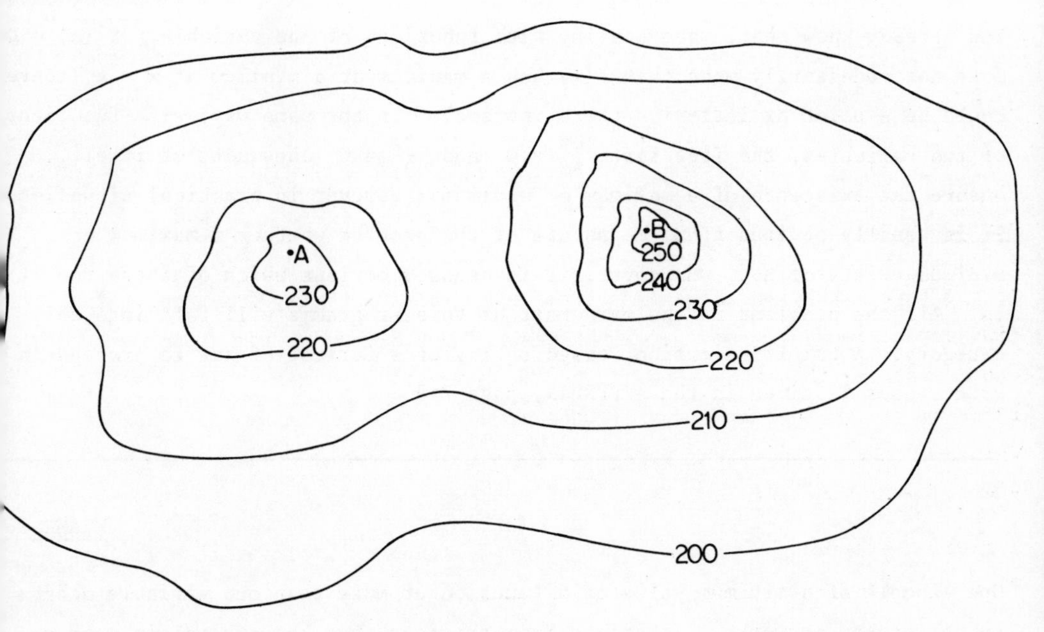

Similar remarks apply to minima.

FRAME 26

If you imagine a vertical section taken through the top of a hill, the summit will be a maximum point on the resulting curve. In particular, if the section is taken perpendicular to Oy, i.e. parallel to the plane Ozx, y will be constant throughout the section and consequently along the curve z will effectively be a function of x only. The slope of this curve is thus given by $\frac{\partial z}{\partial x}$ and at the maximum point $\frac{\partial z}{\partial x} = 0$. In a similar way, by taking a section perpendicular to Ox, it can be seen that $\frac{\partial z}{\partial y}$ is also zero at the maximum point.

Obviously a similar argument also applies to a minimum point. Thus, if $z = f(x,y)$, a necessary condition for either a maximum or a minimum value of z is that both

$$\frac{\partial z}{\partial x} = 0 \quad \text{and} \quad \frac{\partial z}{\partial y} = 0.$$

FRAME 27

You already know that, when dealing with functions of one variable, $f'(a) = 0$ does not necessarily mean that $f(x)$ has a maximum or a minimum at $x = a$ (there could be a point of inflexion there instead). In the same way, with functions of two variables, the fact that $\frac{\partial z}{\partial x} = 0$ and $\frac{\partial z}{\partial y} = 0$ does not, of itself, ensure the existence of a maximum or minimum. However in practical situations it is usually obvious from the nature of the problem whether a maximum or minimum exists or not. Moreover, it is usually obvious which of these two it is. All the problems in the main part of this programme will fall into this category. A brief indication, based on Taylor's Series, of how to proceed in other cases is given in the APPENDIX, page 7:35.

FRAME 28

One example of a minimum value of a function of more than one variable occurs in resistance networks. It arises from the fact that the current in such a network distributes itself so that the total heat generated is a minimum.

The diagram illustrates one such network, a given current i entering and

FRAME 28 continued

leaving at opposite corners as shown. The problem is to find the currents in all the branches. The application of Kirchhoff's first law to the junctions effectively reduces the problem to that of finding just i_1 and i_2. Now, using the fact that the heat generated per unit time in a resistance R by a current I is RI^2, the heat generated per unit time in the network is given by

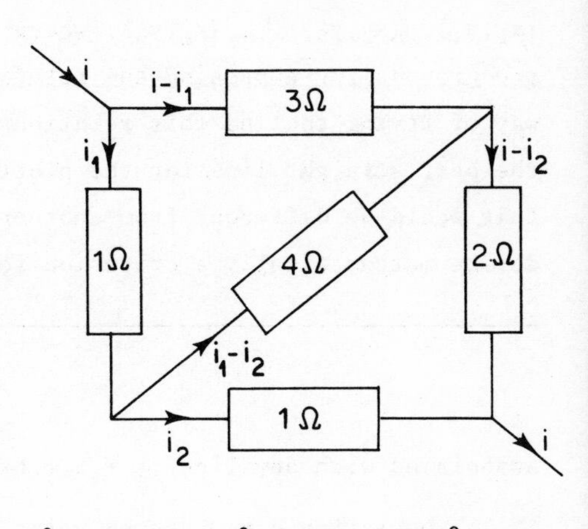

$$H = i_1{}^2 + i_2{}^2 + 2(i - i_2)^2 + 3(i - i_1)^2 + 4(i_1 - i_2)^2$$

For H to have a minimum value, $\dfrac{\partial H}{\partial i_1} = 0$ and $\dfrac{\partial H}{\partial i_2} = 0$. Use these two equations to find i_1 and i_2 in terms of i.

****

28A

$$\frac{\partial H}{\partial i_1} = 2(8i_1 - 4i_2 - 3i) \qquad \frac{\partial H}{\partial i_2} = -2(4i_1 - 7i_2 + 2i)$$

$$i_1 = \frac{29}{40}i \qquad i_2 = \frac{7}{10}i$$

FRAME 29

Another example occurs in curve fitting by the method of least squares. You may already have met this technique if you have studied statistics or numerical methods. The simplest case is that of finding the straight line which best fits a set of experimental data.

Suppose values of y, corresponding to a given series of values of x, have been measured and are therefore subject to error. When plotted they will be represented by a set of points. Let these points be

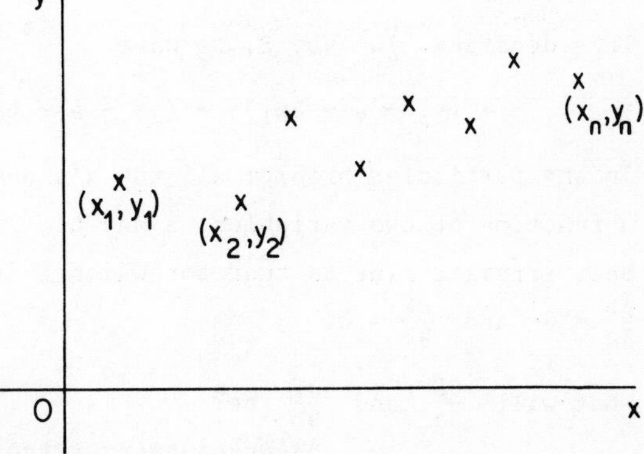

FRAME 29 continued

(x_1,y_1), (x_2,y_2) ... (x_n,y_n), where n is the number of observations. If there are grounds for supposing the relationship between x and y to be linear, one way of trying to find this relationship would be to draw what appears to be the best straight line for the plotted points. However, one person's idea of this would be different from another's. A less hit-and-miss approach is to define mathematically a criterion for deciding which is the best straight line.

FRAME 30

Associated with any line $y = a + bx$ will be a set of y-deviations given by

Deviation d = Observed value of y - Value of y given by the line

For some points this deviation will be positive while for others it will be negative and so the actual distance shown on the diagram for each point is $|d|$. An obvious criterion to adopt would be for $\sum|d|$ to be as small as possible, but this is awkward to deal with mathematically. Instead $\sum d^2$ is used.

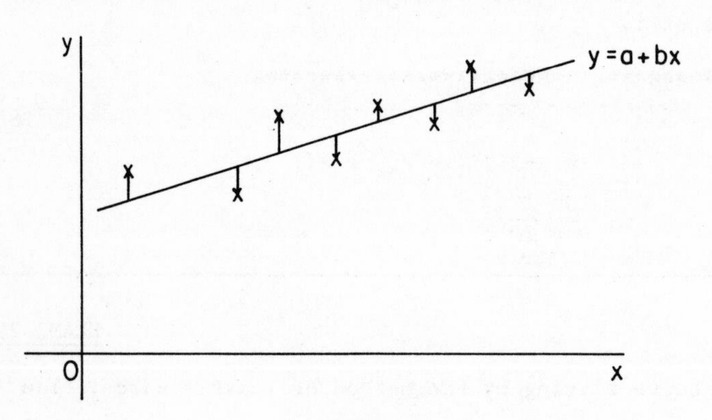

For (x_1,y_1) the value of d is $y_1 - (a + bx_1)$, i.e. $y_1 - a - bx_1$, for (x_2,y_2) it is $y_2 - a - bx_2$, and so on.

Thus denoting $\sum d^2$ by S, we have

$$S = (y_1 - a - bx_1)^2 + (y_2 - a - bx_2)^2 + \ldots + (y_n - a - bx_n)^2$$

In any particular problem all the x's and y's will be fixed and S is therefore a function of two variables, a and b. The least squares method defines the best straight line as that for which S is a minimum. For this we must have $\frac{\partial S}{\partial a} = 0$ and $\frac{\partial S}{\partial b} = 0$.

What will $\frac{\partial S}{\partial a}$ and $\frac{\partial S}{\partial b}$ be?

****

30A

$$\frac{\partial S}{\partial a} = -2(y_1 - a - bx_1) - 2(y_2 - a - bx_2) - \ldots - 2(y_n - a - bx_n)$$

$$\frac{\partial S}{\partial b} = -2x_1(y_1 - a - bx_1) - 2x_2(y_2 - a - bx_2) - \ldots - 2x_n(y_n - a - bx_n)$$

FRAME 31

$\frac{\partial S}{\partial a}$ can be rearranged as

$$-2\{(y_1 + y_2 + \ldots + y_n) - na - b(x_1 + x_2 + \ldots + x_n)\}$$

which, using $\sum$ notation, is

$$-2\left\{\sum_{i=1}^{n} y_i - na - b\sum_{i=1}^{n} x_i\right\}$$

Now rearrange $\frac{\partial S}{\partial b}$ into a similar form.

****

31A

$$-2\left\{\sum_{i=1}^{n} x_i y_i - a\sum_{i=1}^{n} x_i - b\sum_{i=1}^{n} x_i^2\right\}$$

FRAME 32

Putting $\frac{\partial S}{\partial a} = 0$ and $\frac{\partial S}{\partial b} = 0$ leads to

$$na + \left(\sum_{i=1}^{n} x_i\right)b = \sum_{i=1}^{n} y_i$$

$$\left(\sum_{i=1}^{n} x_i\right)a + \left(\sum_{i=1}^{n} x_i^2\right)b = \sum_{i=1}^{n} x_i y_i$$

These are two simultaneous equations (called the normal equations) which can be solved for a and b. As there is only one solution this must give the required minimum.

FRAME 33

Lagrange Multipliers

You have already met examples of maxima and minima involving constraints. One such example is the case of a cylindrical tank (closed at both ends) which is to be constructed so that its capacity V is fixed and its surface area S is a minimum. Here S is a function of the radius r and height h, i.e. $S = 2\pi rh + 2\pi r^2$, but r and h are not independent variables as they are subject to the constraint $\pi r^2 h = V$.

Another, very simple, example of a constraint problem is the following: A farmer wishes to fence off a rectangular area in a field using a given length ℓ of fencing. What should be the dimensions of the rectangle for its area A to be a maximum?

In this case, if x and y are the length and breadth of the rectangle it is necessary to maximise $A(= xy)$ subject to the constraint $2x + 2y = \ell$. The method which you have used previously to solve such a problem was to eliminate one of the variables in A by substitution from the constraint. If, for example, y is eliminated, $A = x(\frac{1}{2}\ell - x)$.

Then $\frac{dA}{dx} = \frac{1}{2}\ell - 2x$ and putting this equal to zero gives $x = \frac{1}{4}\ell$.

From the constraint, $y = \frac{1}{4}\ell$ also and thus the farmer should choose a square.

FRAME 34

An alternative approach that can be used in a problem of this type is that known as the method of LAGRANGE MULTIPLIERS. We shall first illustrate this technique by applying it to the farmer's problem and then, afterwards, you will see the justification for it. Here the function xy has to be maximised subject to the constraint $2x + 2y - \ell = 0$.

We start by inventing the function

$$G = xy + \lambda(2x + 2y - \ell) \quad \text{where } \lambda \text{ is a constant}$$

Then $\frac{\partial G}{\partial x} = y + 2\lambda$ and $\frac{\partial G}{\partial y} = x + 2\lambda$

FRAME 34 continued

Next, $\frac{\partial G}{\partial x}$ and $\frac{\partial G}{\partial y}$ are put equal to zero. The two resulting equations, together with the constraint, then form three simultaneous equations in x, y and λ, i.e.

$$y + 2\lambda = 0 \qquad (34.1)$$

$$x + 2\lambda = 0 \qquad (34.2)$$

$$2x + 2y - \ell = 0 \qquad (34.3)$$

From (34.1) and (34.2), $x = y$

Substitution in (34.3) gives $x = \frac{1}{4}\ell$, $y = \frac{1}{4}\ell$.

FRAME 35

Two things should be noticed about this technique. First, the value of λ is not actually required for its own sake, though it may sometimes help in finding the other variables. The second point is that the method only locates the stationary points, it does not determine their nature. However it is very useful in those practical problems where physical considerations enable you to distinguish between maxima and minima.

You may not think that this method is particularly advantageous in the farmer's problem. It isn't! The problem was chosen to give a simple illustration of the process. You will see later that the value of Lagrange multipliers lies in those problems where the initial elimination of a variable leads to awkward differentiation.

FRAME 36

Now use this method to find the relation between the height and the radius in the tank problem given at the beginning of FRAME 33.

****

36A

$G = 2\pi rh + 2\pi r^2 + \lambda(\pi r^2 h - V)$

$\frac{\partial G}{\partial r} = 0$ and $\frac{\partial G}{\partial h} = 0$ give, respectively,

$$h + 2r + \lambda rh = 0 \qquad (36A.1)$$

$$r(2 + \lambda r) = 0 \qquad (36A.2)$$

and the constraint is $$\pi r^2 h - V = 0 \qquad (36A.3)$$

(36A.2) gives $\lambda r = -2$, as $r = 0$ is not a sensible solution of the problem. Substitution in (36A.1) then gives $h = 2r$. (36A.3) would then have to be used if r and h were required in terms of V.

FRAME 37

In more general terms the type of problem now being considered may be described as maximising or minimising a function $F(x,y)$ subject to a constraint $f(x,y) = 0$.

Using the first method by which the farmer's problem was solved, it is possible to regard F as a function of one variable only, say x. A maximum or minimum for F will then be given by $\frac{dF}{dx} = 0$.

From the earlier programme on Partial Differentiation you know that, if F is a function of two variables x and y which are not independent,

$$\frac{dF}{dx} = \frac{\partial F}{\partial x} + \frac{\partial F}{\partial y}\frac{dy}{dx} \qquad (37.1)$$

where $\frac{dy}{dx}$, in this case, must be obtained from $f(x,y) = 0$.

Differentiating this last equation w.r.t. x gives

$$\frac{\partial f}{\partial x} + \frac{\partial f}{\partial y}\frac{dy}{dx} = 0$$

$$\therefore \quad \frac{dy}{dx} = -\frac{\partial f}{\partial x}\bigg/\frac{\partial f}{\partial y} \qquad (37.2)$$

Thus, combining (37.1) and (37.2), the condition $\frac{dF}{dx} = 0$ is equivalent to

$$\frac{\partial F}{\partial x} - \frac{\partial F}{\partial y}\left(\frac{\partial f}{\partial x}\bigg/\frac{\partial f}{\partial y}\right) = 0$$

$$\text{i.e.} \quad \frac{\partial F}{\partial x}\frac{\partial f}{\partial y} - \frac{\partial F}{\partial y}\frac{\partial f}{\partial x} = 0$$

In the next frame you will see that the method of Lagrange multipliers leads to the same condition.

FRAME 38

Now the function used for G in the examples in FRAMES 34 and 36 was of the form $G = F + \lambda f$.

$\dfrac{\partial G}{\partial x} = 0$ and $\dfrac{\partial G}{\partial y} = 0$ then give, respectively,

$$\frac{\partial F}{\partial x} + \lambda \frac{\partial f}{\partial x} = 0$$

$$\frac{\partial F}{\partial y} + \lambda \frac{\partial f}{\partial y} = 0$$

From the first equation, $\lambda = -\dfrac{\partial F}{\partial x}\Big/\dfrac{\partial f}{\partial x}$

and from the second, $\lambda = -\dfrac{\partial F}{\partial y}\Big/\dfrac{\partial f}{\partial y}$

So, for consistency, $\dfrac{\partial F}{\partial x}\Big/\dfrac{\partial f}{\partial x} = \dfrac{\partial F}{\partial y}\Big/\dfrac{\partial f}{\partial y}$

i.e. $\dfrac{\partial F}{\partial x}\dfrac{\partial f}{\partial y} - \dfrac{\partial F}{\partial y}\dfrac{\partial f}{\partial x} = 0$

As the same condition is obtained by the two methods, they are seen to be equivalent.

FRAME 39

The following problem was worked in Vol I without the use of Lagrange Multipliers. We suggest you now try it making use of this method.

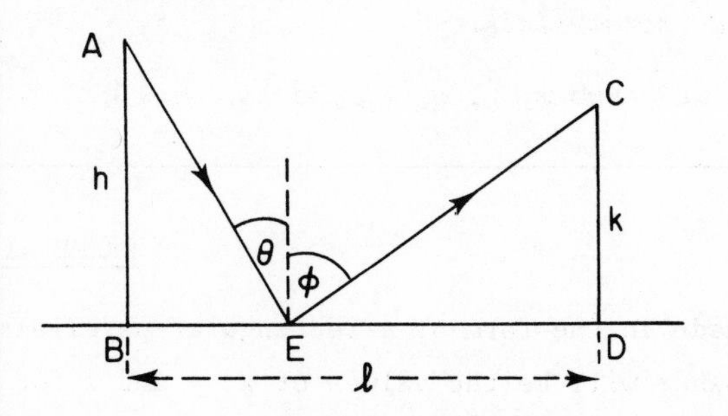

In the diagram a ray of light is to travel from A to C by reflection at a point on the surface BD. The principle of least time states that it will do so in such a way that the time taken is a minimum. Show that, as a consequence of this principle, the point of reflection E is such that $\phi = \theta$.

****

39A

It is necessary to minimise $(h \sec\theta + k \sec\phi)/c$*, where* c *is the speed of light, subject to the constraint* $h \tan\theta + k \tan\phi = \ell$.

$$G = \frac{1}{c}(h \sec\theta + k \sec\phi) + \lambda(h \tan\theta + k \tan\phi - \ell)$$

$\frac{\partial G}{\partial \theta} = 0$ *and* $\frac{\partial G}{\partial \phi} = 0$ *lead to*

$$\sin\theta + c\lambda = 0$$

$$\sin\phi + c\lambda = 0$$

whence $\phi = \theta$

FRAME 40

The method of Lagrange multipliers can be extended to functions of more than two variables with one or more constraints, provided that the number of variables exceeds the number of constraints. Thus if it is desired to minimise or maximise $F(x,y,z)$ subject to the constraint $f(x,y,z) = 0$, the function $G = F + \lambda f$ is formed. This then leads to four equations $\frac{\partial G}{\partial x} = 0$, $\frac{\partial G}{\partial y} = 0$, $\frac{\partial G}{\partial z} = 0$ and the constraint, for the four unknowns x,y,z and λ.

If, on the other hand, it is desired to minimise or maximise $F(x,y,z)$ subject to two constraints $f_1(x,y,z) = 0$ and $f_2(x,y,z) = 0$, the G function is now given by $G = F + \lambda_1 f_1 + \lambda_2 f_2$. What equations would you now expect to have to use to complete the problem and what would be the unknowns in these equations?

****

40A

$\frac{\partial G}{\partial x} = 0$, $\frac{\partial G}{\partial y} = 0$, $\frac{\partial G}{\partial z} = 0$, *and the two constraints.*

These are five equations in the five unknowns x, y, z, λ_1 *and* λ_2.

FRAME 41

A thermal nuclear reactor is to be made in the form of a rectangular parallelepiped whose edges are x, y and z. What will be the values of x, y and z for the reactor to have a minimum volume, subject to the constraint

$$\left(\frac{\pi}{x}\right)^2 + \left(\frac{\pi}{y}\right)^2 + \left(\frac{\pi}{z}\right)^2 = k^2 \quad \text{where } k \text{ is a constant?}$$

****

41A

For simplicity, the constraint can be rewritten $\frac{1}{x^2} + \frac{1}{y^2} + \frac{1}{z^2} = c$, *where* $c = \frac{k^2}{\pi^2}$.

$$G = xyz + \lambda\left(\frac{1}{x^2} + \frac{1}{y^2} + \frac{1}{z^2} - c\right)$$

$$\left.\begin{aligned}\frac{\partial G}{\partial x} &= yz - \frac{2\lambda}{x^3} = 0\\ \frac{\partial G}{\partial y} &= xz - \frac{2\lambda}{y^3} = 0\\ \frac{\partial G}{\partial z} &= xy - \frac{2\lambda}{z^3} = 0\end{aligned}\right\}$$

These three equations give $x = y = z$ *i.e. a cube.*

Substitution in the constraint equation gives the side of the cube to be $\pi\sqrt{3}/k$.

FRAME 42

Miscellaneous Examples

In this frame a collection of miscellaneous examples is given for you to try. Answers are supplied in FRAME 43, together with such working as is considered helpful. Omit Question 10 if you did not read the APPENDIX.

1. If ϕ is a function of the independent variables x, y, z which are changed to independent variables u, v, w by the transformation

$$x = vw/u, \quad y = wu/v, \quad z = uv/w,$$

show that

$$u\frac{\partial\phi}{\partial u} + v\frac{\partial\phi}{\partial v} + w\frac{\partial\phi}{\partial w} = x\frac{\partial\phi}{\partial x} + y\frac{\partial\phi}{\partial y} + z\frac{\partial\phi}{\partial z}$$ (L.U.)

2. If $u = ax + by$, $v = bx - ay$, prove that

(i) $\left(\frac{\partial u}{\partial x}\right)_y \left(\frac{\partial x}{\partial u}\right)_v = \frac{a^2}{a^2 + b^2}$,

(ii) $\left(\frac{\partial y}{\partial v}\right)_x \left(\frac{\partial v}{\partial y}\right)_u = \frac{a^2 + b^2}{a^2}$,

(iii) $\frac{\partial^2 w}{\partial x \partial y} = ab\left(\frac{\partial^2 w}{\partial u^2} - \frac{\partial^2 w}{\partial v^2}\right) + \left(b^2 - a^2\right)\frac{\partial^2 w}{\partial u \partial v}$ where $w = f(x,y) = \phi(u,v)$.

(L.U.)

FRAME 42 continued

3. Find the Jacobian of the transformation in Qu. 1.

4. Assuming that $\int_0^{\pi} \frac{dx}{a - \cos x} = \frac{\pi}{\sqrt{a^2 - 1}}$, $a > 1$, and differentiating under the integral sign deduce the values of

$$\int_0^{\pi} \frac{dx}{(2 - \cos x)^2} \quad \text{and} \quad \int_0^{\pi} \frac{dx}{(5 - 3\cos x)^3} \qquad \text{(C.E.I.)}$$

5. Differentiation under the integral sign can also be used to obtain certain indefinite integrals. Use this technique to find $\int x \cos px\, dx$ given that $\int \sin px\, dx = -\frac{1}{p}\cos px$.

6. Bessel's function of order zero is given by

$$J_0(x) = \frac{1}{\pi}\int_0^{\pi} \cos(x \sin\theta)d\theta$$

Verify, using differentiation under the integral sign, that $y = J_0(x)$ satisfies Bessel's equation of order zero, i.e.

$$\frac{d^2y}{dx^2} + \frac{1}{x}\frac{dy}{dx} + y = 0$$

7. A capacitor is formed from four concentric conducting spherical shells whose radii are a, x, y, b where $a < x < y < b$. The shells are connected in such a way that the capacity C is given by

$$C = \frac{ax}{x - a} + \frac{xy}{y - x} + \frac{yb}{b - y}$$

Find x and y in terms of a and b for C to be a minimum.

FRAME 42 continued

8. It is suspected that n points, representing experimental readings in which y is subject to error, lie approximately on a parabola with equation of the form $y = a + bx + cx^2$. Use a least squares technique similar to that in FRAMES 29-32 to find the normal equations which will give the values of a, b and c for the parabola of best fit.

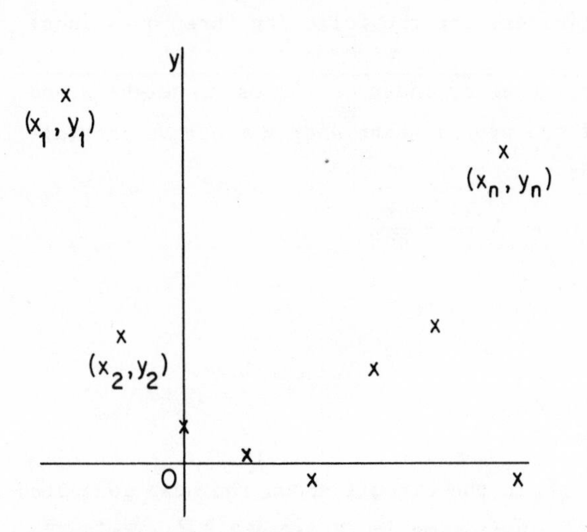

(NOTE: You will need to know how to minimise a function of three independent variables. If z is a function of three such variables u, v and w, maximum or minimum values of z can only occur if $\frac{\partial z}{\partial u} = 0$, $\frac{\partial z}{\partial v} = 0$, $\frac{\partial z}{\partial w} = 0$.)

9. In economics, cost-production problems may involve maximising or minimising a function of more than one variable. For instance, a ball-point pen manufacturer finds that it costs him 50 p to make a pen and 5 p to make a refill. He has a monopoly of the market for both products. The demand equations for the pens and refills are respectively

$$Q_1 = 32\,400 - 120(5P_1 + 2P_2)$$
$$Q_2 = 14\,520 - 40(6P_1 + 5P_2)$$

where Q_1, Q_2 are the expected sales of pens and refills respectively and P_1, P_2 are the corresponding selling prices. Find P_1 and P_2 for maximum net revenue.

10. If $z = (x^2 + y^2)^2 - 2(x^2 - y^2)$ find the points (x,y) at which z is a maximum or a minimum, and state which it is. (L.U.)

FRAME 42 continued

Use the method of Lagrange multipliers for the following three questions:

11. A particle is in a right circular cylinder of radius r, height h and fixed volume. Find r/h if the ground state energy E of the particle is to be a minimum, given that

$$E = k\left(\frac{a^2}{r^2} + \frac{\pi^2}{h^2}\right)$$

where k and a are constants.

12.

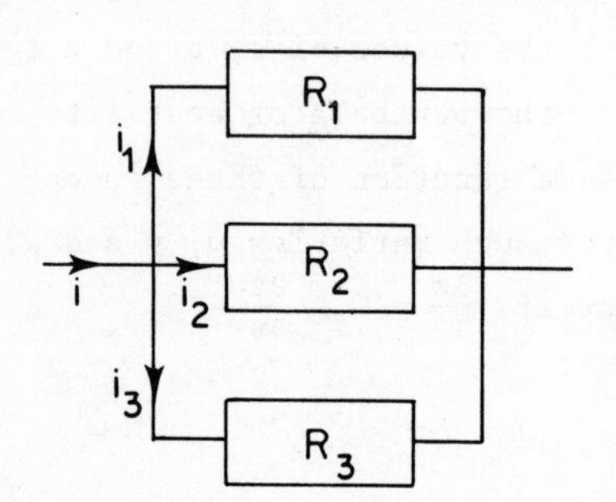

In the circuit shown the heat generated per unit time is $R_1 i_1{}^2 + R_2 i_2{}^2 + R_3 i_3{}^2$. Find i_1, i_2, i_3 in terms of i, R_1, R_2, R_3 if the currents are subject to the constraint $i_1 + i_2 + i_3 = i$.

13. A building is to have a rectangular floor, a flat roof and is to have a volume of 72 000 m^3. The front wall will cost twice as much per unit area as the other walls and the roof $1\frac{1}{2}$ times as much. Neglecting the cost of the floor, find the dimensions of the building for minimum cost.

FRAME 43

Answers to Miscellaneous Examples

1. $$\frac{\partial\phi}{\partial u} = \frac{\partial\phi}{\partial x}\frac{\partial x}{\partial u} + \frac{\partial\phi}{\partial y}\frac{\partial y}{\partial u} + \frac{\partial\phi}{\partial z}\frac{\partial z}{\partial u} = -\frac{vw}{u^2}\frac{\partial\phi}{\partial x} + \frac{w}{v}\frac{\partial\phi}{\partial y} + \frac{v}{w}\frac{\partial\phi}{\partial z}$$

$$u\frac{\partial\phi}{\partial u} = -x\frac{\partial\phi}{\partial x} + y\frac{\partial\phi}{\partial y} + z\frac{\partial\phi}{\partial z}$$

$v\dfrac{\partial\phi}{\partial v}$ and $w\dfrac{\partial\phi}{\partial w}$ are obtained similarly, whence result.

FRAME 43 continued

2. (i) and (ii) $\left(\frac{\partial u}{\partial x}\right)_y$ and $\left(\frac{\partial y}{\partial v}\right)_x$ can be written down immediately.

$\left(\frac{\partial x}{\partial u}\right)_v$ can be obtained after eliminating y between $u = ax + by$ and $v = bx - ay$, or by the method of FRAME 10.

$\left(\frac{\partial v}{\partial y}\right)_u$ can be obtained after eliminating x between $u = ax + by$ and $v = bx - ay$.

(iii) $\frac{\partial w}{\partial x} = a\frac{\partial w}{\partial u} + b\frac{\partial w}{\partial v}$ $\qquad$ $\frac{\partial w}{\partial y} = b\frac{\partial w}{\partial u} - a\frac{\partial w}{\partial v}$

$$\frac{\partial^2 w}{\partial x \partial y} = \left(a\frac{\partial}{\partial u} + b\frac{\partial}{\partial v}\right)\left(b\frac{\partial w}{\partial u} - a\frac{\partial w}{\partial v}\right) = \text{answer}$$

3.
$$\begin{vmatrix} -\frac{vw}{u^2} & \frac{w}{u} & \frac{v}{u} \\ \frac{w}{v} & -\frac{wu}{v^2} & \frac{u}{v} \\ \frac{v}{w} & \frac{u}{w} & -\frac{uv}{w^2} \end{vmatrix} = 4$$

4. $$\int_0^{\pi} \frac{dx}{(a - \cos x)^2} = \frac{\pi a}{(a^2 - 1)^{3/2}}$$

which, when $a = 2$, gives $\int_0^{\pi} \frac{dx}{(2 - \cos x)^2} = \frac{2\pi}{3\sqrt{3}}$

$$\int_0^{\pi} \frac{dx}{(a - \cos x)^3} = \frac{\pi(2a^2 + 1)}{2(a^2 - 1)^{5/2}}$$

which, when $a = \frac{5}{3}$, gives $\int_0^{\pi} \frac{dx}{(5 - 3\cos x)^3} = \frac{59\pi}{2048}$

5. $(\cos px + px \sin px)/p^2$

FRAME 43 continued

6. $$\frac{dy}{dx} = -\frac{1}{\pi}\int_0^\pi \sin\theta \sin(x\sin\theta)d\theta \qquad (43.1)$$

$$= -\frac{1}{\pi}\int_0^\pi x\cos^2\theta\cos(x\sin\theta)d\theta \quad \text{on integration by parts}$$

Differentiating (43.1) w.r.t. x gives

$$\frac{d^2y}{dx^2} = -\frac{1}{\pi}\int_0^\pi \sin^2\theta\cos(x\sin\theta)d\theta$$

$$\frac{d^2y}{dx^2} + \frac{1}{x}\frac{dy}{dx} + y = -\frac{1}{\pi}\int_0^\pi (\sin^2\theta + \cos^2\theta - 1)\cos(x\sin\theta)d\theta = 0$$

7. $$\frac{\partial C}{\partial x} = -\frac{a^2}{(x-a)^2} + \frac{y^2}{(y-x)^2} \qquad \frac{\partial C}{\partial y} = -\frac{x^2}{(y-x)^2} + \frac{b^2}{(b-y)^2}$$

$$\frac{\partial C}{\partial x} = 0 \text{ gives } \frac{y}{y-x} = \frac{a}{x-a}$$

$$\frac{\partial C}{\partial y} = 0 \text{ gives } \frac{x}{y-x} = \frac{b}{b-y}$$

$$x = \frac{3ab}{a+2b} \qquad y = \frac{3ab}{2a+b}$$

8. $$S = \sum_{i=1}^n (y_i - a - bx_i - cx_i^2)^2$$

$$\frac{\partial S}{\partial a} = -2\sum_{i=1}^n (y_i - a - bx_i - cx_i^2)$$

$$\frac{\partial S}{\partial b} = -2\sum_{i=1}^n x_i(y_i - a - bx_i - cx_i^2)$$

$$\frac{\partial S}{\partial c} = -2\sum_{i=1}^n x_i^2(y_i - a - bx_i - cx_i^2)$$

Normal equations are:

$$na + \left(\sum_{i=1}^n x_i\right)b + \left(\sum_{i=1}^n x_i^2\right)c = \sum_{i=1}^n y_i$$

FRAME 43 continued

$$\left(\sum_{i=1}^{n} x_i\right)a + \left(\sum_{i=1}^{n} x_i^2\right)b + \left(\sum_{i=1}^{n} x_i^3\right)c = \sum_{i=1}^{n} x_i y_i$$

$$\left(\sum_{i=1}^{n} x_i^2\right)a + \left(\sum_{i=1}^{n} x_i^3\right)b + \left(\sum_{i=1}^{n} x_i^4\right)c = \sum_{i=1}^{n} x_i^2 y_i$$

9. Net revenue $R = (P_1 - 50)Q_1 + (P_2 - 5)Q_2$

$$= (P_1 - 50)\{32\,400 - 120(5P_1 + 2P_2)\} + (P_2 - 5)\{14\,520 - 40(6P_1 + 5P_2)\}$$

$$\frac{\partial R}{\partial P_1} = 63\,600 - 1200P_1 - 480P_2$$

$$\frac{\partial R}{\partial P_2} = 27\,520 - 480P_1 - 400P_2$$

$\frac{\partial R}{\partial P_1} = 0$ and $\frac{\partial R}{\partial P_2} = 0$ lead to $P_1 = 49$ and $P_2 = 10$.

10. $\frac{\partial z}{\partial x} = 4x(x^2 + y^2 - 1)$ $\qquad$ $\frac{\partial z}{\partial y} = 4y(x^2 + y^2 + 1)$

$\frac{\partial z}{\partial x} = 0$ and $\frac{\partial z}{\partial y} = 0$ give the points $(0,0)$, $(1,0)$, $(-1,0)$.

Minimum at $(1,0)$ and $(-1,0)$.

11. $G = k\left(\frac{a^2}{r^2} + \frac{\pi^2}{h^2}\right) + \lambda(\pi r^2 h - V)$ where V is the fixed volume.

$\frac{\partial G}{\partial r} = -\frac{2ka^2}{r^3} + 2\lambda\pi rh$ $\qquad$ $\frac{\partial G}{\partial h} = -\frac{2k\pi^2}{h^3} + \lambda\pi r^2$

$\frac{\partial G}{\partial r} = 0$ and $\frac{\partial G}{\partial h} = 0$ lead to $\frac{r}{h} = \frac{a}{\pi\sqrt{2}}$

12. $G = R_1 i_1^2 + R_2 i_2^2 + R_3 i_3^2 + \lambda(i_1 + i_2 + i_3 - i)$

$\frac{\partial G}{\partial i_1} = 2R_1 i_1 + \lambda$ $\qquad$ $\frac{\partial G}{\partial i_2} = 2R_2 i_2 + \lambda$ $\qquad$ $\frac{\partial G}{\partial i_3} = 2R_3 i_3 + \lambda$

Putting each of these equal to zero gives $R_1 i_1 = R_2 i_2 = R_3 i_3$

$i_1 = R_2 R_3 i/S$ $\qquad$ $i_2 = R_3 R_1 i/S$ $\qquad$ $i_3 = R_1 R_2 i/S$

where $S = R_1 R_2 + R_2 R_3 + R_3 R_1$.

FRAME 43 continued

13. If x = length of front, y = length of side, z = height,

cost $= 3xz + 2yz + \frac{3}{2}xy$ where the unit is cost per unit area of a side wall

$$G = 3xz + 2yz + \frac{3}{2}xy + \lambda(xyz - 72\,000)$$

$$\frac{\partial G}{\partial x} = 3z + \frac{3}{2}y + \lambda yz \qquad \frac{\partial G}{\partial y} = 2z + \frac{3}{2}x + \lambda xz$$

$$\frac{\partial G}{\partial z} = 3x + 2y + \lambda xy$$

Putting each equal to zero leads to $x = 40$, $y = 60$, $z = 30$.

APPENDIX

FRAME A1

Second Derivative Tests for Maxima and Minima

You have already seen how the sign of the second derivative can help you to distinguish between maximum and minimum values of a function of one variable. When two independent variables are involved, the situation is somewhat more complicated. The following criteria will usually give you the solution to the problem.

Given that $\frac{\partial z}{\partial x} = 0$ and $\frac{\partial z}{\partial y} = 0$, then if

$$\frac{\partial^2 z}{\partial x^2}\frac{\partial^2 z}{\partial y^2} - \left(\frac{\partial^2 z}{\partial x \partial y}\right)^2 > 0 \qquad (A1.1)$$

z has a <u>maximum</u> value if $\frac{\partial^2 z}{\partial x^2}$ and $\frac{\partial^2 z}{\partial y^2}$ are <u>negative</u>,

z has a <u>minimum</u> value if $\frac{\partial^2 z}{\partial x^2}$ and $\frac{\partial^2 z}{\partial y^2}$ are <u>positive</u>.

If you think about the sections mentioned in FRAME 26 you will realise that the statement about the signs of $\frac{\partial^2 z}{\partial x^2}$ and $\frac{\partial^2 z}{\partial y^2}$ is simply an extension of the corresponding result for a function of one variable. However, there is nothing in the corresponding work on one variable that would immediately suggest the inequality (A1.1), but it does follow from Taylor's Series as we shall now show.

FRAME A2

In the programme on Infinite Series in Unit 1, Taylor's Series in one dimension was used to investigate the nature of a stationary value of a function of one variable. There, values of f(x) in the neighbourhood of the stationary point were considered. A similar procedure can be followed for a function of two variables, using Taylor's Series in two dimensions.

Let $x = a$, $y = b$ be a solution of the simultaneous equations $\frac{\partial z}{\partial x} = 0$, $\frac{\partial z}{\partial y} = 0$.

FRAME A2 continued

It is then necessary to investigate the behaviour of the function $z = f(x,y)$ in the neighbourhood of the point (a,b). To do this an approximate expression for f(a+h,b+k) is written down from Taylor's Series, assuming that h and k are sufficiently small for third order and higher terms to be neglected. Can you do this?

****

A2A

$$f(a+h,b+k) \simeq f(a,b) + \frac{h^2}{2} f''_{xx}(a,b) + hk\, f''_{xy}(a,b) + \frac{k^2}{2} f''_{yy}(a,b)$$

FRAME A3

You will now see that whether there is a maximum or minimum at (a,b) depends on the value of the expression

$$\frac{h^2}{2} f''_{xx}(a,b) + hk\, f''_{xy}(a,b) + \frac{k^2}{2} f''_{yy}(a,b)$$

which, for brevity, we shall write as

$$\frac{h^2}{2} f_{xx} + hk\, f_{xy} + \frac{k^2}{2} f_{yy} \qquad (A3.1)$$

What can you say about the value of z at (a,b) if this expression is

(i) positive for all h,k,

(ii) negative for all h,k?

****

A3A

(i) z will have a minimum value

(ii) z will have a maximum value.

FRAME A4

There is, of course, the possibility (albeit remote) that (A3.1) may be zero for some or all values of h and k. In that case one would have to proceed to the next stage in Taylor's Series and the problem becomes too complicated for

FRAME A4 continued

a text of this kind. In what follows it will be assumed, therefore, that (A3.1) is not zero.

Now an alternative way of writing this expression is

$$\frac{1}{2} f_{xx}\left(h^2 + 2hk \frac{f_{xy}}{f_{xx}} + k^2 \frac{f_{yy}}{f_{xx}}\right) \qquad (A4.1)$$

For a minimum value of z this has to be positive and you have already seen that f_{xx} must be positive. Therefore

$$h^2 + 2hk \frac{f_{xy}}{f_{xx}} + k^2 \frac{f_{yy}}{f_{xx}} \qquad (A4.2)$$

must be positive.

Now what can you say about the sign of (A4.2) for z to have a maximum value?

****

A4A

As (A4.1) has to be negative, and f_{xx} is negative, it follows that (A4.2) must be positive.

FRAME A5

Thus, for either a maximum or a minimum, (A4.2) must be positive for all h,k. This expression can be rewritten in the form

(h + ----)^2 + --------

Can you fill in the gaps?

****

A5A

$$\left(h + k\frac{f_{xy}}{f_{xx}}\right)^2 + k^2 \frac{f_{yy}}{f_{xx}} - k^2 \frac{f_{xy}^2}{f_{xx}^2} \qquad (A5A.1)$$

FRAME A6

Now, as h and k vary, the least value of $\left(h + k\dfrac{f_{xy}}{f_{xx}}\right)^2$ is zero.

Thus, if (A5A.1) is to be positive for all h, k, it follows that

$$k^2 \frac{f_{xx}f_{yy} - f_{xy}^{\ 2}}{f_{xx}^{\ 2}} > 0$$

$$\text{i.e.} \quad f_{xx}f_{yy} - f_{xy}^{\ 2} > 0$$

which is (A1.1).

Conversely, if $f_{xx}f_{yy} - f_{xy}^{\ 2} < 0$, there cannot possibly be a maximum or minimum at the point in question.

FRAME A7

Example

Find the maximum and/or minimum values of xye^{2x-3y}.

Writing $z = xye^{2x-3y}$

we have $\dfrac{\partial z}{\partial x} = y(1 + 2x)e^{2x-3y}$ and $\dfrac{\partial z}{\partial y} = x(1 - 3y)e^{2x-3y}$

At what points will both $\dfrac{\partial z}{\partial x}$ and $\dfrac{\partial z}{\partial y}$ be zero?

****

A7A

$\dfrac{\partial z}{\partial x} = 0$ gives $y = 0$ or $x = -\dfrac{1}{2}$

$\dfrac{\partial z}{\partial y} = 0$ gives $x = 0$ when $y = 0$

or $y = \dfrac{1}{3}$ when $x = -\dfrac{1}{2}$

Points are (0,0) and $\left(-\dfrac{1}{2}, \dfrac{1}{3}\right)$

FRAME A8

The derivatives required for testing for maxima or minima at these points are

$$\frac{\partial^2 z}{\partial x^2} = 4y(1 + x)e^{2x-3y} \qquad \frac{\partial^2 z}{\partial y^2} = 3x(3y - 2)e^{2x-3y}$$

$$\frac{\partial^2 z}{\partial x \partial y} = (1 + 2x)(1 - 3y)e^{2x-3y}$$

When more than one point is being tested it is a good idea to set out your working in a table as follows:

	(0,0)	$\left(-\frac{1}{2}, \frac{1}{3}\right)$
$\frac{\partial^2 z}{\partial x^2}$	0	$\frac{2}{3} e^{-2}$
$\frac{\partial^2 z}{\partial y^2}$	0	$\frac{3}{2} e^{-2}$
$\frac{\partial^2 z}{\partial x \partial y}$	1	0
$\frac{\partial^2 z}{\partial x^2} \frac{\partial^2 z}{\partial y^2} - \left(\frac{\partial^2 z}{\partial x \partial y}\right)^2$	-1	e^{-4}

At the point (0,0), there is neither a maximum nor a minimum because $\frac{\partial^2 z}{\partial x^2} \frac{\partial^2 z}{\partial y^2} - \left(\frac{\partial^2 z}{\partial x \partial y}\right)^2$ is negative there. On the other hand, this expression is positive at the point $\left(-\frac{1}{2}, \frac{1}{3}\right)$, telling us that there is either a maximum or a minimum there. As $\frac{\partial^2 z}{\partial x^2}$ and $\frac{\partial^2 z}{\partial y^2}$ are both positive, z has a minimum value, this being $-\frac{1}{6} e^{-2}$.

FRAME A9

Now find the maximum and/or minimum values of $xy(3 - x - y)$.

****

A9A

If $z = xy(3 - x - y)$, $\frac{\partial z}{\partial x} = y(3 - 2x - y)$ *and* $\frac{\partial z}{\partial y} = x(3 - x - 2y)$

Solutions of $\frac{\partial z}{\partial x} = 0$, $\frac{\partial z}{\partial y} = 0$ *are* $(0,0)$, $(0,3)$, $(3,0)$ *and* $(1,1)$.

$\frac{\partial^2 z}{\partial x^2} = -2y$ $\quad \frac{\partial^2 z}{\partial x \partial y} = 3 - 2x - 2y$ $\quad \frac{\partial^2 z}{\partial y^2} = -2x$

	$(0,0)$	$(0,3)$	$(3,0)$	$(1,1)$
$\frac{\partial^2 z}{\partial x^2}$	0	-6	0	-2
$\frac{\partial^2 z}{\partial y^2}$	0	0	-6	-2
$\frac{\partial^2 z}{\partial x \partial y}$	3	-3	-3	-1
$\frac{\partial^2 z}{\partial x^2}\frac{\partial^2 z}{\partial y^2} - \left(\frac{\partial^2 z}{\partial x \partial y}\right)^2$	-9	-9	-9	3

There is a maximum at (1,1) of value 1.

FRAME A10

If $\frac{\partial z}{\partial x} = 0$ and $\frac{\partial z}{\partial y} = 0$ at a point, but there is neither a maximum nor a minimum there, a variety of other situations can arise. Just two of these will be looked at briefly.

One possibility is what is called a "saddle-point", as at A in Fig (i).

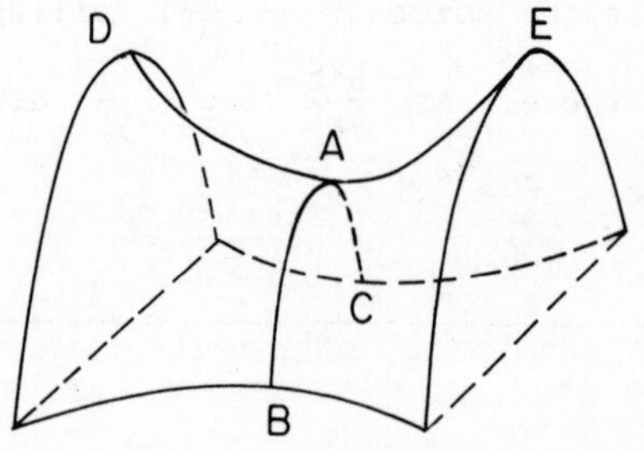

Fig (i)

Here this is a maximum along the section BAC but a minimum along DAE. It is the sort of thing that happens at the top of a pass between two mountains. Such a point could occur between A and B in the diagram in FRAME 25.

FRAME A10 continued

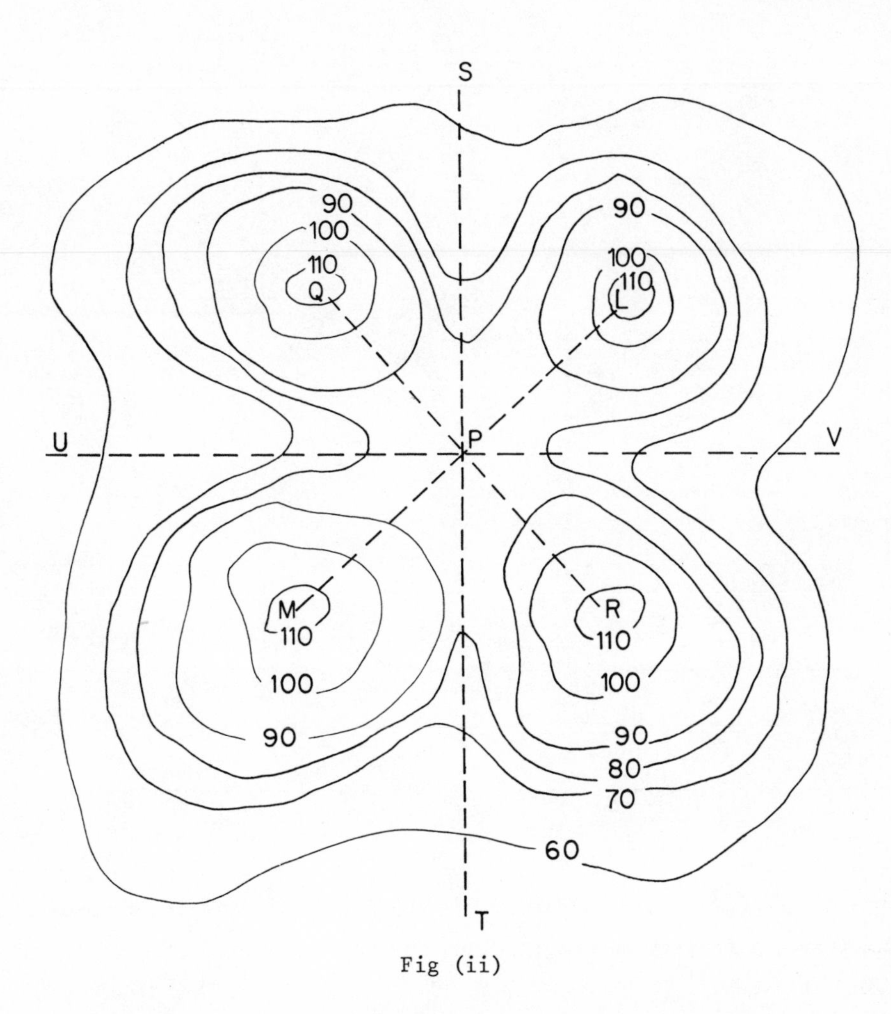

Fig (ii)

The contour map in Fig (ii) shows an even more intriguing possibility. The point P is a maximum to a walker along either of the paths ST or UV but a minimum to a person walking along either of the paths LM or QR.

UNIT 8

MATRICES II

A.C. Bajpai
I.M. Calus
J.A. Fairley

Loughborough University of Technology

INSTRUCTIONS

This programme is divided up into a number of FRAMES which are to be worked *in the order given*. You will be required to participate in many of these frames and in such cases the answers are provided in ANSWER FRAMES, designated by the letter A following the frame number. Steps in the working are given where this is considered helpful. The answer frame is separated from the main frame by a line of asterisks: *********. Keep the answers covered until you have written your own response. If your answer is wrong, go back and try to see why. Do not proceed to the next frame until you have corrected any mistakes in your attempt and are satisfied that you understand the contents up to this point.

Before reading this programme, it is necessary that you are familiar with the following

Prerequisites

The contents of the Matrices I programme in Unit 3 of Vol I.

Vector Algebra, as covered in FRAMES 1-14, pages 3:39 to 3:46.

CONTENTS

FRAME 1

Eigenvalues

We shall start by considering a property of AX which exists in certain cases. Here A is a square matrix and X is a column matrix having the same number of rows as A.

First work out AX in each of the following cases:

(i) $A = \begin{pmatrix} 1 & 2 \\ 2 & -2 \end{pmatrix}$ $X = \begin{pmatrix} 5 \\ -3 \end{pmatrix}$

(ii) $A = \begin{pmatrix} 1 & 2 \\ 2 & -2 \end{pmatrix}$ $X = \begin{pmatrix} 1 \\ -2 \end{pmatrix}$

(iii) $A = \begin{pmatrix} 4 & 6 & 6 \\ 1 & 3 & 2 \\ -1 & -4 & -3 \end{pmatrix}$ $X = \begin{pmatrix} 1 \\ 3 \\ 2 \end{pmatrix}$

(iv) $A = \begin{pmatrix} 4 & 6 & 6 \\ 1 & 3 & 2 \\ -1 & -4 & -3 \end{pmatrix}$ $X = \begin{pmatrix} 3 \\ 1 \\ -1 \end{pmatrix}$

****

1A

(i) $\begin{pmatrix} -1 \\ 16 \end{pmatrix}$ *(ii)* $\begin{pmatrix} -3 \\ 6 \end{pmatrix}$

(iii) $\begin{pmatrix} 34 \\ 14 \\ -19 \end{pmatrix}$ *(iv)* $\begin{pmatrix} 12 \\ 4 \\ -4 \end{pmatrix}$

FRAME 2

Now look carefully at the answers in 1A. In (ii) and (iv) can you spot any relation between them and the corresponding matrix X which does not exist in (i) and (iii)?

****

2A

In (ii) $\begin{pmatrix} -3 \\ 6 \end{pmatrix} = -3 \begin{pmatrix} 1 \\ -2 \end{pmatrix}$ *and in (iv)* $\begin{pmatrix} 12 \\ 4 \\ -4 \end{pmatrix} = 4 \begin{pmatrix} 3 \\ 1 \\ -1 \end{pmatrix}$

i.e., in both cases the elements of AX are constant multiples of those of X and so we can write

$$AX = \lambda X$$

($\lambda = -3$ *in (ii) and* $\lambda = 4$ *in (iv)*).

FRAME 3

As you saw in FRAME 49, page 3:70, Vol I, a vector can be considered as an ordered set of numbers. But this is just what a row or column matrix is, and so we often refer to a row matrix as a row vector and to a column matrix as a column vector.

Thus in cases (ii) and (iv) you will see that, if X and AX are regarded as vectors, then both vectors are in the same direction. This did not happen in (i) even though A was the same as in (ii), nor did it happen in (iii) even though A was the same as in (iv). The problem is, given a square matrix A, how can we find a vector X for which this does happen? This problem is of considerable interest to engineers, who may require the vector X but more often the corresponding value of λ. In the next few frames, we shall describe three practical situations which illustrate this.

FRAME 4

As a first example, we shall consider the oscillations of three flywheels, each of the same moment of inertia I, fixed to a central shaft which is free to rotate in its bearings. (Such oscillatory motion is additional to any rotational motion.) This kind of situation occurs in the study of such problems as the oscillations of the blades of a turbine.

FRAME 4 continued

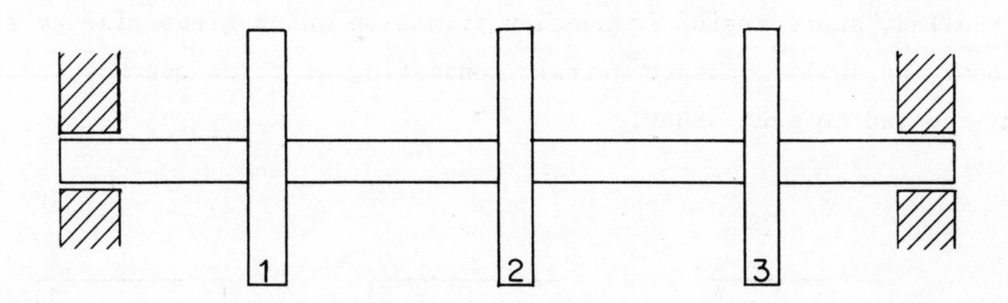

Let θ_1, θ_2, θ_3 be the angular displacements for the oscillatory motion at time t of flywheels 1,2,3 respectively, and let C be the torsional stiffness of the shaft.

Then

$$I\ddot{\theta}_1 = -C(\theta_1 - \theta_2)$$
$$I\ddot{\theta}_2 = C(\theta_1 - \theta_2) - C(\theta_2 - \theta_3)$$
$$I\ddot{\theta}_3 = C(\theta_2 - \theta_3)$$

Now the angle θ turned through in an oscillatory motion is given by $\theta = \theta_{max} \sin \omega t$ and therefore $\ddot{\theta} = -\omega^2\theta$, so writing $\ddot{\theta}_1 = -\omega^2\theta_1$ etc. gives

$$-I\omega^2\theta_1 = -C\theta_1 + C\theta_2$$
$$-I\omega^2\theta_2 = C\theta_1 - 2C\theta_2 + C\theta_3$$
$$-I\omega^2\theta_3 = C\theta_2 - C\theta_3$$

Putting $\frac{I\omega^2}{C} = \lambda$, for simplicity, gives

$$\theta_1 - \theta_2 = \lambda\theta_1$$
$$-\theta_1 + 2\theta_2 - \theta_3 = \lambda\theta_2$$
$$-\theta_2 + \theta_3 = \lambda\theta_3$$

This set of equations can be written in matrix form as

$$\begin{pmatrix} 1 & -1 & 0 \\ -1 & 2 & -1 \\ 0 & -1 & 1 \end{pmatrix}\begin{pmatrix} \theta_1 \\ \theta_2 \\ \theta_3 \end{pmatrix} = \lambda \begin{pmatrix} \theta_1 \\ \theta_2 \\ \theta_3 \end{pmatrix}$$

which is of the form $A\Theta = \lambda\Theta$.

FRAME 5

The so-called "short train" is another situation which gives rise to similar equations. We shall consider a train consisting of three coaches, each of mass m, coupled to each other.

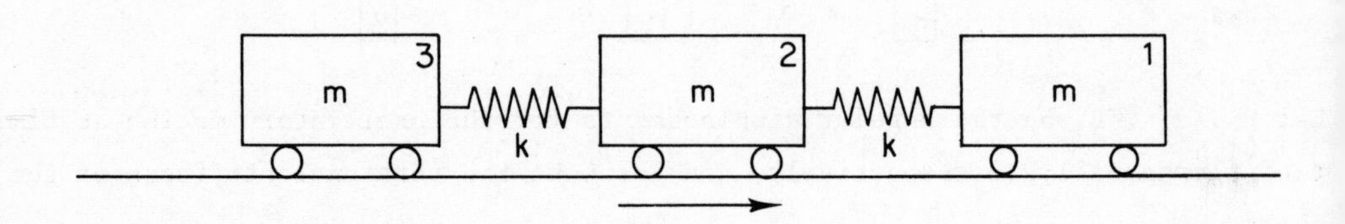

Representing each coupling by a spring of stiffness k, the following equations arise when considering the motion of the coaches within the system:

$$\begin{aligned} m\ddot{x}_1 &= -kx_1 + kx_2 \\ m\ddot{x}_2 &= kx_1 - 2kx_2 + kx_3 \\ m\ddot{x}_3 &= \qquad\quad kx_2 - kx_3 \end{aligned}$$

By a process similar to that used in the last frame these equations become

$$\left.\begin{aligned} x_1 - x_2 \qquad &= \lambda x_1 \\ -x_1 + 2x_2 - x_3 &= \lambda x_2 \\ - x_2 + x_3 &= \lambda x_3 \end{aligned}\right\} \quad (5.1)$$

where $\ddot{x}_1 = -\omega^2 x_1$, etc. and $\lambda = m\omega^2/k$.

Thus in this case

$$\begin{pmatrix} 1 & -1 & 0 \\ -1 & 2 & -1 \\ 0 & -1 & 1 \end{pmatrix}\begin{pmatrix} x_1 \\ x_2 \\ x_3 \end{pmatrix} = \lambda \begin{pmatrix} x_1 \\ x_2 \\ x_3 \end{pmatrix}$$

which is of the form $AX = \lambda X$.

FRAME 6

In electrical network theory the analogous situation is that shown in the circuit below.

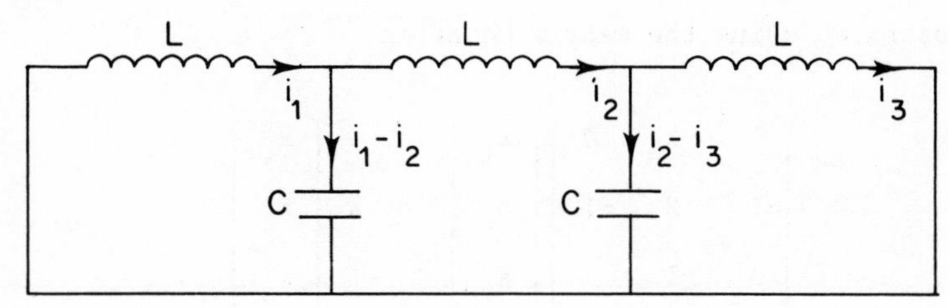

Here the equations are

$$L\frac{d^2 i_1}{dt^2} = -\frac{i_1}{C} + \frac{i_2}{C}$$

$$L\frac{d^2 i_2}{dt^2} = \frac{i_1}{C} - \frac{2i_2}{C} + \frac{i_3}{C}$$

$$L\frac{d^2 i_3}{dt^2} = \frac{i_2}{C} - \frac{i_3}{C}$$

giving, where $\frac{d^2 i_1}{dt^2} = -\omega^2 i_1$ etc. and $\lambda = \omega^2 CL$,

$$\begin{pmatrix} 1 & -1 & 0 \\ -1 & 2 & -1 \\ 0 & -1 & 1 \end{pmatrix}\begin{pmatrix} i_1 \\ i_2 \\ i_3 \end{pmatrix} = \lambda \begin{pmatrix} i_1 \\ i_2 \\ i_3 \end{pmatrix}$$

which is of the form $AJ = \lambda J$, where $J = \{i_1 \quad i_2 \quad i_3\}$.

FRAME 7

You may have noticed that in each of these last three examples the matrix A was the same. This only happened because of the similarity of the three systems. If, for example, the train had had four coaches or the inductances in the three meshes of the network had been different, then different A's would have resulted.

FRAME 8

We shall now see how to solve the equation $AX = \lambda X$ for λ, leaving the method of solution for X until a little later in the programme. The solution for λ will be illustrated using the matrix equation

$$\begin{pmatrix} 1 & -1 & 0 \\ -1 & 2 & -1 \\ 0 & -1 & 1 \end{pmatrix} \begin{pmatrix} x_1 \\ x_2 \\ x_3 \end{pmatrix} = \lambda \begin{pmatrix} x_1 \\ x_2 \\ x_3 \end{pmatrix}$$

of FRAME 5. This is the matrix form of equations (5.1) which can be written as

$$\left.\begin{aligned} (1-\lambda)x_1 - x_2 \qquad\qquad &= 0 \\ -x_1 + (2-\lambda)x_2 - x_3 &= 0 \\ -x_2 + (1-\lambda)x_3 &= 0 \end{aligned}\right\} \qquad (8.1)$$

These are homogeneous linear equations in x_1, x_2, x_3 and an obvious solution is for all x's to be zero. Other solutions exist if

$$\begin{vmatrix} 1-\lambda & -1 & 0 \\ -1 & 2-\lambda & -1 \\ 0 & -1 & 1-\lambda \end{vmatrix} = 0 \qquad (8.2)$$

(See FRAMES 49-54, pages 3:29-3:32, Vol I, if you have forgotten this result.)

Solve this equation for λ.

****

8A

The equation reduces to

$$(1 - \lambda)\lambda(\lambda - 3) = 0$$

giving $\lambda = 1, 0, 3$.

FRAME 9

The equation in λ is called the CHARACTERISTIC EQUATION of the matrix A. The polynomial (in this case a cubic) on the L.H.S. of this equation is called the CHARACTERISTIC POLYNOMIAL. The roots of the characteristic equation, i.e. the values of λ, are sometimes called the CHARACTERISTIC ROOTS or LATENT ROOTS, but the term EIGENVALUES is more frequently used.

FRAME 10

Now look at the determinant in (8.2). Can you see any connection between its elements and those of the matrix A?

****

10A

The elements on the leading diagonal of the determinant are those on the leading diagonal of the matrix, reduced in each case by λ. *The other elements are the same for both the matrix and the determinant.*

FRAME 11

Using the result observed in 10A, find the eigenvalues of the following matrices:

(i) $\begin{pmatrix} 4 & -2 \\ 3 & -3 \end{pmatrix}$ (ii) $\begin{pmatrix} 2 & 6 & 5 & 1 \\ 0 & -3 & 7 & 5 \\ 0 & 0 & 4 & 2 \\ 0 & 0 & 0 & -1 \end{pmatrix}$

****

11A

(i) $\begin{vmatrix} 4-\lambda & -2 \\ 3 & -3-\lambda \end{vmatrix} = 0$ *giving* $\lambda = 3$ *or* -2.

11A continued

(ii)

$$\begin{vmatrix} 2-\lambda & 6 & 5 & 1 \\ 0 & -3-\lambda & 7 & 5 \\ 0 & 0 & 4-\lambda & 2 \\ 0 & 0 & 0 & -1-\lambda \end{vmatrix} = 0 \quad \textit{giving} \quad \lambda = 2,\ -3,\ 4 \text{ or } -1.$$

FRAME 12

Returning to the matrix $\begin{pmatrix} 1 & -1 & 0 \\ -1 & 2 & -1 \\ 0 & -1 & 1 \end{pmatrix}$ of FRAME 8, the sum of the elements on the leading diagonal is $1 + 2 + 1 = 4$, and the sum of the eigenvalues is $1 + 0 + 3 = 4$. It can be proved that, for any square matrix, these two sums are the same. The sum of the elements on the leading diagonal of a square matrix A is called the TRACE of the matrix, sometimes written Tr A.

Now verify that, for each of the matrices in FRAME 11, the trace is equal to the sum of the eigenvalues.

FRAME 13

Returning to equations (8.1), the matrix of the coefficients on the L.H.S.'s is

$$\begin{pmatrix} 1-\lambda & -1 & 0 \\ -1 & 2-\lambda & -1 \\ 0 & -1 & 1-\lambda \end{pmatrix} \qquad (13.1)$$

This can be written in the form

$$\begin{pmatrix} 1 & -1 & 0 \\ -1 & 2 & -1 \\ 0 & -1 & 1 \end{pmatrix} - \lambda \begin{pmatrix} & & \\ & ? & \\ & & \end{pmatrix}$$

Fill in the elements of the unknown matrix.

****

13A

$$\begin{bmatrix} 1 & 0 & 0 \\ 0 & 1 & 0 \\ 0 & 0 & 1 \end{bmatrix} = I$$

FRAME 14

The matrix of coefficients (13.1) can therefore be written as $A - \lambda I$ and the set of equations (8.1) as $(A - \lambda I)X = 0$, which is just another way of writing $AX = \lambda X$.

Why can't we write $AX = \lambda X$ as $(A - \lambda)X = 0$?

****

14A

A is a 3 × 3 matrix and λ is just a single number, so A − λ is meaningless. Only another 3 × 3 matrix can be subtracted from A. λI is such a matrix.

FRAME 15

You will now see that equation (8.2) is equivalent to $|A - \lambda I| = 0$. Although in our demonstration example we have used a particular 3 × 3 matrix for A, the same procedure can be followed for any square matrix A. Taking this general case, the matrix equation $AX = \lambda X$ can be expressed in the form $(A - \lambda I)X = 0$ and the eigenvalues of the matrix A are given by the characteristic equation $|A - \lambda I| = 0$.

FRAME 16

You have seen in FRAMES 4 - 6 how a square matrix (which we called A) can be associated with the oscillations of a physical system, and you now know how to find the eigenvalues of such a matrix. These eigenvalues are important to the engineer as they enable him to calculate the natural frequencies of the system,

FRAME 16 continued

(Of course, as is so often the case, the theoretical model is a simplified version of the actual system. The calculated values will therefore only be approximations to the actual ones.)

If a periodic force is applied to such a system, the amplitude of the resulting oscillations rapidly increases if the frequency of this force is the same as one of the natural frequencies of the system, and this can have disastrous consequences. Perhaps the most spectacular example of this was the collapse, under dynamic wind forces, of the first Tacoma Narrows suspension bridge in the State of Washington in 1940. You may have seen the film of this. While on the subject of bridges, it has long been the custom for soldiers to break step when marching across some types of bridge, to eliminate the risk of their footsteps becoming synchronous with the natural frequency of the structure.

In another field, the aircraft designer must be very conscious of the natural frequencies of various parts of a plane, otherwise, during flight, a state of critical vibration, known as 'flutter', may be induced in, for example, a wing or tail-plane. If this continues, parts of the structure will become stressed beyond their elastic limit and quickly fracture.

A knowledge of the natural frequencies of high-speed rotating shafts, such as in a jet engine's compressor/turbine unit, is also important. A rotating shaft has not only one, but a whole sequence of 'critical speeds', which occur when the natural frequency of the shaft, or one of its harmonics, synchronises with the speed of rotation. The result of sustained running (even for a few seconds) at a critical speed causes the amplitude of the transverse vibration of the shaft to progressively increase. A factor vitally affecting the performance of any turbine is the clearance between the tips of the blades and the internal surface of the casing, efficiency falling rapidly as the clearance increases. So close is the operating clearance in the turbine of a jet engine that critical vibration of the turbine shaft could prove disastrous, as happened only too frequently with early jet aircraft in course of flight. Blade tips moving faster than the speed of sound would make momentary contact with the magnesium turbine casing, thereby causing such a high temperature due to friction that the whole engine would virtually blow up in mid-air.

FRAME 16 continued

Critical vibration of the blades themselves can greatly impair the efficiency of a turbine, as is believed to have been a contributory cause of the teething troubles which afflicted the liner QE2. This may be assumed from press reports of the turbine blades being later stiffened by means of wire lacing, an expedient originally applied to some aero gas-turbines.

FRAME 17

The problem of the determination of eigenvalues is also met in the study of the bending of structures. We shall consider a simple problem to illustrate this - that of a uniform strut of length ℓ under an axial compressive force P at each end. The bending moment equation is

$$EI\frac{d^2y}{dx^2} = -Py$$

where y is the deflection at a distance x from one end and EI is the constant flexural rigidity.

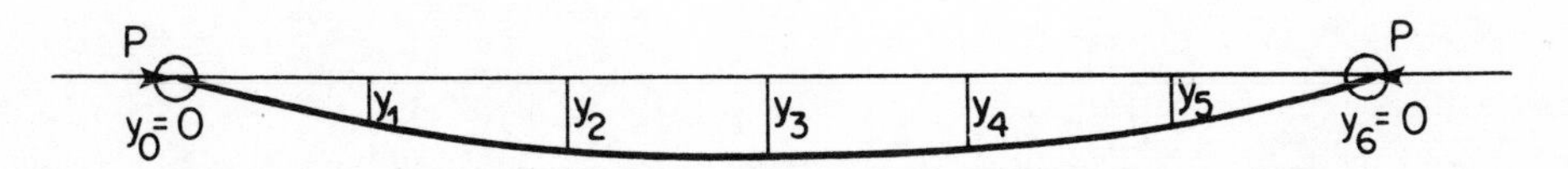

Dividing the length of the strut into six equal intervals, the corresponding deflections can be denoted by y_o, y_1, y_2, ..., y_6 as shown in the diagram (obviously $y_o = y_6 = 0$). Assuming symmetry about the mid-point, $y_4 = y_2$ and $y_5 = y_1$, and so we need only investigate one half of the strut.

An approximate formula for $\frac{d^2y}{dx^2}$ at the point where the deflection is y_r is

$$\frac{y_{r+1} - 2y_r + y_{r-1}}{h^2},$$

using the notation shown in the diagram below.

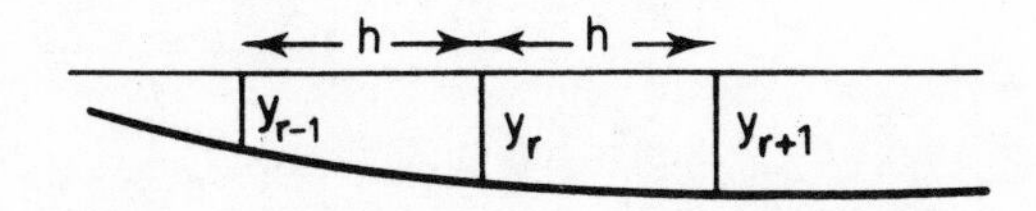

FRAME 17 continued

Thus at the point on the strut where $y = y_2$,

$$\frac{d^2y}{dx^2} \doteqdot \frac{y_3 - 2y_2 + y_1}{(\ell/6)^2}$$

as $h = \frac{\ell}{6}$.

The approximate formula for $\frac{d^2y}{dx^2}$ which we have quoted here comes from the use of finite differences. If you have studied numerical methods you may have met this topic, but if you haven't, it doesn't matter.

Now apply the formula to find approximations for $\frac{d^2y}{dx^2}$ at the points on the strut where $y = y_1$ and $y = y_3$.

****

17A

When $y = y_1$, $\frac{d^2y}{dx^2} \doteqdot \frac{y_2 - 2y_1 + y_0}{\ell^2/36}$ *i.e.* $\frac{y_2 - 2y_1}{\ell^2/36}$ *as* $y_0 = 0$

When $y = y_3$, $\frac{d^2y}{dx^2} \doteqdot \frac{y_4 - 2y_3 + y_2}{\ell^2/36}$ *i.e.* $\frac{2y_2 - 2y_3}{\ell^2/36}$ *as* $y_4 = y_2$

FRAME 18

Using these approximations in the equation $\frac{d^2y}{dx^2} = \frac{-P}{EI}y$ at the points where $y = y_1, y_2, y_3$ gives respectively

$$\frac{y_2 - 2y_1}{\ell^2/36} = \frac{-P}{EI}y_1$$

$$\frac{y_3 - 2y_2 + y_1}{\ell^2/36} = \frac{-P}{EI}y_2$$

$$\frac{2y_2 - 2y_3}{\ell^2/36} = \frac{-P}{EI}y_3$$

FRAME 18 continued

Rearranging these equations and putting $\frac{P\ell^2}{36EI} = \lambda$ gives

$$\left.\begin{aligned} 2y_1 - y_2 \quad\quad &= \lambda y_1 \\ -y_1 + 2y_2 - y_3 &= \lambda y_2 \\ -2y_2 + 2y_3 &= \lambda y_3 \end{aligned}\right\} \quad (18.1)$$

Write these last three equations as a single matrix equation.

****

18A

$$\begin{pmatrix} 2 & -1 & 0 \\ -1 & 2 & -1 \\ 0 & -2 & 2 \end{pmatrix}\begin{pmatrix} y_1 \\ y_2 \\ y_3 \end{pmatrix} = \lambda \begin{pmatrix} y_1 \\ y_2 \\ y_3 \end{pmatrix}$$

which is of the form $AY = \lambda Y$.

FRAME 19

One obvious solution of equations (18.1) is the zero solution, i.e. $y_1 = y_2 = y_3 = 0$. As you have seen, other solutions exist if λ is one of the eigenvalues of the matrix

$$\begin{pmatrix} 2 & -1 & 0 \\ -1 & 2 & -1 \\ 0 & -2 & 2 \end{pmatrix}$$

Now find these eigenvalues.

****

19A

$2 - \sqrt{3}$, 2, $2 + \sqrt{3}$ *or* $0{\cdot}27$, 2, $3{\cdot}73$

CHECK: Sum of eigenvalues = Trace = 6

FRAME 20

Each value of λ gives rise to a corresponding value of P. Substituting in $P = 36EI\lambda/\ell^2$ the values of λ you have just found gives

$$\frac{9{\cdot}72EI}{\ell^2}, \qquad \frac{72EI}{\ell^2}, \qquad \frac{134{\cdot}28EI}{\ell^2}$$

The mathematical interpretation of these results is that the strut can assume a symmetrical curved form due to an axial load P only if P has one of these values.

In practice, a strut may also become deflected due to a lateral force F. It is found that what happens depends on the value of P.

(i) If $P < \frac{9{\cdot}72EI}{\ell^2}$, the strut bends but reassumes its straight form when F is removed.

(ii) If $P = \frac{9{\cdot}72EI}{\ell^2}$, the strut buckles.

The smallest eigenvalue therefore enables us to find the critical buckling load.

Because of the approximate formula used for $\frac{d^2y}{dx^2}$ the critical value found for P by this method is not exact. The accuracy can be improved by increasing the number of intervals. In fact, you may be wondering why we have chosen this approach at all, when the equation $\frac{d^2y}{dx^2} = \frac{-P}{EI}y$ can be solved analytically and the critical buckling load deduced. (It is actually π^2EI/ℓ^2.) However, if the strut were not of uniform cross-section, thus making I variable, the analytical solution would be more difficult, or even impossible, and it is in such cases that the numerical method comes into its own. We did not consider such a case, as the simpler problem serves to illustrate the method more clearly.

FRAME 21

Eigenvectors

Returning to the subject of vibrations, we shall use the oscillations of three masses attached to a taut elastic string to introduce the idea of eigenvectors.

FRAME 21 continued

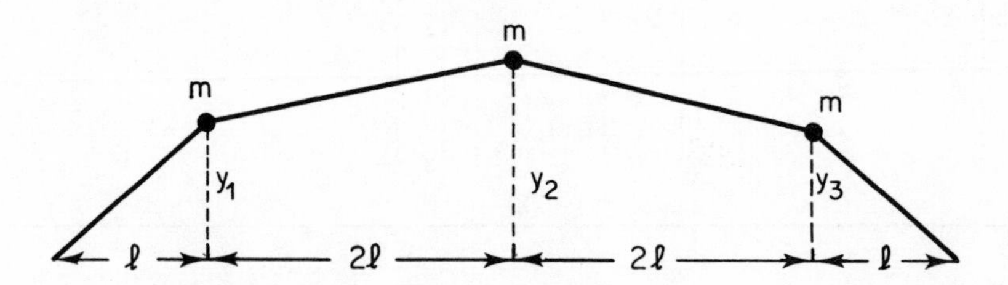

Assuming that the oscillations are of small amplitude so that the tension T in the string can be taken as constant, the equations of motion of the three masses are

$$\left.\begin{aligned} m\ddot{y}_1 &= \frac{-Ty_1}{\ell} + \frac{T(y_2 - y_1)}{2\ell} \\ m\ddot{y}_2 &= \frac{-T(y_2 - y_1)}{2\ell} + \frac{T(y_3 - y_2)}{2\ell} \\ m\ddot{y}_3 &= \frac{-T(y_3 - y_2)}{2\ell} - \frac{Ty_3}{\ell} \end{aligned}\right\} \quad (21.1)$$

Putting $\ddot{y}_1 = -\omega^2 y_1$, etc., as in previous problems, gives

$$\begin{aligned} -m\omega^2 y_1 &= \frac{-3Ty_1}{2\ell} + \frac{Ty_2}{2\ell} \\ -m\omega^2 y_2 &= \frac{Ty_1}{2\ell} - \frac{Ty_2}{\ell} + \frac{Ty_3}{2\ell} \\ -m\omega^2 y_3 &= \frac{Ty_2}{2\ell} - \frac{3Ty_3}{2\ell} \end{aligned}$$

Now write $\dfrac{2\omega^2 m\ell}{T} = \lambda$ and hence obtain these equations in the form $AY = \lambda Y$.

****

<u>21A</u>

$$\left.\begin{aligned} 3y_1 - y_2 \quad &= \lambda y_1 \\ -y_1 + 2y_2 - y_3 &= \lambda y_2 \\ - y_2 + 3y_3 &= \lambda y_3 \end{aligned}\right\} \quad (21A.1)$$

21A continued

i.e.
$$\begin{bmatrix} 3 & -1 & 0 \\ -1 & 2 & -1 \\ 0 & -1 & 3 \end{bmatrix} \begin{bmatrix} y_1 \\ y_2 \\ y_3 \end{bmatrix} = \lambda \begin{bmatrix} y_1 \\ y_2 \\ y_3 \end{bmatrix}$$

FRAME 22

Find the eigenvalues of the matrix A.

****

22A

1, 3, 4.

FRAME 23

When each of these values of λ is substituted in equations (21A.1) a set of three homogeneous equations in y_1, y_2, y_3 is obtained. Remember that the values of λ have been chosen so that each of these sets will have a non-zero solution.

For example, taking $\lambda = 1$ gives

$$\begin{aligned} 3y_1 - y_2 \qquad &= y_1 \\ -y_1 + 2y_2 - y_3 &= y_2 \\ -y_2 + 3y_3 &= y_3 \end{aligned}$$

i.e.
$$2y_1 - y_2 = 0 \qquad (23.1)$$
$$-y_1 + y_2 - y_3 = 0 \qquad (23.2)$$
$$-y_2 + 2y_3 = 0 \qquad (23.3)$$

From (23.1), $y_2 = 2y_1$ and from (23.3), $y_3 = \frac{1}{2}y_2 = y_1$.

If these expressions for y_2 and y_3 are substituted in (23.2) we get $0.y_1 = 0$ which is satisfied for any value of y_1. This is because (23.2) is a linear

FRAME 23 continued

combination of (23.1) and (23.3) and so there are really only two different equations in the three unknowns.

As a result, there is not a unique solution for y_1, y_2, y_3 but an infinite number of possibilities of which the following are a few:

$$y_1 = 1, \quad y_2 = 2, \quad y_3 = 1 \quad ;$$

$$y_1 = -1, \quad y_2 = -2, \quad y_3 = -1 \quad ;$$

$$y_1 = 5, \quad y_2 = 10, \quad y_3 = 5 \quad .$$

Remember that in each case we must have $y_2 = 2y_1$ and $y_3 = y_1$, so that $y_1 : y_2 : y_3$ is always the same. All possible values are included in $\{y_1 \quad y_2 \quad y_3\} = k\{1 \quad 2 \quad 1\}$ where k is any number.

FRAME 24

In the matrix equation $AY = \lambda Y$, the non-zero solution for Y, which is the column vector $\{y_1 \quad y_2 \quad y_3\}$, is called an EIGENVECTOR (or CHARACTERISTIC VECTOR or LATENT VECTOR) of the matrix A.

Corresponding to each eigenvalue there will be a set of eigenvectors. We have found the eigenvectors $k\{1 \quad 2 \quad 1\}$ corresponding to $\lambda = 1$. We suggest that you now find the eigenvectors for $\lambda = 3$ and $\lambda = 4$.

****

24A

$\lambda = 3$ substituted in equations (21A.1) gives

$$\begin{aligned} -y_2 \qquad\quad &= 0 \\ -y_1 - y_2 - y_3 &= 0 \\ -y_2 \qquad\quad &= 0 \end{aligned}$$

i.e. $y_2 = 0$ *and* $y_3 = -y_1$

So $\{y_1 \quad y_2 \quad y_3\}$ *is proportional to* $\{1 \quad 0 \quad -1\}$.

24A continued

$\lambda = 4$ gives

$$\begin{aligned} -y_1 - y_2 \quad &= 0 \\ -y_1 - 2y_2 - y_3 &= 0 \\ -y_2 - y_3 &= 0 \end{aligned}$$

from which $y_2 = -y_1$ *and* $y_3 = y_1$

So the eigenvectors are proportional to $\{1 \quad -1 \quad 1\}$.

FRAME 25

The question now arises as to what interpretation can be given to these eigenvectors.

Corresponding to $\lambda = 1$ there is a free vibration of frequency $\omega/2\pi$, where $\omega^2 = T/2m\ell$, and a mode of oscillation which is such that

$$y_1 : y_2 : y_3 = 1 : 2 : 1,$$

i.e. the displacement of the middle mass is always twice that of either of the two outer ones.

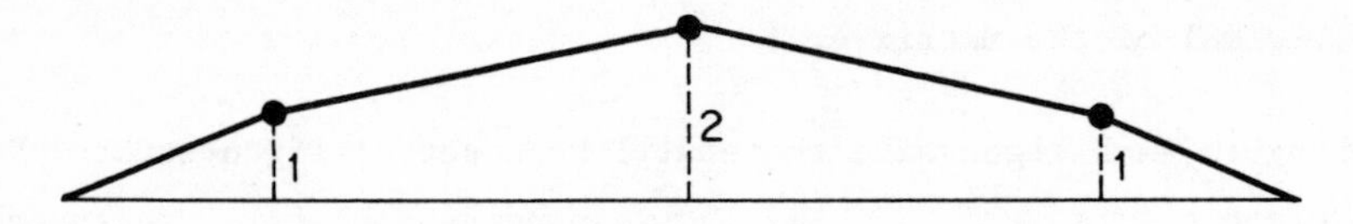

Now state the frequency corresponding to each of the other values of λ and illustrate the mode of oscillation in each case.

****

25A

$\lambda = 3$

Frequency $= \omega/2\pi$ *where* $\omega^2 = 3T/2m\ell$.

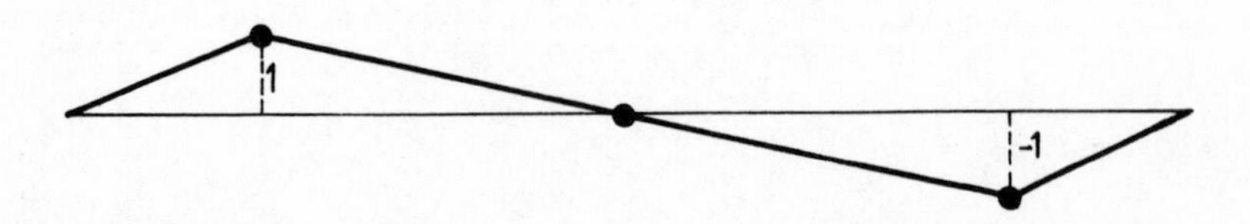

Note that the middle particle remains at rest throughout the motion.

25A continued

$\lambda = 4$

Frequency = $\omega/2\pi$ *where* $\omega^2 = 2T/m\ell$.

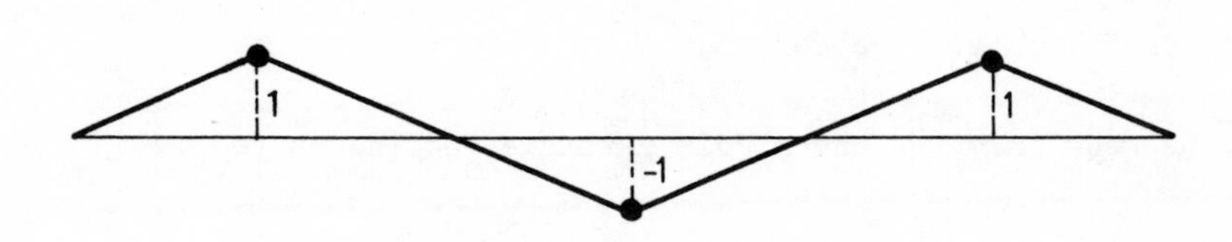

The three modes of oscillation given by $\lambda = 1,3,4$ *are sometimes referred to as the 'normal modes of oscillation'.*

FRAME 26

If you are familiar with the solution of simultaneous differential equations you will realise that equations (21.1) could have been solved by standard methods such as D-operator or Laplace Transform, but the process might have been rather tedious. In fact, the general solution is

$$y_1 = (A\cos ct + B\sin ct) + (C\cos\sqrt{3}ct + E\sin\sqrt{3}ct) + (F\cos 2ct + G\sin 2ct)$$

$$y_2 = 2(A\cos ct + B\sin ct) - (F\cos 2ct + G\sin 2ct)$$

$$y_3 = (A\cos ct + B\sin ct) - (C\cos\sqrt{3}ct + E\sin\sqrt{3}ct) + (F\cos 2ct + G\sin 2ct)$$

where $c = \sqrt{T/2m\ell}$.

You will notice that there are three component oscillations whose frequencies $c/2\pi$, $\sqrt{3}c/2\pi$, $2c/2\pi$ correspond to the three eigenvalues and whose amplitudes correspond to the eigenvectors. For example, the oscillation whose frequency is $c/2\pi$ corresponds to $\lambda = 1$ (check this!) and the ratios of its amplitudes in y_1, y_2, y_3 are 1:2:1.

Any one of the normal modes of vibration can be obtained from the general solution by imposing suitable initial conditions. For example, if at $t = 0$,

FRAME 26 continued

$y_1 = a,\quad y_2 = 0,\quad y_3 = -a\quad$ and $\quad\dot{y}_1 = \dot{y}_2 = \dot{y}_3 = 0\quad$ then

$$y_1 = a\cos\sqrt{3}ct$$

$$y_2 = 0$$

$$y_3 = -a\cos\sqrt{3}ct$$

This is the mode in which the middle particle remains at rest.

FRAME 27

You will remember that, when dealing with vectors, it is sometimes convenient to use a unit vector which is in the same direction as a given vector. The length of the vector $x\underline{i} + y\underline{j} + z\underline{k}$ is $\sqrt{x^2 + y^2 + z^2} = r$, say, so that the unit vector in this direction is

$$\frac{x}{r}\underline{i} + \frac{y}{r}\underline{j} + \frac{z}{r}\underline{k}\ .$$

Note that $\left(\frac{x}{r}\right)^2 + \left(\frac{y}{r}\right)^2 + \left(\frac{z}{r}\right)^2 = 1.$

A similar idea is possible with eigenvectors. Thus in the result at the end of FRAME 23 it is possible to find k such that $y_1^2 + y_2^2 + y_3^2 = 1$. In that particular case the required value would be $\frac{1}{\sqrt{6}}$, i.e. $\frac{1}{\sqrt{1^2 + 2^2 + 1^2}}$.

The corresponding eigenvector would be $\frac{1}{\sqrt{6}}\{1 \quad 2 \quad 1\}$, and this is called a NORMALISED eigenvector.

You should now be able to manage the following example:

Find the eigenvalues of the matrix $\begin{pmatrix} 2 & 6 \\ 1 & -3 \end{pmatrix}$ and corresponding normalised eigenvectors.

****

27A

The characteristic equation is $\lambda^2 + \lambda - 12 = 0$.

The eigenvalues are -4 *and* 3.

$\lambda = -4$ *gives the normalised eigenvector* $\frac{1}{\sqrt{2}}\{1 \quad -1\}$. *If you have written* $\frac{1}{\sqrt{2}}\{-1 \quad 1\}$ *this would be equally acceptable.*

$\lambda = 3$ *gives* $\frac{1}{\sqrt{37}}\{6 \quad 1\}$.

FRAME 28

Diagonalisation of a Matrix

Having obtained the eigenvectors of a matrix A, it is possible to write down a matrix whose columns are these eigenvectors. Thus in the case of the oscillation example just considered this new matrix could be

$$\begin{pmatrix} 1 & 1 & 1 \\ 2 & 0 & -1 \\ 1 & -1 & 1 \end{pmatrix}$$

This matrix is not unique, as the eigenvectors themselves are not unique, as was pointed out in FRAME 23, and also the columns can be written in any order. Such a matrix is called a MODAL MATRIX of A. Denoting it by H, evaluate H^{-1} and thus $H^{-1}AH$. What do you notice about the final result?

****

28A

$$H^{-1} = \frac{1}{6}\begin{pmatrix} 1 & 2 & 1 \\ 3 & 0 & -3 \\ 2 & -2 & 2 \end{pmatrix}$$

$$H^{-1}AH = \begin{pmatrix} 1 & 0 & 0 \\ 0 & 3 & 0 \\ 0 & 0 & 4 \end{pmatrix}$$ *which is a diagonal matrix. The elements on the leading diagonal are the eigenvalues of A, their order corresponding to that in which the eigenvectors were taken to form H.*

FRAME 29

The result observed in the previous answer frame was not just coincidental. If a modal matrix H is formed from a matrix A, the product $H^{-1}AH$ always gives a diagonal matrix D whose non-zero elements are the eigenvalues of A.

We can now write $H^{-1}AH = D$.

From this equation, express A in terms of the other matrices.
(HINT: Begin by pre-multiplying both sides of the equation by H.)

****

29A

$HH^{-1}AH = HD$

i.e. $AH = HD$

Post-multiply each side by H^{-1}

$$AHH^{-1} = HDH^{-1}$$
$$A = HDH^{-1}$$

FRAME 30

Arithmetic with diagonal matrices is very simple. Test this for yourself by finding D^2, D^3 and D^r (where r is a positive integer) when

$$D = \begin{pmatrix} \lambda_1 & 0 & 0 & 0 \\ 0 & \lambda_2 & 0 & 0 \\ 0 & 0 & \lambda_3 & 0 \\ 0 & 0 & 0 & \lambda_4 \end{pmatrix}$$

****

30A

$$D^2 = \begin{pmatrix} {\lambda_1}^2 & 0 & 0 & 0 \\ 0 & {\lambda_2}^2 & 0 & 0 \\ 0 & 0 & {\lambda_3}^2 & 0 \\ 0 & 0 & 0 & {\lambda_4}^2 \end{pmatrix} \qquad D^3 = \begin{pmatrix} {\lambda_1}^3 & 0 & 0 & 0 \\ 0 & {\lambda_2}^3 & 0 & 0 \\ 0 & 0 & {\lambda_3}^3 & 0 \\ 0 & 0 & 0 & {\lambda_4}^3 \end{pmatrix}$$

30A continued

$$D^r = \begin{pmatrix} \lambda_1^r & 0 & 0 & 0 \\ 0 & \lambda_2^r & 0 & 0 \\ 0 & 0 & \lambda_3^r & 0 \\ 0 & 0 & 0 & \lambda_4^r \end{pmatrix}$$

FRAME 31

The results in the last two frames can be used to find powers of a matrix A.

Thus $A^2 = HDH^{-1}HDH^{-1}$

$= HD^2H^{-1}$

$A^3 = HD^2H^{-1}HDH^{-1}$

$= HD^3H^{-1}$

and continuing in this way, one can see that

$A^r = HD^rH^{-1}$

FRAME 32

Quadratic Forms

The next topic which we shall consider in this programme is quadratic forms. Their applications are rather too specialised to be dealt with here so we would just mention that they occur in practical problems such as the optimization of the response of a system involving several variables, and are useful in dynamics for the expression of certain physical quantities such as kinetic or potential energy.

Start by evaluating the matrix product

$$\begin{pmatrix} x_1 & x_2 & x_3 \end{pmatrix} \begin{pmatrix} 2 & 3 & -1 \\ 3 & -4 & -2 \\ -1 & -2 & 1 \end{pmatrix} \begin{pmatrix} x_1 \\ x_2 \\ x_3 \end{pmatrix}$$

****

32A

$2x_1^2 - 4x_2^2 + x_3^2 + 6x_1x_2 - 4x_2x_3 - 2x_3x_1$

FRAME 33

We can thus write X'AX = Q where X = $\{x_1 \quad x_2 \quad x_3\}$, A is the 3 × 3 matrix and Q is the expression in the answer frame. (Technically Q is a 1 × 1 matrix, but having only one element it is usually written without the matrix brackets.)

What do you notice about A and Q?

****

33A

A is symmetric.

All the terms in Q are second order, i.e. quadratic.

The coefficients of x_1^2, x_2^2, x_3^2 *are the elements on the leading diagonal of A.*

The coefficient of x_1x_2 *is twice the element in position 12 in A.*
" " " x_2x_3 *" " " " " " 23 " ".*
" " " x_3x_1 *" " " " " " 31 " ".*

FRAME 34

An expression such as Q is called a QUADRATIC FORM. Such a form would have resulted even if A had not been symmetric, but it is the usual practice to work with a symmetric matrix in this situation.

We now consider the reverse problem of how to express Q in the matrix form X'AX. Bearing in mind the observations made in the last answer frame you should have no difficulty in expressing

$$x_1^2 + 3x_2^2 - 2x_3^2 - 2x_1x_2 + 4x_2x_3 - 6x_3x_1$$

in the form X'AX, where A is symmetric.

****

34A

$$\begin{pmatrix} x_1 & x_2 & x_3 \end{pmatrix}\begin{pmatrix} 1 & -1 & -3 \\ -1 & 3 & 2 \\ -3 & 2 & -2 \end{pmatrix}\begin{pmatrix} x_1 \\ x_2 \\ x_3 \end{pmatrix}$$

The elements on the leading diagonal are the coefficients of x_1^2, x_2^2, x_3^2.
The element in position 12 of A is $\frac{1}{2}$ × *the coefficient of* x_1x_2.
The elements in positions 23 and 31 are obtained in a similar way, and the other elements follow from the symmetry.

FRAME 35

A similar procedure can be followed with quadratic forms where the number of variables is other than 3. Working along the same lines, express the following in a matrix form similar to that used in the last three frames:

(i) $3y_1^2 + y_1y_2 - 2y_2^2$

(ii) $t_1^2 - 2t_2^2 + 3t_4^2 - 4t_1t_2 + 5t_1t_3 + 6t_2t_3 - 4t_2t_4 + 3t_3t_4$

****

35A

(i) $$\begin{pmatrix} y_1 & y_2 \end{pmatrix}\begin{pmatrix} 3 & \frac{1}{2} \\ \frac{1}{2} & -2 \end{pmatrix}\begin{pmatrix} y_1 \\ y_2 \end{pmatrix}$$

(ii) $$\begin{pmatrix} t_1 & t_2 & t_3 & t_4 \end{pmatrix}\begin{pmatrix} 1 & -2 & \frac{5}{2} & 0 \\ -2 & -2 & 3 & -2 \\ \frac{5}{2} & 3 & 0 & \frac{3}{2} \\ 0 & -2 & \frac{3}{2} & 3 \end{pmatrix}\begin{pmatrix} t_1 \\ t_2 \\ t_3 \\ t_4 \end{pmatrix}$$

FRAME 36

The Positive Definite Property

Consider the two quadratic forms $x_1^2 - 4x_1x_2 + 3x_2^2$ and $x_1^2 - 4x_1x_2 + 6x_2^2$. Obviously when $x_1 = x_2 = 0$ the value of each expression is zero. Now by simple algebra they can be written as $(x_1 - 2x_2)^2 - x_2^2$ and $(x_1 - 2x_2)^2 + 2x_2^2$ respectively. Can you spot any property of $(x_1 - 2x_2)^2 + 2x_2^2$, which is not shared by $(x_1 - 2x_2)^2 - x_2^2$, when x_1 and x_2 are given any other real values?

****

36A

As $(x_1 - 2x_2)^2 + 2x_2^2$ is the sum of two squares, it is always positive.

FRAME 37

Next consider the more general expression $x_1^2 - 4x_1x_2 + ax_2^2$, where a is real. Can you say for what range of values of a this will reduce to

(i) the sum of two squares, (ii) the difference of two squares?

****

37A

(i) $a > 4$ *(ii)* $a < 4$

If $a = 4$ the expression reduces to a single square, i.e. $(x_1 - 2x_2)^2$.

FRAME 38

For real values of x_1 and x_2, the following statements can now be made about Q, where $Q = x_1^2 - 4x_1x_2 + ax_2^2$

(i) If $a > 4$, $Q = 0$ when $x_1 = x_2 = 0$ but is otherwise positive.

(ii) If $a = 4$, $Q = 0$ when $x_1 = x_2 = 0$ or when $x_1 = 2x_2$ but is otherwise positive.

(iii) If $a < 4$, $Q = 0$ when $x_1 = x_2 = 0$ or when x_1 and x_2 are in certain fixed ratios, dependent on a, but otherwise can be positive or negative. For example, when $a = 3$, $Q = (x_1 - 2x_2)^2 - x_2^2$ and is zero when $x_1 = x_2 = 0$, $x_1 = 3x_2$ or $x_1 = x_2$.

FRAME 38 continued

You will notice that it is only in case (i) that Q is positive for all values of x_1 and x_2 other than $x_1 = x_2 = 0$. Any Q in this category is said to be POSITIVE DEFINITE. (Any Q in category (ii) is said to be positive semi-definite.)

Now answer the following questions:

1. Which of the following are positive definite?

   (a) $x_1^2 - 2x_1x_2 + 3x_2^2$ (b) $x_1^2 + 2x_1x_2 + x_2^2$

   (c) $x_1^2 + 4x_1x_2 + 2x_2^2$

2. For what range of values of a is $x_1^2 + 6x_1x_2 + ax_2^2$ positive definite?

****

38A

1. (a) 2. $a > 9$

FRAME 39

Now put $x_1^2 - 4x_1x_2 + ax_2^2$ into the form X'AX and find the value of $|A|$. What can you say about the value of $|A|$ corresponding to (i), (ii) and (iii) in the last frame?

****

39A

$$\begin{pmatrix} x_1 & x_2 \end{pmatrix} \begin{pmatrix} 1 & -2 \\ -2 & a \end{pmatrix} \begin{pmatrix} x_1 \\ x_2 \end{pmatrix} \qquad |A| = a - 4$$

(i) $|A| > 0$ *(ii)* $|A| = 0$ *(iii)* $|A| < 0$

You will notice that when Q is positive definite, $|A| > 0$.

FRAME 40

These ideas can easily be extended to functions of more than two real variables.

For example, $x_1^2 + 2x_2^2 + 7x_3^2 - 2x_1x_2 + 4x_1x_3 - 2x_2x_3$ can be written as $(x_1 - x_2 + 2x_3)^2 + (x_2 + x_3)^2 + 2x_3^2$ and, being the sum of three squares, is always positive unless $x_1 = x_2 = x_3 = 0$. Thus it is positive definite.

On the other hand, $x_1^2 + 2x_2^2 + 5x_3^2 - 2x_1x_2 + 4x_1x_3 - 2x_2x_3$ can be expressed as $(x_1 - x_2 + 2x_3)^2 + (x_2 + x_3)^2$ which is zero when $x_1:x_2:x_3 = 3:1:-1$, as well as when $x_1 = x_2 = x_3 = 0$, but is positive otherwise.

Furthermore, $x_1^2 + 2x_2^2 + 4x_3^2 - 2x_1x_2 + 4x_1x_3 - 2x_2x_3$ can be put into the form $(x_1 - x_2 + 2x_3)^2 + (x_2 + x_3)^2 - x_3^2$ which can be positive, zero or negative.

Thus neither of the last two quadratic forms is positive definite.

Now express each of the above three quadratic forms in the form X'AX and find $|A|$ in each case.

****

40A

$$\begin{pmatrix} x_1 & x_2 & x_3 \end{pmatrix}\begin{pmatrix} 1 & -1 & 2 \\ -1 & 2 & -1 \\ 2 & -1 & 7 \end{pmatrix}\begin{pmatrix} x_1 \\ x_2 \\ x_3 \end{pmatrix}, \quad 2$$

$$\begin{pmatrix} x_1 & x_2 & x_3 \end{pmatrix}\begin{pmatrix} 1 & -1 & 2 \\ -1 & 2 & -1 \\ 2 & -1 & 5 \end{pmatrix}\begin{pmatrix} x_1 \\ x_2 \\ x_3 \end{pmatrix}, \quad 0$$

and

$$\begin{pmatrix} x_1 & x_2 & x_3 \end{pmatrix}\begin{pmatrix} 1 & -1 & 2 \\ -1 & 2 & -1 \\ 2 & -1 & 4 \end{pmatrix}\begin{pmatrix} x_1 \\ x_2 \\ x_3 \end{pmatrix}, \quad -1$$

respectively.

FRAME 41

Again you will notice that in the positive definite case, $|A| > 0$. However $|A|$ being >0 does not by itself ensure that a quadratic form is positive definite. For example, $x_1^2 + 2x_3^2 + 2x_1x_2 + 4x_1x_3 + 6x_2x_3$ can be expressed in the matrix form

$$\begin{pmatrix} x_1 & x_2 & x_3 \end{pmatrix} \begin{pmatrix} 1 & 1 & 2 \\ 1 & 0 & 3 \\ 2 & 3 & 2 \end{pmatrix} \begin{pmatrix} x_1 \\ x_2 \\ x_3 \end{pmatrix}$$

and here $|A| > 0$. Yet it can also be expressed in the form

$$(x_1 + x_2 + 2x_3)^2 - (x_2 - x_3)^2 - x_3^2$$

from which it is obvious that it is not positive definite.

FRAME 42

Now, we have seen that $x_1^2 + 2x_2^2 + 7x_3^2 - 2x_1x_2 + 4x_1x_3 - 2x_2x_3$, being positive definite, is positive unless all x's are zero. But one or two x's can be put equal to zero and it will still remain positive. Thus if $x_3 = 0$, $x_1^2 + 2x_2^2 - 2x_1x_2$ must be positive for all x_1, x_2 other than $x_1 = x_2 = 0$ and hence also positive definite. But

$$x_1^2 + 2x_2^2 - 2x_1x_2 = \begin{pmatrix} x_1 & x_2 \end{pmatrix} \begin{pmatrix} 1 & -1 \\ -1 & 2 \end{pmatrix} \begin{pmatrix} x_1 \\ x_2 \end{pmatrix}$$

and hence it is necessary for $\begin{vmatrix} 1 & -1 \\ -1 & 2 \end{vmatrix}$ to be positive, which, of course, it is.

Furthermore, if $x_2 = 0$ as well as x_3, we are only left with x_1^2 which is positive for all $x_1 \neq 0$.

FRAME 43

These results indicate that if

$$\begin{pmatrix} x_1 & x_2 & x_3 \end{pmatrix} \begin{pmatrix} a_{11} & a_{12} & a_{13} \\ a_{21} & a_{22} & a_{23} \\ a_{31} & a_{32} & a_{33} \end{pmatrix} \begin{pmatrix} x_1 \\ x_2 \\ x_3 \end{pmatrix} \qquad (43.1)$$

is to be positive definite we must have

$$a_{11} > 0, \quad \begin{vmatrix} a_{11} & a_{12} \\ a_{21} & a_{22} \end{vmatrix} > 0 \quad \text{and} \quad \begin{vmatrix} a_{11} & a_{12} & a_{13} \\ a_{21} & a_{22} & a_{23} \\ a_{31} & a_{32} & a_{33} \end{vmatrix} > 0 \qquad (43.2)$$

(As explained earlier we always work with A symmetric, so that $a_{12} = a_{21}$, etc.)

Conversely, it can be proved that if conditions (43.2) are all satisfied then (43.1) is positive definite.

Use this determinant test to find whether the following are positive definite:

(i) $x_1^2 + 3x_1x_2 + 2x_2^2$

(ii) $x_1^2 + 2x_2^2 - 3x_3^2 - 2x_1x_2 + 2x_2x_3$

****

43A

Neither

FRAME 44

Extending now to any number of variables, a quadratic form Q in n real variables $x_1, x_2, \ldots, x_n$ is said to be positive definite if it is positive for all values of the variables other than $x_1 = x_2 = \ldots = x_n = 0$. For this to happen it must be possible to express Q as the sum of n squares. If Q is written in the matrix form X'AX, A, chosen to be symmetric, is called the matrix of the quadratic form, and is of size $n \times n$. Q is positive definite if all the determinants indicated in the following matrix diagram are positive.

FRAME 44 continued

$$\begin{vmatrix} a_{11} & a_{12} & a_{13} & \cdot & \cdot & \cdot & a_{1n} \\ a_{21} & a_{22} & a_{23} & \cdot & \cdot & \cdot & \cdot \\ a_{31} & a_{32} & a_{33} & \cdot & \cdot & \cdot & \cdot \\ \cdot & \cdot & \cdot & \cdot & \cdot & \cdot & \cdot \\ \cdot & \cdot & \cdot & \cdot & \cdot & \cdot & \cdot \\ \cdot & \cdot & \cdot & \cdot & \cdot & \cdot & \cdot \\ a_{n1} & \cdot & \cdot & \cdot & \cdot & \cdot & a_{nn} \end{vmatrix}$$

When Q is positive definite the matrix A is also said to be positive definite.

One use of positive definite matrices occurs in the theory of structures. If an elastic structure is subjected to forces and it is required to determine whether the equilibrium is stable, a necessary condition is that what is called the 'local stiffness matrix' shall be positive definite.

An application of positive definiteness which you have already met occurs in the theory of maxima and minima of two independent variables. If, for example, the energy V of a system is expressed in terms of coordinates x and y then in order for the equilibrium of the system to be stable it is necessary that the energy shall be a minimum. This will be so if

$$\frac{\partial^2 V}{\partial x^2} > 0 \quad \text{and} \quad \frac{\partial^2 V}{\partial x^2}\frac{\partial^2 V}{\partial y^2} - \left(\frac{\partial^2 V}{\partial x \partial y}\right)^2 \text{ also } > 0.$$

You will see that these two conditions are equivalent to the statement that

the matrix $\begin{bmatrix} \frac{\partial^2 V}{\partial x^2} & \frac{\partial^2 V}{\partial x \partial y} \\ \frac{\partial^2 V}{\partial x \partial y} & \frac{\partial^2 V}{\partial y^2} \end{bmatrix}$ is positive definite.

FRAME 45

Band and Sparse Matrices

You have already seen (FRAME 30) that arithmetic with diagonal matrices is very simple. However diagonal matrices are very specialised and it is only rarely that these will occur naturally. Other matrices which you have already met, which lead to relatively simple arithmetic, are upper triangular and lower triangular (defined in FRAME 28, page 3:99, of Vol I). A type of matrix, of which the diagonal matrix is a special case, and which does occur naturally (in structural analysis problems, for example), is the band matrix. A square matrix in which all elements that are not zero lie in a band parallel to the leading diagonal is called a BAND MATRIX. This band will be symmetric about the leading diagonal when, as frequently happens in practical problems, the matrix is itself symmetric. The following are some examples of band matrices:

(i) $\begin{bmatrix} 1 & -1 & 0 \\ -1 & 2 & -1 \\ 0 & -1 & 1 \end{bmatrix}$. (which you met in FRAMES 4, 5 and 6)

(ii) $\begin{bmatrix} 1 & -1 & 0 & 0 \\ 0 & -2 & 4 & 0 \\ 0 & 0 & 3 & 0 \\ 0 & 0 & 0 & -1 \end{bmatrix}$ and (iii) $\begin{bmatrix} a & b & c & 0 & 0 \\ d & e & f & g & 0 \\ h & j & k & \ell & m \\ 0 & n & p & q & r \\ 0 & 0 & s & t & u \end{bmatrix}$

Of course, in practice, matrices are usually much larger than the ones illustrated here and it is necessary to use a computer when handling them. One advantage of band matrices is that they can be inverted with greater ease and speed than ordinary matrices of the same size. If the band is narrow in comparison with the size of the matrix there will be many zero elements in the matrix. Any matrix (which may, or may not, be of the band type) which contains a large proportion of zero elements is said to be SPARSE. Such matrices occur in linear programming, engineering structures and electrical systems, including power generation and distribution. Computer programs have been developed which exploit the special nature of sparse matrices.

FRAME 46

Miscellaneous Examples

In this frame a collection of miscellaneous examples is given for you to try. Answers are supplied in FRAME 47, together with such working as is considered helpful.

1. Find the eigenvalues and eigenvectors of the matrix A, where $A = \begin{pmatrix} 1 & 2 \\ 3 & 2 \end{pmatrix}$.

   Show also that the eigenvalues of A^{-1} are the reciprocals of those of A.

2. Find the eigenvectors of the matrix

$$\begin{pmatrix} 3 & 2 & 2 \\ 2 & 2 & 0 \\ 2 & 0 & 4 \end{pmatrix}$$

   Two properties of a symmetric matrix are:

   (i) The eigenvectors are orthogonal. (Two vectors are orthogonal if the sum of the products of their corresponding elements is zero. In vector algebra terms this is saying that their scalar product is zero. In matrix algebra terms it means that, if X and Y are the two column vectors, then $X'Y = 0$.)

   (ii) The modal matrix of normalised eigenvectors is orthogonal.

   Verify (i) and (ii) for the given matrix.

3. Given that one eigenvalue of the matrix

$$C = \begin{pmatrix} 4 & 1 & -1 \\ 2 & 3 & -1 \\ -2 & 1 & 5 \end{pmatrix}$$

   is 2, find the others. Find also the eigenvectors of C and write down a matrix P such that $P^{-1}CP$ is diagonal and verify the correctness of your choice.

FRAME 46 continued

4. A matrix A is to have eigenvalues 1, 2 and 3 with corresponding eigenvectors {1 2 3}, {1 2 1} and {3 2 1}. Use this information to find A by a method involving the modal matrix. (If you can't see how to start this, turn back to answer frame 29A.)

5. Show that the characteristic equation of A, where A is the matrix $\begin{pmatrix} 1 & 5 \\ 2 & 4 \end{pmatrix}$, is $\lambda^2 - 5\lambda - 6 = 0$.

The Cayley-Hamilton theorem states that a square matrix satisfies its own characteristic equation. Verify that $A^2 - 5A - 6I = 0$.

If this last equation is either pre- or post-multiplied by A^{-1}, it becomes $A - 5I - 6A^{-1} = 0$. Use this result to find the inverse of the matrix A.

6. Find, by a method using determinants, the inequalities which must be satisfied by a and b for

$$2x_1^2 + ax_2^2 + 3x_3^2 - 2x_1x_2 + 2bx_2x_3$$

to be positive definite.

FRAME 47

Answers to Miscellaneous Examples

1. Eigenvalues of A are -1 and 4.

Eigenvectors are $k_1\{1 \quad -1\}$ and $k_2\{2 \quad 3\}$ respectively.

$$A^{-1} = \begin{pmatrix} -\frac{1}{2} & \frac{1}{2} \\ \frac{3}{4} & -\frac{1}{4} \end{pmatrix}$$

FRAME 47 continued

2. Eigenvalues are 0, 3, 6.

Eigenvectors are $k_1\{2 \quad -2 \quad -1\}$, $k_2\{1 \quad 2 \quad -2\}$ and $k_3\{2 \quad 1 \quad 2\}$.

Modal matrix of normalised eigenvectors $\begin{pmatrix} \frac{2}{3} & \frac{1}{3} & \frac{2}{3} \\ -\frac{2}{3} & \frac{2}{3} & \frac{1}{3} \\ -\frac{1}{3} & -\frac{2}{3} & \frac{2}{3} \end{pmatrix}$

Denoting this matrix by P, it is necessary to show that $P' = P^{-1}$.
You will realise that this is equivalent to showing that $PP' = I$.

3. 4 and 6

$k_1\{1 \quad -1 \quad 1\} \qquad k_2\{1 \quad 1 \quad 1\} \qquad k_3\{1 \quad 1 \quad -1\}$

$P = \begin{pmatrix} 1 & 1 & 1 \\ -1 & 1 & 1 \\ 1 & 1 & -1 \end{pmatrix}$ is one possibility.

4. One modal matrix H is $\begin{pmatrix} 1 & 1 & 3 \\ 2 & 2 & 2 \\ 3 & 1 & 1 \end{pmatrix}$

D is then $\begin{pmatrix} 1 & 0 & 0 \\ 0 & 2 & 0 \\ 0 & 0 & 3 \end{pmatrix}$ and $H^{-1} = -\frac{1}{4}\begin{pmatrix} 0 & 1 & -2 \\ 2 & -4 & 2 \\ -2 & 1 & 0 \end{pmatrix}$

$$A = \frac{1}{2}\begin{pmatrix} 7 & -1 & -1 \\ 2 & 4 & -2 \\ 1 & 1 & 1 \end{pmatrix}$$

FRAME 47 continued

5.
$$A^{-1} = \frac{1}{6}\begin{pmatrix} -4 & 5 \\ 2 & -1 \end{pmatrix}$$

6.
$$X'AX = \begin{pmatrix} x_1 & x_2 & x_3 \end{pmatrix}\begin{pmatrix} 2 & -1 & 0 \\ -1 & a & b \\ 0 & b & 3 \end{pmatrix}\begin{pmatrix} x_1 \\ x_2 \\ x_3 \end{pmatrix}$$

$2a > 1$ and $2b^2 < 6a - 3$

UNIT 9

FOURIER SERIES and PARTIAL DIFFERENTIAL EQUATIONS

A.C. Bajpai
I.M. Calus
J.A. Fairley

Loughborough University of Technology

INSTRUCTIONS

This Unit comprises two programmes:

(a) Fourier Series
(b) Partial Differential Equations for Technologists

Each programme is divided up into a number of FRAMES which are to be worked *in the order given*. You will be required to participate in many of these frames and in such cases the answers are provided in ANSWER FRAMES, designated by the letter A following the frame number. Steps in the working are given where this is considered helpful. The answer frame is separated from the main frame by a line of asterisks: *********. Keep the answers covered until you have written your own response. If your answer is wrong, go back and try to see why. Do not proceed to the next frame until you have corrected any mistakes in your attempt and are satisfied that you understand the contents up to this point.

Before reading these programmes, it is necessary that you are familiar with the following

Prerequisites

For (a): Integration, as covered in FRAMES 17-40, pages 2:11 to 2:26 of Vol I.

Mean values, as covered in FRAME 53, page 2:106.

For (b): The contents of the programme on Partial Differentiation in Unit 1.

Second order differential equations, as covered in FRAMES 1-31, pages 5:51 to 5:67.

The contents of (a).

CONTENTS

Instructions

FOURIER SERIES

FRAME 1

Introduction

You have probably already met examples of the use of Maclaurin's series where a function f(x) is expressed as a series of powers of x,

$$\text{e.g.} \quad \ln(1 + x) = x - \frac{x^2}{2} + \frac{x^3}{3} - \frac{x^4}{4} + \ldots$$

We do this because it is often easier to deal with powers of x than with the original functions.

In some problems, e.g. those dealing with oscillations, it is more convenient to use a series of sines. As sines are periodic, such a series can only represent a periodic function. In the next frame we shall remind you of some of the features of periodic functions.

FRAME 2

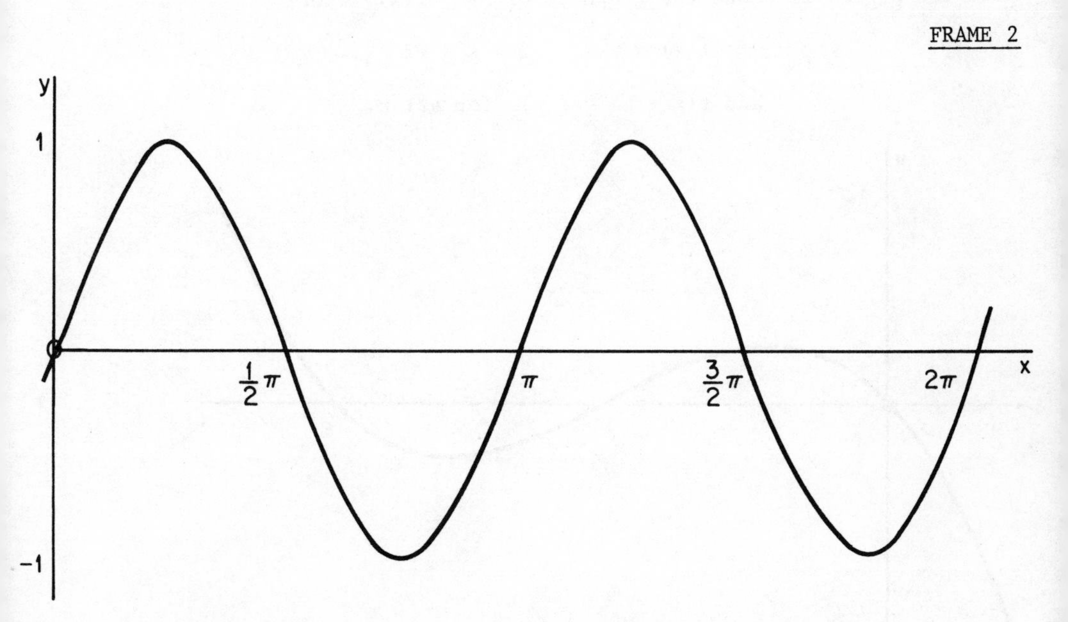

This is the graph of $y = \sin 2x$. The period is π.

Note that $\sin 2x = \sin 2(x + \pi)$ for all values of x, i.e. increasing x by π does not change the value of y.

FRAME 2 continued

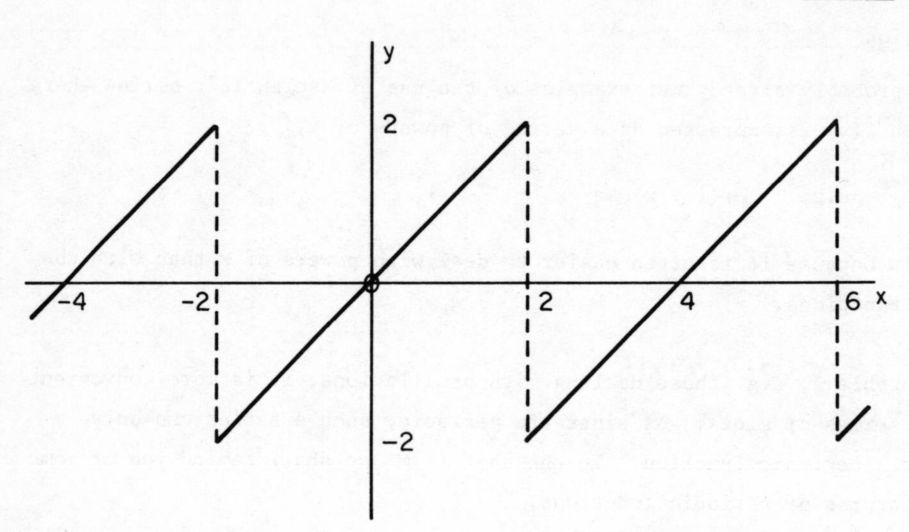

Here the period is 4 and the graph is of $y = f(x)$ where

$$f(x) = x \qquad -2 < x < +2$$

$$\text{and } f(x + 4) = f(x) \quad \text{for all } x.$$

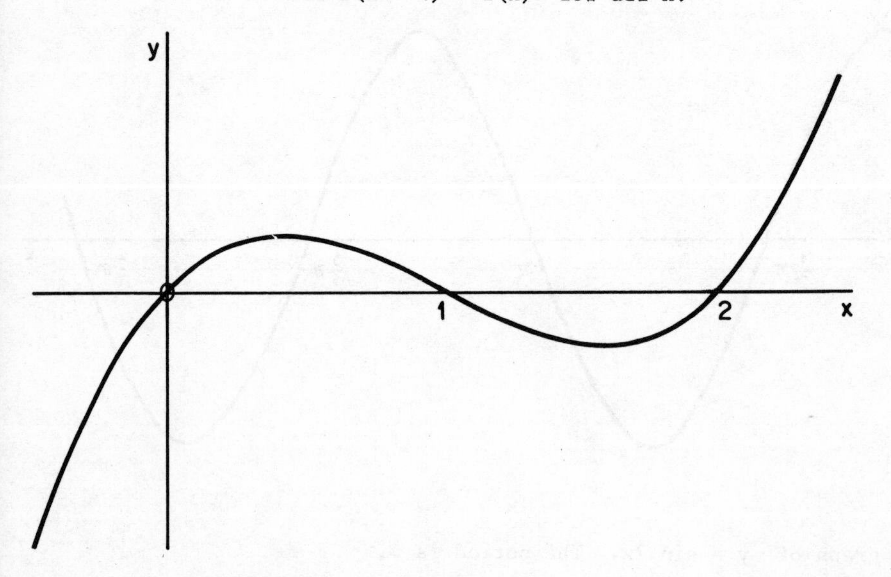

This is the graph of $y = x(x - 1)(x - 2)$ and in this case the function is not periodic.

FRAME 2 continued

State whether each of the following is periodic. If it is, give the period and a set of equations which define the function.

(i)

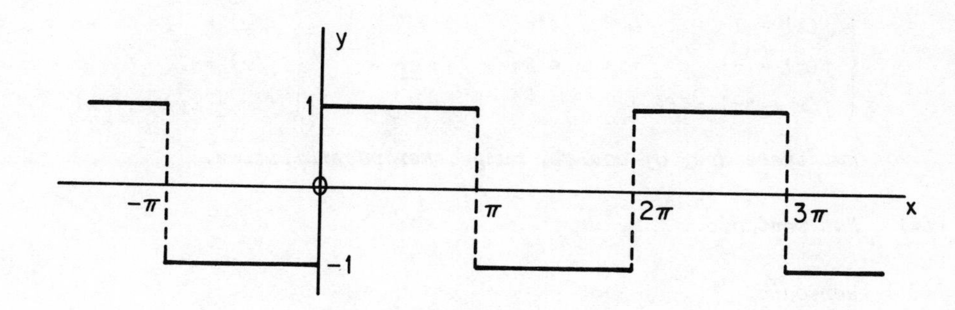

(ii)

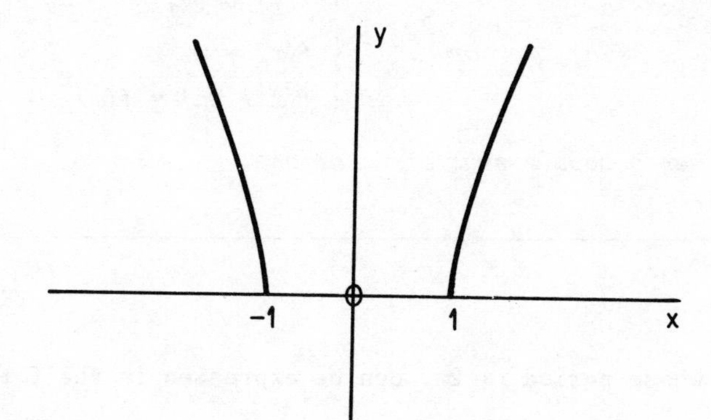

(iii)

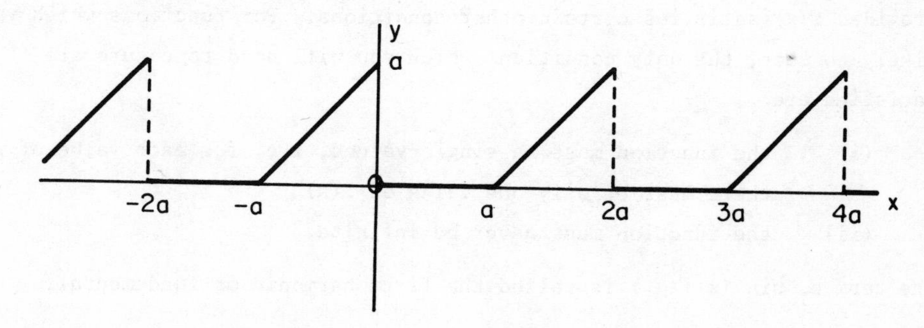

****

2A

(i) *Periodic.*

Period 2π.

The most obvious are

$$\begin{cases} f(x) = 1 & 0 < x < \pi \\ f(x) = -1 & \pi < x < 2\pi \\ f(x + 2\pi) = f(x) \end{cases} \quad \underline{or} \quad \begin{cases} f(x) = -1 & -\pi < x < 0 \\ f(x) = 1 & 0 < x < \pi \\ f(x + 2\pi) = f(x) \end{cases}$$

but there are, of course, many other possibilities.

(ii) *Not periodic.*

(iii) *Periodic.*

Period $2a$.

$$\begin{cases} f(x) = 0 & 0 < x < a \\ f(x) = x - a & a < x < 2a \\ f(x + 2a) = f(x) \end{cases} \quad \underline{or} \quad \begin{cases} f(x) = x + a & -a < x < 0 \\ f(x) = 0 & 0 < x < a \\ f(x + 2a) = f(x) \end{cases}$$

Again, there are other possible sets of equations.

FRAME 3

A periodic function $f(x)$, whose period is 2π, can be expressed in the form

$$f(x) = c_o + c_1 \sin (x + \alpha_1) + c_2 \sin (2x + \alpha_2) + c_3 \sin (3x + \alpha_3) + \ldots \quad (3.1)$$

provided $f(x)$ satisfies certain other conditions. For functions which you are likely to meet, the only conditions which you will need to ensure are satisfied are:

(i) the function must be single-valued, i.e. for each value of x there must be only one value of $f(x)$,

(ii) the function must never be infinite.

The term $c_1 \sin (x + \alpha_1)$ is called the first harmonic or fundamental.

The terms $c_2 \sin (2x + \alpha_2)$, $c_3 \sin (3x + \alpha_3)$ etc. are called the second harmonic, third harmonic, etc. A series such as that on the R.H.S. of (3.1) is called a FOURIER SERIES.

FRAME 3 continued

If the period of f(x) is not 2π, a conversion to a function whose period is 2π can be made by a suitable change of variable, as will be shown later in the programme.

FRAME 4

The series in (3.1) can be expressed in an alternative form as follows:

$$\sin (x + \alpha_1) = \sin x \cos \alpha_1 + \cos x \sin \alpha_1$$

and so we can write $c_1 \sin (x + \alpha_1) = a_1 \cos x + b_1 \sin x$

where $a_1 = c_1 \sin \alpha_1$ and $b_1 = c_1 \cos \alpha_1$.

Converting the other terms in a similar way, the series becomes

$$f(x) = \tfrac{1}{2}a_o + a_1 \cos x + a_2 \cos 2x + a_3 \cos 3x + \ldots$$
$$+ b_1 \sin x + b_2 \sin 2x + b_3 \sin 3x + \ldots \quad (4.1)$$

The reason for writing c_o as $\tfrac{1}{2}a_o$ will become clear later on.

The problem now, in any particular case, is to find the coefficients $a_o, a_1, a_2, \ldots.., b_1, b_2, \ldots..,$ or, what is equivalent, the values of $c_o, c_1, c_2, \ldots.., \alpha_1, \alpha_2, \ldots$ In practice, it is easier to calculate the a's and the b's, and the c's and the α's can then be found from them, if the series is required in the form (3.1). The form (4.1) may, however, be suitable in itself.

Before proceeding to the method of calculating the coefficients, we shall, in the next few frames, give some illustrations showing that the idea of such a series for a periodic function is feasible and how it can be useful to an engineer or applied scientist.

FRAME 5

Let us consider the series

$$\frac{1}{2} + \frac{2}{\pi}\left(\sin x + \frac{1}{3}\sin 3x + \frac{1}{5}\sin 5x + \frac{1}{7}\sin 7x + \ldots\right) \tag{5.1}$$

and draw the graphs representing:

(i) $y = \frac{1}{2}$

(ii) $y = \frac{1}{2} + \frac{2}{\pi}\sin x$

(iii) $y = \frac{1}{2} + \frac{2}{\pi}(\sin x + \frac{1}{3}\sin 3x)$

(iv) $y = \frac{1}{2} + \frac{2}{\pi}(\sin x + \frac{1}{3}\sin 3x + \frac{1}{5}\sin 5x)$

These graphs, from $x = 0$ to 2π, are:

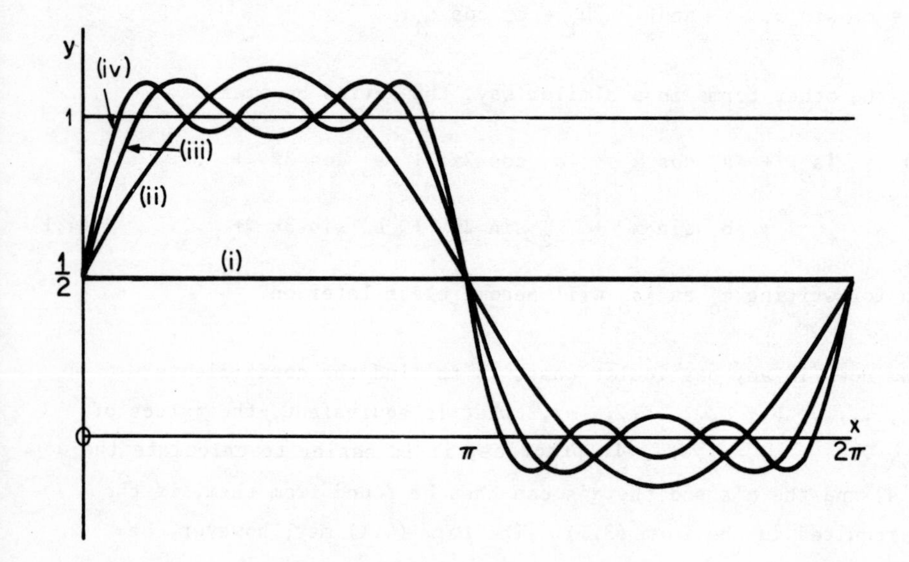

As (ii), (iii) and (iv) are periodic, each with period 2π, the continuation of their graphs for other values of x will simply be repetitions of the parts we have shown.

You will notice that, as the number of terms is increased, the oscillations about $y = 1$ and $y = 0$ get smaller and take place over a greater width. If we now considerably increase the number of terms in the sum, the graph will look like this:

FRAME 5 continued

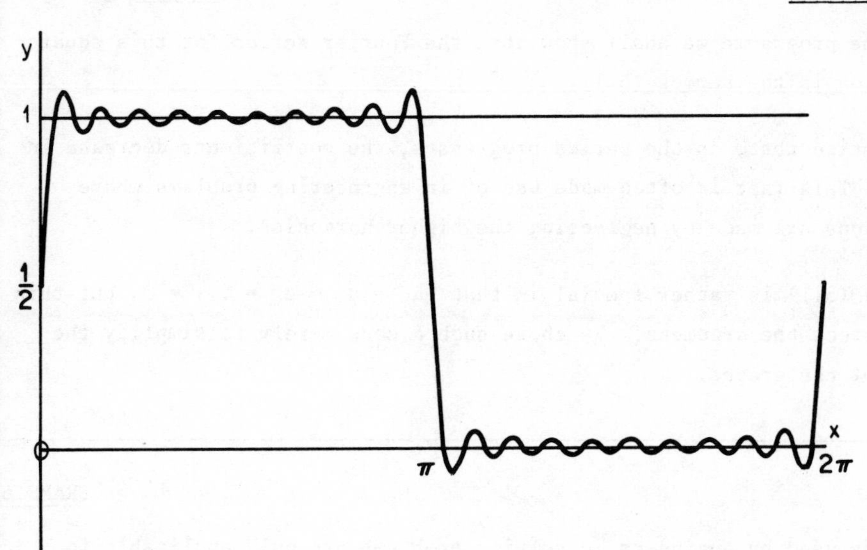

It is therefore reasonable to suggest that if we increase the number of terms indefinitely, the square-wave function shown will be produced.

This has equations

$$\begin{cases} f(x) = 1 & 0 < x < \pi \\ f(x) = 0 & \pi < x < 2\pi \\ f(x + 2\pi) = f(x) & \end{cases}$$

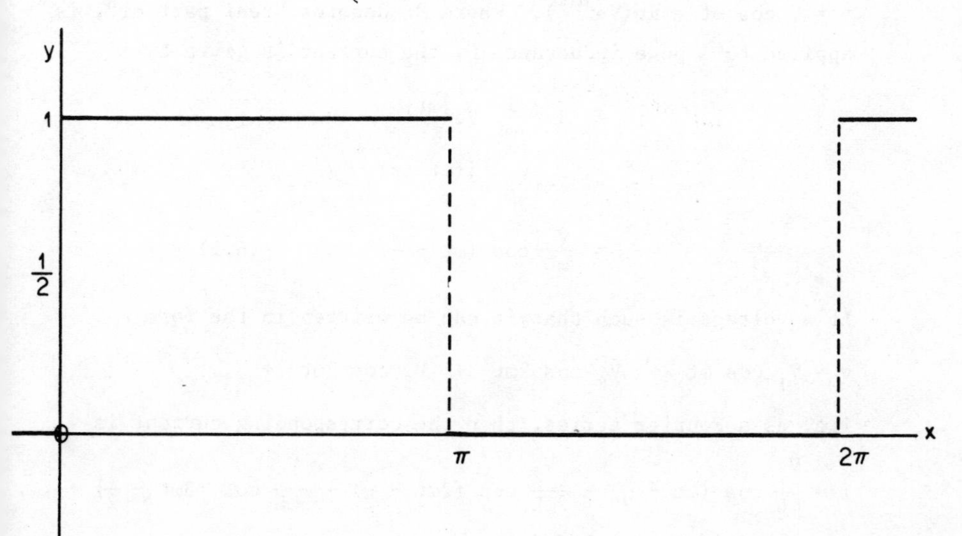

FRAME 5 continued

Later in the programme we shall show that the Fourier series for this square-wave function is the series (5.1).

You will notice that, as the series progresses, the coefficients decrease in magnitude. This fact is often made use of in engineering problems where approximations are made by neglecting the higher harmonics.

(The series (5.1) is rather special in that $a_1 = a_2 = a_3 = \ldots = 0$, but this does not affect the argument. We chose such a case merely to simplify the sketching of the graphs.)

FRAME 6

Some methods used by engineers in solving problems are only applicable to sinusoidal functions and consequently they often find it helpful to use Fourier series to express other functions in terms of sines (and/or cosines). In the field of electrical engineering two such cases are:

(i) The use of complex numbers in dealing with a.c. circuits depends on functions being sinusoidal. For example, if a voltage $v = V \cos \omega t = \mathrm{Re}(Ve^{j\omega t})$, where Re denotes "real part of", is applied to a pure inductance L, the current is given by

$$\begin{aligned} i &= \mathrm{Re}\{\frac{1}{j\omega L} Ve^{j\omega t}\} \\ &= \mathrm{Re}\{\frac{V}{\omega L} e^{j(\omega t - \frac{1}{2}\pi)}\} \\ &= \frac{V}{\omega L} \cos (\omega t - \frac{\pi}{2}) \qquad (6.1) \end{aligned}$$

If a voltage is such that it can be written in the form

$$v = V_1 \cos \omega t + V_2 \cos 2\omega t + V_3 \cos 3\omega t + \ldots$$

i.e. as a Fourier series, then the corresponding current is

$$i = \frac{V_1}{\omega L} \cos (\omega t - \frac{\pi}{2}) + \frac{V_2}{2\omega L} \cos (2\omega t - \frac{\pi}{2}) + \frac{V_3}{3\omega L} \cos (3\omega t - \frac{\pi}{2}) + \ldots$$

by applying the result (6.1) to each term.

FRAME 6 continued

(ii) If the average value of the power in a circuit over time T is required, T being the period of the applied voltage, we have to evaluate

$$\frac{1}{T}\int_{o}^{T} vi\ dt$$

The integration involved is simplified if we can write the voltage and current in the forms

$$v = V_1 \cos(\omega t + \alpha_1) + V_2 \cos(2\omega t + \alpha_2) + \ldots$$

$$i = I_1 \cos(\omega t + \alpha_1 + \phi_1) + I_2 \cos(2\omega t + \alpha_2 + \phi_2) + \ldots$$

as then the integrals of all terms of the form

$V_m \cos(m\omega t + \alpha_m) I_n \cos(n\omega t + \alpha_n + \phi_n)$, where $m \neq n$, are zero.

We shall be proving a simpler version of this result in FRAME 13. Consequently, we are only left with integrals of terms of the form

$$V_n \cos(n\omega t + \alpha_n) I_n \cos(n\omega t + \alpha_n + \phi_n)$$

which are very simple to evaluate.

You will notice that in these two cases, t is used instead of x as the independent variable is time. This is usually the case in electrical examples, but in other applications the independent variable is distance, generally represented by x. For uniformity within the programme, we shall work in terms of x.

You will also notice that multiples of ωt, rather than t, appear in these series. This is because the period of the waveforms is $2\pi/\omega$ instead of 2π.

FRAME 7

Fourier series are also useful in tackling some beam problems, as in the examples shown here and in the next frames.

If the load on a beam varies sinusoidally so also does the deflection and a formula for this is known. If the load can be expressed as the sum of such

FRAME 7 continued

terms then this formula can be applied to each term in turn. One such loading for which this can be done is the so-called patch loading, i.e. loading of the form

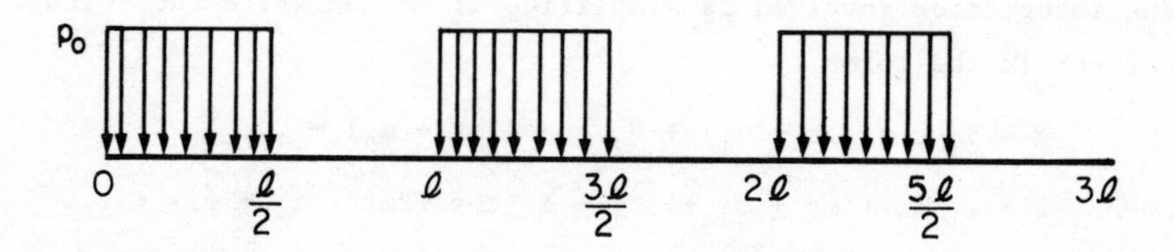

taking the length of the beam in this case as 3ℓ.

This can be expressed as a series of sine terms and later we shall see that it is

$$\frac{1}{2} p_o + \frac{2}{\pi} p_o \left(\sin \frac{2\pi x}{\ell} + \frac{1}{3} \sin \frac{6\pi x}{\ell} + \frac{1}{5} \sin \frac{10\pi x}{\ell} + \ldots\right) \qquad (7.1)$$

Multiples of $\frac{2\pi x}{\ell}$ occur here because the period is ℓ.

The calculation of the deflection due to the constant term is as straightforward as it is for the sine terms.

FRAME 8

Any value of x can be substituted in (7.1) but the practical interpretation only exists for $0 \leq x \leq 3\ell$ i.e. (7.1) represents the waveform

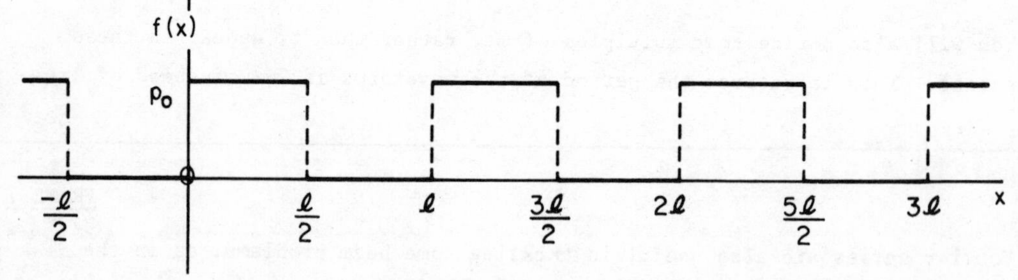

but we are only interested in the section between 0 and 3ℓ.

If the loading on a beam is not periodic it may still be helpful to be able to represent the deflection curve by means of a Fourier series.

FRAME 8 continued

Suppose, for example, we have a cantilever whose deflection curve is as in the diagram.

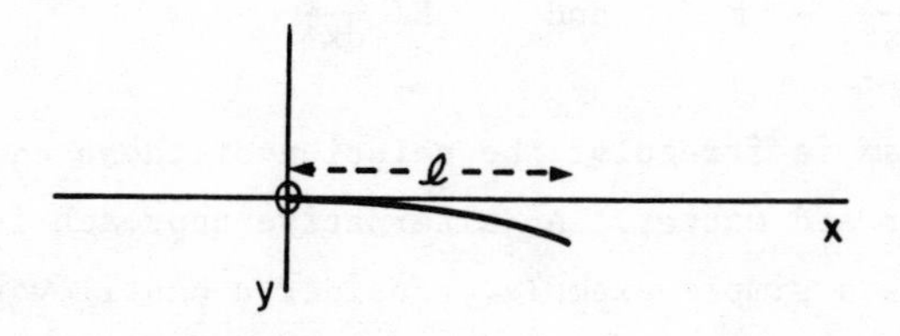

Some of the waveforms of which this is part are shown below.

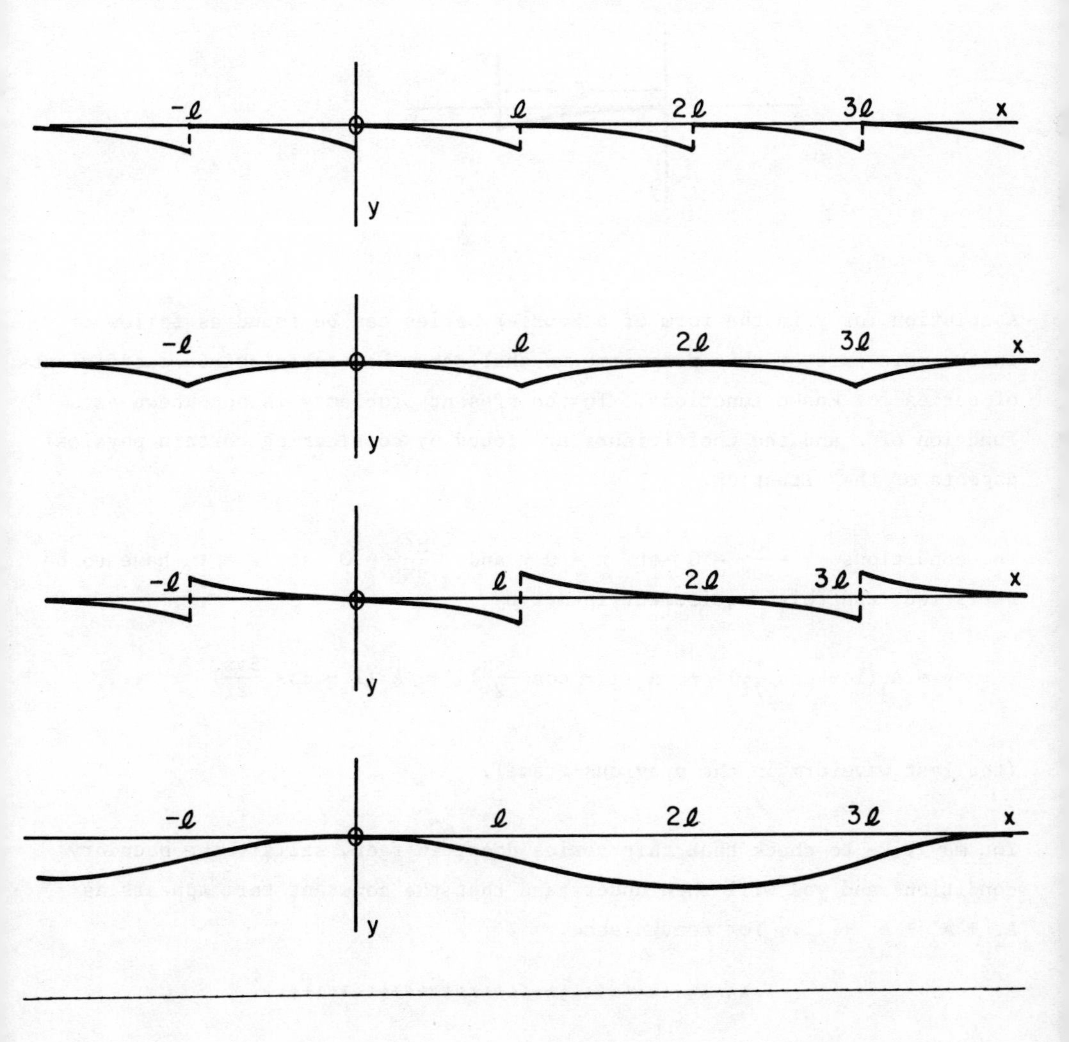

FRAME 9

You are probably familiar with the standard beam equations

$$EI\,\frac{d^2y}{dx^2} = M \qquad \text{and} \qquad EI\,\frac{d^4y}{dx^4} = w$$

If the loading on the beam is irregular the solution of these equations may well not be a straightforward matter. An alternative approach is by a method using Fourier series. As a simple example, consider a cantilever loaded as shown.

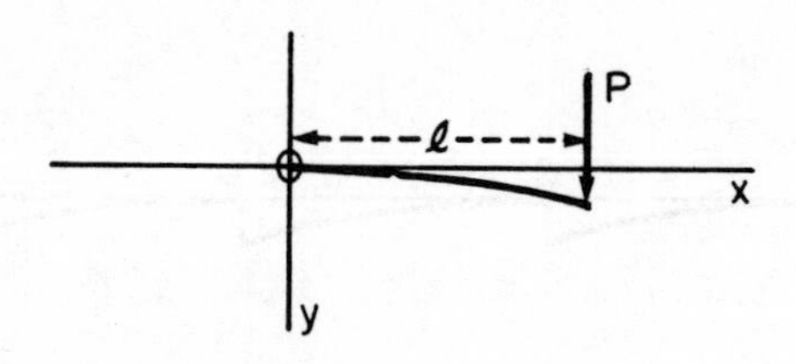

A solution for y in the form of a Fourier series can be found as follows: In the main part of this programme we shall show how to calculate coefficients of series for known functions. In the present problem y is not known as a function of x and the coefficients are found by considering certain physical aspects of the situation.

The conditions $y = \frac{dy}{dx} = 0$ at $x = 0$ and $\frac{d^2y}{dx^2} = 0$ at $x = \ell$ have to be satisfied, and this requirement is met by

$$y = A_1(1 - \cos\frac{\pi x}{2\ell}) + A_3(1 - \cos\frac{3\pi x}{2\ell}) + A_5(1 - \cos\frac{5\pi x}{2\ell}) + \dots$$

(the last waveform in the previous frame).

You may like to check that this series does, in fact, satisfy the boundary conditions and you will then understand that the constant term appears as $A_1 + A_3 + A_5 + \dots$ for convenience.

****

9A

$$\frac{dy}{dx} = A_1\frac{\pi}{2\ell}\sin\frac{\pi x}{2\ell} + A_3\frac{3\pi}{2\ell}\sin\frac{3\pi x}{2\ell} + A_5\frac{5\pi}{2\ell}\sin\frac{5\pi x}{2\ell} + \dots$$

$$\frac{d^2y}{dx^2} = A_1\frac{\pi^2}{4\ell^2}\cos\frac{\pi x}{2\ell} + A_3\frac{9\pi^2}{4\ell^2}\cos\frac{3\pi x}{2\ell} + A_5\frac{25\pi^2}{4\ell^2}\cos\frac{5\pi x}{2\ell} + \dots$$

which can be written as $y'' = C_1\cos\frac{\pi x}{2\ell} + C_3\cos\frac{3\pi x}{2\ell} + C_5\cos\frac{5\pi x}{2\ell} + \dots$

It is then easily verified that the conditions are satisfied.

FRAME 10

In this particular problem the coefficients are found by making use of the principle of virtual work. In order to do so, we require to know the energy stored in the beam. This is given by

$$U = \frac{EI}{2}\int_0^\ell (y'')^2\,dx$$

$$= \frac{EI}{2}\int_0^\ell \left(C_1\cos\frac{\pi x}{2\ell} + C_3\cos\frac{3\pi x}{2\ell} + C_5\cos\frac{5\pi x}{2\ell} + \dots\right)^2 dx$$

from answer frame 9A.

As in the case of the average power example in FRAME 6, the integration is very simple because all the integrals

$$\int_0^\ell \cos\frac{\pi x}{2\ell}\cos\frac{3\pi x}{2\ell}\,dx, \quad \int_0^\ell \cos\frac{\pi x}{2\ell}\cos\frac{5\pi x}{2\ell}\,dx, \quad \int_0^\ell \cos\frac{3\pi x}{2\ell}\cos\frac{5\pi x}{2\ell}\,dx \text{ etc}$$

are zero, leaving only

$$\frac{EI}{2}\int_0^\ell \left(C_1^{\,2}\cos^2\frac{\pi x}{2\ell} + C_3^{\,2}\cos^2\frac{3\pi x}{2\ell} + C_5^{\,2}\cos^2\frac{5\pi x}{2\ell} + \dots\right)dx$$

This is easily evaluated and the coefficients are then found by applying the principle of virtual work. The equation of the deflection curve thus obtained is

$$y = \frac{32P\ell^3}{EI\pi^4}\left\{\left(1 - \cos\frac{\pi x}{2\ell}\right) + \frac{1}{3^4}\left(1 - \cos\frac{3\pi x}{2\ell}\right) + \frac{1}{5^4}\left(1 - \cos\frac{5\pi x}{2\ell}\right) + \dots\right\}$$

FRAME 11

Another application of Fourier series is their use in fitting boundary conditions to the solutions of partial differential equations.

One example is the case of a taut string, one point of which is pulled aside and then released. The string oscillates and the displacement of any point depends both on its distance from one end and on time. The equation of motion is

$$\frac{\partial^2 y}{\partial x^2} = \frac{1}{c^2}\frac{\partial^2 y}{\partial t^2}$$

where c is a constant and y is the displacement of a point distant x from one end at time t. This is a partial differential equation as y is a function of both x and t. One of the boundary conditions is that when $t = 0$ the shape of the string is

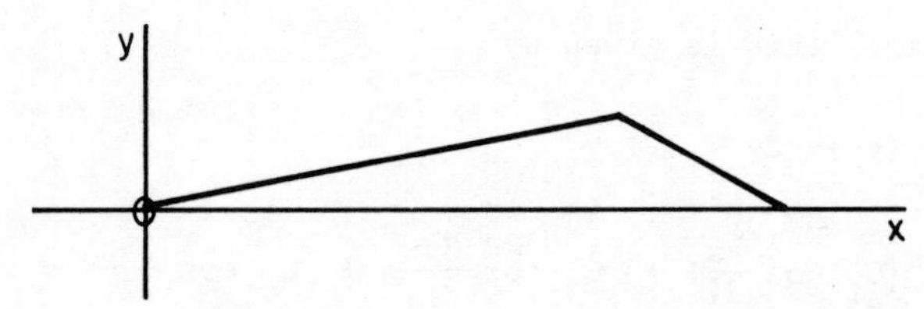

and it is found that in order to apply this boundary condition it is necessary to express this shape as a Fourier series.

Other examples of practical problems requiring a similar technique are certain cases of rod oscillations, heat conduction and voltages in an electric cable. The solution of such partial differential equations is dealt with in the next programme in this volume.

FRAME 12

Finally we would mention one or two applications of Fourier series in medical science.

A musical note consists of a fundamental and higher harmonics, the latter being the overtones, and, in the study of the physiology of hearing, Helmholtz's theory assumes that different fibres in the inner ear are excited only by those

FRAME 12 continued

harmonics to which they are tuned.

An electric current is produced by the activity of the heart. The variation of the voltage involved is periodic and can be measured by an electrocardiograph. The resulting waveform can be analysed into a Fourier series and the coefficients will show marked changes if disease or damage is present.

Human temperature fluctuates periodically, and can thus be represented by a Fourier series (see FRAME 53 Miscellaneous Examples, No. 8).

FRAME 13

Integrals required for calculating the coefficients

The calculation of the coefficients a_o, a_1, a_2, ..., b_1, b_2, ... will require the values of certain definite integrals, which we will now find. In these integrals both m and n are integers.

Two of the integrals that will be required are $\int_{-\pi}^{\pi} \cos nx \, dx$ and $\int_{-\pi}^{\pi} \sin nx \, dx$.

As you know, the integral of a sine or cosine over a complete period is zero. This is obvious from a graph - if you are not already satisfied on this point a quick sketch should convince you. The period of cos nx is $2\pi/n$, and therefore in integrating from $-\pi$ to π we are integrating over n complete periods.

Hence $\int_{-\pi}^{\pi} \cos nx \, dx = 0.$

What is the period of sin nx and what is the value of $\int_{-\pi}^{\pi} \sin nx \, dx$?

****

13A

The period of sin nx is $2\pi/n$.

$\int_{-\pi}^{\pi} \sin nx \, dx = 0$ *as the integral is over n complete periods.*

FRAME 14

The other integrals that will be needed are $\int_{-\pi}^{\pi} \cos mx \cos nx\, dx$,

$\int_{-\pi}^{\pi} \sin mx \sin nx\, dx$ and $\int_{-\pi}^{\pi} \sin mx \cos nx\, dx$.

For the first one, we have

$$\int_{-\pi}^{\pi} \cos mx \cos nx\, dx = \tfrac{1}{2}\int_{-\pi}^{\pi} \{\cos (m + n)x + \cos (m - n)x\}dx$$

$= 0$ if $m \neq n$, as both cosines are integrated over a number of complete periods.

But if $m = n$, the integral becomes $\tfrac{1}{2}\int_{-\pi}^{\pi} (\cos 2nx + 1)dx$

$$= \pi$$

Now work out the other two integrals, not forgetting to consider whether $m = n$ must be treated as a special case.

****

14A

$$\int_{-\pi}^{\pi} \sin mx \sin nx\, dx = \tfrac{1}{2}\int_{-\pi}^{\pi} \{\cos (m - n)x - \cos (m + n)x\}dx$$

$$\begin{cases} = 0 & \text{if } m \neq n \\ = \pi & \text{if } m = n \end{cases}$$

$$\int_{-\pi}^{\pi} \sin mx \cos nx\, dx = \tfrac{1}{2}\int_{-\pi}^{\pi} \{\sin (m + n)x + \sin (m - n)x\}dx$$

$= 0$ *in all cases.*

FRAME 15

The integrals worked out in the last two frames are now summarised for easy reference.

$$\int_{-\pi}^{\pi} \cos nx\, dx = 0 \qquad (15.1)$$

$$\int_{-\pi}^{\pi} \sin nx\, dx = 0 \qquad (15.2)$$

FRAME 15 continued

$$\int_{-\pi}^{\pi} \cos mx \cos nx \, dx = \begin{cases} 0 & m \neq n \quad (15.3) \\ \pi & m = n \quad (15.4) \end{cases}$$

$$\int_{-\pi}^{\pi} \sin mx \sin nx \, dx = \begin{cases} 0 & m \neq n \quad (15.5) \\ \pi & m = n \quad (15.6) \end{cases}$$

$$\int_{-\pi}^{\pi} \sin mx \cos nx \, dx = 0 \quad (15.7)$$

If we integrate these functions between other limits, we find that the values are the same so long as the limits differ by 2π

e.g. 0 to 2π, $\frac{-\pi}{2}$ to $\frac{3\pi}{2}$ etc.

FRAME 16

Calculation of the coefficients

We are now ready to find formulae for calculating the a's and b's.

First, we will show how to find a_o.

Now
$$f(x) = \tfrac{1}{2}a_o + a_1 \cos x + a_2 \cos 2x + \ldots + a_n \cos nx + \ldots$$
$$+ b_1 \sin x + b_2 \sin 2x + \ldots + b_n \sin nx + \ldots \quad (16.1)$$

and integrating both sides from $-\pi$ to π gives

$$\int_{-\pi}^{\pi} f(x) \, dx = \int_{-\pi}^{\pi} \tfrac{1}{2}a_o \, dx$$

The integrals of all the other terms will be zero, as each is a case of either (15.1) or (15.2).

We now get
$$\int_{-\pi}^{\pi} f(x) \, dx = \tfrac{1}{2}a_o \cdot 2\pi$$

$$\therefore \quad a_o = \frac{1}{\pi} \int_{-\pi}^{\pi} f(x) \, dx$$

Note that
$$\tfrac{1}{2}a_o = \frac{1}{2\pi} \int_{-\pi}^{\pi} f(x) \, dx$$

$$= \text{mean value of } f(x) \text{ over the range } -\pi \text{ to } \pi.$$

FRAME 17

The next problem is to find the coefficients of the cosine terms.
To find a_1, we shall multiply both sides of (16.1) by $\cos x$ and integrate from $-\pi$ to π. This gives

$$\int_{-\pi}^{\pi} f(x) \cos x \, dx = \int_{-\pi}^{\pi} a_1 \cos^2 x \, dx$$

Once again, the integrals of all the other terms will be zero, as each is a case of (15.1), (15.3) or (15.7). You should check this.

Now using (15.4) we have

$$\int_{-\pi}^{\pi} f(x) \cos x \, dx = a_1 \pi$$

$$\therefore \quad a_1 = \frac{1}{\pi} \int_{-\pi}^{\pi} f(x) \cos x \, dx$$

Now, how do you suggest that we should find $a_2, a_3, \ldots, a_n$?

****

17A

To find a_2, a_3, etc. multiply both sides of (16.1) by $\cos 2x$, $\cos 3x$, etc. respectively and integrate between $-\pi$ and π.

In general, to find a_n multiply both sides of (16.1) by $\cos nx$ and integrate between $-\pi$ and π.

FRAME 18

Turning now to the case of the general term we suggest that you find the formula for a_n.

****

18A

$$\int_{-\pi}^{\pi} f(x) \cos nx \, dx = \int_{-\pi}^{\pi} a_n \cos^2 nx \, dx$$

$$= a_n \pi$$

$$\therefore \quad a_n = \frac{1}{\pi} \int_{-\pi}^{\pi} f(x) \cos nx \, dx$$

FRAME 19

If we put $n = 2, 3, \ldots$ in the formula for a_n, we get

$$a_2 = \frac{1}{\pi}\int_{-\pi}^{\pi} f(x)\cos 2x\,dx$$

$$a_3 = \frac{1}{\pi}\int_{-\pi}^{\pi} f(x)\cos 3x\,dx$$

and so on. You will notice that putting $n = 1$ gives a_1, of course. Also, putting $n = 0$ gives the formula for a_o obtained in FRAME 16 so that the general formula for a_n also covers this case. Now you will understand why we wrote c_o as $\frac{1}{2}a_o$, rather than a_o, in (4.1).

We are now left with the problem of finding the coefficients of the sine terms i.e. the b's. Proceeding directly to the general case, can you suggest a method for finding the formula for b_n?

****

19A

To find b_n, multiply both sides of (16.1) by $\sin nx$ and integrate from $-\pi$ to π.

FRAME 20

By a method similar to those used previously, now find the formula for b_n.

****

20A

$$\int_{-\pi}^{\pi} f(x)\sin nx\,dx = \int_{-\pi}^{\pi} b_n \sin^2 nx\,dx$$

$$= b_n\,\pi$$

$$\therefore\; b_n = \frac{1}{\pi}\int_{-\pi}^{\pi} f(x)\sin nx\,dx$$

FRAME 21

Summarising, we have the following result:

If a periodic function f(x), of period 2π, can be represented by

$$f(x) = \tfrac{1}{2}a_o + a_1 \cos x + a_2 \cos 2x + \dots + a_n \cos nx + \dots$$
$$+ b_1 \sin x + b_2 \sin 2x + \dots + b_n \sin nx + \dots$$

then $a_n = \frac{1}{\pi}\int_{-\pi}^{\pi} f(x) \cos nx \, dx$

and $b_n = \frac{1}{\pi}\int_{-\pi}^{\pi} f(x) \sin nx \, dx.$

You will notice that a_n is twice the mean value of $f(x) \cos nx$ over a period and b_n is twice the mean value of $f(x) \sin nx$ over a period.

NOTE: As the integrals used in arriving at the formulae for a_n and b_n have the same values for any interval of 2π (see FRAME 15), we could use any such interval instead of $-\pi$ to π. Thus we could also have, for instance,

$$a_n = \frac{1}{\pi}\int_o^{2\pi} f(x) \cos nx \, dx$$

$$b_n = \frac{1}{\pi}\int_o^{2\pi} f(x) \sin nx \, dx$$

An alternative way of writing the series for f(x), which is more compact, is to use the $\sum$ notation, i.e.

$$f(x) = \tfrac{1}{2}a_o + \sum_{n=1}^{\infty} a_n \cos nx + \sum_{n=1}^{\infty} b_n \sin nx$$

FRAME 22

As a first example, let us consider the following waveform, which is of period 2π:

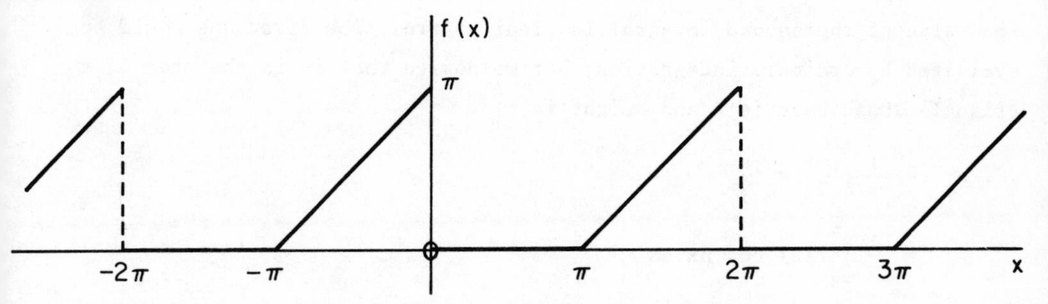

Can you give a set of equations which would define this function?

****

22A

Two possible sets are:

$$\left.\begin{array}{lll} f(x) & = x + \pi & -\pi < x < 0 \\ f(x) & = 0 & 0 < x < \pi \\ f(x + 2\pi) & = f(x) & \end{array}\right\} \qquad (22A.1)$$

$$\left.\begin{array}{lll} f(x) & = 0 & 0 < x < \pi \\ f(x) & = x - \pi & \pi < x < 2\pi \\ f(x + 2\pi) & = f(x) & \end{array}\right\} \qquad (22A.2)$$

FRAME 23

Using the set of equations (22A.1), the Fourier series for the wave-form would be found as follows:

$$a_o = \frac{1}{\pi}\int_{-\pi}^{\pi} f(x)\, dx$$

$$= \frac{1}{\pi}\left[\int_{-\pi}^{o} f(x)\, dx + \int_{o}^{\pi} f(x)\, dx\right]$$

The interval has to be split up in this way, because of the different f(x) in the two halves.

FRAME 23 continued

$$\therefore a_o = \frac{1}{\pi}\left[\int_{-\pi}^{o}(x + \pi)dx + \int_{o}^{\pi} 0 . dx\right]$$

The value of the second integral is clearly zero. The first one could be evaluated by ordinary integration, but we notice that it is the area of a triangle whose base is π and height is π.

$$\therefore a_o = \frac{1}{\pi} . \frac{\pi^2}{2} = \frac{\pi}{2}$$

$$a_n = \frac{1}{\pi}\int_{-\pi}^{\pi} f(x) \cos nx \, dx$$

$$= \frac{1}{\pi}\left[\int_{-\pi}^{o}(x + \pi)\cos nx \, dx + \int_{o}^{\pi} 0 . \cos nx \, dx\right]$$

$$= \frac{1}{\pi}\int_{-\pi}^{o}(x + \pi) \cos nx \, dx + 0$$

$$= \frac{1}{\pi}\left\{\left[(x + \pi)\frac{1}{n} \sin nx\right]_{-\pi}^{o} - \int_{-\pi}^{o} 1 . \frac{1}{n} \sin nx \, dx\right\} \quad \text{on integrating by parts}$$

$$= \frac{1}{\pi}\left\{0 - \left[-\frac{1}{n^2} \cos nx\right]_{-\pi}^{o}\right\}$$

$$= \frac{1}{\pi n^2}\left\{1 - \cos(-n\pi)\right\}$$

$$= \frac{1}{\pi n^2}(1 - \cos n\pi)$$

Now $\cos n\pi = 1$ if n is even, but -1 if n is odd.

$$\therefore a_n = \begin{cases} 0 & \text{if n is even} \\ \frac{2}{\pi n^2} & \text{if n is odd} \end{cases}$$

$$b_n = \frac{1}{\pi}\int_{-\pi}^{\pi} f(x) \sin nx \, dx$$ and the evaluation of this is left to you.

****

23A

$$b_n = \frac{1}{\pi}\int_{-\pi}^{0} (x + \pi)\sin nx\, dx$$

$$= \frac{1}{\pi}\left\{\left[-(x+\pi)\frac{1}{n}\cos nx\right]_{-\pi}^{0} - \int_{-\pi}^{0} -\frac{1}{n}\cos nx\, dx\right\}$$

$$= \frac{1}{\pi}\left\{\frac{-\pi}{n} + \frac{1}{n^2}\left[\sin nx\right]_{-\pi}^{0}\right\}$$

$$= -\frac{1}{n}$$

FRAME 24

The Fourier series for the waveform is, therefore,

$$\frac{\pi}{4} + \frac{2}{\pi}(\cos x + \frac{1}{3^2}\cos 3x + \frac{1}{5^2}\cos 5x + \ldots)$$

$$- (\sin x + \frac{1}{2}\sin 2x + \frac{1}{3}\sin 3x + \frac{1}{4}\sin 4x + \ldots)$$

Alternatively, using the $\sum$ notation, this could be written as

$$\frac{\pi}{4} + \frac{2}{\pi}\sum_{k=1}^{\infty}\frac{1}{(2k-1)^2}\cos(2k-1)x - \sum_{n=1}^{\infty}\frac{1}{n}\sin nx$$

We suggest that you now verify that the same values of a_o, a_n, and b_n, and hence the same series, are obtained if the set of equations (22A.2) is used.

****

24A

$$a_o = \frac{1}{\pi}\int_{0}^{2\pi} f(x)\, dx$$

$$= \frac{1}{\pi}\int_{\pi}^{2\pi} (x - \pi)dx$$

$$= \frac{\pi}{2}$$

$$a_n = \frac{1}{\pi}\int_{0}^{2\pi} f(x)\cos nx\, dx$$

24A continued

$$= \frac{1}{\pi}\int_{\pi}^{2\pi} (x - \pi)\cos nx\, dx$$

$$= \begin{cases} 0 & \text{if } n \text{ is even} \\ \dfrac{2}{\pi n^2} & \text{if } n \text{ is odd} \end{cases}$$

$$b_n = \frac{1}{\pi}\int_{0}^{2\pi} f(x)\sin nx\, dx$$

$$= \frac{1}{\pi}\int_{\pi}^{2\pi} (x - \pi)\sin nx\, dx$$

$$= -\frac{1}{n}$$

FRAME 25

In FRAME 5 we quoted the Fourier series for the square-wave function shown below.

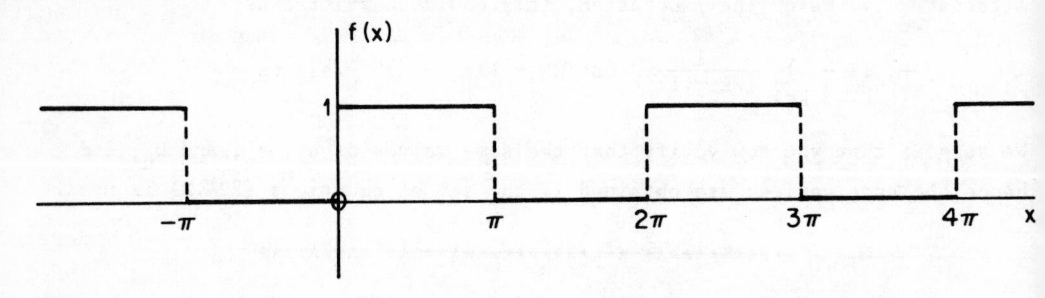

Now that we know how to calculate the coefficients, we can actually obtain the series.

Whether the interval for integration is taken as $-\pi$ to π or 0 to 2π, the formulae for the coefficients will become

$$a_o = \frac{1}{\pi}\int_{0}^{\pi} 1\, dx = \frac{1}{\pi} \times \text{area of rectangle}$$

$$a_n = \frac{1}{\pi}\int_{0}^{\pi} 1 \,.\, \cos nx\, dx$$

FRAME 25 continued

$$b_n = \frac{1}{\pi}\int_0^{\pi} 1 \, . \, \sin nx \, dx$$

Now evaluate the coefficients and hence obtain the series.

****

25A

$$a_o = \frac{1}{\pi} \, . \, \pi = 1$$

$$a_n = \frac{1}{\pi}\left[\frac{1}{n} \sin nx\right]_0^{\pi} = 0$$

$$b_n = \frac{1}{\pi}\left[-\frac{1}{n} \cos nx\right]_0^{\pi}$$

$$= \frac{1}{\pi n}(1 - \cos n\pi)$$

$$= \begin{cases} 0 & \text{if } n \text{ is even} \\ \frac{2}{\pi n} & \text{if } n \text{ is odd} \end{cases}$$

$\therefore$ *The series is* $\frac{1}{2} + \frac{2}{\pi}(\sin x + \frac{1}{3}\sin 3x + \frac{1}{5}\sin 5x + \frac{1}{7}\sin 7x + \ldots)$

FRAME 26

Odd and Even Functions

In the series just obtained some of the coefficients turned out to be zero. If we could tell in advance that this was going to happen, we could reduce the work done in calculating the coefficients.

It is particularly useful, in this context, to be able to recognise an odd or an even function. It may be helpful to remind you, at this stage, of the definition and properties of odd and even functions.

The function f(x) is defined as being even if $f(-x) = f(x)$
" " " " " " " odd " $f(-x) = -f(x)$

FRAME 26 continued

From the definition, any even power of x is seen to be an even function, likewise $\cos x$ and, more generally, $\cos nx$. You will remember that the Maclaurin series for $\cos x$ (and other even functions) contains only even powers of x.

Similarly, any odd power of x is an odd function and so is also $\sin x$, and, more generally, $\sin nx$. The Maclaurin series for $\sin x$ (and other odd functions) contains only odd powers of x.

State which of the following are (a) even, (b) odd, (c) neither.

(i) $x \cos x$

(ii) $x^2 \cos 5x$

(iii) $(x + \pi) \cos x$

(iv) $\sin x \sin 2x$

****

26A

(i)

$$f(x) = x \cos x$$

$$f(-x) = (-x) \cos (-x)$$

$$= -x \cos x$$

$$= -f(x)$$

$\therefore$ *$x \cos x$ is an odd function*

(ii) Even

(iii) Neither

(iv) Even

You will notice that the product of two even functions, as in (ii), or of two odd functions, as in (iv), is even, and that the product of an even function and an odd function, as in (i), is odd.

FRAME 27

It follows from the definition of an even function that its graph is symmetrical about the y-axis. Similarly the graph of an odd function is symmetrical about the origin. This should be obvious to you if you consider the graphs of $y = x^2$ (an even function) and $y = x^3$ (an odd function).

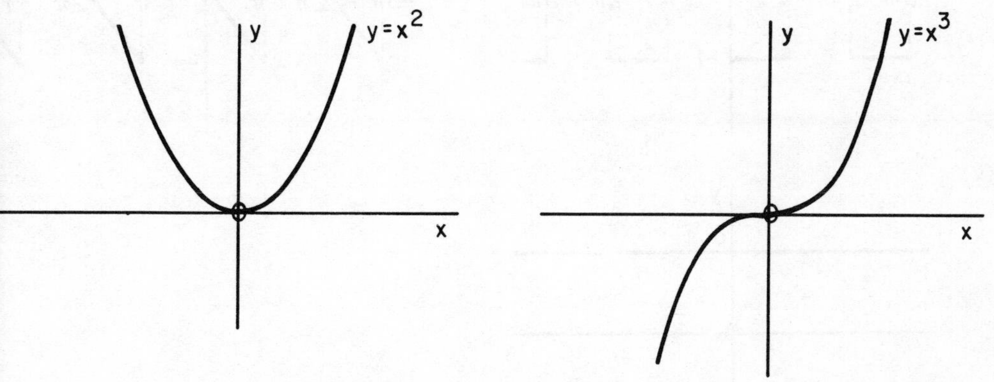

The graphs of $y = \cos x$ (an even function) and $y = \sin x$ (an odd function) show the same kinds of symmetry.

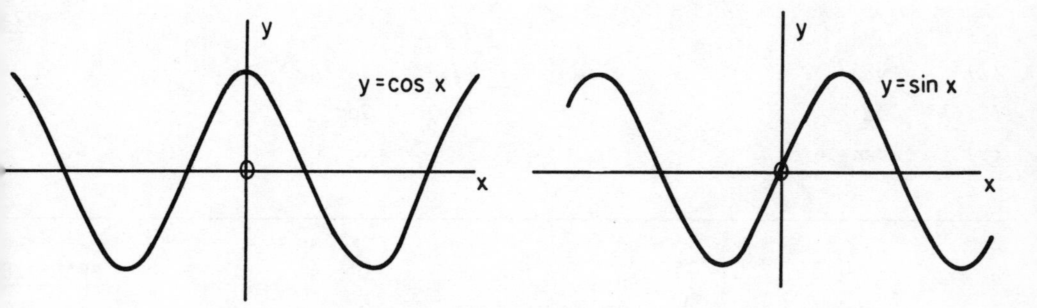

State which of the following graphs represent (a) an even function, (b) an odd function, (c) neither.

(i)

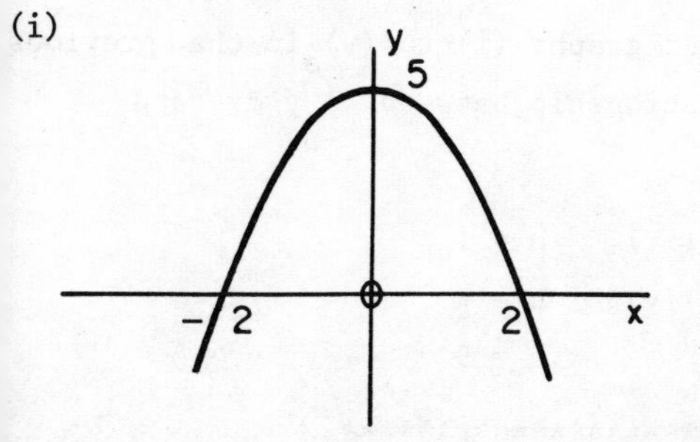

(ii)

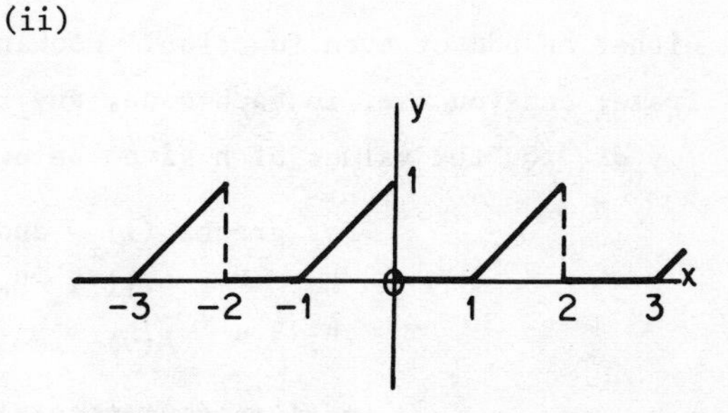

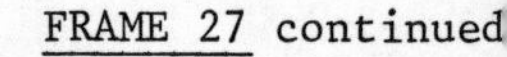

FRAME 27 continued

(iii)

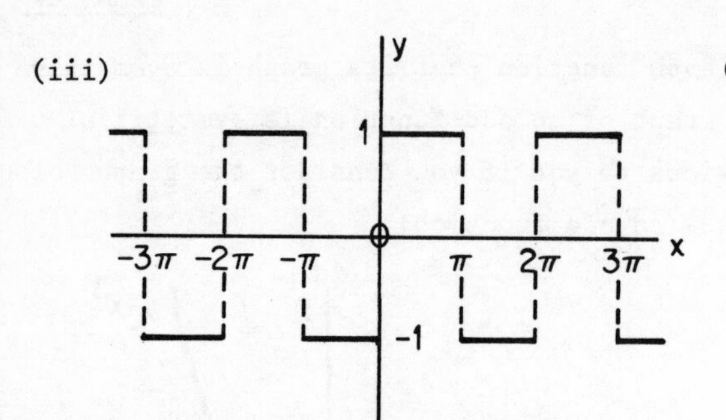

(iv)

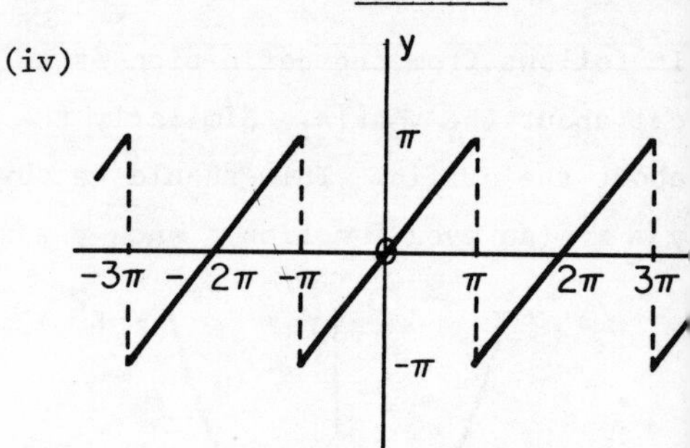

(v)

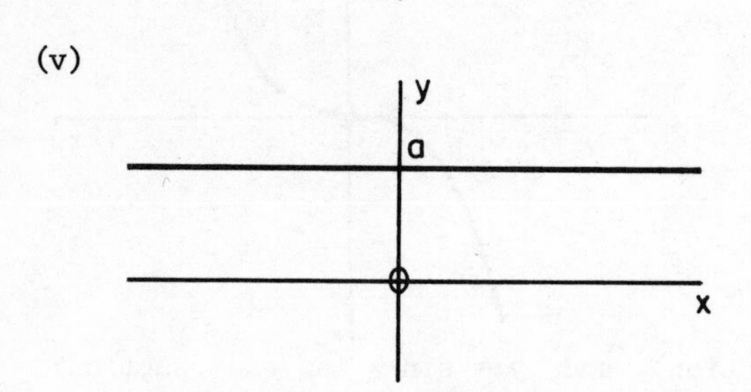

****

27A

(i)	*Even*	*(ii)*	*Neither*
(iii)	*Odd*	*(iv)*	*Odd*
(v)	*Even*		

FRAME 28

Integrals of the form $\int_{-h}^{h} f(x)\, dx$ occur in Fourier series (you have already met them with $h = \pi$) and it will be useful to find out if any conclusions can be drawn about the values of such integrals when $f(x)$ is known to be either an odd or even function. Looking at graphs (i) to (v) in the previous frame, can you see, in each case, any relationship between $\int_{-h}^{0} y\, dx$ and $\int_{0}^{h} y\, dx$ for the values of h given below?

For graphs	(i)	and	(ii)	$h = 1$
" "	(iii)	"	(iv)	$h = \pi$
" "	(v)			$h = 5$

****

28A

(i) $\int_{-1}^{0} y\,dx = \int_{0}^{1} y\,dx$

(ii) *No relationship*

(iii) *and* (iv) $\int_{-\pi}^{0} y\,dx = -\int_{0}^{\pi} y\,dx$

(v) $\int_{-5}^{0} y\,dx = \int_{0}^{5} y\,dx$

FRAME 29

The following general conclusions will now be evident:

For any value of h, $\int_{-h}^{0} f(x)\,dx = \int_{0}^{h} f(x)\,dx$ if f(x) is even, as in (i) and (v),

but $\int_{-h}^{0} f(x)\,dx = -\int_{0}^{h} f(x)\,dx$ if f(x) is odd, as in (iii) and (iv).

What, therefore, can you deduce about the value of $\int_{-h}^{h} f(x)\,dx$ when f(x) is

(a) odd, (b) even?

****

29A

(a) *When f(x) is odd,* $\int_{-h}^{h} f(x)\,dx = 0$

(b) *When f(x) is even,* $\int_{-h}^{h} f(x)\,dx = 2\int_{0}^{h} f(x)\,dx$

(It is also $2\int_{-h}^{0} f(x)\,dx$ *but negative limits are less convenient.)*

FRAME 30

Obtaining the Fourier series for f(x) is simplified in those cases where f(x) can be seen to be either even or odd.

The Fourier series for an even function can only contain terms which are themselves even functions, i.e., it will have no sine terms and will be of the form

$$\tfrac{1}{2}a_0 + \sum_{n=1}^{\infty} a_n \cos nx$$

On the other hand, the Fourier series for an odd function can only contain terms which are themselves odd functions, i.e., it will have only sine terms and will be of the form

$$\sum_{n=1}^{\infty} b_n \sin nx$$

FRAME 31

Now, as an illustration, we shall consider the saw-tooth waveform shown in FRAME 27, graph (iv). It is described by the equations

$$f(x) = x \qquad -\pi < x < \pi$$

$$f(x + 2\pi) = f(x)$$

You have already noted that f(x) is an odd function.

$\therefore$ In its Fourier series, $a_0 = a_n = 0$

$$b_n = \frac{1}{\pi}\int_{-\pi}^{\pi} x \sin nx \, dx$$

$$= \frac{2}{\pi}\int_{0}^{\pi} x \sin nx \, dx \quad \text{as} \quad x \sin nx \text{ is an even function}$$

$$= \frac{2}{\pi}\left\{ \left[-\frac{x}{n}\cos nx\right]_0^{\pi} - \int_0^{\pi} -\frac{1}{n}\cos nx \, dx \right\}$$

$$= \frac{2}{\pi}\left\{ -\frac{\pi}{n}\cos n\pi + \frac{1}{n}\left[\frac{1}{n}\sin nx\right]_0^{\pi} \right\}$$

$$= -\frac{2}{n}\cos n\pi$$

FRAME 31 continued

$$= \begin{cases} \dfrac{2}{n} & \text{if n is odd} \\ -\dfrac{2}{n} & \text{if n is even} \end{cases}$$

∴ The Fourier series for this waveform is

$$2(\sin x - \frac{1}{2}\sin 2x + \frac{1}{3}\sin 3x - \frac{1}{4}\sin 4x + \frac{1}{5}\sin 5x - \ldots)$$

Now obtain the Fourier series for the periodic function shown below.

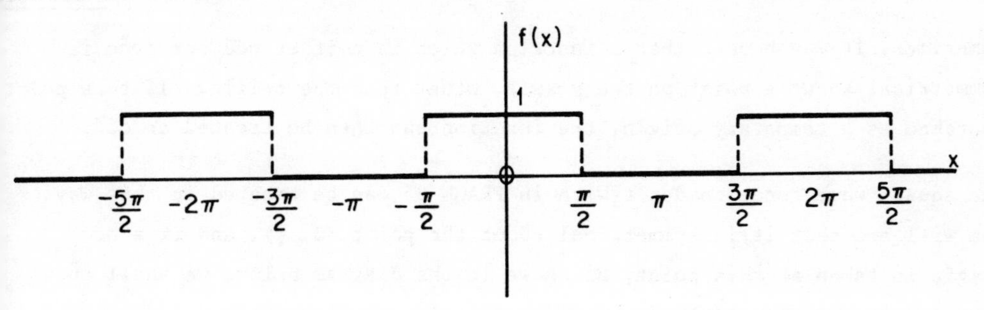

****

31A

This is an even function, so $b_n = 0$.

$$\tfrac{1}{2}a_0 = \text{mean value of } f(x) \text{ over one period}$$

$$= \tfrac{1}{2}$$

$$a_n = \frac{1}{\pi}\int_{-\pi}^{\pi} f(x)\cos nx\,dx$$

$$= \frac{1}{\pi}\int_{-\pi/2}^{\pi/2} 1 \cdot \cos nx\,dx$$

$$= \frac{2}{\pi}\int_{0}^{\pi/2} \cos nx\,dx$$

$$= \frac{2}{\pi}\left[\frac{1}{n}\sin nx\right]_0^{\pi/2}$$

$$= \frac{2}{n\pi}\sin\frac{n\pi}{2}$$

31A continued

$$= \begin{cases} 0 & \text{if } n \text{ is even} \\ \dfrac{2}{n\pi} & \text{if } n \text{ is } 1,\ 5,\ 9 \text{ etc.} \\ -\dfrac{2}{n\pi} & \text{if } n \text{ is } 3,\ 7,\ 11 \text{ etc.} \end{cases}$$

$\therefore$ *The Fourier series is* $\frac{1}{2} + \frac{2}{\pi}(\cos x - \frac{1}{3}\cos 3x + \frac{1}{5}\cos 5x - \frac{1}{7}\cos 7x \,.....)$

FRAME 32

Sometimes, it may happen that a function which is neither odd nor even is symmetrical about a point on the y-axis, other than the origin. If this point is taken as a temporary origin, the function can then be treated as odd.

The square-wave function dealt with in FRAME 25 can be treated in this way. You will see that it is symmetrical about the point $(0, \frac{1}{2})$, and if a new origin is taken at this point, as shown in the diagram below, we shall then have an odd function $\phi(x)$.

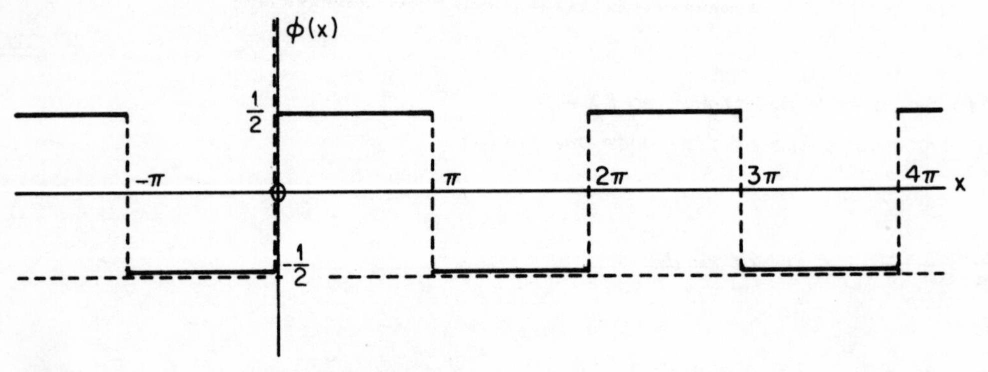

The original axes are shown as -------.

$\phi(x)$ is given by the equations

$$\begin{cases} \phi(x) = -\frac{1}{2} & -\pi < x < 0 \\ \phi(x) = \frac{1}{2} & 0 < x < \pi \\ \phi(x + 2\pi) = \phi(x) & \end{cases}$$

Now find the Fourier series for $\phi(x)$.

****

32A

$\phi(x)$ is odd $\therefore \; a_0 = a_n = 0$

$$b_n = \frac{1}{\pi}\int_{-\pi}^{\pi} \phi(x) \sin nx \, dx$$

$$= \frac{2}{\pi}\int_{0}^{\pi} \tfrac{1}{2} \sin nx \, dx \quad \text{as} \quad \phi(x) \sin nx \quad \text{is an even function}$$

$$= \frac{1}{\pi}\left[-\frac{1}{n}\cos nx\right]_0^{\pi}$$

$$= \frac{1}{\pi n}(1 - \cos n\pi)$$

$$= \begin{cases} 0 & \text{if } n \text{ is even} \\ \frac{2}{\pi n} & \text{if } n \text{ is odd} \end{cases}$$

$$\therefore \; \phi(x) = \frac{2}{\pi}\left(\sin x + \frac{1}{3}\sin 3x + \frac{1}{5}\sin 5x + \frac{1}{7}\sin 7x + \ldots\right)$$

FRAME 33

Changing back to the original origin,

$$f(x) = \frac{1}{2} + \phi(x)$$

$$= \frac{1}{2} + \frac{2}{\pi}\left(\sin x + \frac{1}{3}\sin 3x + \frac{1}{5}\sin 5x + \frac{1}{7}\sin 7x + ..\right)$$

You will notice that this time we only had to work out b_n, and that the value of $\frac{1}{2}a_0$ is the distance of the new origin above the original one.

FRAME 34

Half-range Fourier Series

In FRAMES 8 - 11 we saw that some problems arise where the function is not periodic within the interval of definition but it is useful to represent it by a Fourier series. This difficulty was overcome by using a function which was periodic and which coincided with the given function over the interval of

FRAME 34 continued

definition. The period of the function we choose must obviously be greater than, or equal to, the given interval. Taking the interval as half a period makes it possible to obtain a simple series by defining the function for the other half period in such a way that it is either odd, thus giving a series of sines only, or even, giving a series of cosines only (including the term $\frac{1}{2}a_o$ as this can be regarded as $\frac{1}{2}a_o \cos 0x$).

For example, if we have the function defined between 0 and π as shown in the diagram

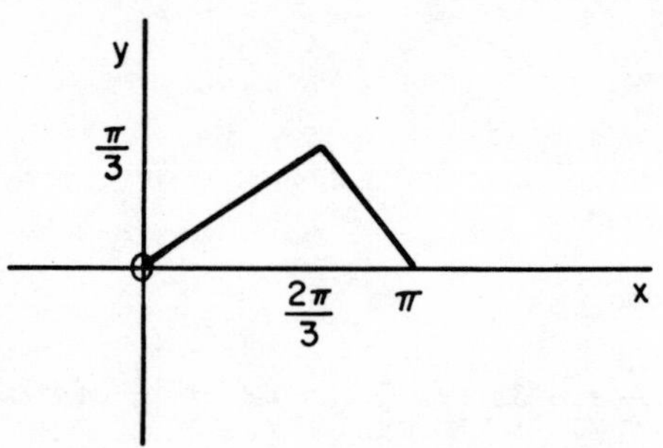

either of the following waveforms can be chosen to represent it.

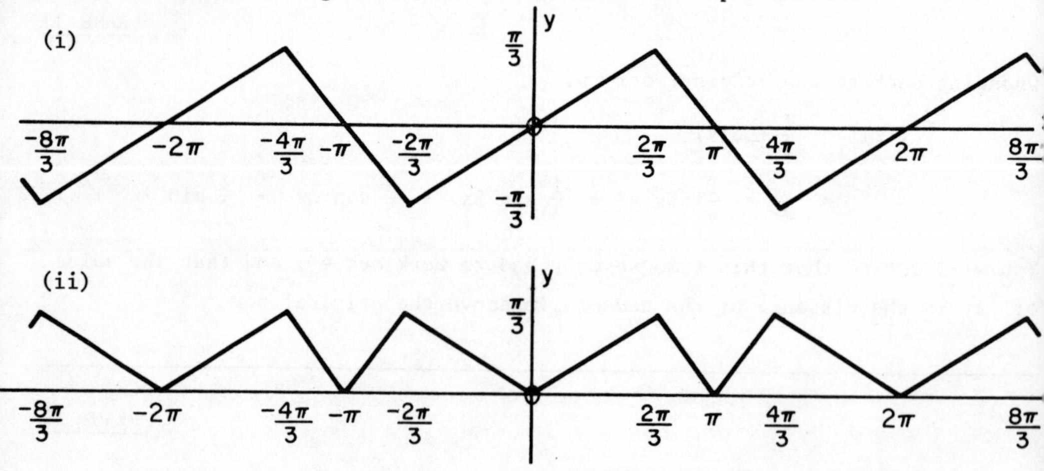

(i) is an odd waveform and its series will have sines only (called a half-range Fourier sine series).

(ii) is an even waveform and its series will have cosines only (called a half -range Fourier cosine series).

FRAME 35

In half-range Fourier series, the calculation of the coefficients is simplified by the assumption of either an odd or an even function to represent the required function over the given interval.

If we assume an odd function f(x), then $a_o = a_n = 0$

$$\text{and} \quad b_n = \frac{1}{\pi}\int_{-\pi}^{\pi} f(x) \sin nx \, dx$$

$$= \frac{2}{\pi}\int_{0}^{\pi} f(x) \sin nx \, dx \quad \text{as} \quad f(x) \sin nx \quad \text{is even.}$$

If we assume an even function f(x), then

$$b_n = 0$$

$$a_o = \frac{1}{\pi}\int_{-\pi}^{\pi} f(x)\, dx = \frac{2}{\pi}\int_{0}^{\pi} f(x)\, dx$$

$$\text{and} \quad a_n = \frac{1}{\pi}\int_{-\pi}^{\pi} f(x) \cos nx \, dx$$

$$= \frac{2}{\pi}\int_{0}^{\pi} f(x) \cos nx \, dx \quad \text{as} \quad f(x) \cos nx \text{ is even.}$$

FRAME 36

Turning now to the example we used in FRAME 34, let us calculate the series for Case (i).

Between 0 and π the function is defined by the equations

$$\begin{cases} f(x) = \frac{1}{2}x & 0 < x < \frac{2\pi}{3} \\ f(x) = \pi - x & \frac{2\pi}{3} < x < \pi \end{cases}$$

$$b_n = \frac{2}{\pi}\int_{0}^{\pi} f(x) \sin nx \, dx$$

FRAME 36 continued

$$= \frac{2}{\pi}\left\{\int_0^{2\pi/3} \tfrac{1}{2}x \sin nx \, dx + \int_{2\pi/3}^{\pi} (\pi - x) \sin nx \, dx\right\}$$

Now show that this becomes $\frac{3}{\pi n^2}\sin\frac{2n\pi}{3}$ and then write down the first few terms of the series.

****

36A

$$b_n = \frac{2}{\pi}\left\{\frac{1}{2}\left[-\frac{x}{n}\cos nx\right]_0^{2\pi/3} - \frac{1}{2}\int_0^{2\pi/3} -\frac{1}{n}\cos nx \, dx + \left[-\frac{\pi - x}{n}\cos nx\right]_{2\pi/3}^{\pi} - \int_{2\pi/3}^{\pi}\frac{1}{n}\cos nx \, dx\right\}$$

$$= \frac{2}{\pi}\left\{-\frac{\pi}{3n}\cos\frac{2n\pi}{3} + \frac{1}{2n^2}\sin\frac{2n\pi}{3} + \frac{\pi}{3n}\cos\frac{2n\pi}{3} + \frac{1}{n^2}\sin\frac{2n\pi}{3}\right\}$$

$$= \frac{3}{\pi n^2}\sin\frac{2n\pi}{3}$$

$$= \begin{cases} 3\sqrt{3}/2\pi n^2 & \text{if } n = 1, 4, 7, \ldots \\ -3\sqrt{3}/2\pi n^2 & \text{if } n = 2, 5, 8, \ldots \\ 0 & \text{if } n = 3, 6, 9, \ldots \end{cases}$$

$$\therefore \quad f(x) = \frac{3\sqrt{3}}{2\pi}\left(\sin x - \frac{1}{2^2}\sin 2x + \frac{1}{4^2}\sin 4x - \frac{1}{5^2}\sin 5x + \ldots\right)$$

FRAME 37

A function f(x) is defined between 0 and π by the equation $f(x) = \pi - x$. Sketch the waveforms which will result if this is represented by a half-range Fourier series involving

(i) sines,

(ii) cosines,

and obtain the series in each case.

****

37A

(i)

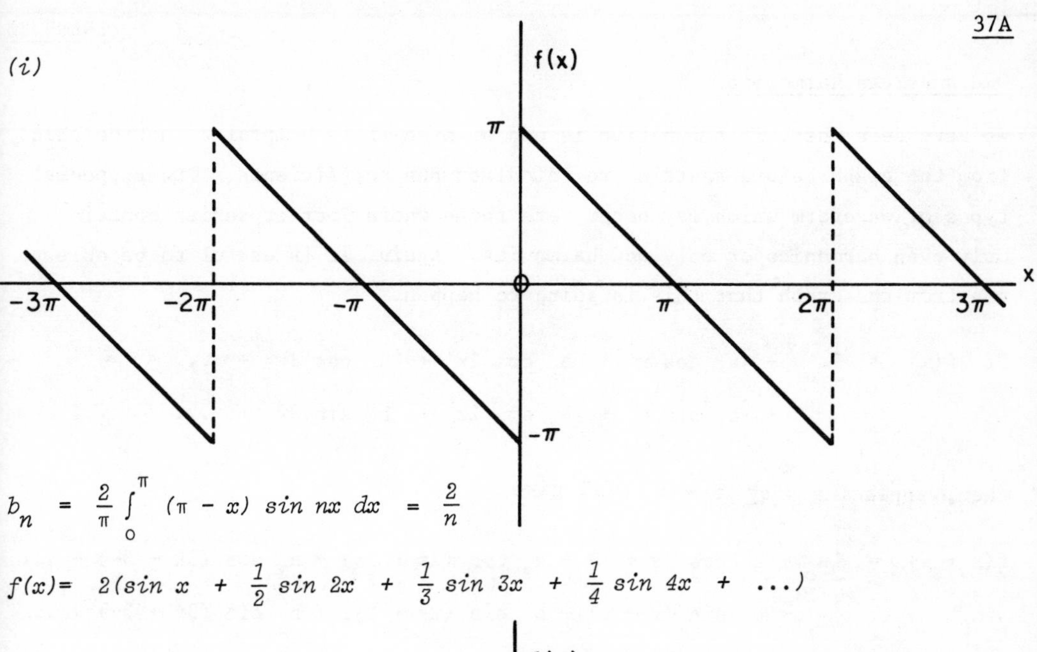

$$b_n = \frac{2}{\pi}\int_0^\pi (\pi - x)\sin nx\,dx = \frac{2}{n}$$

$$f(x) = 2\left(\sin x + \frac{1}{2}\sin 2x + \frac{1}{3}\sin 3x + \frac{1}{4}\sin 4x + \ldots\right)$$

(ii)

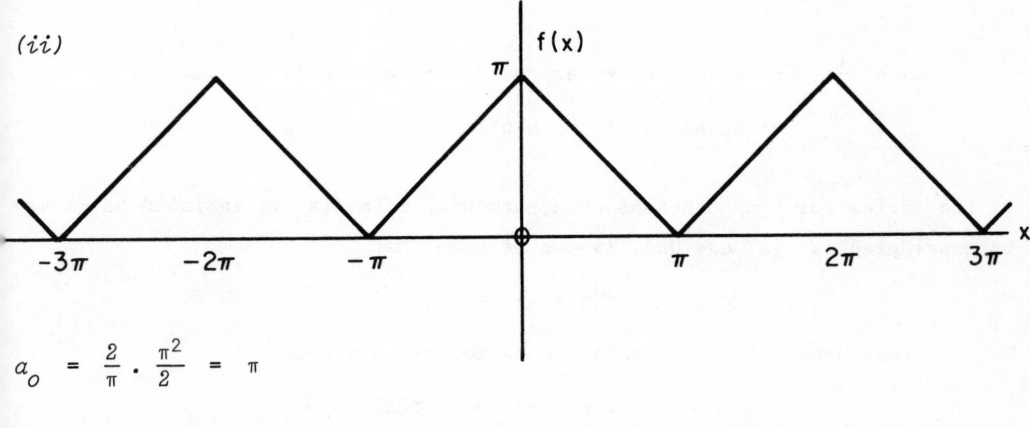

$$a_0 = \frac{2}{\pi}\cdot\frac{\pi^2}{2} = \pi$$

$$a_n = \frac{2}{\pi}\int_0^\pi (\pi - x)\cos nx\,dx$$

$$= \frac{2}{\pi n^2}(1 - \cos n\pi)$$

$$= \begin{cases} 0 & \text{if } n \text{ is even} \\ \frac{4}{\pi n^2} & \text{if } n \text{ is odd} \end{cases}$$

$$f(x) = \frac{\pi}{2} + \frac{4}{\pi}\left(\cos x + \frac{1}{9}\cos 3x + \frac{1}{25}\cos 5x + \ldots\right)$$

FRAME 38

Odd and Even Harmonics

We have seen that, if a function is odd or even, it is helpful to notice this from the graph before starting to calculate the coefficients. Other special types of waveform which may occur are those whose Fourier series contain only even harmonics or only odd harmonics. Again, it is useful to be able to see from the graph that this is going to happen.

If

$$f(x) = \frac{1}{2}a_o + a_1 \cos x + a_2 \cos 2x + a_3 \cos 3x + \ldots$$
$$+ b_1 \sin x + b_2 \sin 2x + b_3 \sin 3x + \ldots$$

then, replacing x by $x + \pi$ will give

$$f(x + \pi) = \frac{1}{2}a_o + a_1 \cos (x + \pi) + a_2 \cos (2x + 2\pi) + a_3 \cos (3x + 3\pi) + \ldots$$
$$+ b_1 \sin (x + \pi) + b_2 \sin (2x + 2\pi) + b_3 \sin (3x + 3\pi) + \ldots$$

$$= \frac{1}{2}a_o - a_1 \cos x + a_2 \cos 2x - a_3 \cos 3x + \ldots$$
$$- b_1 \sin x + b_2 \sin 2x - b_3 \sin 3x + \ldots$$

If the series for f(x) contains even harmonics only ($\frac{1}{2}a_o$ is included as it may be considered as $\frac{1}{2}a_o \cos 0x$), it can be seen that

$$f(x + \pi) = f(x)$$

On the other hand, if it contains odd harmonics only then

$$f(x + \pi) = - f(x)$$

An example of each case is shown below.

EVEN HARMONICS ONLY

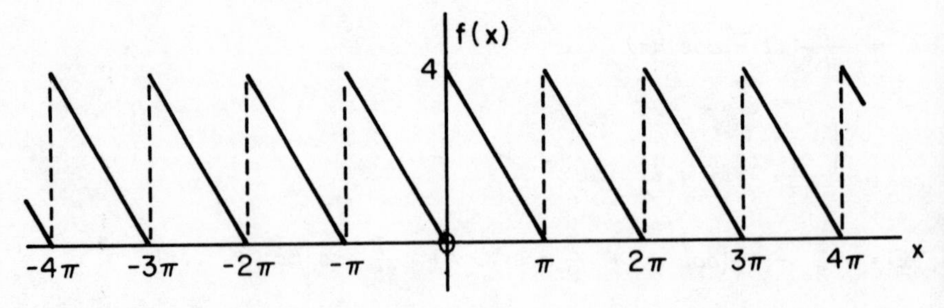

FRAME 38 continued

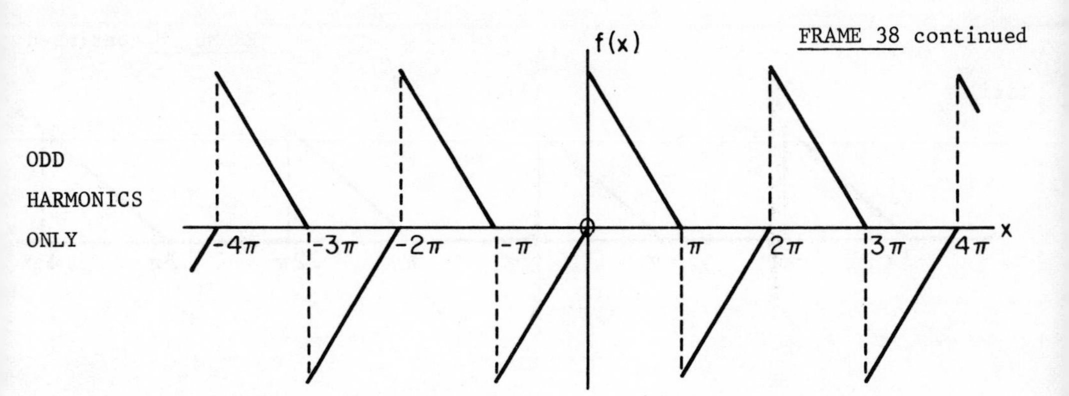

You will probably have realised that, in the first case, the function, as well as repeating itself at intervals of 2π, also repeats itself at intervals of π, as is implied by the relation $f(x + \pi) = f(x)$. This could also be taken as an example of the more general case where the period is other than 2π, which we shall shortly be considering.

Now state whether the series representing the following functions will contain (a) only even harmonics, (b) only odd harmonics, (c) both even and odd harmonics.

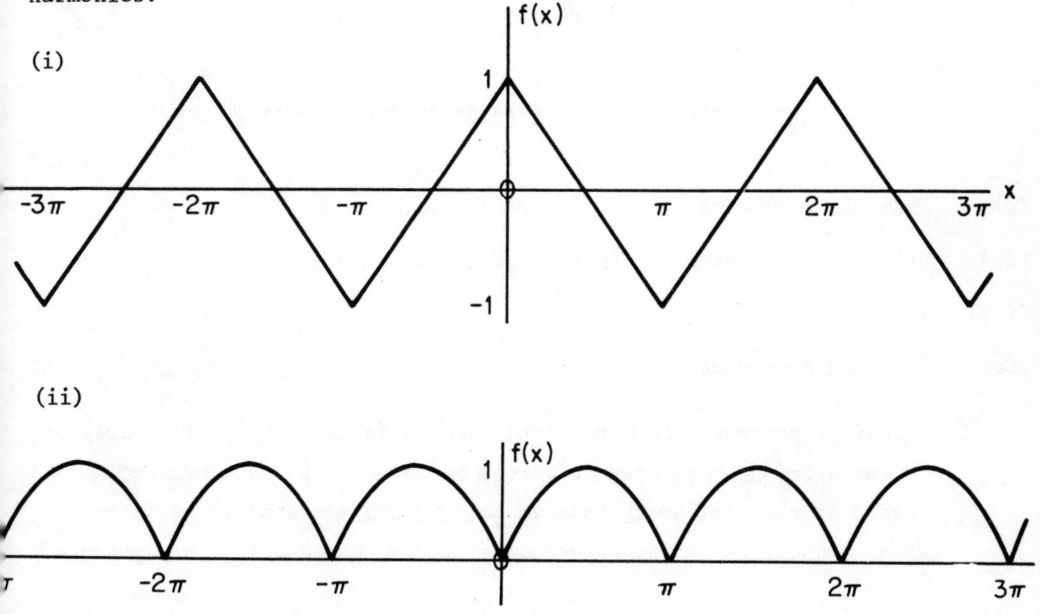

FRAME 38 continued

(iii)

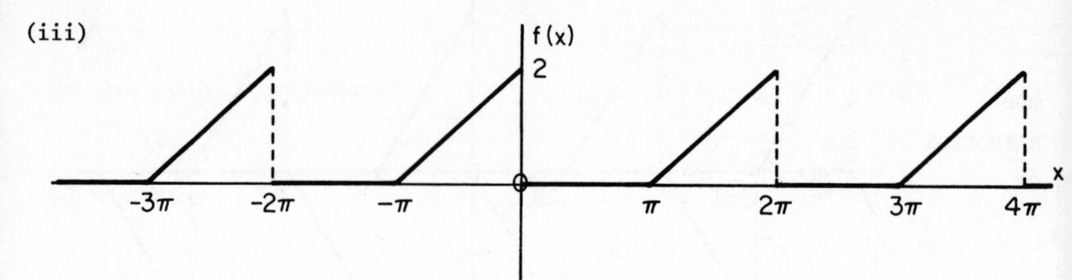

(iv)

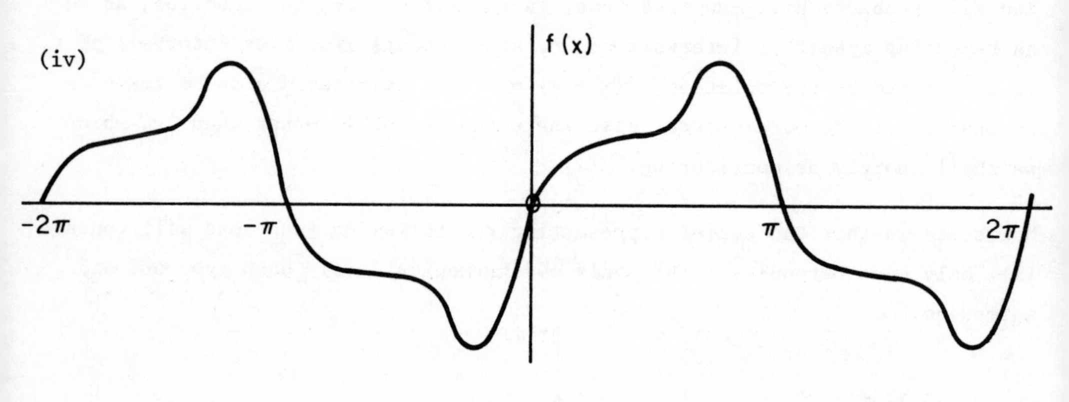

****

38A

(i) *Only odd harmonics.*

(ii) *Only even harmonics. (The rectified sine wave.)*

(iii) *Both.*

(iv) *Only odd harmonics.*

A voltage generated by a rotating machine is an example of a waveform in engineering whose Fourier series contains only odd harmonics. As a result of the similarity between the N and S magnetic poles in any such machine, the voltage generated always has this type of symmetry.

FRAME 39

Sum of a Fourier series at a point of discontinuity

If you substitute a particular value of x into a Fourier series, the sum which results is the corresponding value of f(x).

If, however, this is done at a point of discontinuity $x = k$, say, the sum is the average of the two limiting values of f(x) as $x \to k$ from the left and right respectively.

The series for the waveform

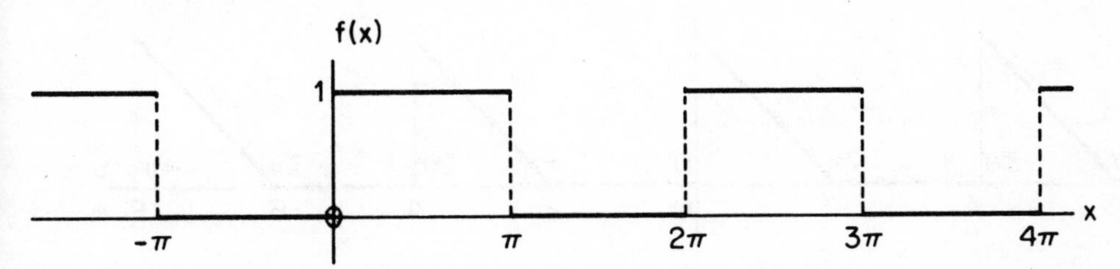

is $\frac{1}{2} + \frac{2}{\pi}\left(\sin x + \frac{1}{3}\sin 3x + \frac{1}{5}\sin 5x + \frac{1}{7}\sin 7x + \ldots\ldots\right)$ (See FRAME 25)

Substituting $x = \pi$, for instance, in this series will give the value $\frac{1}{2}$ which is $\frac{1}{2}(1 + 0)$.

FRAME 40

General Period

So far we have only considered functions of period 2π, but as we saw in FRAMES 6 - 11, most practical problems involve the use of functions with a different period. We shall now look at an example of a function whose period is not 2π - the waveform is shown below.

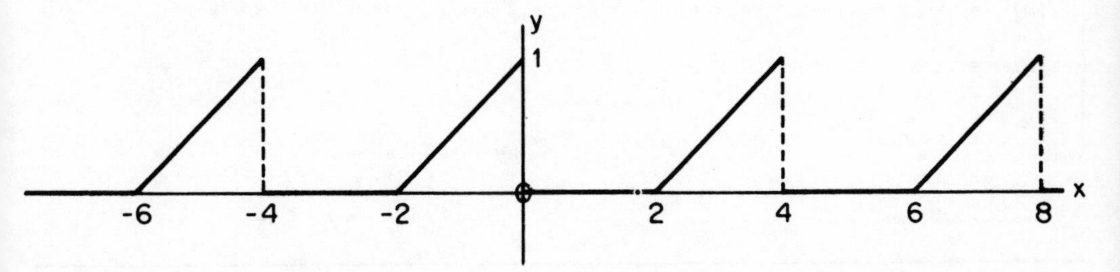

FRAME 40 continued

The equation of this is $y = f(x)$ where

$$\begin{cases} f(x) = \frac{1}{2}(x + 2) & -2 < x < 0 \\ f(x) = 0 & 0 < x < 2 \\ f(x + 4) = f(x) & \text{i.e. the period is 4.} \end{cases}$$

This can, in fact, be converted to the straightforward case where the period is 2π by a change in the horizontal scale as illustrated below.

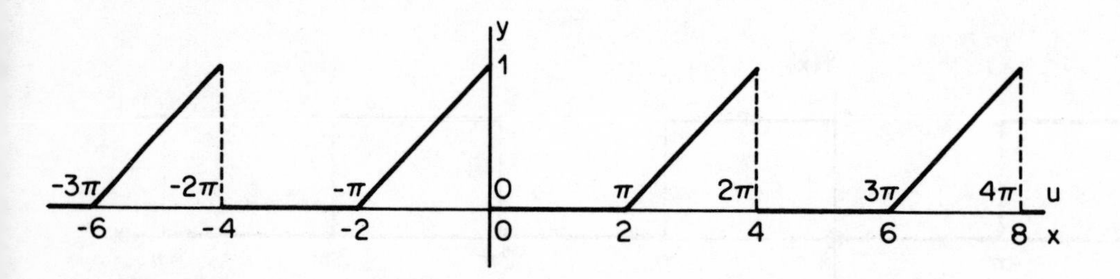

We now treat u as a new independent variable, and hence the equation of the waveform can be written in the form $y = F(u)$.

From the graph write down:

(i) The equations which define $F(u)$. (Ignore the x-scale while you are doing this.)

(ii) The equation which gives u in terms of x.

****

40A

$$\begin{cases} F(u) = \frac{1}{\pi}(u + \pi) & -\pi < u < 0 \\ F(u) = 0 & 0 < u < \pi \\ F(u + 2\pi) = F(u) \end{cases}$$

$$u = \frac{\pi x}{2}$$

FRAME 41

We now have a function F(u) which has period 2π. Show that the Fourier series for this is

$$F(u) = \frac{1}{4} + \frac{2}{\pi^2}\left(\cos u + \frac{1}{3^2}\cos 3u + \frac{1}{5^2}\cos 5u + \ldots\right)$$
$$- \frac{1}{\pi}\left(\sin u + \frac{1}{2}\sin 2u + \frac{1}{3}\sin 3u + \ldots\right)$$

****

41A

$$a_o = \frac{1}{\pi}\cdot\frac{\pi}{2} = \frac{1}{2}$$

$$a_n = \frac{1}{\pi}\int_{-\pi}^{\pi} F(u)\cos nu\, du$$

$$= \frac{1}{\pi}\int_{-\pi}^{0} \frac{1}{\pi}(u + \pi)\cos nu\, du$$

$$= \frac{1}{\pi^2 n^2}(1 - \cos n\pi)$$

$$= \begin{cases} 0 & \text{if } n \text{ is even} \\ \frac{2}{\pi^2 n^2} & \text{if } n \text{ is odd} \end{cases}$$

$$b_n = \frac{1}{\pi}\int_{-\pi}^{\pi} F(u)\sin nu\, du$$

$$= \frac{1}{\pi}\int_{-\pi}^{0} \frac{1}{\pi}(u + \pi)\sin nu\, du$$

$$= -\frac{1}{\pi n}$$

$$\therefore F(u) = \frac{1}{4} + \frac{2}{\pi^2}\left(\cos u + \frac{1}{3^2}\cos 3u + \frac{1}{5^2}\cos 5u + \ldots\right)$$
$$- \frac{1}{\pi}\left(\sin u + \frac{1}{2}\sin 2u + \frac{1}{3}\sin 3u + \ldots\right)$$

FRAME 42

Now $F(u) = y = f(x)$ and $u = \frac{\pi x}{2}$, so the Fourier series for $f(x)$ is

$$f(x) = \frac{1}{4} + \frac{2}{\pi^2}\left(\cos\frac{\pi x}{2} + \frac{1}{3^2}\cos\frac{3\pi x}{2} + \frac{1}{5^2}\cos\frac{5\pi x}{2} + \dots\right)$$

$$- \frac{1}{\pi}\left(\sin\frac{\pi x}{2} + \frac{1}{2}\sin\pi x + \frac{1}{3}\sin\frac{3\pi x}{2} + \dots\right)$$

Having seen how to deal with an example of this type, in the next frame we shall consider the general case with period 2ℓ, where ℓ can take any value.

FRAME 43

When the period is 2ℓ, we again take a new horizontal scale, but this time chosen so that $u = 2\pi$ corresponds to $x = 2\ell$.

-3π	-2π	$-\pi$	0	π	2π	3π	u
-3ℓ	-2ℓ	$-\ell$	0	ℓ	2ℓ	3ℓ	x

Write down the equation giving u in terms of x.

****

43A

$$u = \frac{\pi x}{\ell}$$

FRAME 44

$y = f(x)$ now becomes $y = F(u)$ with period 2π, and the Fourier series for $F(u)$ will be

$$\tfrac{1}{2}a_o + a_1 \cos u + a_2 \cos 2u + a_3 \cos 3u + \dots$$

$$+ b_1 \sin u + b_2 \sin 2u + b_3 \sin 3u + \dots$$

where $a_o = \frac{1}{\pi}\int_{-\pi}^{\pi} F(u)\,du$

$$a_n = \frac{1}{\pi}\int_{-\pi}^{\pi} F(u)\cos nu\,du$$

FRAME 44 continued

$$b_n = \frac{1}{\pi}\int_{-\pi}^{\pi} F(u) \sin nu \, du$$

The series and formulae for the coefficients can be written in terms of x, using $u = \frac{\pi x}{\ell}$, to give the Fourier series for f(x).

$$\text{Thus } f(x) = \tfrac{1}{2}a_o + a_1 \cos\frac{\pi x}{\ell} + a_2 \cos\frac{2\pi x}{\ell} + a_3 \cos\frac{3\pi x}{\ell} + \ldots$$

$$+ b_1 \sin\frac{\pi x}{\ell} + b_2 \sin\frac{2\pi x}{\ell} + b_3 \sin\frac{3\pi x}{\ell} + \ldots$$

where $a_o = \frac{1}{\pi}\int_{-\ell}^{\ell} f(x) \frac{\pi}{\ell}\, dx$ as $du = \frac{\pi}{\ell}\, dx$ and $x = \pm\, \ell$ corresponds to $u = \pm\, \pi$

$$= \frac{1}{\ell}\int_{-\ell}^{\ell} f(x)\, dx$$

$$a_n = \frac{1}{\ell}\int_{-\ell}^{\ell} f(x) \cos\frac{n\pi x}{\ell}\, dx$$

$$b_n = \frac{1}{\ell}\int_{-\ell}^{\ell} f(x) \sin\frac{n\pi x}{\ell}\, dx$$

These can be regarded as standard formulae for the coefficients for the general period, so that it is not necessary to make the u substitution every time.

You will recall that in FRAME 21 we pointed out that for a function of period 2π, the integrals used in evaluating the coefficients can be taken over <u>any</u> interval of 2π.

Similarly, here the integrals can be taken over <u>any</u> interval of 2ℓ. For example, we could use limits 0 and 2ℓ instead of $-\ell$ and ℓ.

FRAME 45

As an example, consider the waveform shown below.

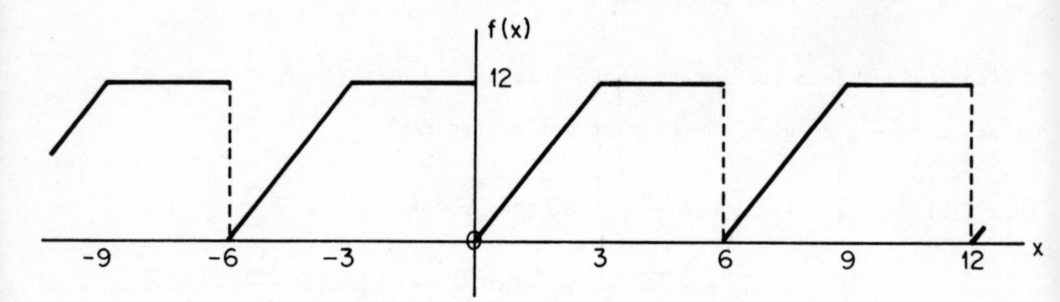

First, write down the equations defining the function and state the value of ℓ in this case.

****

45A

$$\begin{cases} f(x) = 12 & -3 < x < 0 \\ f(x) = 4x & 0 < x < 3 \\ f(x+6) = f(x) \end{cases} \qquad \text{OR} \qquad \begin{cases} f(x) = 4x & 0 < x < 3 \\ f(x) = 12 & 3 < x < 6 \\ f(x+6) = f(x) \end{cases}$$

The period is 6, so the value of ℓ is 3.

FRAME 46

Now write down, but do not evaluate, the formulae for a_o, a_n and b_n for this particular waveform, using the standard results obtained in FRAME 43.

****

46A

$$a_o = \frac{1}{3}\left\{\int_{-3}^{0} 12\,dx + \int_{0}^{3} 4x\,dx\right\} \qquad \left(= \frac{1}{3} \times \text{area under graph between } -3 \text{ and } +3\right)$$

$$a_n = \frac{1}{3}\left\{\int_{-3}^{0} 12\cos\frac{n\pi x}{3}\,dx + \int_{0}^{3} 4x\cos\frac{n\pi x}{3}\,dx\right\}$$

$$b_n = \frac{1}{3}\left\{\int_{-3}^{0} 12\sin\frac{n\pi x}{3}\,dx + \int_{0}^{3} 4x\sin\frac{n\pi x}{3}\,dx\right\}$$

46A continued

OR

$$a_o = \frac{1}{3}\left\{\int_0^3 4x\,dx + \int_3^6 12dx\right\} \qquad (= \frac{1}{3} \times \text{area under graph between 0 and 6})$$

$$a_n = \frac{1}{3}\left\{\int_0^3 4x\cos\frac{n\pi x}{3}\,dx + \int_3^6 12\cos\frac{n\pi x}{3}\,dx\right\}$$

$$b_n = \frac{1}{3}\left\{\int_0^3 4x\sin\frac{n\pi x}{3}\,dx + \int_3^6 12\sin\frac{n\pi x}{3}\,dx\right\}$$

FRAME 47

Proceeding to the evaluation of the coefficients,

$$a_o = \frac{1}{3} \times 54 = 18$$

$$a_n = \frac{4}{3}\left\{\int_{-3}^{0} 3\cos\frac{n\pi x}{3}\,dx + \int_0^3 x\cos\frac{n\pi x}{3}\,dx\right\}$$

$$= \frac{4}{3}\left\{\frac{9}{n\pi}\left[\sin\frac{n\pi x}{3}\right]_{-3}^{0} + \left[x\,.\,\frac{3}{n\pi}\sin\frac{n\pi x}{3}\right]_0^3 - \int_0^3 \frac{3}{n\pi}\sin\frac{n\pi x}{3}\,dx\right\}$$

$$= \frac{4}{n\pi}\,.\,\frac{3}{n\pi}\left[\cos\frac{n\pi x}{3}\right]_0^3$$

$$= \frac{12}{n^2\pi^2}(\cos n\pi - 1)$$

$$= \begin{cases} 0 & \text{if n is even} \\ -\dfrac{24}{n^2\pi^2} & \text{if n is odd} \end{cases}$$

Now show that $b_n = -\dfrac{12}{n\pi}$ and write down the series for the waveform.

****

47A

$$9 - \frac{24}{\pi^2}\left(\cos\frac{\pi x}{3} + \frac{1}{3^2}\cos\pi x + \frac{1}{5^2}\cos\frac{5\pi x}{3} + \dots\right)$$

$$- \frac{12}{\pi}\left(\sin\frac{\pi x}{3} + \frac{1}{2}\sin\frac{2\pi x}{3} + \frac{1}{3}\sin\pi x + \dots\right)$$

FRAME 48

You will now easily see that the idea of a half-range Fourier series can be extended to a half-range other than π. This will entail the modification of the results in FRAME 44 for period 2ℓ, as in FRAME 35 for period 2π.

Thus, for an odd function, $a_o = a_n = 0$

and $b_n = \frac{2}{\ell}\int_0^{\ell} f(x) \sin \frac{n\pi x}{\ell}\, dx$

Now write down the corresponding results for an even function.

****

48A

$$b_n = 0$$

$$a_o = \frac{2}{\ell}\int_0^{\ell} f(x)\, dx$$

$$a_n = \frac{2}{\ell}\int_0^{\ell} f(x) \cos \frac{n\pi x}{\ell}\, dx$$

FRAME 49

Now try the following example.

Find the half-range Fourier sine series for x^2 in the interval $0 < x < 3$, and sketch the waveform represented by this series over several periods.

****

49A

$$b_n = \frac{2}{3}\int_0^{3} x^2 \sin \frac{n\pi x}{3}\, dx$$

$$= \frac{2}{3}\left\{\left[-\frac{3x^2}{n\pi} \cos \frac{n\pi x}{3}\right]_0^3 - \int_0^3 2x\left(\frac{-3}{n\pi}\right) \cos \frac{n\pi x}{3}\, dx\right\}$$

$$= \frac{2}{3}\left\{-\frac{27}{n\pi} \cos n\pi + \frac{6}{n\pi}\left[\frac{3x}{n\pi} \sin \frac{n\pi x}{3}\right]_0^3 - \frac{6}{n\pi}\int_0^3 \frac{3}{n\pi} \sin \frac{n\pi x}{3}\, dx\right\}$$

$$= \frac{2}{3}\left\{-\frac{27}{n\pi} \cos n\pi - \frac{18}{n^2\pi^2}\left[-\frac{3}{n\pi} \cos \frac{n\pi x}{3}\right]_0^3\right\}$$

49A continued

$$= \frac{2}{3}\left\{-\frac{27}{n\pi}\cos n\pi + \frac{54}{n^3\pi^3}(\cos n\pi - 1)\right\}$$

$$= \begin{cases} -\dfrac{18}{n\pi} & \text{if } n \text{ is even} \\[2ex] \dfrac{18}{n\pi} - \dfrac{72}{n^3\pi^3} & \text{if } n \text{ is odd} \end{cases}$$

The series is $18\left\{\left(\frac{1}{\pi} - \frac{4}{\pi^3}\right)\sin\frac{\pi x}{3} - \frac{1}{2\pi}\sin\frac{2\pi x}{3} + \left(\frac{1}{3\pi} - \frac{4}{3^3\pi^3}\right)\sin \pi x - \frac{1}{4\pi}\sin\frac{4\pi x}{3} + \ldots\right\}$

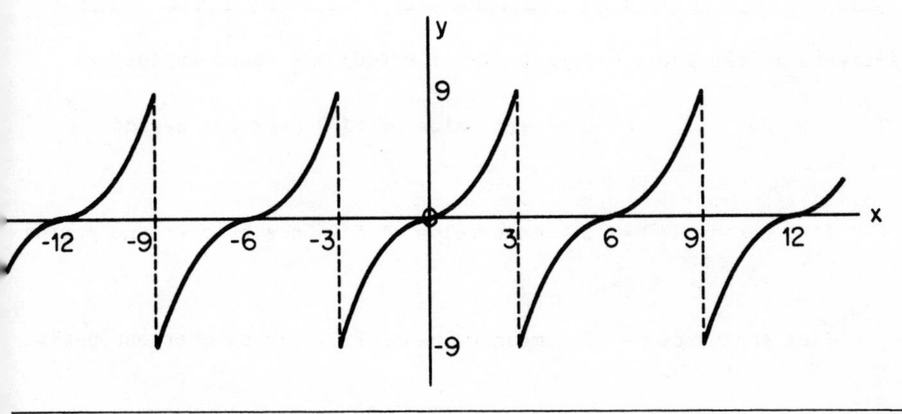

FRAME 50

Numerical Harmonic Analysis

The formulae obtained for calculating the coefficients in previous frames involved integrals whose evaluation depended on knowing the equation for y in terms of x. However, this may not be known and in its place we are only given or can measure numerical values of y for certain values of x. In such cases it is necessary to modify the formulae for the coefficients so that a numerical method can be used for their evaluation.

As an example:

If an alternating voltage is applied to a circuit consisting of a rectifier, a resistance and an inductance in series, the current is of the form

FRAME 50 continued

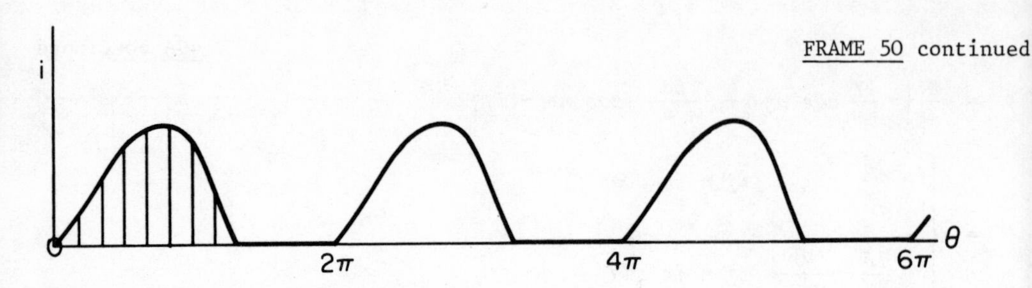

In a particular case the values of i, measured at 12 equidistant values of θ covering a period of 2π, were:

θ	0	π/6	π/3	π/2	2π/3	5π/6	π	7π/6	4π/3	3π/2	5π/3	11π/6
i	0	2·3	5·5	8·9	10·8	11·4	9·9	4·8	0	0	0	0

The coefficients in the Fourier series for $i = f(\theta)$ are found as follows:

$$a_o = \frac{1}{\pi}\int_0^{2\pi} f(\theta)\,d\theta = 2 \times \text{mean value of } f(\theta) \text{ over one period}$$

$$a_n = \frac{1}{\pi}\int_0^{2\pi} f(\theta)\cos n\theta\,d\theta = 2 \times \text{mean value of } f(\theta)\cos n\theta \text{ over one period}$$

$$b_n = \frac{1}{\pi}\int_0^{2\pi} f(\theta)\sin n\theta\,d\theta = 2 \times \text{mean value of } f(\theta)\sin n\theta \text{ over one period}$$

$f(\theta)$ is not known, so an approximation to the mean value of $f(\theta)$ is found by taking the mean of the 12 values of i at the tabulated θ's,

$$\text{i.e.} \quad \frac{\sum i}{12}$$

Similarly the mean value of $f(\theta)\cos n\theta$ is taken as

$$\frac{\sum i \cos n\theta}{12}$$

and that of $f(\theta)\sin n\theta$ as

$$\frac{\sum i \sin n\theta}{12}.$$

Thus $a_o = 2 \times \frac{\Sigma i}{12} = \frac{\Sigma i}{6}$, $\quad a_1 = \frac{\Sigma\, i \cos\theta}{6}$, $\quad b_1 = \frac{\Sigma\, i \sin\theta}{6}$ etc.

The calculation is best set out in a tabular form. For convenience, we shall change to θ in degrees.

FRAME 50 continued

θ^o	i	cos θ	i cos θ	sin θ	i sin θ	cos 2θ	i × cos 2θ	sin 2θ	i × sin 2θ
0	0	1	0	0	0	1	0	0	0
30	2·3	0·866	1·99	0·5	1·15	0·5	1·15	0·866	1·99
60	5·5	0·5	2·75	0·866	4·76	-0·5	-2·75	0·866	4·76
90	8·9	0	0	1	8·9	-1	-8·9	0	0
120	10·8	-0·5	-5·4	0·866	9·35	-0·5	-5·4	-0·866	-9·35
150	11·4	-0·866	-9·87	0·5	5·7	0·5	5·7	-0·866	-9·87
180	9·9	-1	-9·9	0	0	1	9·9	0	0
210	4·8	-0·866	-4·16	-0·5	-2·4	0·5	2·4	0·866	4·16
240	0	-0·5	0	-0·866	0	-0·5	0	0·866	0
270	0	0	0	-1	0	-1	0	0	0
300	0	0·5	0	-0·866	0	-0·5	0	-0·866	0
330	0	0·866	0	-0·5	0	0·5	0	-0·866	0
$\sum$	53·6		-24·59		27·46		2·10		-8·31
$\Sigma/6$	8·9		-4·1		4·6		0·3		-1·4

$a_o = 8·9$ $a_1 = -4·1$ $b_1 = 4·6$ $a_2 = 0·3$ $b_2 = -1·4$

Now continue the table to find a_3, b_3, a_4 and b_4.

****

50A

$a_3 = -0·8$ $b_3 = 0·0$ $a_4 = 0·2$ $b_4 = 0·1$

FRAME 51

Further calculation gives $a_5 = -0·1$, $b_5 = -0·1$

Thus the series is

4·5 - 4·1 cos θ + 0·3 cos 2θ - 0·8 cos 3θ + 0·2 cos 4θ - 0·1 cos 5θ ..

+ 4·6 sin θ - 1·4 sin 2θ + 0·1 sin 4θ - 0·1 sin 5θ ....

This can also be expressed in the form

FRAME 51 continued

$$4{\cdot}5 + 6{\cdot}2 \sin(\theta - 40^{\circ}) + 1{\cdot}4 \sin(2\theta + 168^{\circ}) + 0{\cdot}8 \sin(3\theta - 90^{\circ})$$
$$+ 0{\cdot}2 \sin(4\theta + 27^{\circ}) + 0{\cdot}1 \sin(5\theta - 135^{\circ}) + \ldots$$

You will notice that the amplitudes of successive harmonics decrease so that the higher harmonics contribute very little to the total.

The approximate method we have used for calculating the a's and b's becomes less accurate as the harmonics get higher. The only way of improving the accuracy of the method is to decrease the interval between the tabular values, which means, of course, increasing the number of readings. As a general guide, if the period is divided into k intervals the method can only be relied upon to estimate the first $\left(\frac{k}{2} - 1\right)$ harmonics.

In the present problem, for instance, with 12 intervals we can only expect to go to the fifth harmonic, and even then we would usually only consider the amplitudes of the fourth and fifth harmonics relative to that of the fundamental instead of treating each as being accurate in itself.

In previous frames you have seen what happens to the Fourier series in certain special cases (e.g. odd and even functions) and also how to deal with a period other than 2π. The corresponding results for numerical harmonic analysis will be obvious to you if you remember that it is simply a matter of replacing integration by summation. You will meet some of these cases in the Miscellaneous Examples in FRAME 53.

FRAME 52

Summary

Before giving you some miscellaneous examples to try we will summarise briefly the main results, for easy reference.

NOTE: In each case the formula for a_o can be obtained by putting n = 0 in that for a_n.

FRAME 52 continued

Period 2π

$$f(x) = \tfrac{1}{2}a_o + \sum_{n=1}^{\infty} a_n \cos nx + \sum_{n=1}^{\infty} b_n \sin nx$$

Full range: f(x) defined for $-\pi < x < \pi$

$$a_n = \frac{1}{\pi}\int_{-\pi}^{\pi} f(x) \cos nx \, dx$$

$$b_n = \frac{1}{\pi}\int_{-\pi}^{\pi} f(x) \sin nx \, dx$$

Half-range: f(x) defined for $0 < x < \pi$

Cosine series: $$a_n = \frac{2}{\pi}\int_{0}^{\pi} f(x) \cos nx \, dx$$

$$b_n = 0$$

Sine series: $$a_n = 0$$

$$b_n = \frac{2}{\pi}\int_{0}^{\pi} f(x) \sin nx \, dx$$

General Period 2ℓ

$$f(x) = \tfrac{1}{2}a_o + \sum_{n=1}^{\infty} a_n \cos \frac{n\pi x}{\ell} + \sum_{n=1}^{\infty} b_n \sin \frac{n\pi x}{\ell}$$

Full range: f(x) defined for $-\ell < x < \ell$

$$a_n = \frac{1}{\ell}\int_{-\ell}^{\ell} f(x) \cos \frac{n\pi x}{\ell} \, dx$$

$$b_n = \frac{1}{\ell}\int_{-\ell}^{\ell} f(x) \sin \frac{n\pi x}{\ell} \, dx$$

Half-range: f(x) defined for $0 < x < \ell$

Cosine series: $$a_n = \frac{2}{\ell}\int_{0}^{\ell} f(x) \cos \frac{n\pi x}{\ell} \, dx$$

$$b_n = 0$$

FRAME 52 continued

Sine Series: $a_n = 0$

$$b_n = \frac{2}{\ell}\int_0^{\ell} f(x)\sin\frac{n\pi x}{\ell}\,dx$$

Numerical Harmonic Analysis

For period 2π: $a_n = 2 \times$ Mean value of $f(x)\cos nx$ over one period
$b_n = 2 \times$ " " " $f(x)\sin nx$ " " "

FRAME 53

Miscellaneous Examples

In this frame a collection of miscellaneous examples is given for you to try. Answers are supplied in FRAME 54, together with such working as is considered helpful. A rough sketch of the waveform is generally advisable.

1. Find the Fourier series for the function defined by:

$$\begin{cases} f(x) = e^x & -\pi < x < \pi \\ f(x + 2\pi) = f(x) & \end{cases}$$

NOTE: $\int e^{ax}\sin bx\,dx = \dfrac{e^{ax}(a\sin bx - b\cos bx)}{a^2 + b^2}$

and $\int e^{ax}\cos bx\,dx = \dfrac{e^{ax}(a\cos bx + b\sin bx)}{a^2 + b^2}$

2. Find the Fourier sine series for the trapezoidal wave defined for the half-range 0 to π as shown in the diagram.

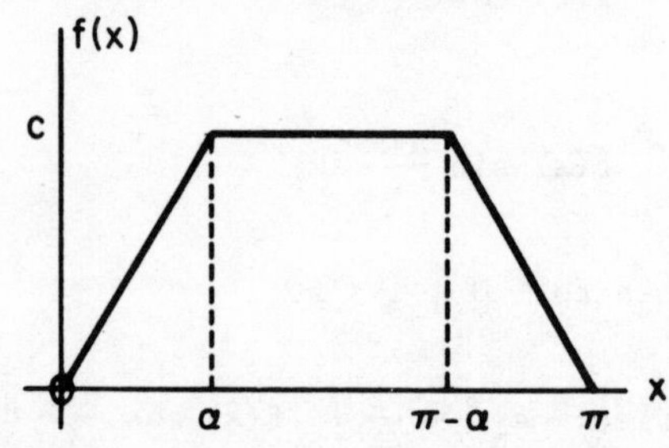

This wave is important in determining the flux distribution in the air gap of an alternator.

FRAME 53 continued

3. On a long beam, patches of constant loading p_0 of length $k\ell$ alternate with patches of length $(1 - k)\ell$ free of pressure, as shown in the diagram.

Find the Fourier analysis for the loading. By putting $k = \frac{1}{2}$ in this series, verify the result (7.1).

4. Find a Fourier series to represent the rectified sine wave $e = E|\sin \omega t|$. (This is the output voltage when the e.m.f. $E \sin \omega t$ acts on a full-wave rectifier.)

5. Find (i) a half-range Fourier sine series, and
   (ii) a half-range Fourier cosine series,

   to represent the function defined between 0 and T as $y = 1 + t/T$.

6. A function $f(x)$ of period 2ℓ is defined by

$$f(x) = -\frac{k}{\ell}(\ell + x) \qquad -\ell < x < 0$$

$$f(x) = \frac{k}{\ell}(\ell - x) \qquad 0 < x < \ell$$

Find the Fourier series which represents this.

The following three examples are all numerical. As the technique involved is similar in each case, it is suggested that you try only one of them, choosing one which is nearest to your own particular interest.

7. Referring to FRAME 38, Figure (iv), the voltage produced by one such machine is as shown in the table, the period being divided into twelve equal intervals.

FRAME 53 continued

θ^o	0	30	60	90	120	150	180	210	240	270	300	330
v	137	164	265	325	156	-54	-137	-164	-265	-325	-156	54

Analyse this waveform as far as the fifth harmonic.

8. A person's temperature is a function of the time of day and his temperature curve normally repeats itself daily. Thus his temperature is the same at, for example, 2 pm every day. The following values were extracted from a one-hourly temperature chart.

Time	Mid-night	2 am	4 am	6 am	8 am	10 am	Noon	2 pm	4 pm	6 pm	8 pm	10 pm
Temp °C	36·61	36·50	36·45	36·54	36·69	36·76	36·94	37·13	37·35	37·25	36·98	36·86

As in the example in FRAME 50 the table shows 12 readings taken at equal intervals over one cycle (midnight corresponds to $\theta = 0$, 2 am to $\theta = \pi/6$ etc).

Find a Fourier series as far as the third harmonic to represent the temperature curve.

9. The table shows the displacement x mm of a sliding piece from a fixed reference point for every 30° of rotation of the crank.

θ	0	30	60	90	120	150	180	210	240	270	300	330
x	298	356	373	337	254	155	80	51	60	93	147	221

Find a Fourier series for x as far as the third harmonic.

FRAME 54

Answers to Miscellaneous Examples

1.

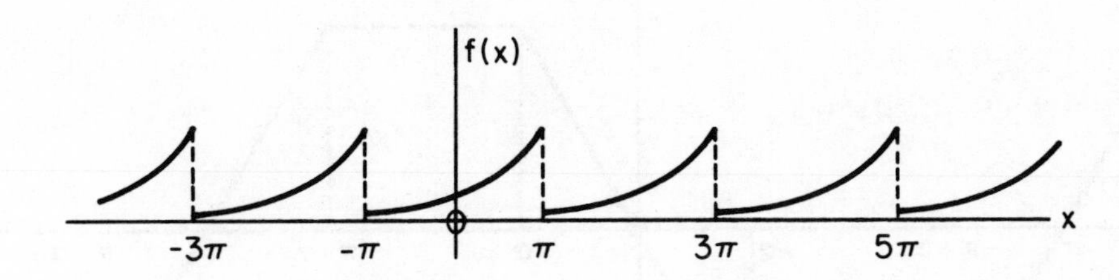

$$a_o = \frac{1}{\pi}\int_{-\pi}^{\pi} e^x\,dx = \frac{1}{\pi}(e^{\pi} - e^{-\pi}) = \frac{2\sinh\pi}{\pi}$$

$$a_n = \frac{1}{\pi}\int_{-\pi}^{\pi} e^x \cos nx\,dx$$

$$= \frac{1}{\pi}\left[\frac{e^x}{1+n^2}(\cos nx + n\sin nx)\right]_{-\pi}^{\pi}$$

$$= \frac{1}{\pi(1+n^2)}\left[e^{\pi}\cos n\pi - e^{-\pi}\cos n\pi\right]$$

$$= \frac{e^{\pi} - e^{-\pi}}{\pi(1+n^2)}\cos n\pi$$

$$= \begin{cases} \dfrac{2\sinh\pi}{\pi(1+n^2)} & \text{if n is even} \\ \dfrac{-2\sinh\pi}{\pi(1+n^2)} & \text{if n is odd} \end{cases}$$

$$b_n = \frac{1}{\pi}\int_{-\pi}^{\pi} e^x \sin nx\,dx$$

$$= \begin{cases} \dfrac{-2n\sinh\pi}{\pi(1+n^2)} & \text{if n is even} \\ \dfrac{2n\sinh\pi}{\pi(1+n^2)} & \text{if n is odd} \end{cases}$$

$$\therefore f(x) = \frac{2\sinh\pi}{\pi}\left\{\frac{1}{2} - \frac{1}{2}\cos x + \frac{1}{5}\cos 2x - \frac{1}{10}\cos 3x + \ldots + \frac{1}{2}\sin x - \frac{2}{5}\sin 2x + \frac{3}{10}\sin 3x - \ldots\right\}$$

FRAME 54 continued

2.

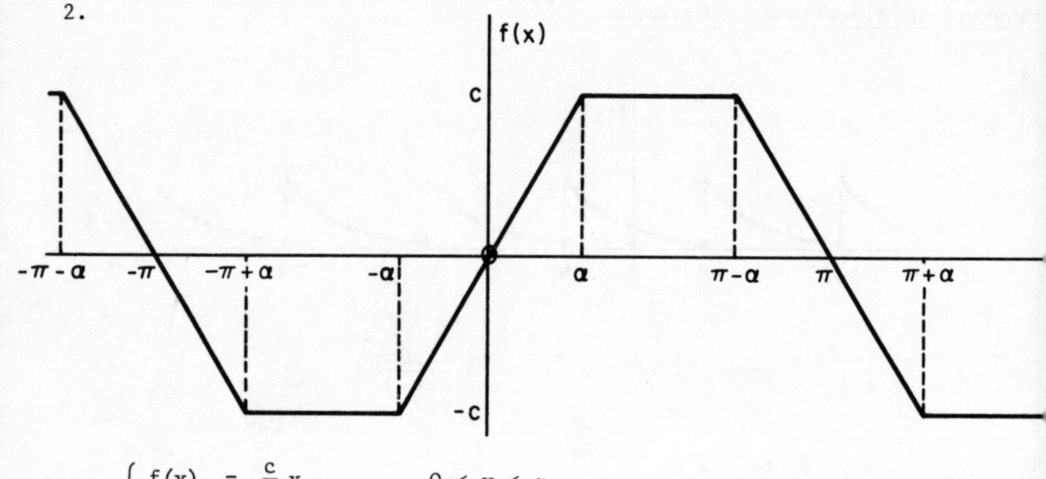

$$\begin{cases} f(x) = \frac{c}{\alpha} x & 0 < x < \alpha \\ f(x) = c & \alpha < x < \pi-\alpha \\ f(x) = \frac{c}{\alpha}(\pi - x) & \pi-\alpha < x < \pi \end{cases}$$

There will only be odd harmonics.

$$b_n = \frac{2}{\pi}\left[\int_0^{\alpha} \frac{c}{\alpha} x \sin nx\, dx + \int_{\alpha}^{\pi-\alpha} c \sin nx\, dx + \int_{\pi-\alpha}^{\pi} \frac{c}{\alpha}(\pi - x) \sin nx\, dx\right]$$

$$= \frac{2c}{\pi n^2 \alpha}\left[\sin n\alpha + \sin n(\pi - \alpha)\right]$$

$$= \begin{cases} 0 & \text{if n is even} \\ \dfrac{4c \sin n\alpha}{\pi n^2 \alpha} & \text{if n is odd} \end{cases}$$

$$f(x) = \frac{4c}{\pi\alpha}\left(\sin\alpha \sin x + \frac{\sin 3\alpha}{3^2} \sin 3x + \frac{\sin 5\alpha}{5^2} \sin 5x + \dots\right)$$

3.

$$\begin{cases} f(x) = p_o & 0 < x < k\ell \\ f(x) = 0 & k\ell < x < \ell \\ \text{Period } \ell \end{cases}$$

$$a_o = \frac{2}{\ell} \times p_o k\ell = 2kp_o$$

FRAME 54 continued

$$a_n = \frac{2}{\ell}\int_0^{k\ell} p_o \cos\frac{2n\pi x}{\ell}\,dx = \frac{p_o}{n\pi}\sin 2nk\pi$$

$$b_n = \frac{2}{\ell}\int_0^{k\ell} p_o \sin\frac{2n\pi x}{\ell}\,dx = \frac{p_o}{n\pi}(1 - \cos 2nk\pi)$$

$$f(x) = p_o k + \frac{p_o}{\pi}\Big[\sin 2k\pi \cos\frac{2\pi x}{\ell} + \frac{1}{2}\sin 4k\pi\cos\frac{4\pi x}{\ell} + \frac{1}{3}\sin 6k\pi\cos\frac{6\pi x}{\ell} + ..$$

$$+ (1-\cos 2k\pi)\sin\frac{2\pi x}{\ell} + \frac{1}{2}(1-\cos 4k\pi)\sin\frac{4\pi x}{\ell} + \frac{1}{3}(1-\cos 6k\pi)\sin\frac{6\pi x}{\ell} + ..\Big]$$

Putting $k = \frac{1}{2}$ gives the series

$$\frac{1}{2}p_o + \frac{2p_o}{\pi}\left[\sin\frac{2\pi x}{\ell} + \frac{1}{3}\sin\frac{6\pi x}{\ell} + \frac{1}{5}\sin\frac{10\pi x}{\ell} + \ldots\right]$$

which was quoted for the patch loading in FRAME 7.

4.

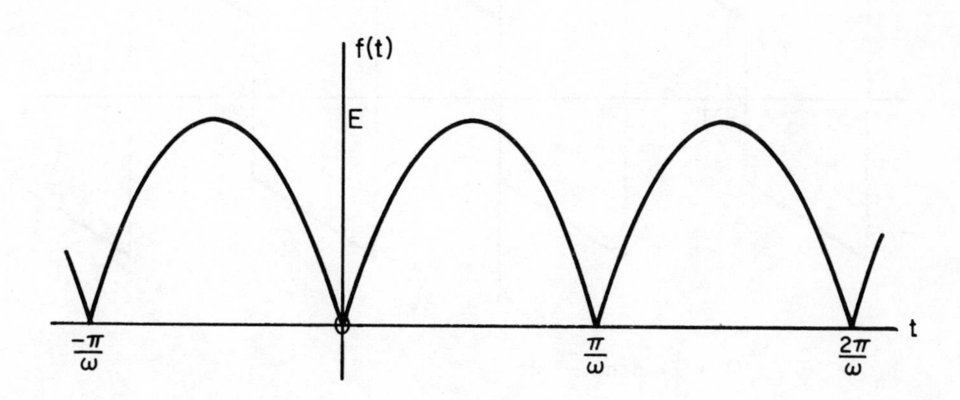

This is an even function, period π/ω.

$$a_o = \frac{2\omega}{\pi}\int_0^{\pi/\omega} E\sin\omega t\,dt$$

$$= 4E/\pi$$

FRAME 54 continued

$$a_n = \frac{2\omega}{\pi}\int_0^{\pi/\omega} E \sin \omega t \cos 2n\omega t \, dt$$

$$= \frac{E\omega}{\pi}\int_0^{\pi/\omega} \{\sin (2n + 1)\omega t - \sin (2n - 1)\omega t\} dt$$

$$= -\frac{4E}{\pi}\frac{1}{(2n - 1)(2n + 1)}$$

Did you integrate by parts? If so you were using an unnecessarily complicated method.

$$e = \frac{2E}{\pi} - \frac{4E}{\pi}\left(\frac{1}{1.3}\cos 2\omega t + \frac{1}{3.5}\cos 4\omega t + \frac{1}{5.7}\cos 6\omega t + \ldots\right)$$

5. (i)

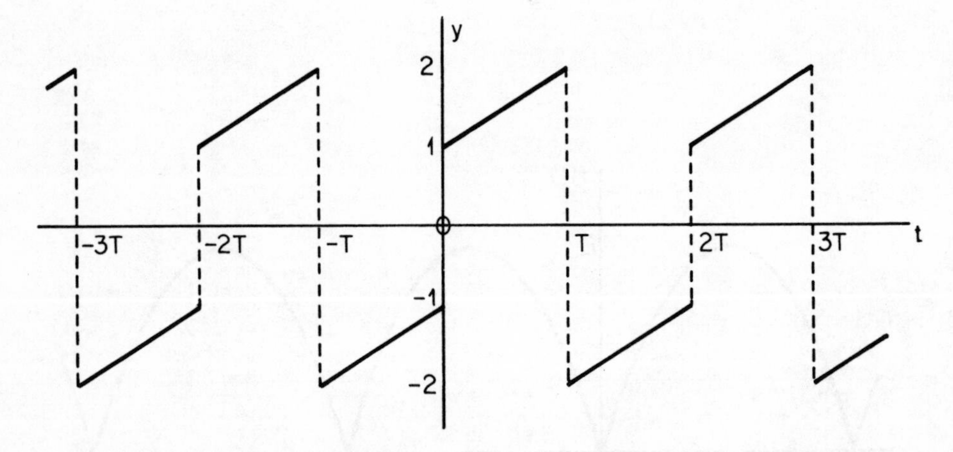

$$b_n = \frac{2}{T}\int_0^T \left(1 + \frac{t}{T}\right) \sin \frac{n\pi t}{T} \, dt$$

$$= \frac{2}{n\pi}(1 - 2 \cos n\pi)$$

$$= \begin{cases} \dfrac{-2}{n\pi} & \text{if n is even} \\ \dfrac{6}{n\pi} & \text{if n is odd} \end{cases}$$

$$y = \frac{2}{\pi}\left(3 \sin \frac{\pi t}{T} - \frac{1}{2}\sin \frac{2\pi t}{T} + \frac{3}{3}\sin \frac{3\pi t}{T} - \frac{1}{4}\sin \frac{4\pi t}{T} + \frac{3}{5}\sin \frac{5\pi t}{T} - \ldots\right)$$

FRAME 54 continued

(ii)

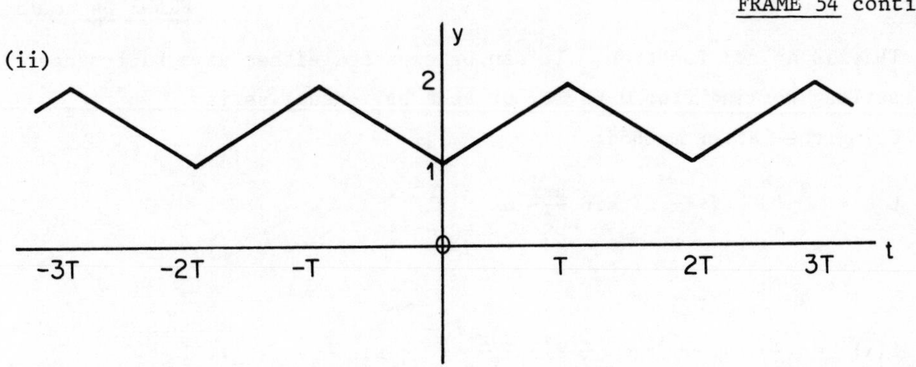

$$a_o = \frac{2}{T}\int_o^T (1 + \frac{t}{T})dt$$

$$= 3$$

$$a_n = \frac{2}{T}\int_o^T (1 + \frac{t}{T}) \cos \frac{n\pi t}{T}\, dt$$

$$= \frac{2}{\pi^2 n^2}(\cos n\pi - 1)$$

$$= \begin{cases} 0 & \text{if n is even} \\ \frac{-4}{\pi^2 n^2} & \text{if n is odd} \end{cases}$$

$$y = \frac{3}{2} - \frac{4}{\pi^2}\left(\cos \frac{\pi t}{T} + \frac{1}{3^2}\cos \frac{3\pi t}{T} + \frac{1}{5^2}\cos \frac{5\pi t}{T} + \ldots\right)$$

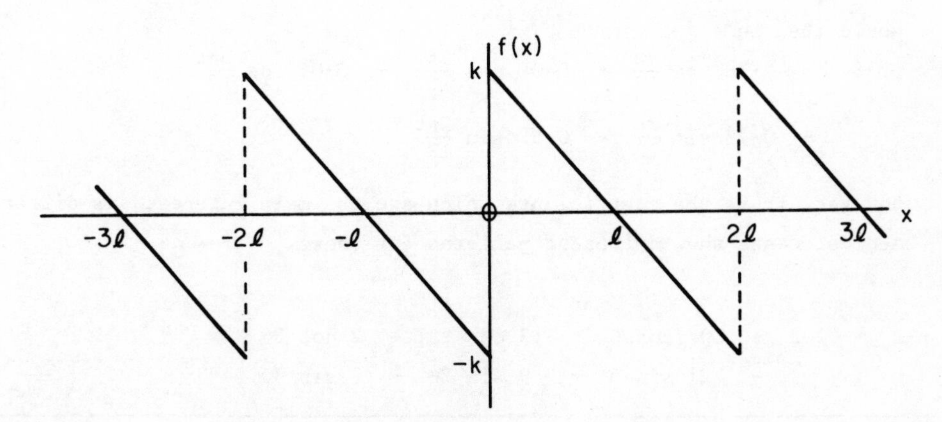

FRAME 54 continued

This is an odd function. It can be expanded either as a full-range series, working from 0 to 2ℓ, or as a half-range series. Using the latter method:

$$b_n = \frac{2}{\ell}\int_0^{\ell} \frac{k}{\ell}(\ell - x)\sin\frac{n\pi x}{\ell}\,dx$$

$$= \frac{2k}{n\pi}$$

$$f(x) = \frac{2k}{\pi}\left(\sin\frac{\pi x}{\ell} + \frac{1}{2}\sin\frac{2\pi x}{\ell} + \frac{1}{3}\sin\frac{3\pi x}{\ell} + \ldots\right)$$

7. As $f(\theta + 180^\circ) = -f(\theta)$, only the odd harmonics are present so it is only necessary to calculate a_1, a_3, a_5, b_1, b_3 and b_5.

$$v = 127\cos\theta + 9\cos 3\theta + \cos 5\theta$$
$$+ 248\sin\theta - 72\sin 3\theta + 5\sin 5\theta$$
$$= 279\sin(\theta + 27^\circ) + 72\sin(3\theta + 173^\circ) + 5\sin(5\theta + 11^\circ)$$

Did you notice that the values for $v\cos\theta$, $v\sin\theta$, $v\cos 3\theta$ etc. for $\theta = 180$ to 330 repeat those for $\theta = 0$ to 150? This always happens if only the odd harmonics are present and consequently only the table from 0 to 150 is really needed.

8. $$36\cdot84 - 0\cdot18\cos\theta - 0\cdot06\cos 2\theta + 0\cdot05\cos 3\theta$$
$$- 0\cdot35\sin\theta + 0\cdot02\sin 2\theta$$

This could be given in terms of t (hours measured from midnight) and would then be

$$36\cdot84 - 0\cdot18\cos\frac{\pi t}{12} - 0\cdot06\cos\frac{2\pi t}{12} + 0\cdot05\cos\frac{3\pi t}{12}$$
$$- 0\cdot35\sin\frac{\pi t}{12} + 0\cdot02\sin\frac{2\pi t}{12}$$

However, it is the coefficients which are the main interest, as different medical cases show different patterns for these.

9. $$x = 202 + 107\cos\theta - 13\cos 2\theta + 2\cos 3\theta$$
$$+ 121\sin\theta + 9\sin 2\theta - \sin 3\theta$$

PARTIAL DIFFERENTIAL EQUATIONS FOR TECHNOLOGISTS

Although there are many different types of partial differential equations with a variety of methods of solution, it is mainly one particular kind which occurs in engineering and applied science. This programme will concentrate on the solution of such equations by one method (separation of variables), this method being of the most value in practice.

FRAME 1

Introduction

If y is a function of one variable x, an equation involving its derivatives will be an ordinary differential equation,

$$\text{e.g.} \qquad \frac{d^2y}{dx^2} + 6\frac{dy}{dx} - 2y = \sin 3x \qquad (1.1)$$

If, however, y is a function of two variables x and t, any derivatives with respect to x or t will be partial and so an equation involving them will be a partial differential equation,

$$\text{e.g.} \qquad \frac{\partial^2 y}{\partial t^2} + 6\frac{\partial y}{\partial t} = 4\frac{\partial^2 y}{\partial x^2} \qquad (1.2)$$

The same is true for a function of more than two variables, so that we might have

$$\frac{\partial^2\theta}{\partial x^2} + \frac{\partial^2\theta}{\partial y^2} + \frac{\partial^2\theta}{\partial z^2} = \frac{1}{h^2}\frac{\partial\theta}{\partial t} \qquad (1.3)$$

where θ is a function of x, y, z and t (h is a constant).

The solution of (1.1) involves finding y as a function of x, that of (1.2) y as a function of x and t and that of (1.3) θ as a function of x, y, z and t.

This programme will be restricted to the solution of partial differential equations involving two independent variables only.

An example of such an equation is (1.2). Like all differential equations it has an unlimited number of solutions. One of these is

$$y = Ae^{-3t}\sin\frac{3x}{2}$$

You might like to verify this.

****

1A

$$\frac{\partial y}{\partial t} = -3Ae^{-3t}\sin\frac{3x}{2}$$

$$\frac{\partial^2 y}{\partial t^2} = 9Ae^{-3t}\sin\frac{3x}{2}$$

$$\frac{\partial^2 y}{\partial x^2} = -\frac{9}{4}Ae^{-3t}\sin\frac{3x}{2}$$

$$\frac{\partial^2 y}{\partial t^2} + 6\frac{\partial y}{\partial t} = -9Ae^{-3t}\sin\frac{3x}{2} = 4\frac{\partial^2 y}{\partial x^2}$$

FRAME 2

We shall now mention a few cases where partial differential equations arise in practice.

One example is that of the transverse vibration of a tightly stretched string.

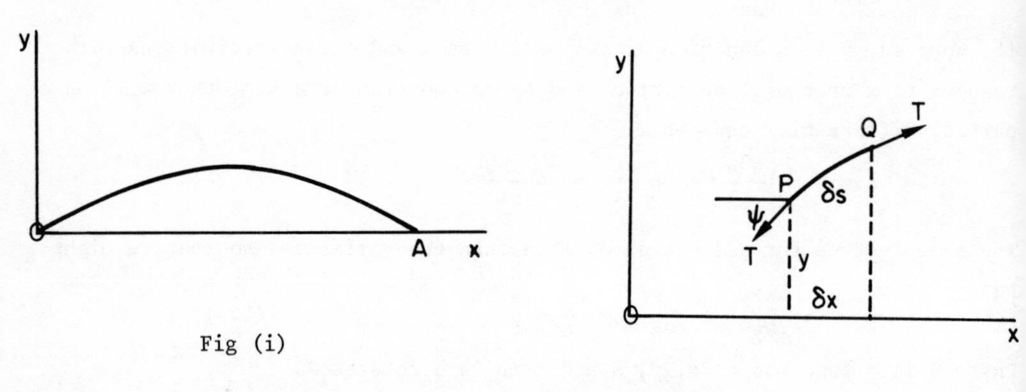

Fig (i)

Fig (ii)

Suppose that the ends of the string are fixed at the two points O and A and that the string is set in motion in the plane Oxy. At a subsequent time t let the string be displaced as shown in Fig (i). It is assumed that the amplitude of the oscillations is small. Therefore the change in length of the string during the motion is small enough for the tension T to be assumed constant. Fig (ii) shows a small element PQ of the string on an enlarged scale.

The acceleration of this element is $\frac{\partial^2 y}{\partial t^2}$, the partial derivative being used as y is a function of both x and t. Hence the force required to produce this acceleration is $\rho\delta s \frac{\partial^2 y}{\partial t^2}$ where ρ is the mass per unit length of the string.

The actual force producing the motion is the difference between the components parallel to Oy of T at the points P and Q, i.e. the increase in the component of T in this direction as x increases by an amount δx.

At P, this component of T is $T \sin \psi$.

Now $\sin \psi = \psi - \frac{\psi^3}{3!} \dots$ and $\tan \psi = \psi + \frac{\psi^3}{3} \dots$, and as ψ is small, powers of ψ can be neglected.

FRAME 2 continued

Hence $\sin\psi \simeq \tan\psi = \dfrac{\partial y}{\partial x}$ and the component of T at P is $T\dfrac{\partial y}{\partial x}$.

The increase in this as x increases by δx is $\dfrac{\partial}{\partial x}\left(T\dfrac{\partial y}{\partial x}\right)\delta x$

$$= T\frac{\partial^2 y}{\partial x^2}\,\delta x$$

$\therefore$ The equation of motion is $\rho\,\delta s\,\dfrac{\partial^2 y}{\partial t^2} = T\dfrac{\partial^2 y}{\partial x^2}\,\delta x$

Now $\delta x = \delta s \cos\psi$

$\simeq \delta s$ as $\cos\psi \simeq 1$ if powers of ψ are ignored.

$$\therefore \rho\frac{\partial^2 y}{\partial t^2} = T\frac{\partial^2 y}{\partial x^2}$$

This is usually written $\dfrac{\partial^2 y}{\partial t^2} = c^2\dfrac{\partial^2 y}{\partial x^2}$ where $c^2 = T/\rho$ and is constant.

FRAME 3

Another example occurs in the study of simple fluid flow.

Suppose a non-viscous fluid of unit depth is streaming over a horizontal plane and axes Ox, Oy are taken anywhere in the plane. Let the velocity of the fluid at any point whose horizontal coordinates are x and y have components u and v parallel to Ox and Oy.

Consider an imaginary rectangle with sides δx and δy, and centre (x,y) as shown in the diagram.

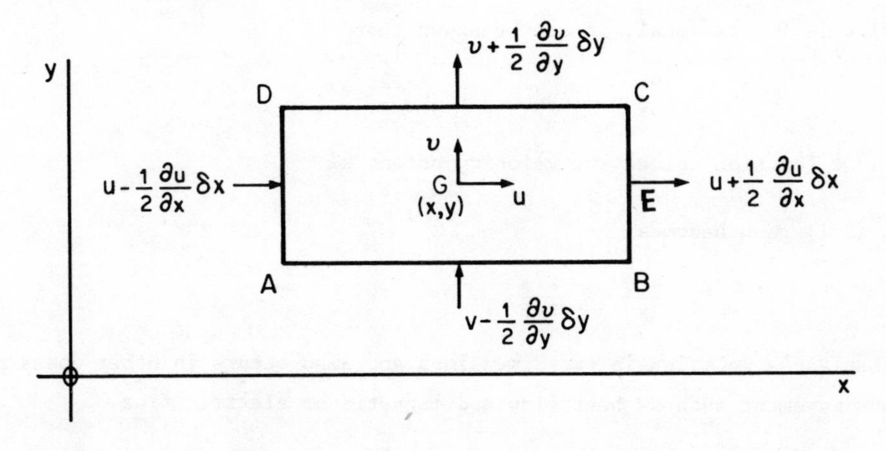

FRAME 3 continued

The rate of increase of u with respect to x is $\frac{\partial u}{\partial x}$.

$\therefore$ The increase in u in going from G to E is $\frac{\partial u}{\partial x} \cdot \frac{1}{2}\delta x$ and hence the velocity component at E parallel to Ox is $u + \frac{1}{2}\frac{\partial u}{\partial x}\delta x$.

Similarly the velocity components at the mid-points of the other sides are as shown in the diagram.

Thus in unit time the volume of fluid crossing over BC is approximately $\left(u + \frac{1}{2}\frac{\partial u}{\partial x}\delta x\right)\delta y$ and over AD is $\left(u - \frac{1}{2}\frac{\partial u}{\partial x}\delta x\right)\delta y$, giving between them a net loss of $\left(u + \frac{1}{2}\frac{\partial u}{\partial x}\delta x\right)\delta y - \left(u - \frac{1}{2}\frac{\partial u}{\partial x}\delta x\right)\delta y$

$$= \frac{\partial u}{\partial x}\delta x\delta y$$

Similarly, for the flow over AB and DC, the net loss is $\frac{\partial v}{\partial y}\delta y\delta x$.

$\therefore$ Total net loss $= \left(\frac{\partial u}{\partial x} + \frac{\partial v}{\partial y}\right)\delta x\delta y$.

If there is no sink or source within the rectangle, and the fluid is assumed to be of constant density, this must be zero.

$$\therefore \frac{\partial u}{\partial x} + \frac{\partial v}{\partial y} = 0 \qquad (3.1)$$

If the flow is irrotational, it can be shown that

$$u = \frac{\partial \phi}{\partial x} \quad \text{and} \quad v = \frac{\partial \phi}{\partial y}$$

where ϕ is a function called the velocity potential.

Equation (3.1) then becomes

$$\frac{\partial^2 \phi}{\partial x^2} + \frac{\partial^2 \phi}{\partial y^2} = 0$$

This is Laplace's Equation in two dimensions and also occurs in other cases of continuous movement such as heat flow and magnetic or electric flux.

FRAME 4

Some more examples are listed below. However, in these cases, only the final differential equation will be quoted, the derivation being omitted.

(i) If the sides of a bar of constant cross-sectional area are insulated and heat flows along the bar then the temperature θ at a point distant x from one end at time t satisfies the equation

$$\frac{\partial^2\theta}{\partial x^2} = \frac{1}{h^2}\frac{\partial\theta}{\partial t}$$

where h is a constant.

(ii) If a uniform beam vibrates transversely then the displacement y at a distance x from one end at time t satisfies the equation

$$EI\frac{\partial^4 y}{\partial x^4} + m\frac{\partial^2 y}{\partial t^2} = 0$$

(iii) If a transmission line consists of two parallel wires, the voltage drop v across the wires and the current i in the line, at a distance x from the sending end at time t, satisfy the equations

$$CL\frac{\partial^2 v}{\partial t^2} + (CR + LG)\frac{\partial v}{\partial t} + RGv = \frac{\partial^2 v}{\partial x^2}$$

and $$CL\frac{\partial^2 i}{\partial t^2} + (CR + LG)\frac{\partial i}{\partial t} + RGi = \frac{\partial^2 i}{\partial x^2}$$

where C, L, R and G are respectively the capacitance, inductance, resistance and leakance per unit length.

FRAME 5

Solution by direct integration

You will remember that the simplest ordinary differential equations such as $\frac{d^2y}{dx^2} = f(x)$ can be solved by direct integration.

For instance, if $\frac{d^2y}{dx^2} = 2$

FRAME 5 continued

$$\text{then} \quad \frac{dy}{dx} = 2x + A$$

$$\text{and} \quad y = x^2 + Ax + B$$

where A and B are constants, each of which can take any value. Here, $2x + A$ is the most general function that gives 2 when differentiated with respect to x and $x^2 + Ax + B$ is the most general function that gives $2x + A$. $y = x^2 + Ax + B$ is the general solution of the differential equation $\frac{d^2y}{dx^2} = 2$.

In a similar manner certain partial differential equations can be solved by direct integration,

e.g. $\frac{\partial^2 y}{\partial x^2} = 2$ where y is a function of x and t.

Then, $\frac{\partial y}{\partial x} = 2x + f(t)$ because $2x + f(t)$ is the most general function that gives 2 when differentiated partially with respect to x.

In the same way, $y = x^2 + x\,f(t) + g(t)$ and this is the general solution, $f(t)$ and $g(t)$ being arbitrary functions of t (which may contain constants).

Now find the general solution of

$$\frac{\partial^2 y}{\partial x \partial t} = 2$$

****

5A

$$\frac{\partial y}{\partial t} = 2x + f(t)$$

$$y = 2xt + \int f(t)\,dt + g(x)$$

$$= 2xt + F(t) + g(x)$$

$f(t)$ being an arbitrary function of t, $\int f(t)\,dt$ is also an arbitrary function, which we have called $F(t)$.

FRAME 6

Returning to the equation $\frac{d^2y}{dx^2} = 2$ in FRAME 5, if additional information is given, e.g. when $x = 0$, $y = -3$ and $\frac{dy}{dx} = 1$, it is possible to determine the values of A and B.

In the case of the partial differential equation $\frac{\partial^2 y}{\partial x^2} = 2$ (see FRAME 5 again) it is possible to find suitable expressions for f(t) and g(t) if additional information is available.

For instance, if $y = \sin t$ for all t when $x = 0$ and $y = x^2 + 2x$ for all x when $t = 0$, substitution in the general solution

$$y = x^2 + x\,f(t) + g(t) \quad \text{gives}$$

$$\begin{aligned} \sin t &= g(t) \\ \text{and} \quad x^2 + 2x &= x^2 + x\,f(0) + g(0) \\ \therefore \quad 2x &= x\,f(0) \quad \text{as} \quad g(0) = 0 \\ \therefore \quad f(0) &= 2 \end{aligned}$$

Thus f(t) is any function of t which has the value 2 when $t = 0$. Write down a few expressions for f(t) which satisfy this condition.

****

6A

Some possible expressions are:

$$2, \qquad 2 + t, \qquad 2\cos t, \qquad 2e^t, \qquad 2\cosh t.$$

There are, of course, many more.

FRAME 7

As any of these expressions for f(t) is suitable, one solution is

$$y = x^2 + 2x + \sin t$$

Another is $y = x^2 + 2x\cos t + \sin t$, and so on.
In FRAME 5 you found the general solution of $\frac{\partial^2 y}{\partial x \partial t} = 2$. Now see if you can find suitable expressions for F(t) and g(x) so that $y = 0$ when $x = 0$ for all t, and $y = x^2$ when $t = 0$ for all x.

****

7A

$y = 0$ when $x = 0$ $\therefore\ 0 = F(t) + g(0)$ (7A.1)

$y = x^2$ when $t = 0$ $\therefore\ x^2 = F(0) + g(x)$ (7A.2)

Equation (7A.2) is satisfied by $g(x) = x^2 - A$, $F(0) = A$ *where A is a constant.*

Then $g(0) = -A$ *and (7A.1) gives* $F(t) = A$

$$\therefore\quad y = 2xt + A + x^2 - A$$
$$= 2xt + x^2$$

You will notice that, in this case, although there are many possibilities for F(t) and g(x), there is only one solution for y.

FRAME 8

Two simple cases where partial differential equations are solved by direct integration are:

(i) The solution of a certain type of ordinary differential equation, i.e. first order exact (see the first programme in Unit 5 of Vol I).

(ii) Conjugate functions (see the second programme in Unit 12 of this volume).

Taking first an example of (i), the equation

$$(2xy - \sin y)dx + (x^2 + 3y^2 - x\cos y)dy = 0 \qquad (8.1)$$

is exact, i.e. it is of the form $\frac{\partial u}{\partial x}\,dx + \frac{\partial u}{\partial y}\,dy = 0$. This is equivalent to $du = 0$, the solution of which is $u = c$.

Here, $\frac{\partial u}{\partial x} = 2xy - \sin y$ and $\frac{\partial u}{\partial y} = x^2 + 3y^2 - x\cos y$

Integrating, these two equations give respectively

$$u = x^2y - x\sin y + f(y)$$

$$\text{and } u = x^2y + y^3 - x\sin y + g(x)$$

FRAME 8 continued

Comparing these, you will see that these two requirements are met by $f(y) = y^3$, $g(x) = 0$. The solution of (8.1) is

$$x^2y - x \sin y + y^3 = c$$

(The technique used here is slightly different from that used in the programme mentioned above.)

Now solve the exact equation

$$(2x - e^y)dx + (2 - xe^y)dy = 0$$

****

8A

$$\frac{\partial u}{\partial x} = 2x - e^y \quad \textit{and} \quad \frac{\partial u}{\partial y} = 2 - xe^y$$

$u = x^2 - xe^y + f(y)$ *and*

$u = 2y - xe^y + g(x)$

Comparing, $f(y) = 2y$ *and* $g(x) = x^2$

The solution is $x^2 - xe^y + 2y = c$

FRAME 9

Two functions u and v, of x and y, are called conjugate if they satisfy the equations

$$\frac{\partial u}{\partial x} = \frac{\partial v}{\partial y} \quad \text{and} \quad \frac{\partial u}{\partial y} = -\frac{\partial v}{\partial x}$$

Such functions occur in the study of heat flow, electrostatics and fluid motion. In heat flow, the equations $u = \alpha$, $v = \beta$, represent the isothermal lines and flux lines respectively; in electrostatics they represent equipotential lines and flux lines, and in fluid motion equipotential lines and stream lines. Given one of these functions, it is possible to find the other. If $v = 2xy$, see if you can find u.

****

9A

$$\frac{\partial u}{\partial x} = 2x \quad \text{and} \quad \frac{\partial u}{\partial y} = -2y$$

$$u = x^2 + f(y)$$

$$u = -y^2 + g(x)$$

$$\therefore u = x^2 - y^2$$

An arbitrary constant can be added to the right hand side.

FRAME 10

An exponential trial solution

You will remember that an exponential trial solution is often useful when dealing with ordinary differential equations.

For example, a trial solution of the form $y = Ae^{mx}$ is the basis of the auxiliary equation when solving

$$a\frac{d^2y}{dx^2} + b\frac{dy}{dx} + cy = 0 \qquad (10.1)$$

(A full treatment of this is given in the second programme in Unit 5 of Vol I.)

A similar technique can sometimes be used in solving partial differential equations.

We shall illustrate this by finding a solution for equation (1.2)

$$\text{i.e.} \quad \frac{\partial^2 y}{\partial t^2} + 6\frac{\partial y}{\partial t} = 4\frac{\partial^2 y}{\partial x^2}$$

which is analogous to (10.1) in that it is linear with constant coefficients.

Taking as a trial solution $y = Ae^{mx+nt}$

$$\frac{\partial y}{\partial t} = nAe^{mx+nt} \qquad \frac{\partial^2 y}{\partial t^2} = n^2Ae^{mx+nt}$$

$$\frac{\partial y}{\partial x} = mAe^{mx+nt} \qquad \frac{\partial^2 y}{\partial x^2} = m^2Ae^{mx+nt}$$

FRAME 10 continued

$\therefore$ This is a solution if $n^2Ae^{mx+nt} + 6nAe^{mx+nt} = 4m^2Ae^{mx+nt}$

i.e. if $n^2 + 6n = 4m^2$

i.e. if $m = \pm \frac{1}{2}\sqrt{n(n+6)}$

A few possible solutions, obtained by giving different values to n, are

$$y = Ae^{2x+2t}$$

$$y = Ae^{-2x-8t}$$

$$y = Ae^{(3/2)ix-3t}$$

$$y = Ae^{-(3/2)ix-3t}$$

Now say which of the following are also solutions:

(i) $y = Ae^{2x-8t}$

(ii) $y = Ae^{4x}$

(iii) $y = Ae^{-6t}$

(iv) $y = Ae^{-2x+2t}$

****

10A

(i), (iii) and (iv) are solutions.
(ii) is not a solution.

FRAME 11

As the equation (1.2) is linear with constant coefficients, the sum of any two or more solutions is also a solution. For example, each of the following is also a solution.

$$y = A_1e^{2x+2t} + A_2e^{2x-8t}$$

$$y = A_1e^{-2x+2t} + A_2e^{-(3/2)ix-3t} + A_3e^{-6t}$$

$$y = A_1e^{(3/2)ix-3t} + A_2e^{-(3/2)ix-3t}$$

FRAME 11 continued

You will realise that A_1, A_2, A_3 have been used instead of A because the coefficients need not all be the same.

The last solution can also be written as

$$y = e^{-3t}\left(A_1 e^{3ix/2} + A_2 e^{-3ix/2}\right)$$

$$= e^{-3t}\left(B_1 \cos\frac{3x}{2} + B_2 \sin\frac{3x}{2}\right)$$

If B_1, which is an arbitrary constant, is put equal to zero, the solution given in FRAME 1 is obtained.

Find the relation between m and n for $y = Ae^{mx+nt}$ to be a solution of $\frac{\partial^2 y}{\partial x^2} + 2\frac{\partial^2 y}{\partial x \partial t} = y$. Use this relation to say which of the following are solutions:

(i) $Ae^{2x-\frac{3}{4}t}$

(ii) Ae^{-x}

(iii) Ae^{-2x+t}

(iv) $A_1 e^{x} + A_2 e^{i(x-t)}$

(v) $A_1 e^{-\frac{1}{2}x-\frac{3}{4}t} + A_2 e^{3x+it} + A_3 e^{-x}$

****

11A

$m^2 + 2mn = 1$

(i), (ii) and (iv) are solutions.

(iii) and (v) are not solutions.

FRAME 12

The method of separation of variables

The trial solution $y = Ae^{mx+nt}$ can also be written as $y = Ae^{mx}e^{nt}$ which is the product of a function of x and a function of t, i.e. it is of the form $X(x)T(t)$. It is often simpler to start with this as a trial solution rather than the more specific exponential function.

As a first example, we shall find a solution of

$$\frac{\partial^2 y}{\partial x^2} + \frac{\partial^2 y}{\partial t^2} = 0 \qquad (12.1)$$

which satisfies the conditions $y = \sin t$ when $x = 0$ for all t, and $y \to 0$ as $x \to \infty$.

Let a trial solution be $y = X(x)T(t)$, which for convenience we shall write as $y = XT$, where it is understood that X is a function of x only and T is a function of t only.

Then $\dfrac{\partial y}{\partial x} = T\dfrac{dX}{dx}$ and $\dfrac{\partial^2 y}{\partial x^2} = T\dfrac{d^2X}{dx^2}$,

$\dfrac{\partial y}{\partial t} = X\dfrac{\partial T}{\partial t}$ and $\dfrac{\partial^2 y}{\partial t^2} = X\dfrac{d^2T}{dt^2}$.

Thus $y = XT$ is a solution of (12.1) if

$$T\frac{d^2X}{dx^2} + X\frac{d^2T}{dt^2} = 0$$

Separating the variables, this can be written

$$\frac{1}{X}\frac{d^2X}{dx^2} = -\frac{1}{T}\frac{d^2T}{dt^2} \qquad (12.2)$$

What would the left-hand side become in each of the following cases?

(i) $X = 4e^{2x}$

(ii) $X = x^2$

(iii) $X = xe^{-x}$

(iv) $X = A\cos 3x + B\sin 3x$

****

12A

(i) 4 *(ii)* $\frac{2}{x^2}$

(iii) $\frac{x-2}{x}$ *(iv)* -9

FRAME 13

Whilst it would be possible to find T such that $-\frac{1}{T}\frac{d^2T}{dt^2} = 4$ or -9, it is impossible for $-\frac{1}{T}\frac{d^2T}{dt^2}$ to be equal to $\frac{2}{x^2}$ or $\frac{x-2}{x}$, as T contains t only. Consequently you will see that (12.2) can only be satisfied if $\frac{1}{X}\frac{d^2X}{dx^2}$ is a constant (and therefore $-\frac{1}{T}\frac{d^2T}{dt^2}$ also).

Thus we can write $\frac{1}{X}\frac{d^2X}{dx^2} = -\frac{1}{T}\frac{d^2T}{dt^2} = \lambda$

i.e. $\frac{1}{X}\frac{d^2X}{dx^2} = \lambda$ and $-\frac{1}{T}\frac{d^2T}{dt^2} = \lambda$

$$\frac{d^2X}{dx^2} - \lambda X = 0 \quad \text{and} \quad \frac{d^2T}{dt^2} + \lambda T = 0$$

The problem of solving the original differential equation (12.1) has now been reduced to that of solving two ordinary differential equations. Write down the general solutions of these when

(i) $\lambda = k^2$

(ii) $\lambda = -k^2$

where k is real.

****

13A

(i) $X = A_1e^{kx} + B_1e^{-kx}$

$T = A_2 \cos kt + B_2 \sin kt$

(ii) $X = A_1 \cos kx + B_1 \sin kx$

$T = A_2e^{kt} + B_2e^{-kt}$

FRAME 14

For it to be possible for y to equal sin t when $x = 0$, the first set of solutions in FRAME 13A must be chosen.

Thus $y = (A_1 e^{kx} + B_1 e^{-kx})(A_2 \cos kt + B_2 \sin kt)$

Does the condition $y \to 0$ as $x \to \infty$ tell us anything about any of the constants?

****

14A

Yes. $A_1 = 0$, *as otherwise* $y \to \infty$ *as* $x \to \infty$.

FRAME 15

Now we are left with

$$y = B_1 e^{-kx}(A_2 \cos kt + B_2 \sin kt)$$

$$= e^{-kx}(A \cos kt + B \sin kt) \quad \text{where } A = B_1 A_2 \text{ and } B = B_1 B_2$$

Also, $y = \sin t$ when $x = 0$.

What further information does this give us about the constants?

****

15A

$A = 0$, $k = 1$ *and* $B = 1$

∴ *The solution is* $y = e^{-x} \sin t$.

FRAME 16

In FRAME 2 we derived a partial differential equation giving the displacement of a tightly stretched vibrating string, i.e.

$$\frac{\partial^2 y}{\partial t^2} = c^2 \frac{\partial^2 y}{\partial x^2}$$

Suppose it is required to find y in terms of x and t if the motion is started

FRAME 16 continued

by displacing a stretched string of length ℓ into the form $y = a \sin \pi x/\ell$, and then releasing it.

Taking $y = XT$ as a trial solution, proceed to the stage corresponding to (12.2) in the previous example.

****

16A

$$X \frac{d^2T}{dt^2} = c^2 T \frac{d^2X}{dx^2}$$

$$\frac{1}{T}\frac{d^2T}{dt^2} = \frac{c^2}{X}\frac{d^2X}{dx^2} \qquad \left(\text{or} \quad \frac{1}{c^2T}\frac{d^2T}{dt^2} = \frac{1}{X}\frac{d^2X}{dx^2}\right)$$

FRAME 17

For the same reason as in the previous example, we can now write

$$\frac{1}{T}\frac{d^2T}{dt^2} = \frac{c^2}{X}\frac{d^2X}{dx^2} = \lambda \quad \text{(a constant)} \qquad (17.1)$$

Which sign should be given to λ in this case and why?

****

17A

λ must be negative as the solution of $\frac{c^2}{X}\frac{d^2X}{dx^2} = \lambda$ has to be trigonometric in order to accommodate the condition $y = a \sin \pi x/\ell$ when $t = 0$.

FRAME 18

To ensure that λ is negative, put $\lambda = -k^2$ where k is real. (17.1) then yields

$$\frac{1}{T}\frac{d^2T}{dt^2} = -k^2 \quad \text{and} \quad \frac{c^2}{X}\frac{d^2X}{dx^2} = -k^2$$

Now find the general solutions of these two equations.

****

18A

$$T = A_1 \cos kt + B_1 \sin kt$$

$$X = A_2 \cos \frac{kx}{c} + B_2 \sin \frac{kx}{c}$$

FRAME 19

Thus $y = \left(A_2 \cos \frac{kx}{c} + B_2 \sin \frac{kx}{c}\right)\left(A_1 \cos kt + B_1 \sin kt\right)$

The first condition is that when $x = 0$, $y = 0$ for all t.

What information does this give about any of the constants?

****

19A

$$A_2 = 0$$

FRAME 20

Now, replacing B_2A_1 by A and B_2B_1 by B, we can write

$y = \sin \frac{kx}{c}(A \cos kt + B \sin kt)$.

The next condition is that when $t = 0$, $\frac{\partial y}{\partial t} = 0$ for all x as the string is initially at rest.

Use this to find further information about the constants.

****

20A

$$\frac{\partial y}{\partial t} = \sin \frac{kx}{c}(-Ak \sin kt + Bk \cos kt)$$

$$B = 0.$$

FRAME 21

Now $y = A \sin \frac{kx}{c} \cos kt$

The final condition is that when $t = 0$, $y = a \sin \pi x/\ell$.

Use this to write down the required solution for y.

****

<u>21A</u>

$$A = a$$

$$\frac{k}{c} = \frac{\pi}{\ell}$$

$$y = a \sin \frac{\pi x}{\ell} \cos \frac{\pi c t}{\ell}$$

<u>FRAME 22</u>

As another example we will consider the following problem:

The current i in a cable satisfies the equation

$$\frac{\partial^2 i}{\partial x^2} = \frac{1}{c}\frac{\partial i}{\partial t} + i$$

A solution of this equation is required which will satisfy the conditions $i = 0$ when $x = \ell$, and $\frac{\partial i}{\partial t} = -ae^{-2ct}$ when $x = 0$, both for all values of t.

First, let $i = XT$ and show that it is possible to separate the variables to give the ordinary differential equations

$$\frac{d^2X}{dx^2} - \lambda X = 0$$

$$\text{and} \quad \frac{dT}{dt} + k(1 - \lambda)T = 0$$

****

<u>22A</u>

$$T\frac{d^2X}{dx^2} = \frac{X}{c}\frac{dT}{dt} + XT$$

$$\frac{1}{X}\frac{d^2X}{dx^2} = \frac{1}{cT}\frac{dT}{dt} + 1 = \lambda$$

$$\text{giving} \quad \frac{d^2X}{dx^2} - \lambda X = 0 \qquad (22A.1)$$

$$\text{and} \quad \frac{dT}{dt} + c(1 - \lambda)T = 0 \qquad (22A.2)$$

FRAME 23

Now complete the solution.

HINT: First solve equation (22A.2); the value of λ should then be apparent from the boundary conditions.

****

23A

The solution of (22A.2) is

$$T = Ae^{-c(1-\lambda)t}$$

The condition $\frac{\partial i}{\partial t} = -ae^{-2ct}$ *when* $x = 0$ *requires the exponent in* T *to be* $-2ct$.

$\therefore -c(1 - \lambda) = -2c$

$\lambda = -1$

Thus, $T = Ae^{-2ct}$

and (22A.1) becomes $\frac{d^2X}{dx^2} + X = 0$

giving $X = B \cos x + C \sin x$

$\therefore i = e^{-2ct}(B_1 \cos x + C_1 \sin x)$

$\frac{\partial i}{\partial t} = -2ce^{-2ct}(B_1 \cos x + C_1 \sin x)$

Now, $\frac{\partial i}{\partial t} = -ae^{-2ct}$ *when* $x = 0$

$\therefore -a = -2cB_1$

$\therefore B_1 = \frac{a}{2c}$

$i = e^{-2ct}(\frac{a}{2c} \cos x + C_1 \sin x)$

Also $i = 0$ *when* $x = \ell$

$\therefore 0 = \frac{a}{2c} \cos \ell + C_1 \sin \ell$

$C_1 = -\frac{a}{2c} \cot \ell$

$i = \frac{a}{2c} e^{-2ct}(\cos x - \cot \ell \sin x)$

You will see that this can be written in the form

$$i = \frac{ae^{-2ct} \sin (\ell - x)}{2c \sin \ell}$$

FRAME 24

Examples involving the use of Fourier Series

We shall now consider the solution of equation (12.1)

$$\text{i.e.} \quad \frac{\partial^2 y}{\partial x^2} + \frac{\partial^2 y}{\partial t^2} = 0$$

which is valid for $0 \leqq x \leqq \pi$ for the following boundary conditions:

(a) $y = 0$ when $x = 0$ } for all t

(b) $y = 0$ when $x = \pi$ }

(c) $y \to 0$ when $t \to \infty$

(d) $y = 1$ when $t = 0$ for $0 < x < \pi$

Separation of the variables (see FRAMES 12 and 13) yields

$$\frac{d^2X}{dx^2} - \lambda X = 0 \quad \text{and} \quad \frac{d^2T}{dt^2} + \lambda T = 0$$

Which set of solutions in 13A is compatible with condition (c)?

Does this condition give you any information about the constants in the set you choose?

****

24A

Set (ii) with $A_2 = 0$ (assuming k positive).

Then $T = B_2 e^{-kt}$ which $\to 0$ as $t \to \infty$.

FRAME 25

Thus $y = e^{-kt}(A \cos kx + B \sin kx)$

Now apply condition (a).

****

25A

$A = 0$

FRAME 26

We now have

$$y = Be^{-kt} \sin kx \qquad (26.1)$$

Applying condition (b) gives $0 = B \sin k\pi$.

What can you conclude about k?

****

26A

k must be an integer. ($B = 0$ *would give the solution* $y = 0$, *which would contradict condition (d).*)

FRAME 27

The only condition not used so far is (d). Substituting it in (26.1) gives

$$1 = B \sin kx \quad \text{for all } x \text{ between } 0 \text{ and } \pi.$$

This is impossible as B and k are constants, so a solution of the form (26.1) cannot be made to satisfy condition (d). To overcome this difficulty we proceed as follows.

It is known from 26A that k must be an integer, and so

$$y = B_1e^{-t} \sin x, \quad y = B_2e^{-2t} \sin 2x, \quad y = B_3e^{-3t} \sin 3x, \quad \text{etc.}$$

are all solutions of $\frac{\partial^2 y}{\partial x^2} + \frac{\partial^2 y}{\partial t^2} = 0$ which satisfy conditions (a), (b) and (c).

A more general solution would be

$$y = B_1e^{-t} \sin x + B_2e^{-2t} \sin 2x + B_3e^{-3t} \sin 3x + \ldots \qquad (27.1)$$

Verify that this solution satisfies conditions (a), (b) and (c).

FRAME 28

Applying condition (d) to this solution gives

$$1 = B_1 \sin x + B_2 \sin 2x + B_3 \sin 3x + \dots \quad \text{for } 0 < x < \pi$$

We now have a feasible solution if the R.H.S. of this equation is the half-range Fourier sine series for 1,

$$\text{i.e. if } B_n = \frac{2}{\pi}\int_0^{\pi} 1 \,.\, \sin nx \, dx$$

$$= \frac{2}{\pi}\left[-\frac{1}{n}\cos nx\right]_0^{\pi}$$

$$= \frac{2}{n\pi}(1 - \cos n\pi)$$

$$= \begin{cases} 0 & \text{if } n \text{ is even} \\ \dfrac{4}{n\pi} & \text{if } n \text{ is odd} \end{cases}$$

(A detailed treatment of Fourier Series is given in the first programme of this Unit.)

The solution (27.1) then becomes

$$y = \frac{4}{\pi}(e^{-t} \sin x + \frac{1}{3} e^{-3t} \sin 3x + \frac{1}{5} e^{-5t} \sin 5x + \dots)$$

FRAME 29

A stretched string of length ℓ, fixed at both ends, is set oscillating by displacing the mid-point a distance a perpendicular to the length of the string and releasing it from rest. The differential equation of the motion, derived in FRAME 2, is

$$\frac{\partial^2 y}{\partial t^2} = c^2 \frac{\partial^2 y}{\partial x^2}$$

First of all, write down the four boundary conditions which the solution has to satisfy.

****

29A

(i) $y = 0$ *when* $x = 0$ *for all* t

(ii) $y = 0$ *when* $x = \ell$ *for all* t

(iii) $\dfrac{\partial y}{\partial t} = 0$ *when* $t = 0$ *for all* x

(iv) $y = 2ax/\ell$ *from* $x = 0$ *to* $x = \frac{1}{2}\ell$
$y = (2a/\ell)(\ell - x)$ *from* $x = \frac{1}{2}\ell$ *to* $x = \ell$ } *when* $t = 0$

as the shape of the string when released is

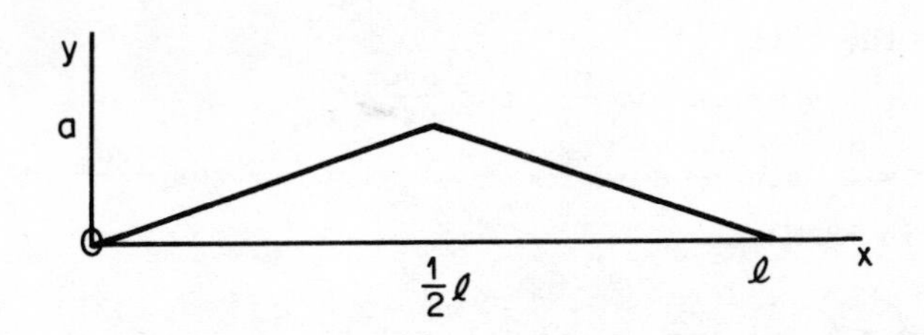

FRAME 30

Separation of the variables (see FRAMES 16 and 17) gives

$$\frac{d^2T}{dt^2} - \lambda T = 0 \quad \text{and} \quad \frac{d^2X}{dx^2} - \frac{\lambda}{c^2} X = 0$$

What must be the sign of λ?

****

30A

Negative.

For the string to oscillate, the solution must be trigonometric in t.
Also, condition (iv) suggests that a Fourier series will be needed and this implies that the solution must be trigonometric in x.

FRAME 31

Putting $\lambda = -k^2$ and using conditions (i) and (iii), derive the form of solution $y = A \sin \dfrac{kx}{c} \cos kt$, and show that condition (ii) requires $k = n\pi c/\ell$, where n is an integer.

****

31A

The derivation of the solution $y = A \sin \frac{kx}{c} \cos kt$ *was given in FRAMES 18 - 21*

$y = 0$ *when* $x = \ell$ *gives* $0 = A \sin \frac{k\ell}{c} \cos kt$

$\therefore \frac{k\ell}{c}$ *must be a multiple of* π

i.e. $\frac{k\ell}{c} = n\pi$ *where n is an integer.*

FRAME 32

The solution is now of the form

$$y = A \sin \frac{n\pi x}{\ell} \cos \frac{n\pi ct}{\ell}$$

or, more generally, $y = A_1 \sin \frac{\pi x}{\ell} \cos \frac{\pi ct}{\ell} + A_2 \sin \frac{2\pi x}{\ell} \cos \frac{2\pi ct}{\ell} + \ldots$
using the ideas given in FRAME 27.

Putting $t = 0$ gives $y = A_1 \sin \frac{\pi x}{\ell} + A_2 \sin \frac{2\pi x}{\ell} + \ldots$
and this must represent the function defined in condition (iv) for the range $0 < x < \ell$.

Find the coefficients of the terms in the half-range Fourier sine series for this function and complete the solution.

****

32A

$$A_n = \frac{2}{\ell}\left\{ \int_0^{\frac{1}{2}\ell} \frac{2ax}{\ell} \sin \frac{n\pi x}{\ell}\, dx + \int_{\frac{1}{2}\ell}^{\ell} \frac{2a}{\ell}(\ell - x) \sin \frac{n\pi x}{\ell}\, dx \right\}$$

$$= \frac{4a}{\ell^2}\left\{ \left[-\frac{\ell x}{n\pi} \cos \frac{n\pi x}{\ell} + \frac{\ell^2}{n^2\pi^2} \sin \frac{n\pi x}{\ell} \right]_0^{\frac{1}{2}\ell} \right.$$

$$\left. + \left[-\frac{\ell(\ell - x)}{n\pi} \cos \frac{n\pi x}{\ell} - \frac{\ell^2}{n^2\pi^2} \sin \frac{n\pi x}{\ell} \right]_{\frac{1}{2}\ell}^{\ell} \right\}$$

$$= \begin{cases} 0 & \text{if } n \text{ is even} \\ \dfrac{8a}{\pi^2 n^2} & \text{if } n = 1, 5, 9 \text{ etc.} \\ -\dfrac{8a}{\pi^2 n^2} & \text{if } n = 3, 7, 11 \text{ etc.} \end{cases}$$

$$y = \frac{8a}{\pi^2}\left[\sin \frac{\pi x}{\ell} \cos \frac{\pi ct}{\ell} - \frac{1}{3^2} \sin \frac{3\pi x}{\ell} \cos \frac{3\pi ct}{\ell} + \frac{1}{5^2} \sin \frac{5\pi x}{\ell} \cos \frac{5\pi ct}{\ell} \ldots \right]$$

FRAME 33

Miscellaneous Examples

In this frame a collection of miscellaneous examples is given for you to try. Answers are supplied in FRAME 34, together with such working as is considered helpful.

1. Find a solution of $\frac{\partial^2 y}{\partial x \partial t} = \sin t$ which is such that $y = 2 + t$ when $x = 0$, and $y = 2$ when $t = 0$.

2. A metal rod AB, length ℓ, is placed with A at $(0,0)$ and B at $(\ell,0)$. The temperature θ satisfies the equation

$$\frac{\partial \theta}{\partial t} = k \frac{\partial^2 \theta}{\partial x^2} \quad \text{where } k > 0.$$

At time t the end A has temperature $\theta_o e^{-\omega t}$ $(\omega > 0)$ while B is always kept at zero temperature. Assuming a solution of the type $\theta = Xe^{-\omega t}$, where X is a function of x only, show that the temperature at the mid-point of the rod is

$$\tfrac{1}{2}\theta_o e^{-\omega t} \sec \tfrac{1}{2}\sqrt{\frac{\omega}{k}}\,\ell$$

NOTE: Sometimes the form of the solution is partly known or given. In this example $\theta = Xe^{-\omega t}$ is taken as the trial solution rather than the more general $\theta = XT$. Only one ordinary differential equation (that for X in this case) then arises.

3. If ϕ is the temperature at any point in a plane the equation for steady heat flow when expressed in polar coordinates is

$$\frac{\partial^2 \phi}{\partial r^2} + \frac{1}{r}\frac{\partial \phi}{\partial r} + \frac{1}{r^2}\frac{\partial^2 \phi}{\partial \theta^2} = 0$$

Find, by the method of separation of variables, a solution to this equation, the solution to involve θ trigonometrically. If $\phi = 0$ when $\theta = 0$ for all values of r and $\phi \to 0$ as $r \to \infty$ for all values of θ, show that the solution reduces to the form

$$\phi = Cr^{-n} \sin n\theta$$

(HINT: In the ordinary differential equation for R in terms of r, either use the trial solution $R = ar^m$, where m is to be determined, or substitute $r = e^u$.)

FRAME 33 continued

4. A rod of length ℓ radiates heat according to the law

$$\frac{\partial \phi}{\partial t} = K \frac{\partial^2 \phi}{\partial x^2} - h\phi$$

h and K being positive constants. By means first of the substitution $\phi = e^{-ht} W$, where W is a function of x and t, and then separation of variables, find a solution of this equation given the conditions

(a) $\phi = 0$ when $x = 0$,

(b) $\phi = 0$ when $x = \ell$,

(c) $\phi \to 0$ as $t \to \infty$.

5. The equation of motion for small transverse horizontal oscillations of a thin rod is

$$m \frac{\partial^2 y}{\partial t^2} + EI \frac{\partial^4 y}{\partial x^4} = 0$$

Find a solution of this equation if it is known to be of the form $y = XT$ and T involves real trigonometric functions only. If the rod is of length ℓ and is clamped horizontally at both ends, show that $\cos n\ell \cosh n\ell = 1$ where $n^4 = mp^2/EI$ and $2\pi/p$ is the period of the oscillation.

6. Find a solution of $\frac{\partial^2 y}{\partial x^2} + c^2 \frac{\partial^2 y}{\partial t^2} = 0$ that satisfies the conditions:

(a) $y = 0$ when $x = 0$

(b) $y \to 0$ when $t \to \infty$

(c) $y = a \sin 2cx + 2a \sin cx$ when $t = 0$.

7. For an insulated rod of constant cross-sectional area, the heat conduction equation is $\frac{\partial \theta}{\partial t} = h^2 \frac{\partial^2 \theta}{\partial x^2}$. A rod of length ℓ has one end (at $x = 0$) kept at 0°C and the other (at $x = \ell$) is kept at 100°C until steady state conditions prevail. The temperature of the hot end is suddenly reduced to zero. Find the expression for the temperature θ as a function of x and t, measuring t from the time when the end at $x = \ell$ has its temperature changed.

FRAME 33 continued

8. In the steady flow of heat in the x,y plane, the temperature θ satisfies $\frac{\partial^2\theta}{\partial x^2} + \frac{\partial^2\theta}{\partial y^2} = 0$. A square plate has its edges along the lines $x = 0$, $x = \pi$, $y = 0$, $y = \pi$. The edges along $x = 0$ and $x = \pi$ are kept at zero temperature. The edge $y = 0$ is insulated and the edge $y = \pi$ is kept at $\theta = 100$. Show that

$$\theta = \frac{400}{\pi}\left(\frac{\sin x \cosh y}{\cosh \pi} + \frac{\sin 3x \cosh 3y}{3 \cosh 3\pi} + \frac{\sin 5x \cosh 5y}{5 \cosh 5\pi} + \ldots\right)$$

9. The potential v at a distance x along a certain cable of length ℓ satisfies the equation $\frac{\partial^2 v}{\partial x^2} = rc\,\frac{\partial v}{\partial t}$ where r is the resistance/unit length and c the capacitance/unit length, both r and c being constant. The cable is at a uniform potential V throughout its length but at $t = 0$ both ends are suddenly reduced to zero potential. Find

   (i) v as a function of x and t

   (ii) the total electrostatic energy at any time in the cable in terms of V, R, C and t where R and C are its total resistance and capacitance.
   (The E.S. energy at any time is $\int_0^\ell \frac{1}{2}cv^2\,dx$.)

10. A sheet of copper at 1080°C is suddenly immersed in a bath of oil at its boiling point, which is 180°C. The sheet is 2 cm thick. Assuming that the sheet is of large enough dimensions for the heat losses from the four thin edges to be ignored, the temperature θ at a distance x from one surface of the sheet after time t satisfies the partial differential equation

$$\frac{\partial\theta}{\partial t} = \alpha\,\frac{\partial^2\theta}{\partial x^2}$$

where α is the thermal diffusivity (for copper this is 0·88 cm^2/s). Calculate the temperature at the centre of the sheet after 2 seconds.
(HINT: Substitute $\theta = \psi + 180$ in the differential equation and then find a solution for ψ to fit the boundary conditions.)

FRAME 34

Answers to Miscellaneous Examples

1. $y = -x \cos t + F(t) + g(x)$

$y = 2 + t$ when $x = 0$ $\therefore 2 + t = F(t) + g(0)$

$y = 2$ when $t = 0$ $\therefore 2 = -x + F(0) + g(x)$

$$\text{i.e. } F(t) + g(0) = 2 + t$$
$$F(0) + g(x) = 2 + x$$

One pair of functions is $F(t) = 2 + t$, $g(x) = x$.

$\therefore y = -x \cos t + 2 + t + x$.

More generally we could write $F(t) = c + t$ (c a constant) and $g(x) = x + 2 - c$. The same y would result.

2. Let $\theta = Xe^{-\omega t}$, then

$$k\frac{d^2X}{dx^2} + \omega X = 0$$

$$\therefore \theta = e^{-\omega t}\left(A \cos \sqrt{\frac{\omega}{k}}\, x + B \sin \sqrt{\frac{\omega}{k}}\, x\right)$$

$$= e^{-\omega t}(A \cos px + B \sin px) \quad \text{if we write p for } \sqrt{\frac{\omega}{k}}$$

When $x = 0$, $\theta = \theta_o e^{-\omega t}$

$\therefore \theta_o e^{-\omega t} = e^{-\omega t} A$ or $A = \theta_o$

$$\theta = e^{-\omega t}(\theta_o \cos px + B \sin px)$$

When $x = \ell$, $\theta = 0$

$\therefore \theta_o \cos p\ell + B \sin p\ell = 0$ or $B = -\theta_o \cot p\ell$

$$\therefore \theta = \theta_o e^{-\omega t}(\cos px - \cot p\ell \sin px)$$
$$= \theta_o e^{-\omega t} \sin p(\ell - x) \operatorname{cosec} p\ell$$

When $x = \frac{1}{2}\ell$, $\theta = \theta_o e^{-\omega t} \sin \frac{1}{2}p\ell \operatorname{cosec} p\ell$

$$= \tfrac{1}{2}\theta_o e^{-\omega t} \sec \tfrac{1}{2}p\ell$$
$$= \tfrac{1}{2}\theta_o e^{-\omega t} \sec \tfrac{1}{2}\sqrt{\frac{\omega}{k}}\,\ell$$

3. Let $\phi = RT$ (here, T is being used for a function of θ only),

then

$$r^2 \frac{d^2R}{dr^2} + r\frac{dR}{dr} - \lambda R = 0 \quad \text{and} \quad \frac{d^2T}{d\theta^2} + \lambda T = 0$$

As θ is trigonometric, put $\lambda = k^2$, then

$$r^2 \frac{d^2R}{dr^2} + r\frac{dR}{dr} - k^2 R = 0 \quad \text{and} \quad \frac{d^2T}{d\theta^2} + k^2 T = 0$$

The latter gives $T = A\cos k\theta + B\sin k\theta$.

The former gives either

(i) $m^2 - k^2 = 0$, when the trial solution is used, leading to

$m = \pm k$

and $R = A_1 r^k + B_1 r^{-k}$

or

(ii) $\frac{d^2R}{du^2} - k^2 R = 0$ on putting $r = e^u$.

$$R = A_1 e^{ku} + B_1 e^{-ku}$$

$$= A_1 r^k + B_1 r^{-k}$$

Hence $\phi = (A_1 r^k + B_1 r^{-k})(A\cos k\theta + B\sin k\theta)$

$\phi = 0$ when $\theta = 0$ requires $A = 0$

$$\phi = (A_2 r^k + B_2 r^{-k})\sin k\theta$$

$\phi \to 0$ as $r \to \infty$ requires $A_2 = 0$

$$\phi = B_2 r^{-k}\sin k\theta$$

which, if $k = n$ and $B_2 = C$ is of the form

$$\phi = Cr^{-n}\sin n\theta.$$

FRAME 34 continued

4. If $\phi = e^{-ht}W$, the given equation becomes

$$\frac{\partial W}{\partial t} = K\frac{\partial^2 W}{\partial x^2}$$

Let $W = XT$, then

$$\frac{dT}{dt} - \lambda KT = 0, \quad \frac{d^2X}{dx^2} - \lambda X = 0$$

Put $\lambda = -n^2$. If you have tried $\lambda = n^2$, you will have found that it is impossible to satisfy both conditions (a) and (b).

$$\phi = e^{-(h+n^2K)t}(A\cos nx + B\sin nx)$$

(c) is already satisfied

(a) gives $A = 0$

$$\phi = Be^{-(h+n^2K)t}\sin nx$$

(b) gives $n\ell = k\pi$ where k is an integer

$$\phi = Be^{-(h+k^2\pi^2K/\ell^2)t}\sin\frac{k\pi x}{\ell}$$

Insufficient information is given in this example for B and k to be found

An alternative method of solution is to do a direct separation of variables on the original equation without first getting the simpler W equation.

5. Let $y = XT$, then

$$\frac{d^2T}{dt^2} - \lambda T = 0, \quad \frac{d^4X}{dx^4} + k^4\lambda X = 0 \quad \text{where} \quad k^4 = \frac{m}{EI}$$

Put $\lambda = -c^4$ as T is to be trigonometric, then

$$\frac{d^2T}{dt^2} + c^4T = 0, \quad \frac{d^4X}{dx^4} - k^4c^4X = 0$$

The former gives $T = A_1\cos c^2t + B_1\sin c^2t$ and the latter has auxiliary equation $\mu^4 - k^4c^4 = 0$ if the trial solution $X = Ae^{\mu x}$ is used.

FRAME 34 continued

i.e. $\mu = kc, \ -kc, \ ikc$ and $-ikc$

giving $X = Ae^{kcx} + Be^{-kcx} + C\cos kcx + E\sin kcx$

or $X = A\cosh kcx + B\sinh kcx + C\cos kcx + E\sin kcx$

which is probably the more useful form here in view of what we are asked to prove.

$$y = (A_1\cos c^2t + B_1\sin c^2t)(A\cosh kcx + B\sinh kcx + C\cos kcx + E\sin kcx)$$

The conditions are (a) when $x = 0, \ y = 0$

(b) when $x = \ell, \ y = 0$

(c) when $x = 0, \ \dfrac{\partial y}{\partial x} = 0$

(d) when $x = \ell, \ \dfrac{\partial y}{\partial x} = 0$

(a) gives $A + C = 0$.

(b) gives $A\cosh kc\ell + B\sinh kc\ell + C\cos kc\ell + E\sin kc\ell = 0$

$$\frac{\partial y}{\partial x} = (A_1\cos c^2t + B_1\sin c^2t)kc(A\sinh kcx + B\cosh kcx - C\sin kcx + E\cos kcx)$$

and so (c) gives $B + E = 0$.

(d) gives $A\sinh kc\ell + B\cosh kc\ell - C\sin kc\ell + E\cos kc\ell = 0$.

Substituting for C and E in the equations given by (b) and (d),

$$A(\cosh kc\ell - \cos kc\ell) + B(\sinh kc\ell - \sin kc\ell) = 0$$
$$A(\sinh kc\ell + \sin kc\ell) + B(\cosh kc\ell - \cos kc\ell) = 0$$

The condition for these to have a non-zero solution for A and B is

$$\begin{vmatrix} \cosh kc\ell - \cos kc\ell & \sinh kc\ell - \sin kc\ell \\ \sinh kc\ell + \sin kc\ell & \cosh kc\ell - \cos kc\ell \end{vmatrix} = 0^*$$

or $(\cosh kc\ell - \cos kc\ell)^2 - (\sinh^2 kc\ell - \sin^2 kc\ell) = 0$

or $\cosh kc\ell \cos kc\ell = 1$

Let $kc = n$ then $\cosh n\ell \cos n\ell = 1$

and $\dfrac{EI}{m} = \dfrac{c^4}{n^4}$ or $n^4 = \dfrac{mc^4}{EI} = \dfrac{mp^2}{EI}$ if $c^2 = p$.

The time period of the oscillations is $2\pi/c^2$, i.e. $2\pi/p$.

* If you are unfamiliar with this result, solve each of the preceding equations for A/B and equate the two values.

FRAME 34 continued

6. Let $y = XT$, then

$$\frac{d^2X}{dx^2} - \lambda X = 0 \quad \text{and} \quad c^2\frac{d^2T}{dt^2} + \lambda T = 0$$

Put $\lambda = -k^2$ (X must be trigonometric)

then $y = (A\cos kx + B\sin kx)(A_1e^{kt/c} + B_1e^{-kt/c})$

Condition (a) requires $A = 0$, then

$$y = (A_2e^{kt/c} + B_2e^{-kt/c})\sin kx$$

Condition (b) requires $A_2 = 0$, thus

$$y = B_2e^{-kt/c}\sin kx$$

To satisfy (c) it is necessary to use two values of k. If $k = c$, a solution is $y = B_2e^{-t}\sin cx$ and if $k = 2c$, another solution is $y = B_3e^{-2t}\sin 2cx$.

Hence $y = B_2e^{-t}\sin cx + B_3e^{-2t}\sin 2cx$ is also a solution.

Now $y = a\sin 2cx + 2a\sin cx$ when $t = 0$

$\therefore\ a\sin 2cx + 2a\sin cx \equiv B_2\sin cx + B_3\sin 2cx$

giving $B_2 = 2a$, $B_3 = a$.

Hence $y = 2ae^{-t}\sin cx + ae^{-2t}\sin 2cx$.

You will notice that this example is intermediate between that in FRAME 16 where one value of k is required and that in FRAME 24 where it must take an infinite number of values.

FRAME 34 continued

7. Let $\theta = XT$, then

$$\frac{dT}{dt} - \lambda h^2 T = 0, \qquad \frac{d^2X}{dx^2} - \lambda X = 0$$

Put $\lambda = -n^2$ (+ n^2 would make $\theta \to \infty$ as $t \to \infty$)

then $\theta = e^{-h^2n^2t}(A \cos nx + B \sin nx)$

When $x = 0$, $\theta = 0$ $\therefore A = 0$

$$\theta = Be^{-h^2n^2t} \sin nx$$

When $x = \ell$, $\theta = 0$ $\therefore n\ell = k\pi$ where k is an integer

$$\theta = Be^{-h^2k^2\pi^2t/\ell^2} \sin \frac{k\pi x}{\ell}$$

When $t = 0$, $\theta = 100x/\ell$ which requires the more general solution

$$\theta = b_1 e^{-h^2\pi^2t/\ell^2} \sin \frac{\pi x}{\ell} + b_2 e^{-4h^2\pi^2t/\ell^2} \sin \frac{2\pi x}{\ell} + b_3 e^{-9h^2\pi^2t/\ell^2} \sin \frac{3\pi x}{\ell} + \ldots$$

Then when $t = 0$, $\theta = b_1 \sin \frac{\pi x}{\ell} + b_2 \sin \frac{2\pi x}{\ell} + b_3 \sin \frac{3\pi x}{\ell} + \ldots$

For the required distribution

$$b_k = \frac{2}{\ell}\int_0^{\ell} \frac{100x}{\ell} \sin \frac{k\pi x}{\ell}\, dx$$

$$= \begin{cases} \dfrac{200}{k\pi} & \text{when k is odd} \\ -\dfrac{200}{k\pi} & \text{when k is even} \end{cases}$$

$$\theta = \frac{200}{\pi}\left(e^{-h^2\pi^2t/\ell^2} \sin \frac{\pi x}{\ell} - \frac{1}{2} e^{-4h^2\pi^2t/\ell^2} \sin \frac{2\pi x}{\ell} + \frac{1}{3} e^{-9h^2\pi^2t/\ell^2} \sin \frac{3\pi x}{\ell} - \frac{1}{4} e^{-16h^2\pi^2t/\ell^2} \sin \frac{4\pi x}{\ell} + \ldots\right)$$

FRAME 34 continued

8. Let $\theta = XY$, then

$$\frac{d^2X}{dx^2} - \lambda X = 0, \qquad \frac{d^2Y}{dy^2} + \lambda Y = 0$$

Put $\lambda = -k^2$ (When $y = \pi$, $\theta = 100$. Insertion of this condition will leave a function of x. This will probably have to be a Fourier Series, thus requiring trigonometric terms. $\therefore$ λ is expected to be negative.)

$$\theta = (A \cos kx + B \sin kx)(A_1 \cosh ky + B_1 \sinh ky)$$

$$\theta = 0 \text{ when } x = 0 \quad \therefore \quad A = 0$$

so $\theta = \sin kx \,(A_2 \cosh ky + B_2 \sinh ky)$

$$\frac{\partial\theta}{\partial y} = k \sin kx \,(A_2 \sinh ky + B_2 \cosh ky)$$

$\frac{\partial\theta}{\partial y} = 0$ when $y = 0$ $\therefore$ $B_2 = 0$ (This is the mathematical interpretation of the edge $y = 0$ being insulated.)

Now $\theta = A_2 \sin kx \cosh ky$.

$\theta = 0$ when $x = \pi$ $\therefore$ $\sin k\pi = 0$ i.e. k is an integer

$\theta = 100$ when $y = \pi$ cannot be satisfied by the present solution so take the more general form

$$\theta = a_1 \sin x \cosh y + a_2 \sin 2x \cosh 2y + a_3 \sin 3x \cosh 3y + \dots$$

$\theta = 100$ when $y = \pi$ now gives

$$100 = a_1 \sin x \cosh \pi + a_2 \sin 2x \cosh 2\pi + a_3 \sin 3x \cosh 3\pi + \dots$$

$$= c_1 \sin x + c_2 \sin 2x + c_3 \sin 3x + \dots$$

where $c_k = a_k \cosh k\pi$

This can be satisfied by a half-range Fourier sine series for 100 $(0 < x < \pi)$

Then $c_k = \frac{2}{\pi}\int_0^\pi 100 \sin kx \, dx$

$$= \begin{cases} 0 & \text{if k is even} \\ \frac{400}{\pi k} & \text{if k is odd} \end{cases}$$

$$a_k = \begin{cases} 0 & \text{if k is even} \\ \frac{400}{\pi k \cosh k\pi} & \text{if k is odd} \end{cases}$$

$$\theta = \frac{400}{\pi}\left(\frac{\sin x \cosh y}{\cosh \pi} + \frac{\sin 3x \cosh 3y}{3 \cosh 3\pi} + \frac{\sin 5x \cosh 5y}{5 \cosh 5\pi} + \dots\right)$$

FRAME 34 continued

9. Let $v = XT$, then if $rc = p^2$

$$\frac{d^2X}{dx^2} - \lambda h^2 X = 0, \qquad \frac{dT}{dt} - \lambda T = 0$$

Put $\lambda = -n^2$ (+ n^2 would make $v \to \infty$ as $t \to \infty$)

$$v = e^{-n^2 t}(A \cos pnx + B \sin pnx)$$

When $x = 0$, $v = 0$ $\therefore A = 0$

$$v = Be^{-n^2 t} \sin pnx$$

When $x = \ell$, $v = 0$ $\therefore pn\ell = k\pi$ where k is an integer

$$n = \frac{k\pi}{p\ell}$$

$$v = Be^{-k^2\pi^2 t/p^2\ell^2} \sin \frac{k\pi x}{\ell}$$

or, more generally,

$$v = b_1 e^{-\pi^2 t/p^2\ell^2} \sin \frac{\pi x}{\ell} + b_2 e^{-4\pi^2 t/p^2\ell^2} \sin \frac{2\pi x}{\ell} + \ldots$$

When $t = 0$,

$$v = b_1 \sin \frac{\pi x}{\ell} + b_2 \sin \frac{2\pi x}{\ell} + \ldots$$

but $v = V$ when $t = 0$

$$b_k = \frac{2}{\ell}\int_0^\ell V \sin \frac{k\pi x}{\ell}\, dx$$

$$= \begin{cases} 0 & \text{when k is even} \\ \dfrac{4V}{k\pi} & \text{when k is odd} \end{cases}$$

$$v = \frac{4V}{\pi}\left(e^{-\pi^2 t/p^2\ell^2} \sin \frac{\pi x}{\ell} + \frac{1}{3} e^{-9\pi^2 t/p^2\ell^2} \sin \frac{3\pi x}{\ell} + \frac{1}{5} e^{-25\pi^2 t/p^2\ell^2} \sin \frac{5\pi x}{\ell} + \ldots\right)$$

ES energy $= \int_0^\ell \frac{1}{2}cv^2\, dx$ and only the terms involving the square of a sine contribute to the integral.

The mth term in v is $\dfrac{4V}{(2m-1)\pi} e^{-(2m-1)^2\pi^2 t/p^2\ell^2} \sin \dfrac{(2m-1)\pi x}{\ell}$

FRAME 34 continued

The energy in the cable due to this term is

$$\int_0^{\ell} \frac{1}{2} c \frac{16V^2}{(2m-1)^2\pi^2} e^{-2(2m-1)^2\pi^2 t/p^2\ell^2} \sin^2 \frac{(2m-1)\pi x}{\ell} dx$$

$$= \frac{8cV^2}{(2m-1)^2\pi^2} e^{-2(2m-1)^2\pi^2 t/p^2\ell^2} \int_0^{\ell} \sin^2 \frac{(2m-1)\pi x}{\ell} dx$$

$$= \frac{8cV^2}{(2m-1)^2\pi^2} e^{-2(2m-1)^2\pi^2 t/p^2\ell^2} \frac{1}{2}\ell$$

$$= \frac{4CV^2}{(2m-1)^2\pi^2} e^{-2(2m-1)^2\pi^2 t/RC}$$

Total energy at this time $= \dfrac{4CV^2}{\pi^2} \displaystyle\sum_{m=1}^{\infty} \frac{1}{(2m-1)^2} e^{-2(2m-1)^2\pi^2 t/RC}$.

10. $$\frac{\partial\psi}{\partial t} = \alpha \frac{\partial^2\psi}{\partial x^2}$$

Let $\psi = XT$, then

$$\frac{dT}{dt} - \lambda\alpha T = 0 \quad \text{and} \quad \frac{d^2X}{dx^2} - \lambda X = 0$$

Put $\lambda = -k^2$ (ψ will not increase with time).

$$\psi = e^{-k^2\alpha t}(A \cos kx + B \sin kx)$$

$x = 0,\ \psi = 0 \quad \therefore\ A = 0$

$$\psi = Be^{-k^2\alpha t} \sin kx$$

$x = 2,\ \psi = 0 \quad \therefore\ \sin 2k = 0$

$k = n\pi/2$ where n is an integer

$$\psi = Be^{-n^2\pi^2\alpha t/4} \sin \frac{n\pi x}{2}$$

or, more generally,

$$\psi = B_1 e^{-\pi^2\alpha t/4} \sin \frac{\pi x}{2} + B_2 e^{-\pi^2\alpha t} \sin \pi x + B_3 e^{-9\pi^2\alpha t/4} \sin \frac{3\pi x}{2} + \ldots$$

For $t = 0,\ \psi = B_1 \sin \dfrac{\pi x}{2} + B_2 \sin \pi x + B_3 \sin \dfrac{3\pi x}{2} + \ldots$

At $t = 0,\ \psi = 900$

$$\therefore\ B_n = \int_0^2 900 \sin \frac{n\pi x}{2} dx$$

FRAME 34 continued

$$= \begin{cases} 0 & \text{if } n \text{ is even} \\ \dfrac{3600}{n\pi} & \text{if } n \text{ is odd} \end{cases}$$

$$\psi = \frac{3600}{\pi}\left(e^{-\pi^2\alpha t/4}\sin\frac{\pi x}{2} + e^{-9\pi^2\alpha t/4}\sin\frac{3\pi x}{2} + \ldots\right)$$

When $x = 1$, $t = 2$, $\alpha = 0{\cdot}88$

$$\psi = \frac{3600}{\pi}\, e^{-4\cdot 34} \quad + \text{negligible terms})$$

$$\simeq 15$$

The temperature at the centre of the sheet after 2 seconds = 195°C.

UNIT 10

LINE, SURFACE AND VOLUME INTEGRALS

A.C. Bajpai
I.M. Calus
J.A. Fairley

Loughborough University of Technology

INSTRUCTIONS

This Unit comprises four programmes:

(a) Three-dimensional Coordinate Geometry
(b) Line Integrals
(c) Surface Integrals
(d) Volume Integrals

Each programme is divided up into a number of FRAMES which are to be worked *in the order given*. You will be required to participate in many of these frames and in such cases the answers are provided in ANSWER FRAMES, designated by the letter A following the frame number. Steps in the working are given where this is considered helpful. The answer frame is separated from the main frame by a line of asterisks: *********. Keep the answers covered until you have written your own response. If your answer is wrong, go back and try to see why. Do not proceed to the next frame until you have corrected any mistakes in your attempt and are satisfied that you understand the contents up to this point.

Before reading these programmes, it is necessary that you are familiar with the following

Prerequisites

For (a): The contents of the Coordinate Systems programme, in Unit 1 of Vol I.

Vector Algebra, as covered in FRAMES 1-39, pages 3:39 to 3:64.

For (b): Integration, as covered in FRAMES 17-24, pages 2:11 to 2:17.

The contents of the Partial Differentiation programme in Unit 1.

Representation of points in three dimensions, as covered in FRAME 1 of (a).

For (c): Integration, as for (b).

For (d): Integration, as for (b).

The contents of (a).

CONTENTS

FRAMES		PAGE
	VOLUME INTEGRALS	

THREE-DIMENSIONAL COORDINATE GEOMETRY

FRAME 1

Cartesian Coordinates

In two dimensions a point P can be specified by giving the two distances x and y as shown in Fig (i).

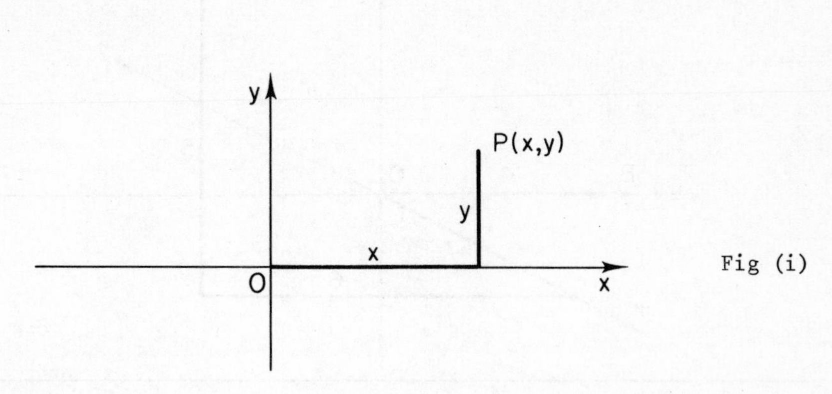

Fig (i)

Similarly, in three dimensions a point P can be specified by giving the three distances x, y and z as shown in Fig (ii), where Ox, Oy and Oz are mutually perpendicular.

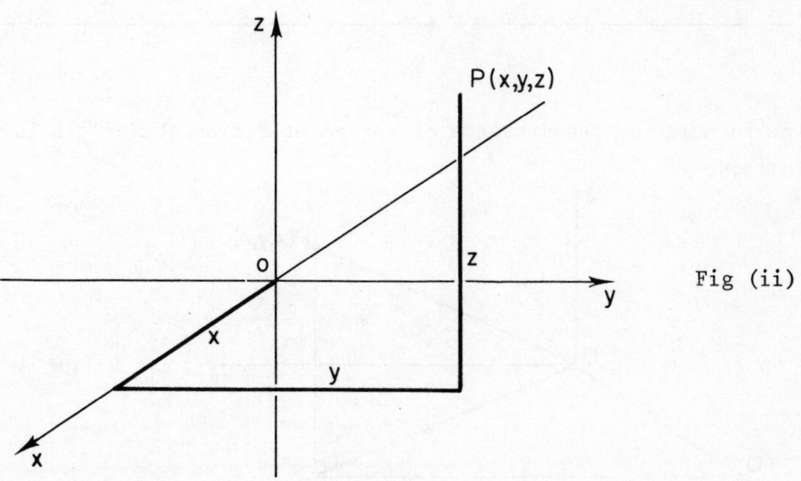

Fig (ii)

FRAME 1 continued

The positive directions for x,y,z are conventionally those indicated by the arrows. Thus, in Fig (iii), A is the point (7,6,5) and B is the point (0,-4,0).

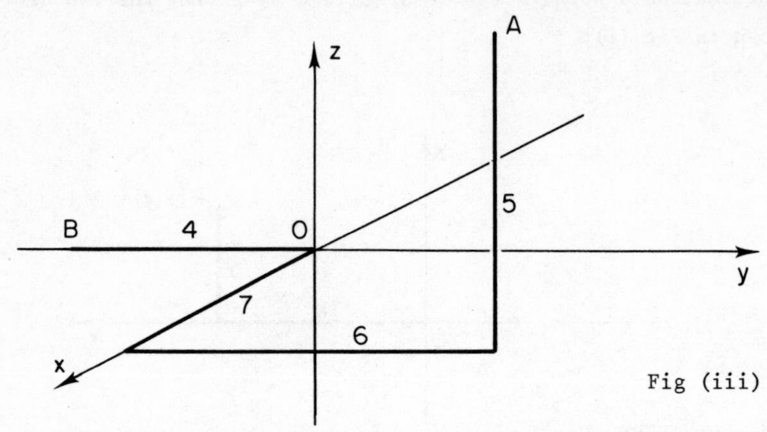

Fig (iii)

FRAME 2

You will see that the three axes form three planes, Oxy, Oyz and Ozx, called the coordinate planes. These planes divide space up into eight octants. That octant for which x,y,z are all positive is called the positive octant.

FRAME 3

The formula for the distance of any point P from the origin is found as follows:

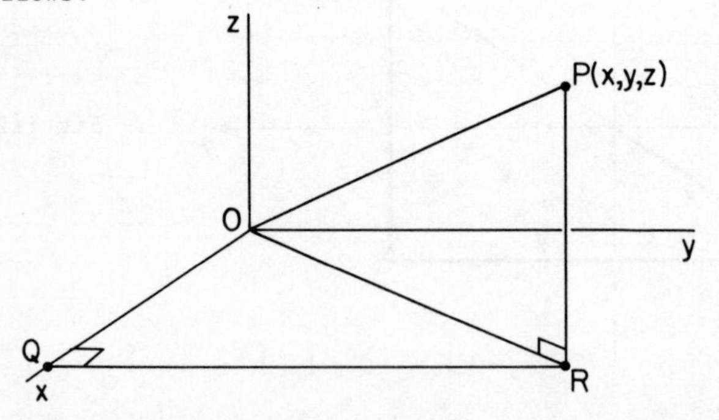

$$OP^2 = OR^2 + RP^2$$
$$= OQ^2 + QR^2 + RP^2$$
$$= x^2 + y^2 + z^2$$

$$OP = \sqrt{x^2 + y^2 + z^2}$$

What is the length of OA in Fig (iii) of FRAME 1?

****

3A

$\sqrt{110}$

FRAME 4

In the following diagram, P_1 is the point (x_1, y_1, z_1) and P_2 is (x_2, y_2, z_2). What are the distances P_1Q, QR, P_1R, RP_2 and P_1P_2?

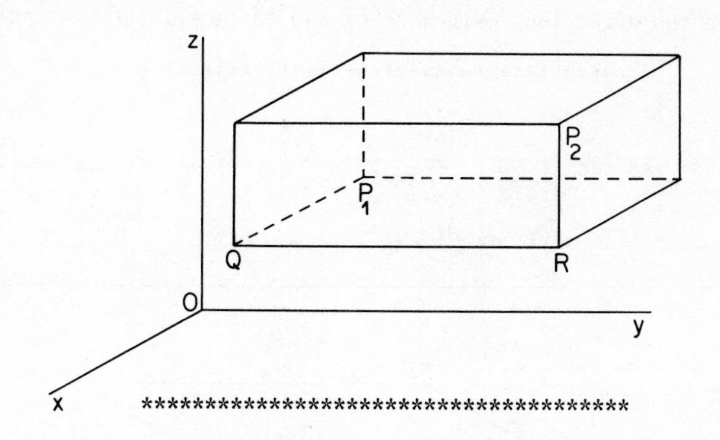

****

4A

$x_2 - x_1$, $y_2 - y_1$, $\sqrt{(x_2 - x_1)^2 + (y_2 - y_1)^2}$, $z_2 - z_1$, $\sqrt{(x_2 - x_1)^2 + (y_2 - y_1)^2 + (z_2 - z_1)^2}$

FRAME 5

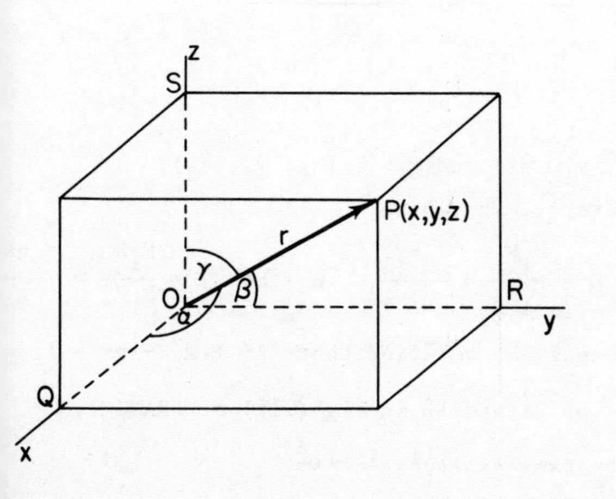

The direction of $\overrightarrow{OP}$ is determined by the three angles α, β, γ which it makes with Ox, Oy, Oz respectively.

$\cos\alpha$, $\cos\beta$, $\cos\gamma$ are known as the DIRECTION COSINES of $\overrightarrow{OP}$ and are usually denoted by ℓ, m, n respectively. Just as the coordinates of P are written

FRAME 5 continued

(x,y,z), so the direction cosines of $\overrightarrow{OP}$ are written $[\ell,m,n]$.

From the diagram you will see that

$$\cos\alpha = \frac{OQ}{OP} = \frac{x}{r} \qquad \cos\beta = \frac{OR}{OP} = \frac{y}{r} \qquad \cos\gamma = \frac{OS}{OP} = \frac{z}{r}$$

It follows that $\ell^2 + m^2 + n^2 = \frac{x^2}{r^2} + \frac{y^2}{r^2} + \frac{z^2}{r^2}$

$= 1$ (using result in FRAME 3).

Now write down the direction cosines of $\overrightarrow{OA}$ and $\overrightarrow{OB}$ in Fig (iii) of FRAME 1.

****

5A

$\overrightarrow{OA}$ *has direction cosines* $\left[\frac{7}{\sqrt{110}}, \frac{6}{\sqrt{110}}, \frac{5}{\sqrt{110}}\right]$, *often written as* $\left[\frac{7,6,5}{\sqrt{110}}\right]$

$\overrightarrow{OB}$ " " " $[0, -1, 0]$

FRAME 6

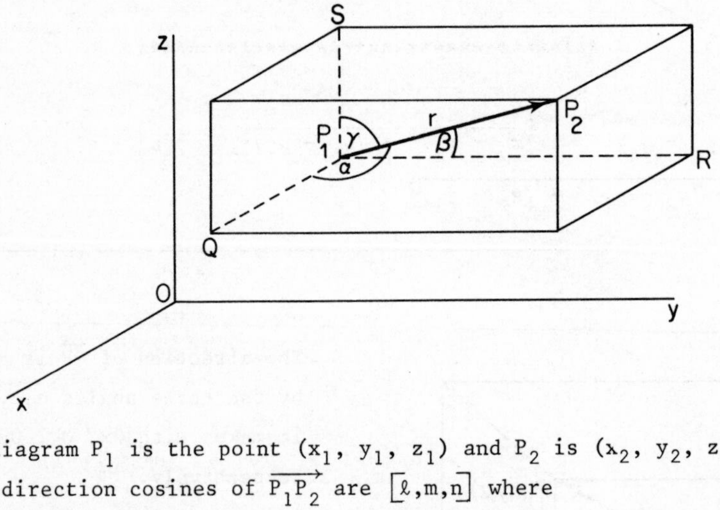

In this diagram P_1 is the point (x_1, y_1, z_1) and P_2 is (x_2, y_2, z_2).

Here the direction cosines of $\overrightarrow{P_1P_2}$ are $[\ell,m,n]$ where

$$\ell = \cos\alpha = \frac{P_1Q}{P_1P_2} = \frac{x_2-x_1}{r}, \quad m = \cos\beta = \frac{P_1R}{P_1P_2} = \frac{y_2-y_1}{r}, \quad n = \cos\gamma = \frac{P_1S}{P_1P_2} = \frac{z_2-z_1}{r}$$

Using the last result in 4A it can easily be verified that $\ell^2 + m^2 + n^2 = 1$.

Now write down the direction cosines of $\overrightarrow{BA}$ and $\overrightarrow{AB}$ in Fig (iii) of FRAME 1.

****

6A

$\overrightarrow{BA}$ $\left[\dfrac{7,10,5}{\sqrt{174}}\right]$

$\overrightarrow{AB}$ $\left[\dfrac{-7,-10,-5}{\sqrt{174}}\right]$

FRAME 7

The ratios $x_2-x_1 : y_2-y_1 : z_2-z_1$ are called the DIRECTION RATIOS of the line P_1P_2 of FRAME 6. Note that $k(x_2-x_1) : k(y_2-y_1) : k(z_2-z_1)$, where k is a constant (positive or negative), are also direction ratios of P_1P_2. Thus the direction ratios of both $\overrightarrow{AB}$ and $\overrightarrow{BA}$ in 6A can be taken as 7:10:5.

FRAME 8

In two dimensions, a single equation such as $y^2 = 4ax$ represents a curve. In three dimensions, a single equation such as $z = x^2 + y^2$ represents a surface. We shall have a look at the equations of a few of the common surfaces, beginning with the plane.

A plane is defined by three non-collinear points A,B,C, say. To find the equation of the plane through these points it is convenient to use a vector approach.

Let $\overrightarrow{OA} = \underline{a}$, $\overrightarrow{OB} = \underline{b}$, $\overrightarrow{OC} = \underline{c}$ and $\overrightarrow{OP} = \underline{r}$ where P is any point in the plane.

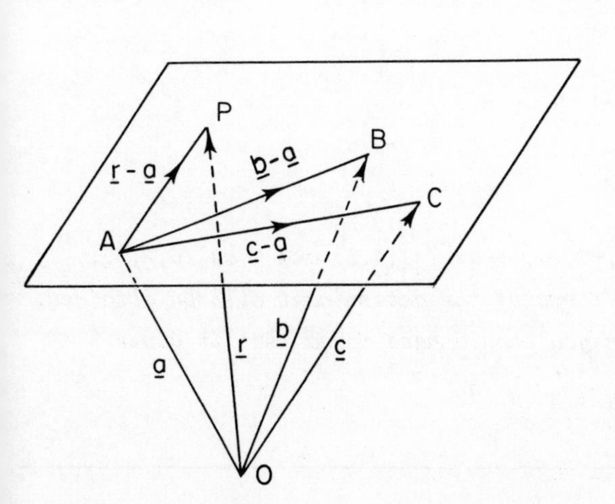

$(\underline{r}-\underline{a})_\wedge(\underline{b}-\underline{a})$ is a vector perpendicular to $\overrightarrow{AP}$ and $\overrightarrow{AB}$, and thus, as P is in the plane ABC, also perpendicular to $\overrightarrow{AC}$.

$\therefore\ (\underline{r}-\underline{a})_\wedge(\underline{b}-\underline{a}).(\underline{c}-\underline{a}) = 0$

Show that in determinant form this becomes

$$\begin{vmatrix} x-a_1 & y-a_2 & z-a_3 \\ b_1-a_1 & b_2-a_2 & b_3-a_3 \\ c_1-a_1 & c_2-a_2 & c_3-a_3 \end{vmatrix} = 0$$

FRAME 8 continued

where $\underline{r} = x\underline{i} + y\underline{j} + z\underline{k}$

$\underline{a} = a_1\underline{i} + a_2\underline{j} + a_3\underline{k}$, etc.

****

8A

$$(\underline{r}-\underline{a})\wedge(\underline{b}-\underline{a}).(\underline{c}-\underline{a}) = (\underline{c}-\underline{a}).(\underline{r}-\underline{a})\wedge(\underline{b}-\underline{a}) \quad \text{as} \quad \underline{p}.\underline{q} = \underline{q}.\underline{p}$$

$$= (\underline{r}-\underline{a}).(\underline{b}-\underline{a})\wedge(\underline{c}-\underline{a})$$

(see FRAME 39 of the Vector Algebra programme in Unit 3)

$$= \begin{vmatrix} x-a_1 & y-a_2 & z-a_3 \\ b_1-a_1 & b_2-a_2 & b_3-a_3 \\ c_1-a_1 & c_2-a_2 & c_3-a_3 \end{vmatrix}$$

(see FRAME 37 of the same Programme)

FRAME 9

Write down, in determinant form, the equation of the plane passing through the points (2,-1,3), (-1,4,3) and (0,2,5). Simplify the result by expanding the determinant.

****

9A

One possible determinant form is

$$\begin{vmatrix} x-2 & y+1 & z-3 \\ -3 & 5 & 0 \\ -2 & 3 & 2 \end{vmatrix} = 0$$

This is obtained by taking A as (2,-1,3), B as (-1,4,3) and C as (0,2,5). If the points are interchanged, other forms of the determinant will be obtained. Whichever one you have written down you should have found that it gives

$$10x + 6y + z = 17$$

FRAME 10

You will notice that the equation $10x + 6y + z = 17$ is linear in x, y and z, and this is generally the case with the equation of a plane. In special cases the equation may be of a simpler form. If the constant term is zero, as in $10x + 6y + z = 0$, through what point must all such planes pass?

****

10A

Through the origin.

FRAME 11

If one of the coefficients (of x, y or z) is zero, the plane is parallel to one of the coordinate axes. For example, as shown in the diagram, $2x + 3z = 6$ is parallel to Oy.

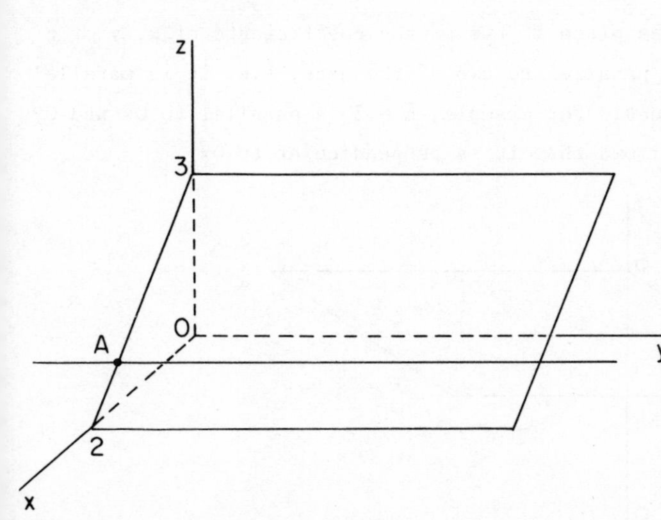

If, for instance, $x = \frac{3}{2}$ and $z = 1$, then P, a point in the plane, can lie anywhere along the line through $A\left(\frac{3}{2}, 0, 1\right)$ parallel to Oy.

A similar thing happens for all other pairs of values of x and z which satisfy $2x + 3z = 6$.

All these parallel lines go to make up the plane.

Note that for a plane parallel to Oy, it is the coefficient of y which is zero.

Can you state what is special about

(i) $z = \frac{1}{2}y + 1$,

(ii) $z = \frac{1}{2}y$?

****

11A

(i) It is parallel to Ox.

(ii) It passes through Ox, as shown in the diagram.

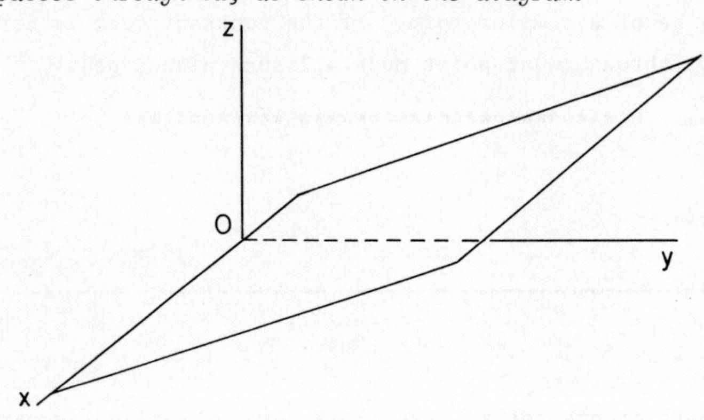

FRAME 12

A further simplification takes place if two of the coefficients of x, y or z are zero. The plane is then parallel to two of the axes, i.e. it is parallel to one of the coordinate planes. For example, z = 5 is parallel to Ox and Oy and therefore to Oxy. It follows that it is perpendicular to Oz.

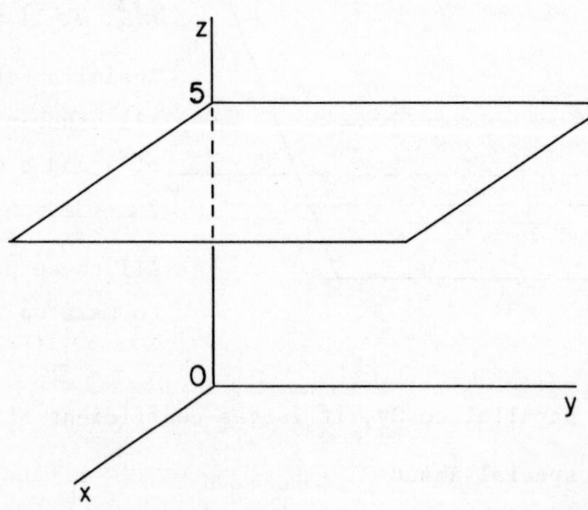

Can you describe the following planes:

(i) x = 3 (ii) y = -4 (iii) z = 0?

****

12A

(i) *It is parallel to Oyz (or perpendicular to Ox) and makes an intercept of 3 on the x-axis.*

(ii) *It is parallel to Ozx (or perpendicular to Oy) and makes an intercept of -4 on the y-axis.*

(iii) *It is the plane Oxy.*

FRAME 13

Turning next to the sphere, this is defined if its centre and radius are known. The equation takes the simplest form when the centre is at the origin.

If P (x,y,z) is any point on the surface of such a sphere, radius R, we must have

$$OP^2 = R^2$$

$$\text{i.e.} \quad x^2 + y^2 + z^2 = R^2$$

What would be the equation of a sphere whose centre is at (α,β,γ) and whose radius is R?

****

13A

$$(x - \alpha)^2 + (y - \beta)^2 + (z - \gamma)^2 = R^2$$

FRAME 14

The last surface we shall mention in Cartesian coordinates is the cylinder. The simplest equation occurs when its axis is along a coordinate axis, and we shall restrict ourselves to this case. For a circular cylinder, radius a, whose axis is along Oz, the equation is $x^2 + y^2 = a^2$. You will notice that, for any pair of values of x and y satisfying $x^2 + y^2 = a^2$, z can take any value, thus giving rise to a line on the surface which is parallel to Oz. Such lines are the generators of the cylinder.

What is the equation of a cylinder whose radius is 3 and whose axis is along Ox?

****

14A

$y^2 + z^2 = 9$

FRAME 15

The intersection of two surfaces is, in general, a curve. Thus, in three dimensions, it is necessary to give the equations of any two surfaces which pass through a curve, in order to define it. A straight line can be defined by the intersection of two planes. A circle could be defined by the intersection of a sphere and a plane, or by the intersection of a cylinder and a plane.

Now answer the following questions:

(i) Which of the following pairs of equations represent the same line?

(a) $\left.\begin{array}{l} z = \frac{1}{2}y \\ z = 0 \end{array}\right\}$ (b) $\left.\begin{array}{l} y = 0 \\ z = 0 \end{array}\right\}$ (c) $\left.\begin{array}{l} z = 0 \\ x = 0 \end{array}\right\}$

(ii) What curve is represented by

(a) $\left.\begin{array}{l} x^2 + y^2 = a^2 \\ z = 0 \end{array}\right\}$ (b) $\left.\begin{array}{l} x^2 + y^2 + z^2 = 9 \\ z = 2 \end{array}\right\}$?

****

15A

(i) (a) and (b) both represent Ox.

(ii) (a) Circle, radius a, centre O, in plane Oxy.

(b) Circle, radius $\sqrt{5}$, centre (0,0,2), in plane $z = 2$.

FRAME 16

Cylindrical Polar Coordinates

Just as in two dimensions polar coordinates are sometimes used instead of Cartesian, so in three dimensions other systems of coordinates are sometimes used. Of these, two common ones will be described here.

FRAME 16 continued

In the first, x and y are replaced by plane polar coordinates and z is left unaltered.

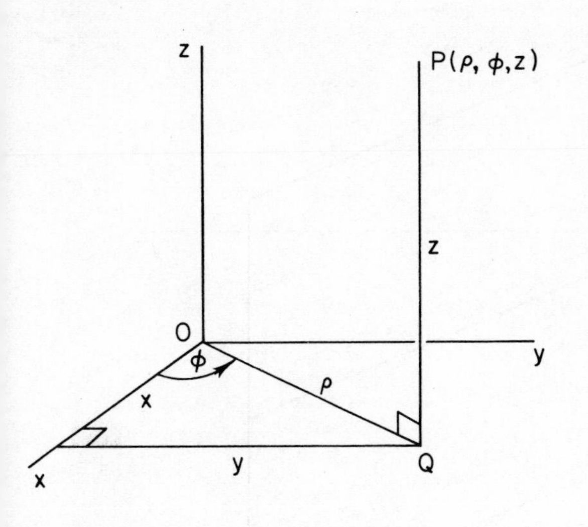

(ρ,ϕ,z) are the CYLINDRICAL POLAR COORDINATES of the point P. Obviously the relationships between these and Cartesian coordinates are as follows:

$$\left.\begin{aligned} x &= \rho \cos \phi \\ y &= \rho \sin \phi \\ z &= z \end{aligned}\right\}$$

$$\left.\begin{aligned} \rho &= \sqrt{x^2 + y^2} \\ \tan \phi &= y/x \\ z &= z \end{aligned}\right\}$$

Note that, in finding ϕ from $\tan \phi = y/x$, you must be careful to observe the quadrant in which Q lies.

Transform the following sets of Cartesian coordinates into cylindrical polars:

(i) $(1,\sqrt{3},2)$

(ii) $(-1,-1,-1)$

(iii) $(-2,0,4)$

****

16A

(i) $(2,\pi/3,2)$

(ii) $(\sqrt{2},5\pi/4,-1)$ *or* $(\sqrt{2},-3\pi/4,-1)$

(iii) $(2,\pi,4)$

FRAME 17

The equations of a few special surfaces will now be mentioned.

ϕ = const. is a plane extending in one direction from the z-axis, e.g. $\phi = \pi/3$

FRAME 17 continued

is the plane illustrated in the diagram.

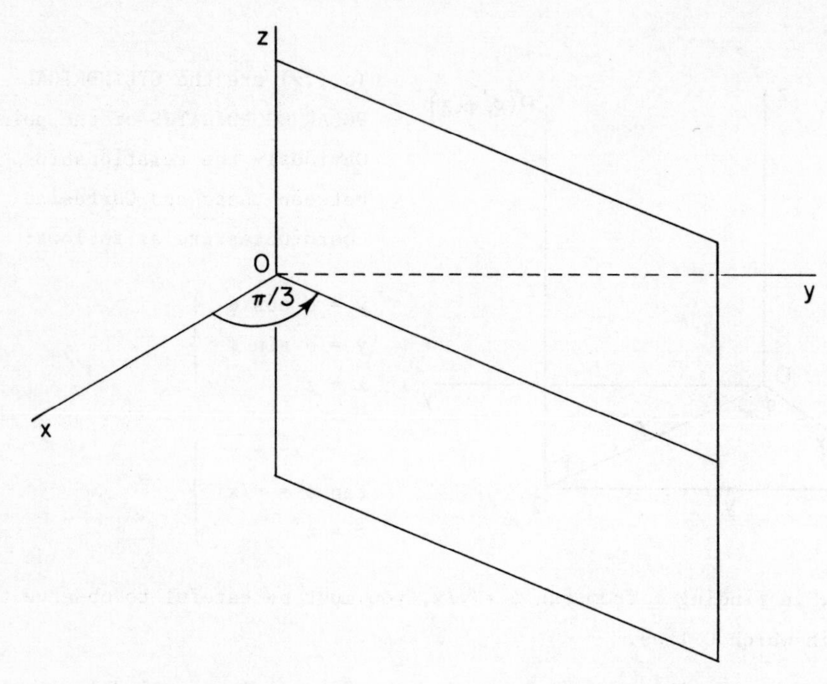

Can you describe the surfaces (i) $\rho = 3$, (ii) $z = -4$?

****

17A

(i) A cylinder with axis along Oz and radius 3.

(ii) A plane parallel to Oxy and 4 units below it.

FRAME 18

Spherical Polar Coordinates

The diagram illustrates the second system of polar coordinates that can be used.

(r,θ,ϕ) are the SPHERICAL POLAR COORDINATES of the point P. The relationships between these and Cartesian coordinates are as follows:

FRAME 18 continued

$$\left.\begin{aligned} x &= r \sin\theta \cos\phi \\ y &= r \sin\theta \sin\phi \\ z &= r \cos\theta \end{aligned}\right\}$$

$$\left.\begin{aligned} r &= \sqrt{x^2 + y^2 + z^2} \\ \tan\theta &= \sqrt{x^2 + y^2}/z \\ \tan\phi &= y/x \end{aligned}\right\}$$

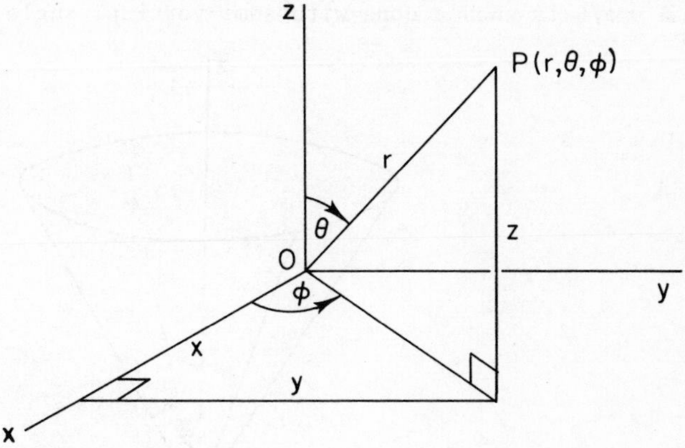

Once again, in finding ϕ it is important to note the quadrant in which Q lies. It is conventional to take θ such that $0 \leq \theta \leq \pi$.

Now transform the following sets of Cartesian coordinates into spherical polars:

(i) $(1, \sqrt{3}, 2)$

(ii) $(-1, -1, -\sqrt{2})$

(iii) $(0, 2, 0)$

****

18A

(i) $(2\sqrt{2}, \pi/4, \pi/3)$

(ii) $(2, 3\pi/4, 5\pi/4)$ *or* $(2, 3\pi/4, -3\pi/4)$

(iii) $(2, \pi/2, \pi/2)$

FRAME 19

The equations of a few special surfaces will now be mentioned.

ϕ = const. is a plane formed in the same way as for cylindrical coordinates.

θ = const. is a cone with vertex at O and axis along Oz, e.g.

$\theta = \pi/6$ is such a cone with semi-vertical angle $\pi/6$ and the vertex downwards.

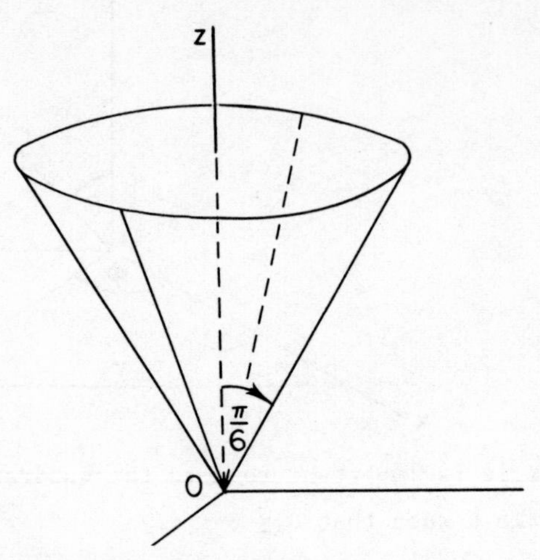

Can you describe the surfaces (i) $r = 5$ (ii) $\theta = 5\pi/6$?

****

19A

(i) *A sphere centre O, radius 5.*

(ii) *A cone with vertex upwards at O, axis along Oz' and semi-vertical angle $\pi/6$.*

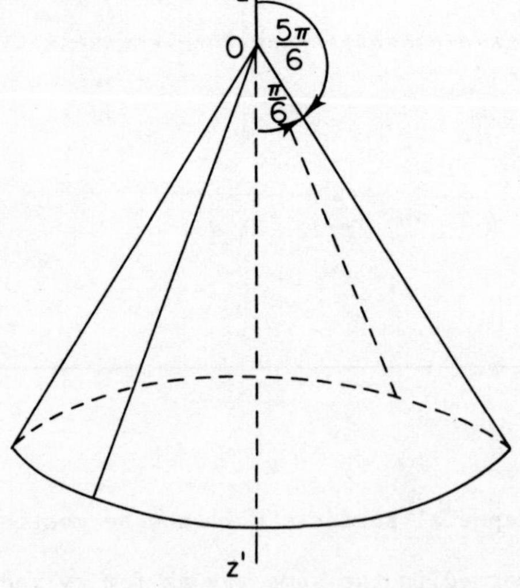

LINE INTEGRALS

FRAME 1

Introduction

Suppose a unit charge is moved along a curve in an electric field of force $\underline{F}$, where $\underline{F}$ represents at any point the strength and direction of the field. In such a case work is done either by or against the field. The magnitude of the work done can be found as follows:

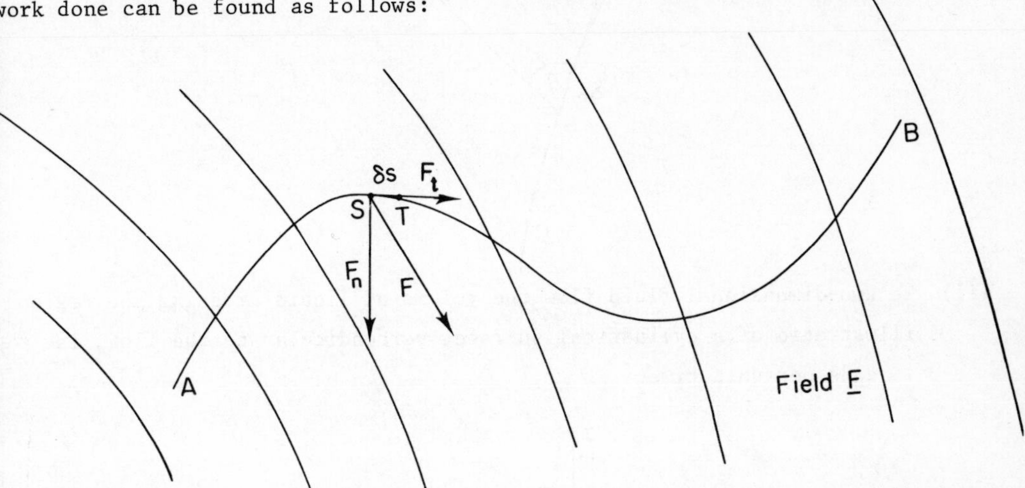

Let ST, length δs, be a small element of the path AB followed. The work done over ST is approximately $F_t \delta s$ where F_t is the component of $\underline{F}$ tangential to the path at S.

The total work done is the limit of the sum of all such quantities from A to B, as $\delta s \to 0$, i.e. $\lim\limits_{\delta s \to 0} \sum F_t \delta s$.

This is $\int F_t ds$ over the curve AB, which can be written more briefly as $\int_{AB} F_t ds$ or $\int_A^B F_t ds$. This is an example of a LINE INTEGRAL. The curve AB can be either two- or three-dimensional.

FRAME 2

Two other examples of line integrals which occur in practice are:

(i) If a current i flows in a plane wire, the magnetic force at a point U is given by $\int_C \frac{i \sin \theta}{r^2} ds$, C being the curve formed by the wire.

FRAME 2 continued

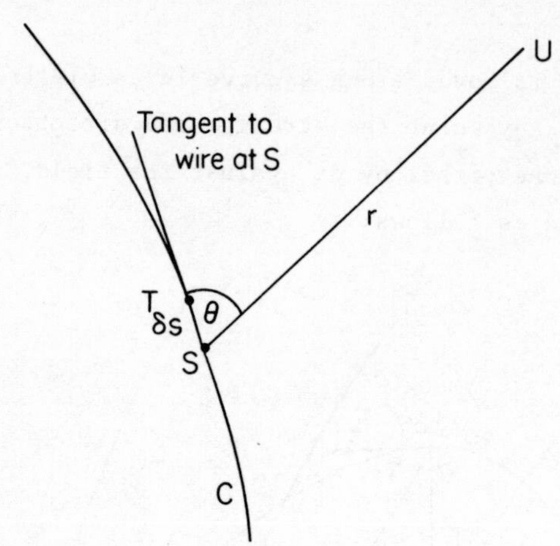

(ii) In two-dimensional fluid flow the volume of liquid crossing the region illustrated of a cylindrical surface, perpendicular to the flow, is $\int_{AB} q_n ds$ per unit time.

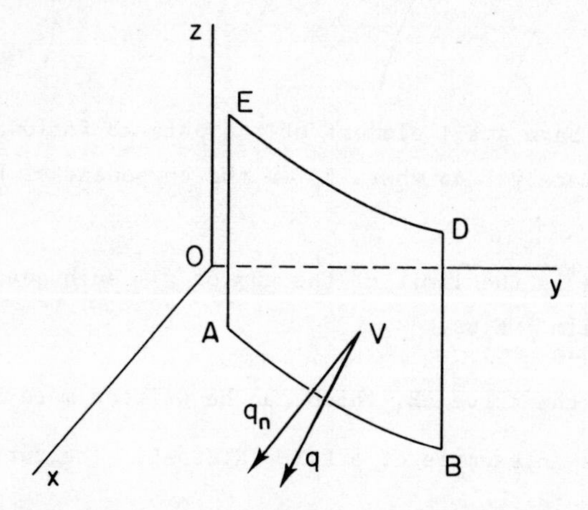

In the diagram, ABDE is part of a band of unit height (BD = AE = 1 unit) of the surface. The motion of the fluid is everywhere parallel to the plane Oxy. q is the velocity of the fluid at any point V and q_n is the component of q perpendicular to the surface at V. (The fact that the flow is two-dimensional means that q is the same at all points on the line through V parallel to Oz.)

FRAME 3

As you will have seen in the preceding examples, a line integral usually arises in the form $\int_C G\,ds$, where G is some function of position on the curve C. However, for evaluation, other forms are necessary.

Taking first the case of two-dimensional fluid flow just mentioned, let u and v be the velocity components parallel to the x- and y-axes respectively and let ST be a small element of the surface AB.

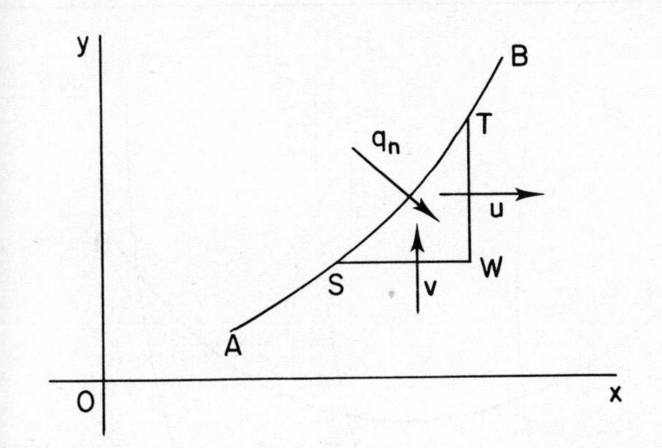

S is the point (x,y)

T is the point $(x + \delta x, y + \delta y)$

$$\text{Flow across ST} + \text{Flow across SW} = \text{Flow across TW}$$

$$\text{Flow across ST} = \text{Flow across TW} - \text{Flow across SW}$$

$$q_n \delta s = u\delta y - v\delta x$$

$$\therefore \int_{AB} q_n ds = \int_{AB} u dy - \int_{AB} v dx,$$

which is usually written as $\int_{AB} u dy - v dx$

Secondly, we consider the case where G is the F_t in the two-dimensional case of FRAME 1. If F_t has components P and Q in the x- and y-directions,

$$F_t \delta s = P\delta x + Q\delta y$$

where δx, δy are the components of δs.

$$\therefore \int_{AB} F_t ds = \int_{AB} P dx + Q dy$$

You will appreciate that P and Q are functions of x and y.

FRAME 3 continued

For the three-dimensional case, the corresponding result is

$$\int_C Pdx + Qdy + Rdz$$

where C is the path followed.

Alternatively, vector methods can be used to express $\int_C F_t ds$ in a form suitable for evaluation and in FRAME 4 this will be done for the three-dimensional case. If you are unfamiliar with vector algebra you can proceed directly to FRAME 5.

FRAME 4

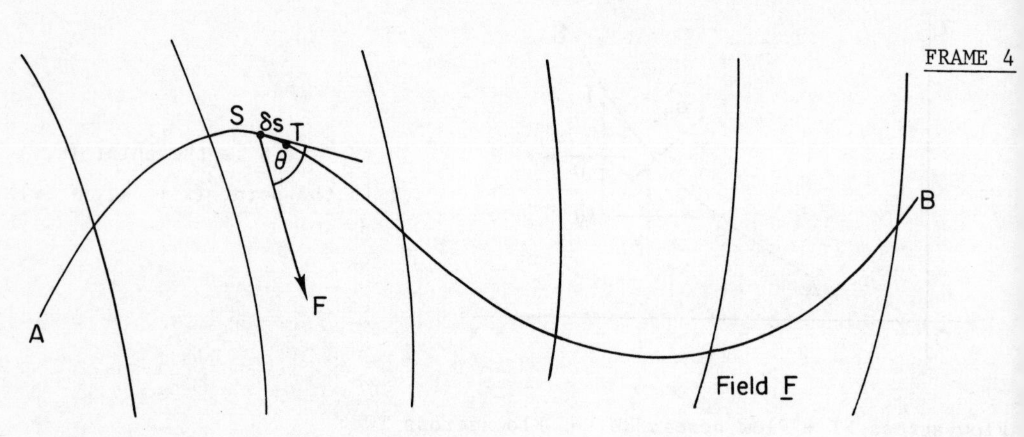

Here the work done is $\int_C F \cos \theta \, ds$ as $F_t = F \cos \theta$

$= \int_C \underline{F}.d\underline{s}$ which, if F_1, F_2, F_3 are the components of $\underline{F}$ parallel to the axes,

$$= \int_C (F_1\underline{i} + F_2\underline{j} + F_3\underline{k}).(dx\underline{i} + dy\underline{j} + dz\underline{k})$$

$$= \int_C F_1dx + F_2dy + F_3dz$$

FRAME 5

Evaluation of Line Integrals in Two Dimensions

We shall now consider the evaluation of integrals of the form

$$\int_C Pdx + Qdy$$

This is best illustrated by an example.

FRAME 5 continued

We shall consider the evaluation of $\int y^2dx + xy\,dy$ from the origin O to the point A(1,1) along each of the following paths:

(i) a straight line

(ii) the parabola $y^2 = x$

(iii) part of the x-axis and the line x = 1.

(i) The equation of OA is y = x.

Hence dy = dx.

Using these relationships

$$\int_{OA} y^2dx + xy\,dy = \int_0^1 x^2dx + x^2dx$$

$$= \int_0^1 2x^2dx$$

$$= \frac{2}{3}$$

Now try part (ii) for yourself.

****

5A

$$y^2 = x$$

$$2y\,dy = dx$$

$$\int_{OPA} y^2dx + xy\,dy = \int_0^1 x\,dx + x\left(\frac{1}{2}\,dx\right)$$

$$= \int_0^1 \frac{3}{2}\,x\,dx$$

$$= \frac{3}{4}$$

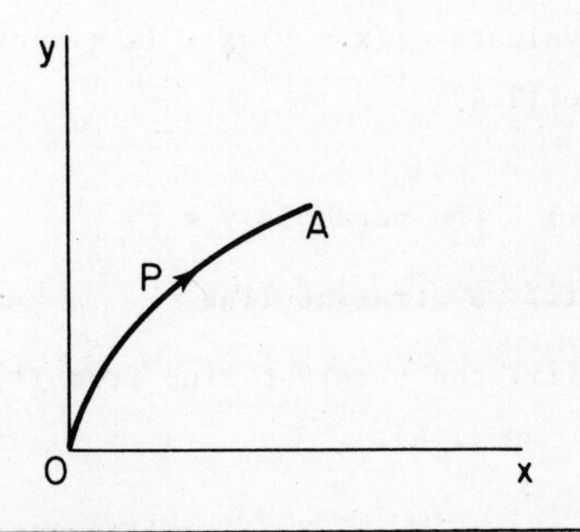

FRAME 6

In (iii) it is necessary to consider the path in two parts OB and BA.

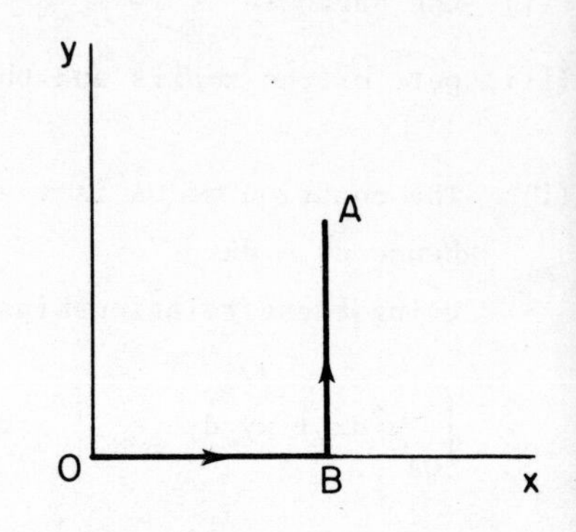

Along OB, $y = 0$ and $dy = 0$

$$\therefore \int_{OB} y^2dx + xy\,dy = 0$$

Along BA, $x = 1$ and $dx = 0$

$$\therefore \int_{BA} y^2dx + xy\,dy = \int_0^1 y\,dy$$

$$= \frac{1}{2}$$

Thus $$\int_{OBA} y^2dx + xy\,dy = 0 + \frac{1}{2}$$

$$= \frac{1}{2}$$

FRAME 7

You will notice that the result depends on the path of integration as well as the end points.

Now try this example.

Evaluate $\int (x - y)dx + (x + y)dy$ along each of the following paths from (1,1) to (2,4):

(i) the parabola $y = x^2$

(ii) a straight line

(iii) the straight line from (1,1) to (1,4) followed by that from (1,4) to (2,4).

****

7A

(i) $\displaystyle\int_{(1,1)}^{(2,4)} (x - y)dx + (x + y)dy$

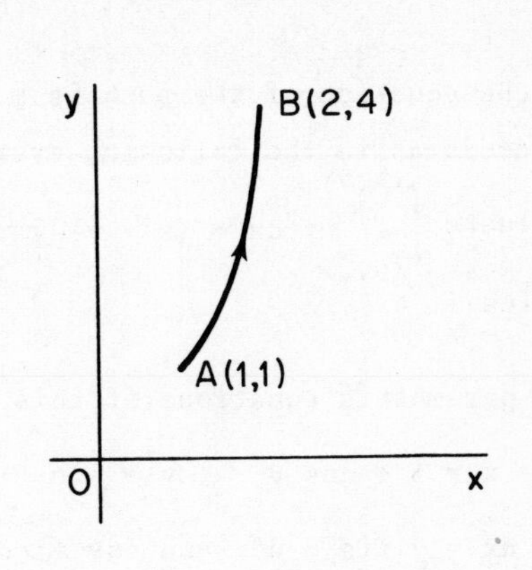

$$= \int_1^2 (x - x^2)dx + (x + x^2)2xdx$$

$$= \int_1^2 (x + x^2 + 2x^3)dx$$

$$= \frac{34}{3}$$

(ii) The equation of the line is $y = 3x - 2$

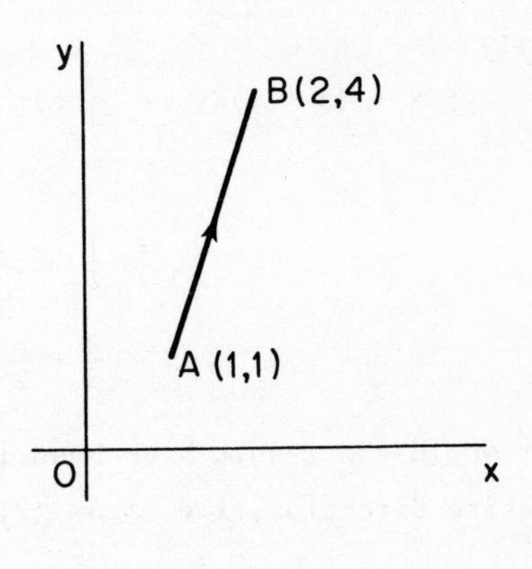

$$\int_{(1,1)}^{(2,4)} (x - y)dx + (x + y)dy$$

$$= \int_1^2 (x - 3x + 2)dx + (x + 3x - 2)3dx$$

$$= \int_1^2 (10x - 4)dx$$

$$= 11$$

(iii) $\displaystyle\int_{(1,1)}^{(2,4)} (x - y)dx + (x + y)dy$

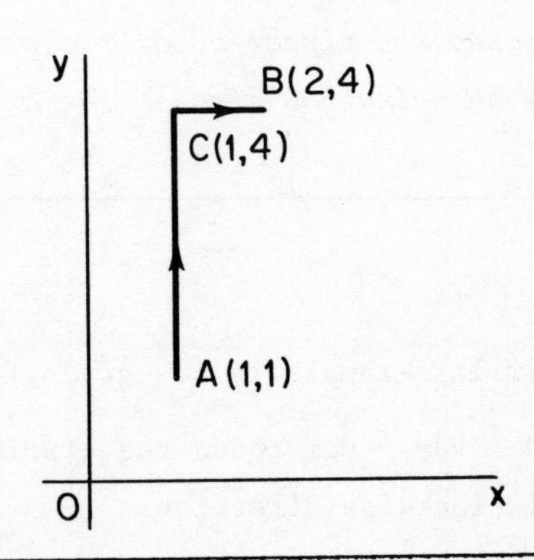

$$= \int_1^4 (1 + y)dy \quad + \quad \int_1^2 (x - 4)dx$$

$$= 10\tfrac{1}{2} - 2\tfrac{1}{2}$$

$$= 8$$

FRAME 8

If the equation of the path is more conveniently given in parametric form, one proceeds as in the following example:

Evaluate $\int_{(0,0)}^{(2,0)} y^2dx + (x + y)dy$ along the upper half of a circle centre (1,0), radius 1.

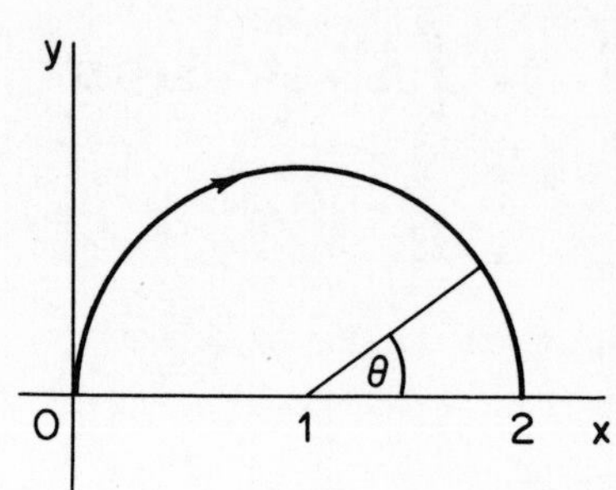

The parametric equations of this circle are

$$x = 1 + \cos\theta \qquad y = \sin\theta$$

$\therefore\ dx = -\sin\theta\, d\theta$ and $dy = \cos\theta\, d\theta$

The point (0,0) corresponds to $\theta = \pi$ and (2,0) to $\theta = 0$.

$$\int_{(0,0)}^{(2,0)} y^2dx + (x + y)dy = \int_{\pi}^{0} \sin^2\theta(-\sin\theta\, d\theta) + (1 + \cos\theta + \sin\theta)\cos\theta\, d\theta$$

$$= \int_{\pi}^{0} (-\sin^3\theta + \cos\theta + \cos^2\theta + \sin\theta\cos\theta)d\theta$$

$$= \frac{4}{3} - \frac{\pi}{2}$$

What would the result have been if the same path had been followed, but in the opposite direction, i.e. from (2,0) to (0,0)?

****

8A

Changing the direction of the path interchanges the limits of the integral, thus changing the sign of the answer to give $\frac{\pi}{2} - \frac{4}{3}$.

FRAME 9

A similar example is now given for you to try.

Find $\int 3xdy - ydx$ round the circle with centre the origin and radius 4, in the anti-clockwise direction.

****

9A

The parametric equations are

$x = 4\cos\theta, \qquad y = 4\sin\theta$

$$\oint 3x\,dy - y\,dx = \int_0^{2\pi} 12\cos\theta(4\cos\theta)d\theta - 4\sin\theta(-4\sin\theta)d\theta$$

$$= \int_0^{2\pi} (48\cos^2\theta + 16\sin^2\theta)d\theta$$

$$= 64\pi$$

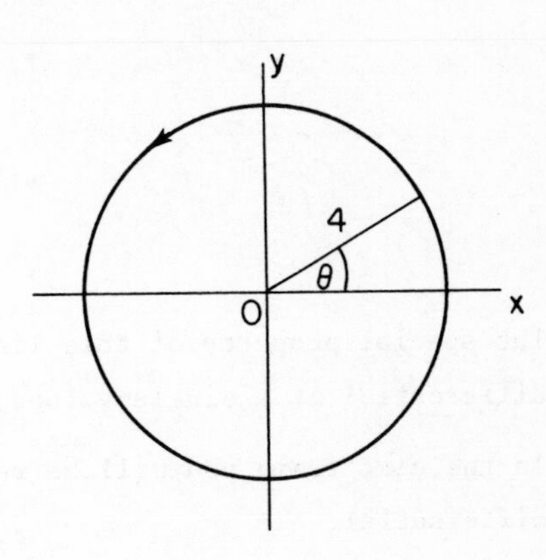

The symbol $\oint$ *is used in line integrals to denote that the integral is taken round a closed curve.*

If you started at a point other than $\theta = 0$ *you should still get the same answer.*

FRAME 10

Dependence on Path of Integration

For each of the line integrals evaluated in FRAMES 5-7 it was noticed that the value depended upon the path of integration. However, we now suggest that you evaluate

$$\int_{(0,1)}^{(1,3)} y^2dx + 2xy\,dy$$

along any path you care to choose and we can be quite confident that when you turn to the answer frame you will find the correct answer to your particular problem.

****

10A

If you didn't get 9 for your answer, you should check your working.
In the next frame we shall let you into the secret of how we were able to predict this answer.

FRAME 11

You will see that if $\phi = xy^2$ then, as $d\phi = \frac{\partial\phi}{\partial x}dx + \frac{\partial\phi}{\partial y}dy$,
in this case $d\phi = y^2dx + 2xy\,dy$

$$\text{so that} \int_{(0,1)}^{(1,3)} y^2dx + 2xy\,dy = \int_{(0,1)}^{(1,3)} d\phi$$

$$= \Big[\phi\Big]_{(0,1)}^{(1,3)}$$

$$= \Big[xy^2\Big]_{(0,1)}^{(1,3)}$$

$$= 9$$

The special property of this line integral is that $y^2dx + 2xy\,dy$ is an exact differential of a single-valued function, i.e. it is $d\phi$ where $\phi = xy^2$.

In the next frame you will be reminded of how to recognise an exact differential.

FRAME 12

The condition for Pdx and Qdy to be equal to $d\phi$ is that

$$P = \frac{\partial\phi}{\partial x} \quad \text{and} \quad Q = \frac{\partial\phi}{\partial y}$$

$$\text{so that} \quad \frac{\partial P}{\partial y} = \frac{\partial^2\phi}{\partial y\partial x} \quad \text{and} \quad \frac{\partial Q}{\partial x} = \frac{\partial^2\phi}{\partial x\partial y}$$

$$\text{and therefore} \quad \frac{\partial P}{\partial y} = \frac{\partial Q}{\partial x} \quad \text{as} \quad \frac{\partial^2\phi}{\partial y\partial x} = \frac{\partial^2\phi}{\partial x\partial y}.$$

In the example of FRAME 10, $P = y^2$ and $Q = 2xy$

$$\frac{\partial P}{\partial y} = 2y \quad \text{and} \quad \frac{\partial Q}{\partial x} = 2y$$

$$\therefore \frac{\partial P}{\partial y} = \frac{\partial Q}{\partial x}$$

Which of the following are exact differentials?

(i) $(y^2 + xy)dx - (x^2 - xy)dy$

(ii) $(2xy - \sin y)dx + (x^2 + 3y^2 - x\cos y)dy$

(iii) $(2x - e^y)dx + (2 - xe^y)dy$

****

12A

(i) is not, (ii) and (iii) are.

FRAME 13

We shall now show how to evaluate $\int (2xy - \sin y)dx + (x^2 + 3y^2 - x\cos y)dy$ along any path from (1,0) to (2,π). You have just shown that the differential is exact, so there must be a function ϕ such that

$$\frac{\partial\phi}{\partial x} = 2xy - \sin y \quad \text{and} \quad \frac{\partial\phi}{\partial y} = x^2 + 3y^2 - x\cos y$$

Integrating, these two equations give respectively

$$\phi = x^2y - x\sin y + f(y)$$

$$\text{and} \quad \phi = x^2y + y^3 - x\sin y + g(x)$$

Comparing these, you will see that these two requirements are met by

$$f(y) = y^3, \qquad g(x) = 0.$$

Thus $\phi = x^2y - x\sin y + y^3$, which is a single-valued function.

(If you need a more detailed treatment of the solution for ϕ, see FRAMES 5-9 in "Partial Differential Equations for Technologists", the second programme in Unit 9.)

$$\therefore \int_{(1,0)}^{(2,\pi)} (2xy - \sin y)dx + (x^2 + 3y^2 - x\cos y)dy = \Big[x^2y - x\sin y + y^3\Big]_{(1,0)}^{(2,\pi)}$$

$$= 4\pi + \pi^3$$

Now find $\int (2x - e^y)dx + (2 - xe^y)dy$ along any path from (0,-2) to (3,0).

****

13A

$\phi = x^2 - xe^y + 2y$

The value of the integral is 10.

FRAME 14

If, in the type of integral considered in the last four frames, the end points coincide (i.e. the path is a closed curve), what would be the value of the integral?

****

14A

0

NOTE: This only occurs for this particular type of integral. You have already seen an example in FRAME 9 where the integral round a closed path is not zero.

FRAME 15

You have now seen that if Pdx + Qdy is an exact differential of a single-valued function,

(i) $\int_{AB}$ Pdx + Qdy is independent of the path followed between A and B

and (ii) $\oint$ Pdx + Qdy = 0 for any path.

These results have physical implications, in that if P and Q are the components of a force the work done in moving between two points against such a force is independent of the path followed. A field of force which has this property is said to be CONSERVATIVE.

FRAME 16

You may have been wondering about the condition 'a single-valued function' which has been specified in FRAMES 10-15. The functions occurring in this Unit do satisfy this condition, so we shall not consider here why it is necessary. An explanation of what happens when the condition is not satisfied can be found in "Advanced Mathematics for Technical Students - Part II" by H.V. Lowry and H.A. Hayden (Longmans) on page 298.

FRAME 17

Evaluation of Line Integrals in Three Dimensions

The method of evaluation in this case is very similar to that for two dimensions. We shall illustrate it by examples.

Find $\int (3x - 2y)dx + (y + 2z)dy - x^2dz$ from (0,0,0) to (1,2,3) along

(i) the curve $x = t, \quad y = 2t^2, \quad z = 3t^3$

(ii) the straight lines from (0,0,0) to (0,2,0), then to (0,2,3) and finally to (1,2,3).

(i) From the parametric equations,

$dx = dt \qquad dy = 4tdt \qquad dz = 9t^2dt$

(0,0,0) corresponds to t = 0 and (1,2,3) to t = 1.

$\therefore$ the integral becomes $\displaystyle\int_0^1 (3t - 4t^2)dt + (2t^2 + 6t^3)4tdt - (t^2)9t^2dt$

$$= \int_0^1 (3t - 4t^2 + 8t^3 + 15t^4)dt$$

$$= 31/6$$

(ii) Along OA, $x = 0 \qquad z = 0$

$dx = 0 \qquad dz = 0$

$\therefore$ the integral becomes $\displaystyle\int_0^2 y\,dy = 2$

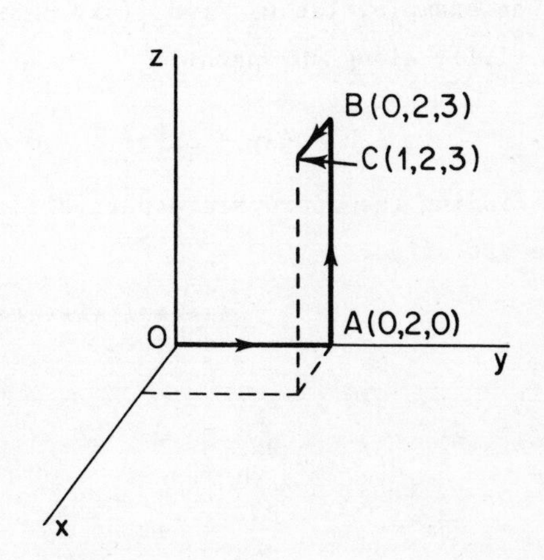

Along AB, $x = 0 \quad y = 2$

Now complete the solution.

****

17A

Along AB, $dx = 0 \quad dy = 0$

$\therefore$ *the integral becomes zero.*

Along BC, $y = 2 \quad z = 3$

$dy = 0 \quad dz = 0$

$\therefore$ *the integral becomes* $\int_0^1 (3x - 4)dx = -5/2$

$\therefore$ *the integral along the complete path* $= 2 + 0 - 5/2$

$= -\frac{1}{2}$

FRAME 18

In a manner similar to the two-dimensional case, the value of $\int Pdx + Qdy + Rdz$ between two points is independent of the path of integration if $Pdx + Qdy + Rdz$ is the exact differential of a single-valued function ϕ.

The condition for this is $P = \frac{\partial\phi}{\partial x} \qquad Q = \frac{\partial\phi}{\partial y} \qquad R = \frac{\partial\phi}{\partial z}$ giving

$$\frac{\partial P}{\partial y} = \frac{\partial Q}{\partial x}, \qquad \frac{\partial Q}{\partial z} = \frac{\partial R}{\partial y}, \qquad \frac{\partial R}{\partial x} = \frac{\partial P}{\partial z} \tag{18.1}$$

As an example, let us find $\int (4xy - 3x^2z^2)dx + 2x^2dy - 2x^3zdz$ from $(0,1,2)$ to $(3,-1,1)$, along any path.

$$P = 4xy - 3x^2z^2 \qquad Q = 2x^2 \qquad R = -2x^3z$$

By finding the appropriate partial derivatives, show that conditions (18.1) are satisfied.

****

18A

$\frac{\partial P}{\partial y} = 4x \qquad \frac{\partial Q}{\partial z} = 0 \qquad \frac{\partial R}{\partial x} = -6x^2z$

$\frac{\partial P}{\partial z} = -6x^2z \qquad \frac{\partial Q}{\partial x} = 4x \qquad \frac{\partial R}{\partial y} = 0$

FRAME 19

Then $\frac{\partial\phi}{\partial x} = 4xy - 3x^2z^2$ gives $\phi = 2x^2y - x^3z^2 + f(y,z)$

$\frac{\partial\phi}{\partial y} = 2x^2$ gives $\phi = 2x^2y + g(z,x)$

and $\frac{\partial\phi}{\partial z} = -2x^3z$ gives $\phi = -x^3z^2 + h(x,y)$

Comparing these you will see that these three requirements are met by

$f(y,z) = 0, \quad g(z,x) = -x^3z^2, \quad h(x,y) = 2x^2y$

Thus $\phi = 2x^2y - x^3z^2$, which is single-valued.

$$\therefore \int_{(0,1,2)}^{(3,-1,1)} (4xy - 3x^2z^2)dx + 2x^2dy - 2x^3z\,dz = \Big[2x^2y - x^3z^2\Big]_{(0,1,2)}^{(3,-1,1)}$$

$$= -45$$

FRAME 20

Miscellaneous Examples

In this frame a collection of miscellaneous examples is given for you to try. Answers are supplied in FRAME 21, together with such working as is considered helpful.

1. Evaluate $\oint 2ydx - xdy$ in the anti-clockwise direction round the parallelogram whose vertices are (0,0), (2,0), (3,1) and (1,1).

2. Evaluate $\oint 2xydx - x^2ydy$ in the anti-clockwise direction round the path ACBOA, where ACB is the straight line $x = 2$ and BOA is the parabola $y^2 = 8x$.

3. Find $\oint y^2dx + xdy$ in the anti-clockwise direction round the ellipse
$$\frac{x^2}{16} + \frac{y^2}{9} = 1$$

4. Show that $\int_{(0,2)}^{(1,5)} (e^x + y)dx + (x + \ln y)dy$ is independent of the path, and find its value.

FRAME 20 continued

5. Evaluate $\int_C (x^2 + y - z)dx + yzdy + (x - y)dz$ where C is

   (i) the arc of the curve $x = t + 2, \quad y = 2t - 1, \quad z = 2t^2 - t$ from (2,-1,0) to (3,1,1),

   (ii) the straight line from (2,-1,0) to (2,-1,1) followed by that to (3,-1,1) and finally by that to (3,1,1).

6. Show that

$$\int_{(-1,2,3)}^{(3,2,-1)} (3x^2 - 3yz + 2xz)dx + (3y^2 - 3xz + z^2)dy + (3z^2 - 3xy + x^2 + 2yz)dz$$

   is independent of the path of integration and evaluate the integral along any path.

7. In three-dimensional fluid flow the line integral over a path AB of the velocity vector is $\int_{AB} udx + vdy + wdz$, where u,v,w are the components of the velocity. Find this if $u = y$, $v = z$, $w = x$ and the path is the curve $x = t$, $y = t^2 - 1$, $z = 2t$ from $t = 0$ to $t = 2$. Velocities are in metres per second, lengths in metres and time t in seconds.

8. In fluid flow the line integral of the velocity vector round a closed path is called the circulation. Find the circulation about the square enclosed by the lines $x = \pm 1$, $y = \pm 1$ in the case of the two-dimensional flow given by $u = x + y$, $v = x^2 - y$. Velocities are in metres per second and lengths in metres.

Omit the following examples if you did not read FRAME 4.

9. If $\underline{F} = (x^2 - y^2)\underline{i} + 2xy\underline{j}$, find the work done by the force $\underline{F}$ in the displacement from (1,0) to (2,2) along the curve $y = x^2 - x$.

10. If $\underline{P} = y^2\underline{i} - x^2\underline{j} + xyz\underline{k}$, evaluate $\int_C \underline{P}.d\underline{s}$ where C is

    (i) the x-axis from -1 to +1

    (ii) the cubical parabola $z = x^3$, $y = 2$ from (0,2,0) to (1,2,1).

    Give a physical interpretation of these results.

FRAME 21

Answers to Miscellaneous Examples

1. Along OA, $y = 0$, $dy = 0$

$$\therefore \int_{OA} 2y\,dx - x\,dy = 0$$

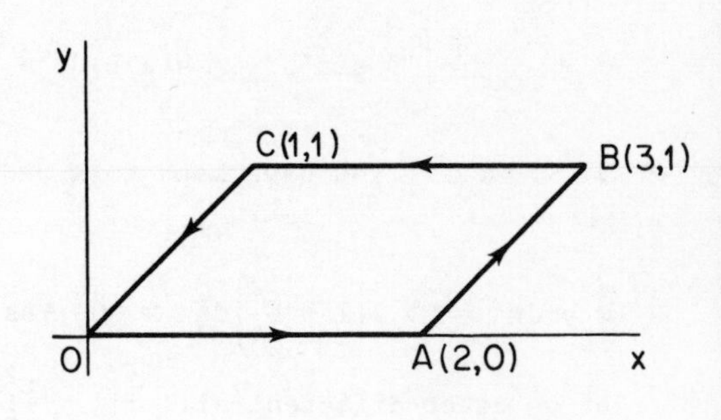

Along AB, $y = x - 2$, $dy = dx$

$$\therefore \int_{AB} = \int_2^3 2(x - 2)dx - x\,dx$$

$$= -\frac{3}{2}$$

Along BC, $y = 1$, $dy = 0$

$$\therefore \int_{BC} = \int_3^1 2dx$$

$$= -4$$

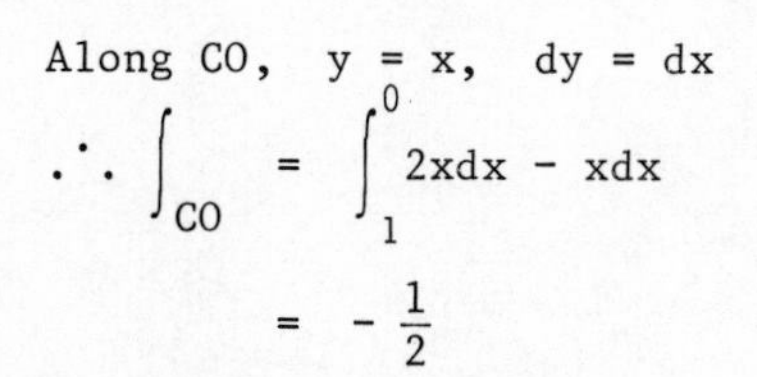

Along CO, $y = x$, $dy = dx$

$$\therefore \int_{CO} = \int_1^0 2x\,dx - x\,dx$$

$$= -\frac{1}{2}$$

$$\therefore \text{Total integral} = 0 - \frac{3}{2} - 4 - \frac{1}{2} = -6.$$

2. Along ACB, $x = 2$, $dx = 0$

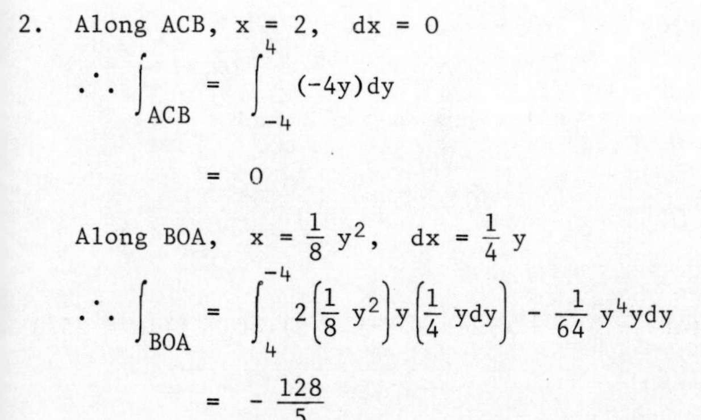

$$\therefore \int_{ACB} = \int_{-4}^{4} (-4y)dy$$

$$= 0$$

Along BOA, $x = \frac{1}{8}y^2$, $dx = \frac{1}{4}y$

$$\therefore \int_{BOA} = \int_4^{-4} 2\left(\frac{1}{8}y^2\right)y\left(\frac{1}{4}y\,dy\right) - \frac{1}{64}y^4 y\,dy$$

$$= -\frac{128}{5}$$

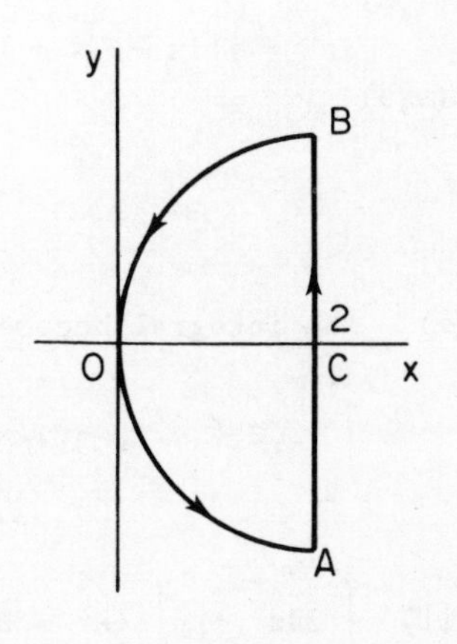

$$\therefore \text{Total integral} = 0 - \frac{128}{5} = -\frac{128}{5}.$$

FRAME 21 continued

3. The easiest solution uses parametric equations,

i.e., $x = 4\cos\theta$, $y = 3\sin\theta$ giving

$$\int_0^{2\pi} 9\sin^2\theta(-4\sin\theta)d\theta + 4\cos\theta(3\cos\theta)d\theta = 12\pi$$

If, however, you have kept to x and y, a possible solution is:

$\oint y^2dx = \oint 9\left(1 - \frac{x^2}{16}\right)dx = 0$ as this is the integral round a closed pat

of an exact differential. $\left[\left(1 - \frac{x^2}{16}\right)dx = d\left(x - \frac{x^3}{48}\right)\right]$.

$$\oint x dy = \int_{-3}^{+3} +4\sqrt{1 - y^2/9}\,dy + \int_{+3}^{-3} -4\sqrt{1 - y^2/9}\,dy$$

$$= 2\int_{-3}^{+3} 4\sqrt{1 - y^2/9}\,dy$$

$$= 12\pi.$$

4. $\frac{\partial P}{\partial y} = 1 = \frac{\partial Q}{\partial x}$

$$\int_{(0,2)}^{(1,5)} (e^x + y)dx + (x + \ln y)dy = \Big[e^x + xy + y\ln y - y\Big]_{(0,2)}^{(1,5)}$$

$$= e + 1 + 5\ln 5 - 2\ln 2$$

5. (i) The integral becomes

$$\int_0^1 (-t^2 + 7t + 3)dt + (4t^3 - 4t^2 + t)2dt + (-t + 3)(4t - 1)dt = 26/3$$

(ii) $\int_0^1 3dz + \int_2^3 (x^2 - 2)dx + \int_{-1}^1 y\,dy = 22/3$

FRAME 21 continued

6. $\dfrac{\partial P}{\partial y} = -3z = \dfrac{\partial Q}{\partial x}$

$\dfrac{\partial Q}{\partial z} = -3x + 2z = \dfrac{\partial R}{\partial y}$

$\dfrac{\partial R}{\partial x} = -3y + 2x = \dfrac{\partial P}{\partial z}$

$$\text{Integral} = \Big[x^3 + y^3 + z^3 - 3xyz + x^2z + z^2y\Big]_{(-1,2,3)}^{(3,2,-1)} = -28$$

7. $$\int_{AB} udx + vdy + wdz = \int_{AB} ydx + zdy + xdz$$

$$= \int_0^2 (t^2 - 1)dt + 2t(2t)dt + t2dt$$

$$= 15\tfrac{1}{3}\ \text{m}^2/\text{s}$$

8. $$\text{Circulation} = \oint (x + y)dx + (x^2 - y)dy$$

$$= \oint (xdx - ydy) + \oint (ydx + x^2dy)$$

The first integral is zero round a closed curve as it is the integral of an exact differential.

$$\therefore \text{Circulation} = \oint ydx + x^2dy$$

$$\int_{AB} = \int_{-1}^{1} dy = 2$$

$$\int_{BC} = \int_{1}^{-1} dx = -2$$

$$\int_{CD} = \int_{1}^{-1} dy = -2$$

$$\int_{DA} = \int_{-1}^{1} (-1)dx = -2$$

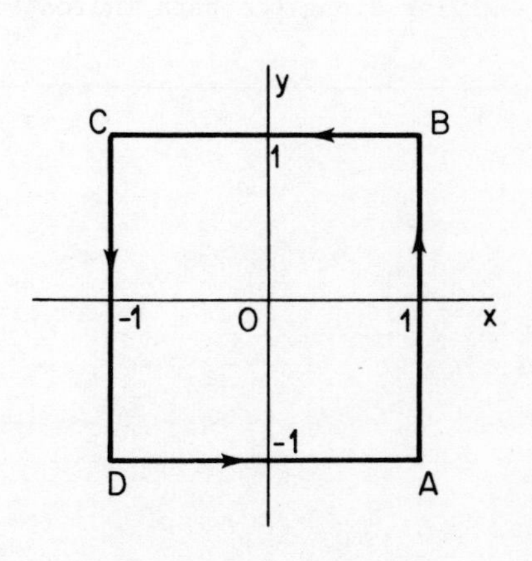

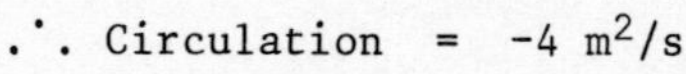

$$\therefore \text{Circulation} = -4\ \text{m}^2/\text{s}$$

FRAME 21 continued

NOTE (of interest to students of fluid dynamics):

If the motion of a fluid is irrotational and the line integral of the velocity (called the velocity potential in this case) is single-valued, the circulation in every closed circuit in the fluid is zero. It follows, therefore, that $\int udx + vdy$ (in the case of two-dimensional flow) is exact, so that $u = \frac{\partial\phi}{\partial x}$ and $v = \frac{\partial\phi}{\partial y}$ where ϕ is the velocity potential.

9. Work done

$$= \int \underline{F}.d\underline{s}$$

$$= \int (x^2 - y^2)dx + 2xy\,dy$$

$$= \int (2x^3 - x^4)dx + (2x^3 - 2x^2)(2x - 1)dx$$

$$= 8\frac{4}{15} \text{ between the given limits.}$$

10. $\int \underline{P}.d\underline{s} = \int y^2dx - x^2dy + xyz\,dz$

(i) 0

(ii) $\int_0^1 4dx - 0 + \int_0^1 2x^4\,3x^2dx = 4\frac{6}{7}$

In each case this can be interpreted as the work done by a force $\underline{P}$ in moving along the path indicated.

SURFACE INTEGRALS

FRAME 1

Introduction

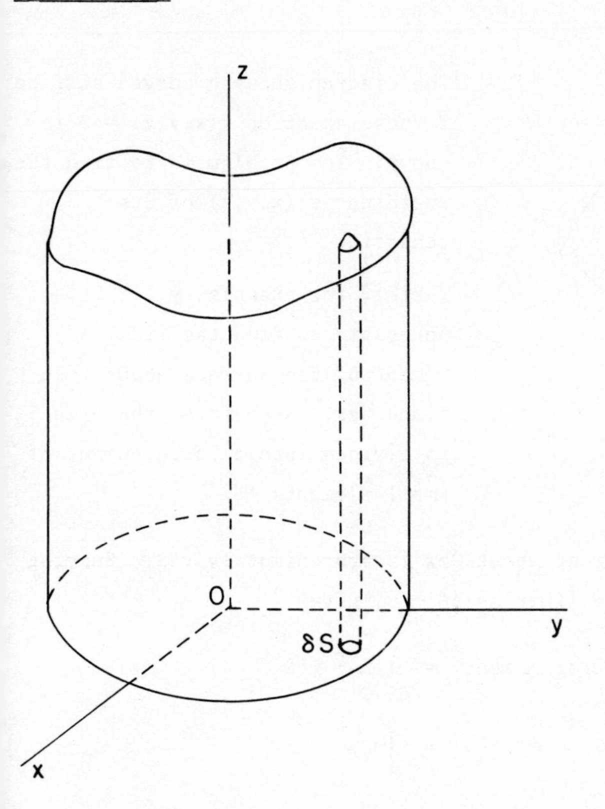

The diagram represents the portion of a circular cylinder cut off by a surface $z = f(x,y)$. The axis of the cylinder is along Oz and its base is in the x-y plane. The problem is to find the volume of this portion. As it is not a solid of revolution, the integration formulae which you have used for finding the volumes of such solids cannot be used in this case. However, we still use a method which divides the volume into a large number of small elements.

We take a small element of area δS in the base of the cylinder and consider the column of the solid standing on this area as base.

Then, volume of column $\simeq z\delta S$.

Summing over all elements δS, and taking the limit as $\delta S \to 0$, gives

$$\text{Total volume} = \lim_{\delta S \to 0} \sum z\delta S$$

This is $\int zdS$ over the surface S which forms the base of the cylinder, sometimes written as $\int_S zdS$. This is an example of a SURFACE INTEGRAL.

FRAME 2

Some other examples will now be used to illustrate the idea of surface integrals.

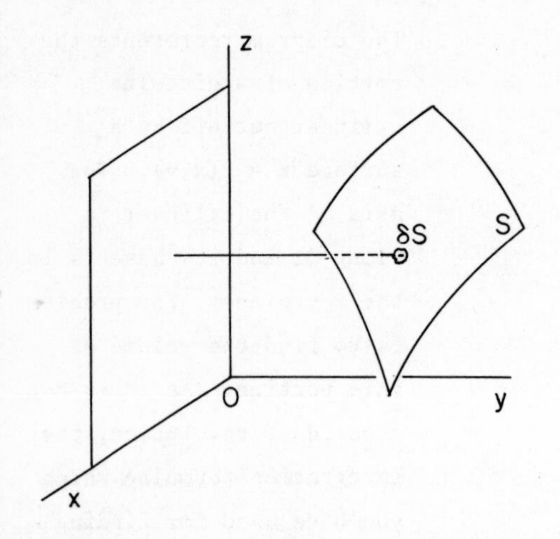

The diagram shows a curved surface S whose equation $f(x,y,z) = 0$ is known. The problem is to find the coordinates $(\bar{x},\bar{y},\bar{z})$ of its centroid.

Taking, for example, $\bar{y}$, it is necessary to find the first moment of the surface about the plane Ozx. As before, the area S is divided into a large number of small elements δS.

The first moment of one such element about Ozx is approximately $y\delta S$. Summing over all elements, and taking the limit as $\delta S \to 0$, gives

$$\text{Total first moment} = \lim_{\delta S \to 0} \sum y\delta S$$

$$= \int y dS$$

$\bar{y}$ is then found from $S\bar{y} = \int y dS$.

If the surface is not a standard one whose area is known, S can be found from $\lim_{\delta S \to 0} \sum \delta S = \int dS$.

$\bar{x}$ and $\bar{z}$ can be found in a similar manner.

FRAME 3

If the surface in FRAME 2 is of uniform density σ, can you write down the moment of inertia of the element δS about Oz and hence obtain an expression for the moment of inertia of the complete surface about Oz?

****

3A

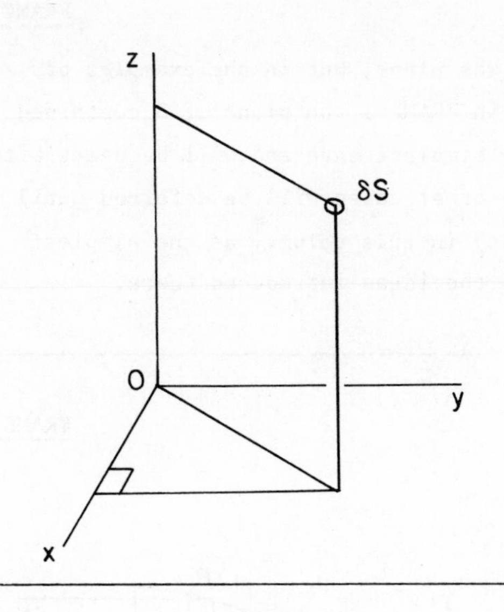

The diagram is intended to show that the distance of the element δS *from Oz is* $\sqrt{x^2 + y^2}$.

Mass of element $= \sigma\delta S$

M. of I. of element about Oz

$$\simeq \sigma\delta S(x^2 + y^2)$$

i.e. $\sigma(x^2 + y^2)\delta S$

M. of I. of surface about Oz

$$= \lim_{\delta S\to 0} \sum \sigma(x^2 + y^2)\delta S$$

$$= \int \sigma(x^2 + y^2)dS$$

FRAME 4

In FRAME 2 of Line Integrals (see page 10:17) the formula for the volume of liquid crossing a cylindrical surface in the case of two-dimensional fluid flow was mentioned. We now consider the volume crossing any surface S when the flow is not two-dimensional, i.e. when q is not the same at all points in a vertical line.

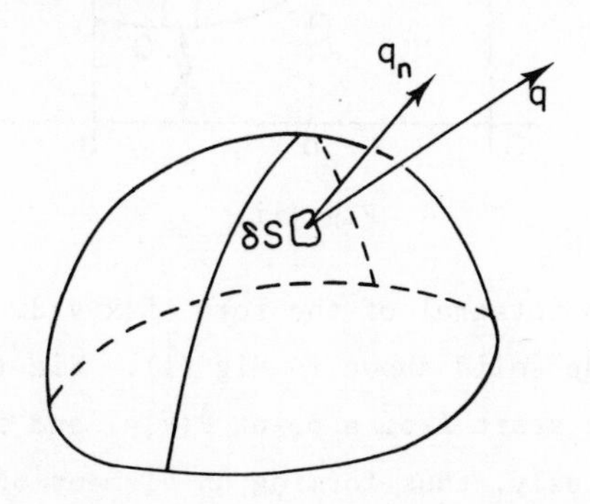

The volume of liquid crossing an element δS of the surface per unit time is approximately $q_n\delta S$, where q_n is the component of the velocity q normal to the surface δS. Hence the volume crossing the complete surface per unit time is $\int q_n dS$.

This is an example of the transfer of a physical quantity across a surface, also known as flux. This idea of flux also occurs in the flow of, for example, a gas, heat, electricity or particles (say from a radioactive source), so you are likely to meet surface integrals if you are concerned with the study of any of these topics.

FRAME 5

In the example in FRAME 1 the surface S was plane, but in the examples of FRAMES 2-4 it was curved. Furthermore, in FRAME 1 the plane of S contained two of the coordinate axes. This is the simplest case and will be dealt with in this programme. Consideration of the other cases will be deferred until you have reached Unit 11 (Vector Analysis) in this volume, as the simplest approach requires a knowledge of some of the ideas introduced there.

FRAME 6

Evaluation in Cartesian Coordinates

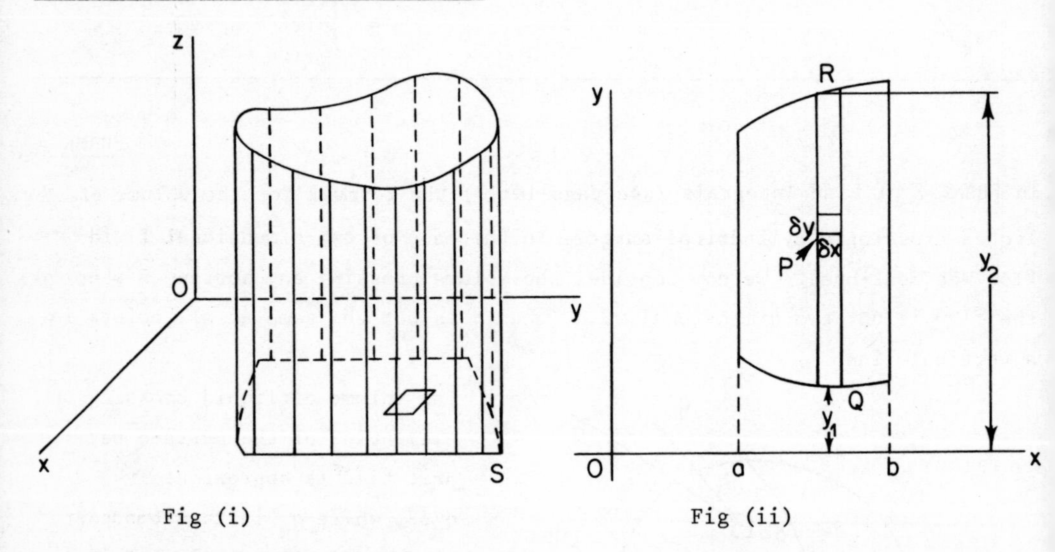

Fig (i) Fig (ii)

We shall now consider how to evaluate an integral of the form $\int f(x,y)dS$ which, if $z = f(x,y)$, could be the volume of the solid shown in Fig (i). Fig (ii) shows the base of this solid. For δS we start from a point $P(x,y)$ and take increments δx and δy in x and y respectively, thus forming an element of area as shown. This gives

$$\delta S = \delta x \delta y$$

The volume of the column standing on this rectangle is approximately $z\delta x\delta y$. Summing over all such rectangles from Q to R and taking the limit as $\delta y \to 0$ gives, approximately, the volume of the slice standing on the strip QR as base,

FRAME 6 continued

i.e. $$\text{Volume of slice} \triangleq \lim_{\delta y \to 0} \sum_{y=y_1}^{y=y_2} z\delta x\delta y$$

$$= \left(\lim_{\delta y \to 0} \sum_{y=y_1}^{y=y_2} z\delta y \right) \delta x \quad \text{as } \delta x \text{ is constant throughout this summation}$$

$$= \left(\int_{y_1}^{y_2} z\,dy \right) \delta x$$

Summing over all such slices from x = a to x = b and taking the limit as $\delta x \to 0$ gives the volume of the complete solid,

i.e. $$\text{Volume of solid} = \lim_{\delta x \to 0} \sum_{x=a}^{x=b} \left(\int_{y_1}^{y_2} z\,dy \right) \delta x$$

$$= \int_a^b \left(\int_{y_1}^{y_2} z\,dy \right) dx$$

Here we have formed the slice by taking the strip QR parallel to Oy. However, as you will see later, it is sometimes easier to form slices by taking strips parallel to Ox. We would then sum first in the x-direction and then in the y.

FRAME 7

As is so often the case when a notation is involved, the way in which double integration is denoted is not unique. As two integrations are involved $\int z\,dS$ is sometimes written as $\iint z\,dS$. Again, some people denote $\int_a^b \left(\int_{y_1}^{y_2} z\,dy \right) dx$ by $\int_a^b \int_{y_1}^{y_2} z\,dy\,dx$, but others use $\int_a^b \int_{y_1}^{y_2} z\,dx\,dy$. Either of these notations may lead to ambiguity,

e.g. does $\int_1^2 \int_3^4 z\,dx\,dy$ mean $\int_1^2 \left(\int_3^4 z\,dx \right) dy$ or $\int_1^2 \left(\int_3^4 z\,dy \right) dx$?

An alternative notation for $\int_a^b \left(\int_{y_1}^{y_2} z\,dy \right) dx$, which avoids this difficulty and which we shall use throughout, is

$$\int_a^b dx \int_{y_1}^{y_2} z\,dy$$

FRAME 8

We shall now show how to evaluate such an integral, taking as an example

$$\int_1^2 dx \int_0^x (x + 2y)dy.$$

The first thing to do is to integrate (x + 2y) with respect to y, treating x as a constant, and you will then get

$$\int_0^x (x + 2y)dy = \Big[xy + y^2\Big]_0^x ,$$

the limits outside the square brackets being those of y. Substituting these limits gives $2x^2$.

$$\text{Thus} \int_1^2 dx \int_0^x (x + 2y)dy = \int_1^2 2x^2dx$$

$$= \left[\frac{2}{3}x^3\right]_1^2$$

$$= \frac{14}{3}$$

FRAME 9

Referring to FRAME 6 we shall now sketch the area, for the present example, corresponding to that shown in Fig (ii). It is first necessary to draw the four boundaries y = 0, y = x, x = 1 and x = 2, these boundaries being given by the limits in the integral. The area defined by them is the trapezium ABCD. An elemental strip QR is shown, corresponding to a similar strip in Fig (ii) of FRAME 6. You will notice that in this case the length of such a strip varies with

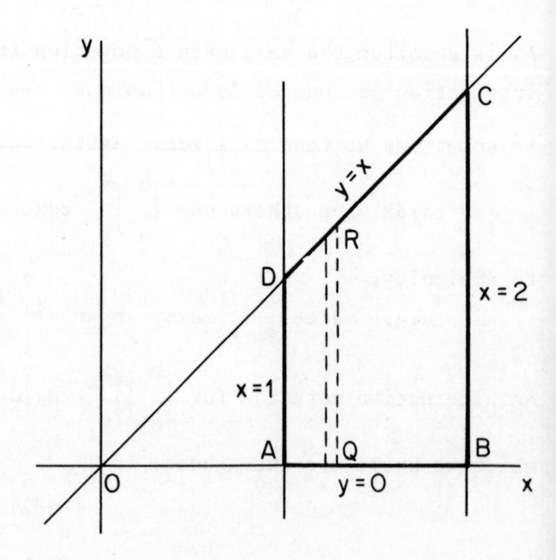

FRAME 9 continued

its position. This is taken account of by the varying value of y_2 (= x). In the general situation, as in FRAME 6, y_1 would also vary, but in the present example it is constant (= 0).

You will notice that the area ABCD was not dependent in any way upon the function x + 2y. Just as $\int_0^3 (x^2 + 2)dx$ can be considered as representing an area under the curve $y = x^2 + 2$, so also $\int_1^2 dx \int_0^x (x + 2y)dy$ can be considered as representing the volume under the surface z = x + 2y standing on the trapezium ABCD as base.

FRAME 10

Now evaluate (i) $\int_0^1 dx \int_{x^2}^4 xy\, dy$,

(ii) $\int_0^2 dx \int_{x^2}^4 xy\, dy$,

and sketch the area over which the integral is taken in each case.

****

10A

(i) $$\int_0^1 dx \int_{x^2}^4 xy\, dy = \int_0^1 \left(8x - \frac{x^5}{2}\right)dx = \frac{47}{12}$$

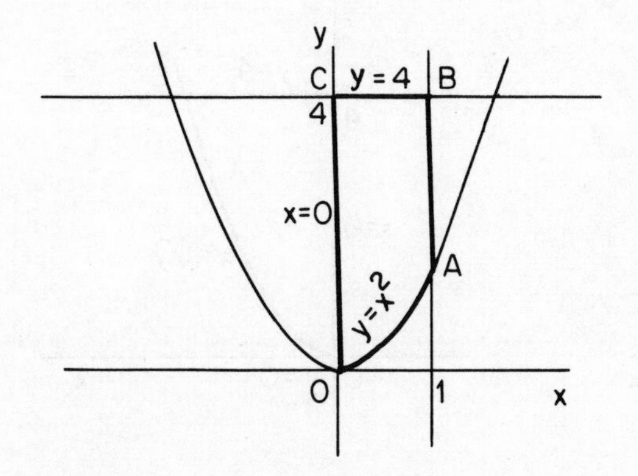

The area is OABC

10A continued

(ii) $\int_0^2 dx \int_{x^2}^4 xy\ dy = \int_0^2 \left(8x - \frac{x^5}{2}\right) dx = \frac{32}{3}$

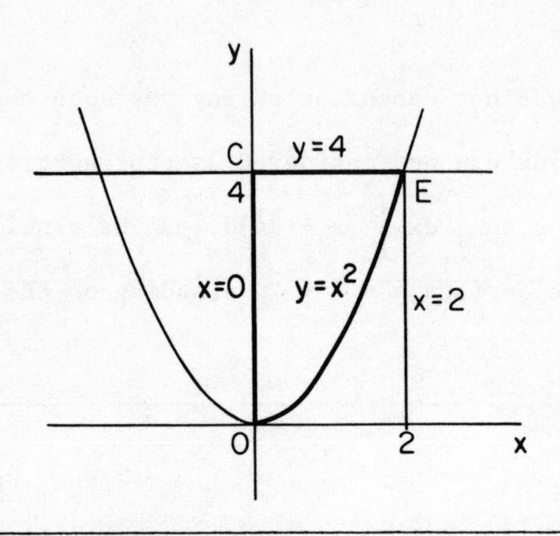

The area is OEC. Comparing this with the previous diagram, you will see that, in moving the right-hand limit of x from 1 to 2, the boundary AB has been reduced to the point E.

FRAME 11

Now verify that $\int_0^4 dy \int_0^{\sqrt{y}} xy\ dx$ gives the same answer as (ii) in the last frame. Can you see why this is so?

****

11A

$\int_0^4 dy \int_0^{\sqrt{y}} xy\ dx = \int_0^4 \frac{y^2}{2} dy = \frac{32}{3}$

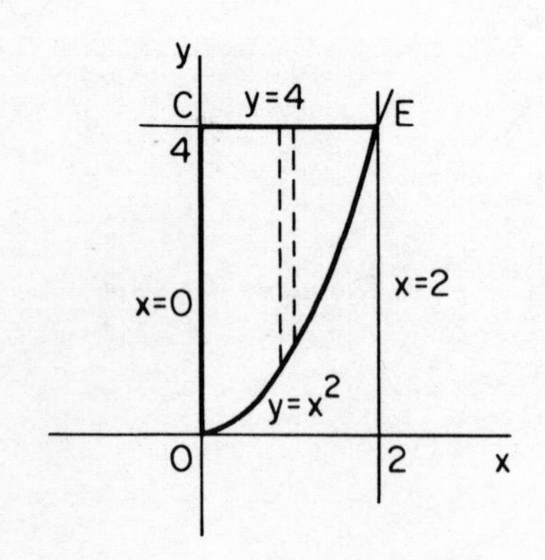

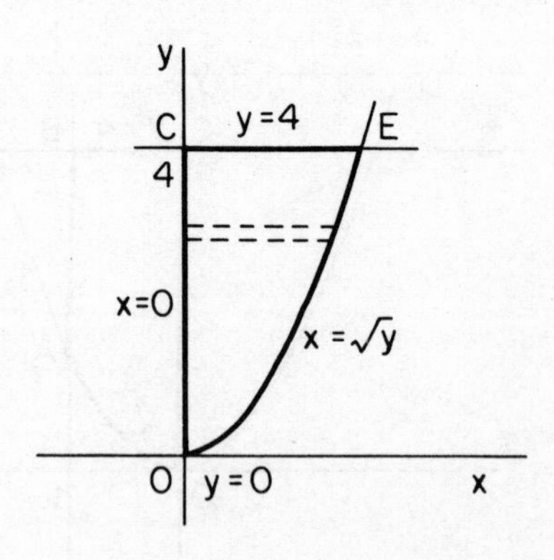

11A continued

The two answers are the same because it is the same function xy that is being integrated, and the area over which the integration is taken is the same in each case. This should be obvious if you compare the two diagrams. When integrating first with respect to x and then with respect to y, the elemental strips are taken parallel to the x-axis.

FRAME 12

In the preceding examples you have seen how to sketch the area over which the integrals were taken. We shall now consider how to find the limits of integration corresponding to a given area. As an illustration, we shall find $\int (x - y)dS$ over the triangle whose vertices are at the points (0,1), (2,1) and (2,3). We strongly recommend that a diagram should first be drawn.

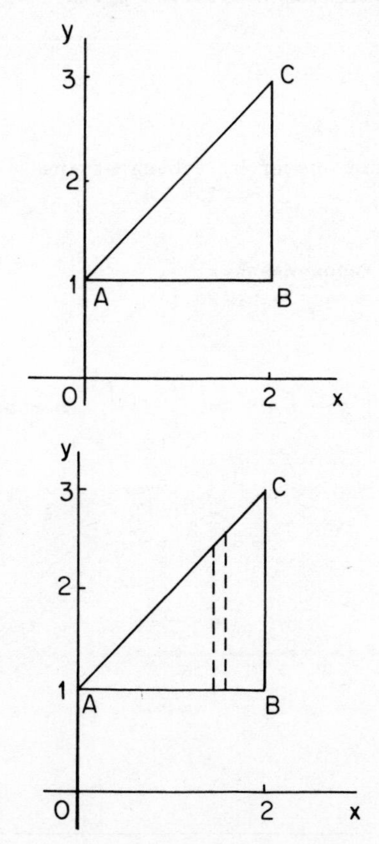

The triangle is ABC. We next have to decide whether to take strips parallel to Ox or to Oy. In this particular example neither has any advantage over the other. We shall take strips parallel to Oy.

Choosing the strips parallel to Oy means that we shall integrate first with respect to y and then with respect to x, so that $\int (x - y)dS$ must be written as

$$\int_?^? dx \int_?^? (x - y)dy.$$

The values of the limits now have to be decided.

First we decide the limits for y by writing down the equations of

FRAME 12 continued

the boundaries formed by the ends of the strips. These boundaries are AB and AC in this example.

It will thus be seen that the range of integration for y is from $y = 1$ to $y = x + 1$.

Integration with respect to x, which corresponds to summation over all strips, is then between $x = 0$ and $x = 2$.

$\int (x - y)dS$ thus becomes $\int_0^2 dx \int_1^{x+1} (x - y)dy$, which on evaluation gives

$\int_0^2 (\frac{1}{2}x^2 - x)dx$, i.e. $-\frac{2}{3}$.

Now, as a check, see if you can obtain the same answer by taking strips parallel to Ox.

****

12A

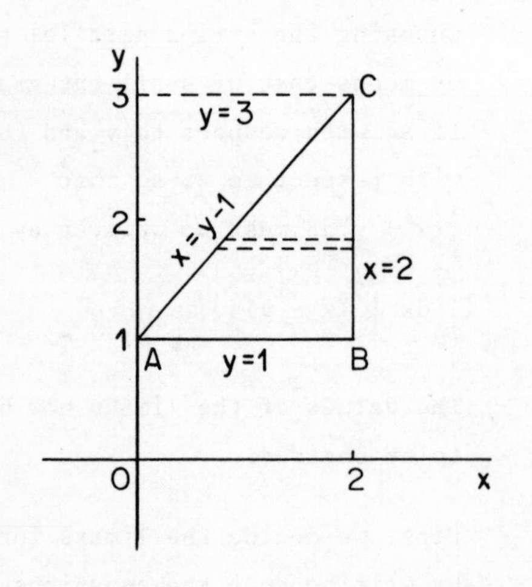

$$\int (x - y)dS = \int_1^3 dy \int_{y-1}^2 (x - y)dx$$

$$= \int_1^3 \left(\frac{3}{2} - 2y + \frac{y^2}{2}\right) dy$$

$$= -\frac{2}{3}$$

FRAME 13

The following problem will now be considered:

Evaluate $\int x^2 dS$ over the trapezium whose vertices are (0,0), (5,0), (3,1) and (0,1).

First sketch the area and then comment on the relative merits of the two ways in which the strips may be taken.

13A

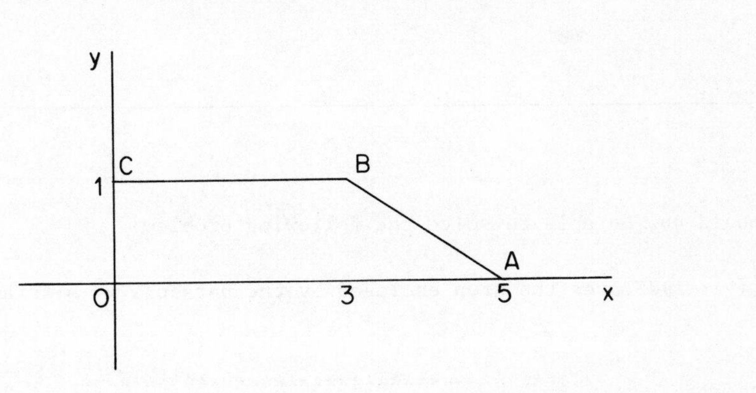

It is easier to take strips parallel to Ox. If strips are taken parallel to Oy, the area must be divided into two parts - the rectangle whose upper boundary is CB and the triangle whose upper boundary is BA.

FRAME 14

Taking the strips parallel to Ox, now write down the limits for x and then those for y.

****

14A

$x = 0$ and $x = 5 - 2y$

$y = 0$ and $y = 1$

FRAME 15

Thus $\int x^2 dS$ becomes $\int_0^1 dy \int_0^{5-2y} x^2 dx$. Evaluate this.

****

15A

$$\int_0^1 dy \int_0^{5-2y} x^2 dx = \int_0^1 \frac{(5-2y)^3}{3} dy$$

$$= \left[-\frac{1}{24}(5-2y)^4\right]_0^1$$

$$= \frac{68}{3}$$

FRAME 16

You should now be able to solve the following problem:

Find $\int (y+1)dS$ over the area enclosed by the parabola $y^2 = 4x$ and the line $x = 1$.

****

16A

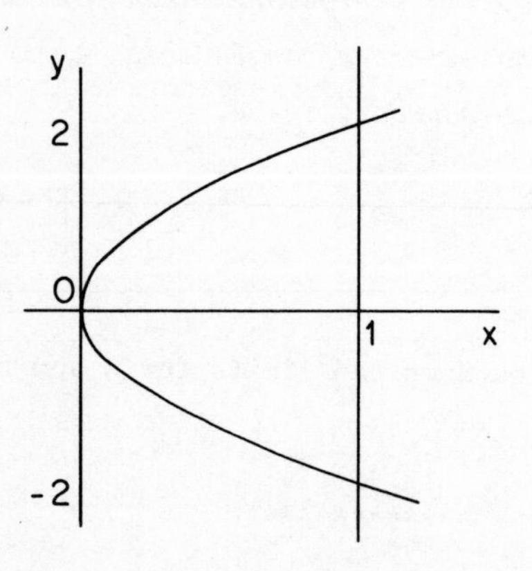

If you took strips parallel to Ox you should have

$$\int_{-2}^{2} dy \int_{y^2/4}^{1} (y+1)dx = \frac{8}{3}$$

If you took strips parallel to Oy you should have

$$\int_0^1 dx \int_{-2\sqrt{x}}^{2\sqrt{x}} (y+1)dy = \frac{8}{3}$$

Either way, if you integrated over the upper half of the area and then doubled your result, you will have obtained an incorrect answer. This is because although the area is symmetrical about Ox, the function y + 1 is not.

FRAME 17

Evaluation in Polar Coordinates

Suppose it is required to evaluate $\int(x^2 + 2y^2)dS$ over the circle $x^2 + y^2 = a^2$. If this is done in Cartesian coordinates it gives, taking strips parallel to Oy, $\int_{-a}^{a} dx \int_{-\sqrt{a^2-x^2}}^{\sqrt{a^2-x^2}} (x^2 + 2y^2)dy$, the evaluation of which is rather troublesome. It is no better if strips are taken parallel to Ox. An easier method is to use polar coordinates.

FRAME 18

The first thing to do is to decide what element of area to take for δS when polar coordinates are used. In FRAME 6 we formed δS by starting from a point $P(x,y)$ and taking increments δx and δy.

$(x, y+\delta y)$ $(x+\delta x, y+\delta y)$
δy
$P(x,y)$ δx $(x+\delta x, y)$

In polar coordinates we start from a point $P(r,\theta)$ and take increments δr and $\delta\theta$. The element of area δS thus obtained is as shown, and you will see that it is the difference between the two sectors whose radii are r and $r + \delta r$.

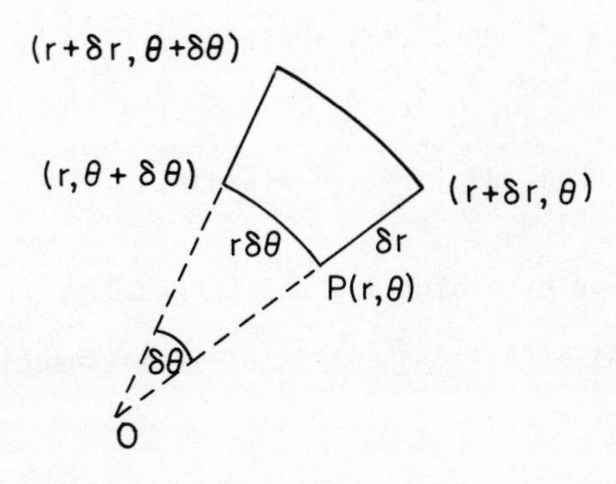

$$\delta S = \tfrac{1}{2}(r + \delta r)^2\delta\theta - \tfrac{1}{2}r^2\delta\theta$$

$$= r\delta r\delta\theta + \tfrac{1}{2}(\delta r)^2\delta\theta$$

FRAME 18 continued

The second term is small compared with the first and in the limit its contribution will be zero. You will note that taking δS as $r\delta r\delta\theta$ is equivalent to taking it to be the area of a rectangle whose sides are $r\delta\theta$ and δr. A similar approach will be adopted when considering volume integrals.

FRAME 19

Returning to the example in FRAME 17, when polar coordinates are used $\int(x^2 + 2y^2)dS$ becomes $\int(r^2\cos^2\theta + 2r^2\sin^2\theta)dS$, i.e. $\int r^2(1 + \sin^2\theta)dS$, which equals $\iint r^2(1 + \sin^2\theta)rdrd\theta$. It is now necessary to find the limits of r and θ in the integration.

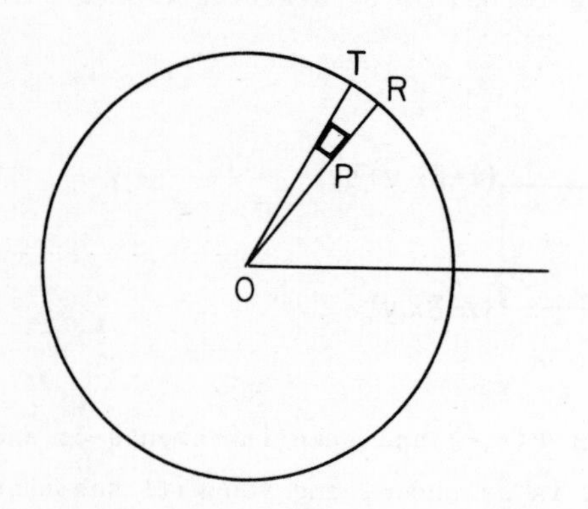

The diagram shows an element δS such as was described in the previous frame. Summing over all such elements in the sector ORT leads to integration with respect to r over the range r = 0 to r = a. Summing over all such sectors in the circle leads to integration with respect to θ over a range of 2π,

e.g. 0 to 2π or $-\pi$ to π.

$$\text{Thus } \int r^2(1 + \sin^2\theta)dS = \int_0^{2\pi} d\theta \int_0^a r^2(1 + \sin^2\theta)rdr$$

$$= \int_0^{2\pi} d\theta \int_0^a r^3(1 + \sin^2\theta)dr$$

Now complete the solution by evaluating the integral.

****

19A

$3\pi a^4/4$

FRAME 20

ow find $\int (x + y)dS$ over the sector OABC of the circle $x^2 + y^2 = 9$.

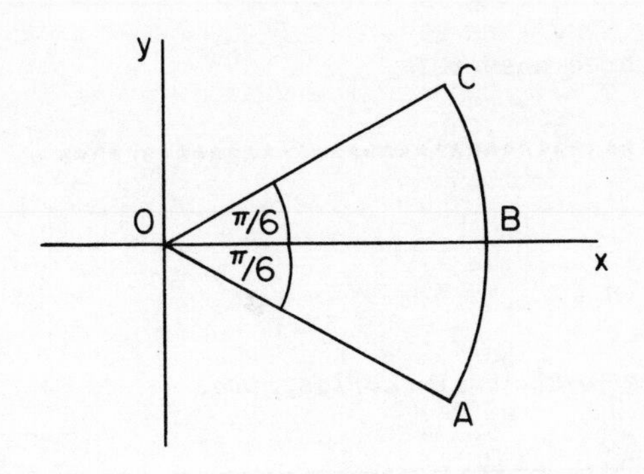

****

20A

$$\int_{-\pi/6}^{\pi/6} d\theta \int_0^3 r^2(\cos\theta + \sin\theta)dr = 9$$

FRAME 21

A Property of $\int_k^\ell dx \int_m^n f(x)g(y)dy$

Certain double integrals have a property which is sometimes useful. In FRAME 19 you found the value of $\int_0^{2\pi} d\theta \int_0^a r^3(1 + \sin^2\theta)dr$.

Now evaluate $\int_0^{2\pi} (1 + \sin^2\theta)d\theta$ and $\int_0^a r^3 dr$. Is there any connection between these answers and that in 19A?

****

21A

3π and $a^4/4$

Yes. The answer in 19A is their product.

FRAME 22

Now find $\int_2^5 dx \int_{-1}^3 \frac{x^2}{y+2}\,dy$, $\int_2^5 x^2 dx$ and $\int_{-1}^3 \frac{1}{y+2}\,dy$. Can you see any relation between the three answers?

****

22A

$\frac{117}{3} \ln 5$, $\frac{117}{3}$ *and* $\ln 5$

The first answer is the product of the last two.

FRAME 23

You will notice that in both $\int_0^{2\pi} d\theta \int_0^a r^3(1 + \sin^2\theta)dr$ and $\int_2^5 dx \int_{-1}^3 \frac{x^2}{y+2}\,dy$ all the limits are constants. Furthermore in each integrand the variables are separable, i.e. the former is of the form $\phi(r)\psi(\theta)$ and the latter of the form $f(x)g(y)$. A double integral which satisfies these conditions can always be written as the product of two simple integrals.

Thus, if k, ℓ, m and n are all constants,

$$\int_k^\ell dx \int_m^n f(x)g(y)dy = \int_k^\ell f(x)dx \times \int_m^n g(y)dy \tag{23.1}$$

Now express $\int_\alpha^\beta d\theta \int_a^b \phi(r)\psi(\theta)dr$ as the product of two simple integrals.

****

23A

$\int_\alpha^\beta \psi(\theta)d\theta \times \int_a^b \phi(r)dr$

FRAME 24

The Integral $\int_0^\infty e^{-x^2}dx$

In the programmes on Infinite Series and Integration in Volume I it was mentioned that the indefinite integral $\int e^{-x^2}dx$ cannot be found by ordinary methods. However it is possible to evaluate the definite integral $\int_0^\infty e^{-x^2}dx$.

What relationship is there between the value of this integral and the value of $\int_0^\infty e^{-y^2}dy$?

****

24A

They are equal. In a definite integral it doesn't matter what letter is used for the variable.

FRAME 25

Denoting $\int_0^\infty e^{-x^2}dx$ by I, it follows that $I = \int_0^\infty e^{-y^2}dy$ and

$$I^2 = \int_0^\infty e^{-x^2}dx \int_0^\infty e^{-y^2}dy.$$

Using the result (23.1) can you see what double integral can represent I^2? Indicate by a sketch the area over which the double integral is taken.

****

25A

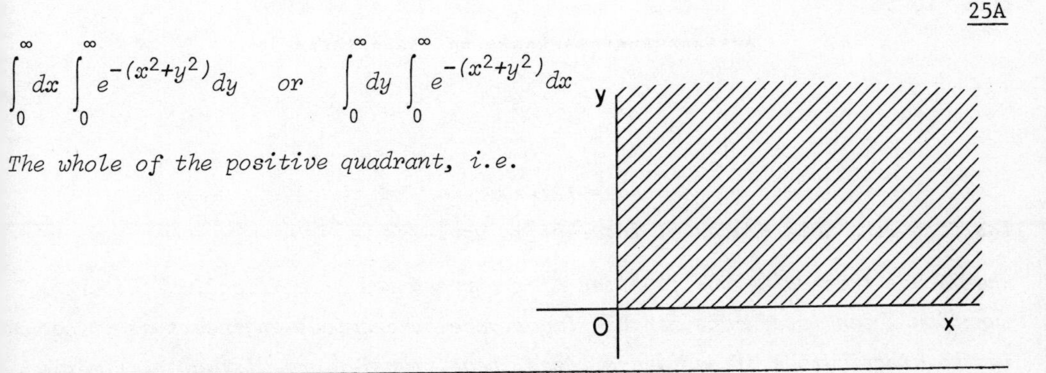

$$\int_0^\infty dx \int_0^\infty e^{-(x^2+y^2)}dy \quad or \quad \int_0^\infty dy \int_0^\infty e^{-(x^2+y^2)}dx$$

The whole of the positive quadrant, i.e.

FRAME 26

Whichever result in 25A is used it can be written as

$$\int e^{-(x^2+y^2)}\,dS$$

taken over the whole of the positive quadrant.

This integral can be evaluated using polar coordinates. See if you can do this.

****

26A

$$\int_0^{\pi/2} d\theta \int_0^{\infty} e^{-r^2} r\,dr = \pi/4$$

FRAME 27

It follows that $I = \sqrt{\pi}/2$

$$\text{i.e.} \quad \int_0^{\infty} e^{-x^2}\,dx = \frac{\sqrt{\pi}}{2}$$

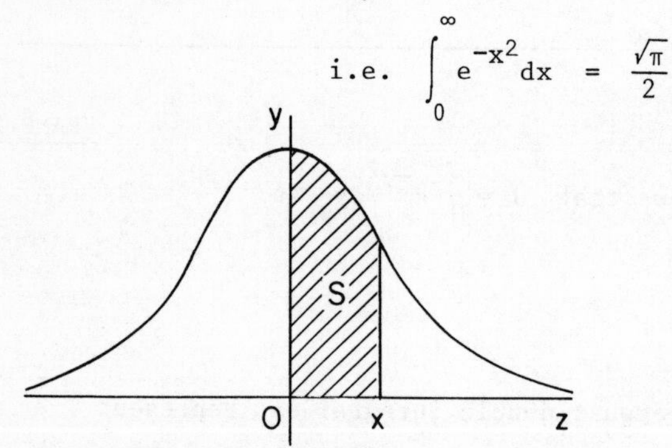

The diagram shows the curve

$$y = Ae^{-z^2}$$

The area S is thus given by

$$\int_0^x Ae^{-z^2}\,dz$$

Note that S is a function of the upper limit x.

What must be the value of A for the area under the right-hand half of the curve to be 1?

****

27A

$$A\int_0^{\infty} e^{-z^2}\,dz = 1$$

$$A = 2/\sqrt{\pi}$$

The following note may be of interest to you:

When A has this value the function of x represented by S is called the error function which is of considerable importance to chemical engineers as it occurs in the distribution of residence times, heat transfer and diffusion of matter.

27A continued

It is denoted by erf x *and thus* $\operatorname{erf} x = \frac{2}{\sqrt{\pi}}\int_0^x e^{-z^2}\,dz$

FRAME 28

In probability theory, $\int_0^u Be^{-\frac{1}{2}t^2}\,dt$ represents the probability of the standardised normal variate taking a value between 0 and u, B being such that $\int_{-\infty}^{\infty} Be^{-\frac{1}{2}t^2}\,dt = 1$. What is the relation between $\int_{-\infty}^{\infty} Be^{-\frac{1}{2}t^2}\,dt$ and $\int_0^{\infty} Be^{-\frac{1}{2}t^2}\,dt$? What substitution will make the latter integral into one you recognise? Find the value of B.

****

28A

As $e^{-\frac{1}{2}t^2}$ *is an even function* $\int_{-\infty}^{\infty} Be^{-\frac{1}{2}t^2}\,dt = 2\int_0^{\infty} Be^{-\frac{1}{2}t^2}\,dt$

The substitution $z = t/\sqrt{2}$ *changes* $\int_0^{\infty} Be^{-\frac{1}{2}t^2}\,dt$ *into* $B\sqrt{2}\int_0^{\infty} e^{-z^2}\,dz$,

whence $B = 1/\sqrt{2\pi}$.

Note that $\frac{1}{\sqrt{2\pi}}\int_0^u e^{-\frac{1}{2}t^2}\,dt = \frac{1}{\sqrt{2\pi}}\int_0^{u/\sqrt{2}} \sqrt{2}\, e^{-z^2}\,dz = \frac{1}{2}\operatorname{erf}\frac{u}{\sqrt{2}}$

FRAME 29

Some Applications of Surface Integrals

As a first example we shall find the product of inertia of a uniform rectangular plate, mass M, sides a and b, about a pair of adjacent sides. (Explanatory note: The product of inertia of an element of mass δm about a pair of rectangular axes is $(\delta m)xy$, where x and y are the distances of δm from the two axes. Products of inertia are used when calculating principal axes of inertia.)

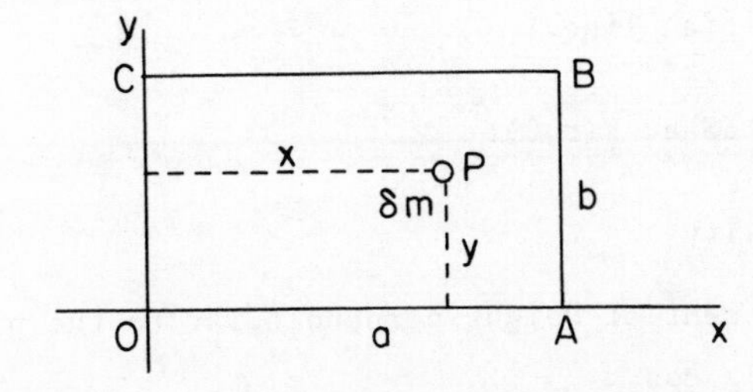

Axes Ox, Oy are taken along the sides about which the product of inertia is required.

FRAME 29 continued

The product of inertia of an element δm at $P(x,y)$ about Ox, Oy is $(\delta m)xy$

$$= (\sigma \delta S)xy \quad \text{where } \sigma \text{ is the surface density}$$

i.e. $\sigma xy \delta S$

Summing over all the elements, the product of inertia of the whole plate is given by

$$\int \sigma xy dS$$

$$= \int_0^a dx \int_0^b \sigma xy dy$$

Evaluate this integral, expressing the answer in terms of M, a and b.

****

29A

$$\frac{\sigma a^2 b^2}{4} = \frac{Mab}{4}$$

FRAME 30

We shall now consider the following example:

A uniform circular plate of mass M and radius a lies on a rough plane which is inclined at an angle α to the horizontal. The coefficient of friction between the plate and the plane is μ. At a point on its circumference the plate is pinned to the plane and lies with the diameter through the pin horizontal. It is required to find the torque preventing the motion of the plate when limiting friction is acting.

The problem is best dealt with using polar coordinates, taking the pole at the pin and the horizontal diameter as initial line.

Dealing first with an element of area δS at $P(r,\theta)$:

Mass = $\sigma \delta S$ where σ is the surface density

Reaction R normal to the plane = Component of weight perpendicular to the plane

$$= (\sigma \delta S)g \cos \alpha$$

FRAME 30 continued

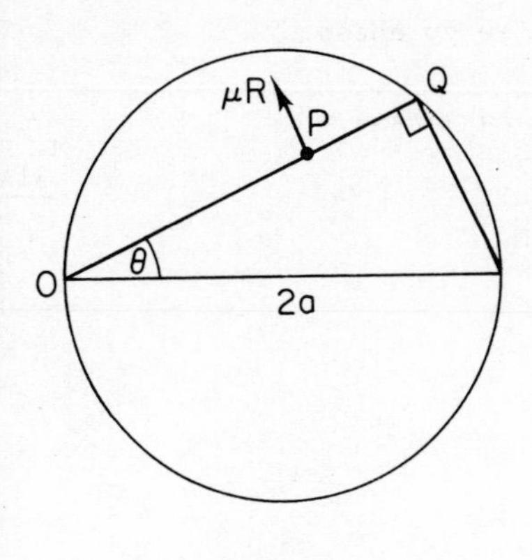

Fig (i)

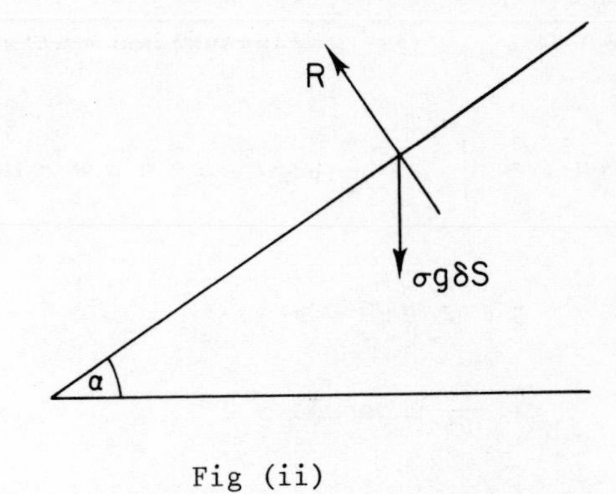

Fig (ii)

When limiting friction is acting, $F = \mu R$

$$= \mu\sigma g \cos\alpha\ \delta S$$

This force acts at P along the plane in a direction perpendicular to OP, as it is opposing a tendency for P to turn about O.

Moment of F about O = $(\mu\sigma g \cos\alpha\ \delta S)r$

Summing now over all the elements, the total frictional torque is

$$\int \mu\sigma g r \cos\alpha\ dS$$

To evaluate this integral it must be written in the form

$$\int_{?}^{?} d\theta \int_{?}^{?} \mu\sigma g r^2 \cos\alpha\ dr,$$

replacing dS by $r dr d\theta$.

By referring to Fig (i), can you insert the four limits into this integral?

****

30A

$$\int_{-\pi/2}^{\pi/2} d\theta \int_{0}^{2a\cos\theta} \mu\sigma g r^2 \cos\alpha\ dr$$

Note: The upper limit for r is the length OQ.

FRAME 31

You should now be able to complete this problem, so go ahead.

****

31A

$$Torque = \int_{-\pi/2}^{\pi/2} \frac{8}{3}\,\mu\sigma g a^3 \cos\alpha \cos^3\theta \, d\theta$$

$$= \frac{32}{9}\,\mu\sigma g a^3 \cos\alpha$$

$$= \frac{32}{9\pi}\,\mu M g a \cos\alpha$$

FRAME 32

The last problem in this section is a fairly straightforward one on finding a volume and you should be able to tackle it without any assistance from us.

A circular hole of radius b is made centrally through a sphere of radius a. Use double integration to find the volume remaining.

****

32A

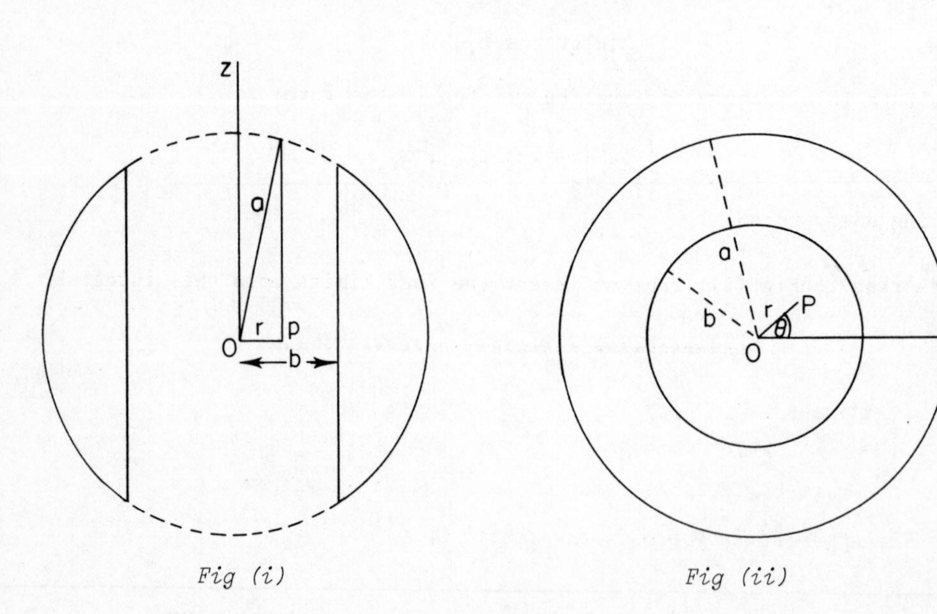

Fig (i) *Fig (ii)*

32A continued

We have taken the origin O at the centre of the sphere, and the z-axis along the axis of the hole. Fig (i) represents a section through the z-axis and Fig (ii) a section through O perpendicular to the z-axis.

We shall find the volume of the hole and subtract it from that of the sphere.

$$\text{Volume of upper half of hole} = \int z\,dS$$

$$= \int \sqrt{a^2 - r^2}\,dS$$

$$= \int_0^{2\pi} d\theta \int_0^b \sqrt{a^2 - r^2}\,r\,dr$$

$$= \int_0^{2\pi} \left[-\frac{1}{3}(a^2 - r^2)^{3/2}\right]_0^b d\theta$$

$$= \int_0^{2\pi} \frac{1}{3}\left\{a^3 - (a^2 - b^2)^{3/2}\right\} d\theta$$

$$= \frac{2\pi}{3}\left\{a^3 - (a^2 - b^2)^{3/2}\right\}$$

$$\text{Volume of remainder} = \frac{4\pi}{3}a^3 - \frac{4\pi}{3}\left\{a^3 - (a^2 - b^2)^{3/2}\right\}$$

$$= \frac{4\pi}{3}(a^2 - b^2)^{3/2}$$

Alternatively, if you have found the volume of the remainder directly it is given by the value of $2\int_0^{2\pi} d\theta \int_b^a \sqrt{a^2 - r^2}\,r\,dr$.

FRAME 33

Miscellaneous Examples

In this frame a collection of miscellaneous examples is given for you to try. Answers are supplied in FRAME 34, together with such working as is considered helpful.

FRAME 33 continued

1. Evaluate $\int_0^3 dx \int_0^{6-2x} 2y(3 - x)dy$ and sketch the area over which this integral is taken.

2. Evaluate $\int 2xy(7x + 12y)dS$ over the area between the parabola $y = x^2$ and the line $y = x$.

3. If you were to try to evaluate $\int_0^1 dy \int_y^1 e^{x^2} dx$ as it stands, you would be faced with the impossibility of finding $\int e^{x^2} dx$. The analytical technique for overcoming this difficulty is that of change of order of integration. You saw in answer frames 10A, 11A and 12A that two double integrals give the same result if in each case the same function is integrated and also the area of integration is the same. Sketch the area over which $\int_0^1 dy \int_y^1 e^{x^2} dx$ is taken and use your sketch to write down the limits for $\int_?^? dx \int_?^? e^{x^2} dy$ to give the same result. Hence find the value of the original integral.

4. Evaluate $\int (a - y)dS$ over the upper half of the circle $x^2 + y^2 = a^2$.

5. A square plate has sides of length 2a and O is the mid-point of one of the sides. The surface density at any point P in the plate is proportional to OP^2. Find the position of the centre of gravity.

6. Find the mean value of $e^{-(x^2+y^2)}$ over the area of the circle whose centre is at the origin and whose radius is 2.

FRAME 34

Answers to Miscellaneous Examples

1. Integral $= \int_0^3 4(3 - x)^3 dx = 81$

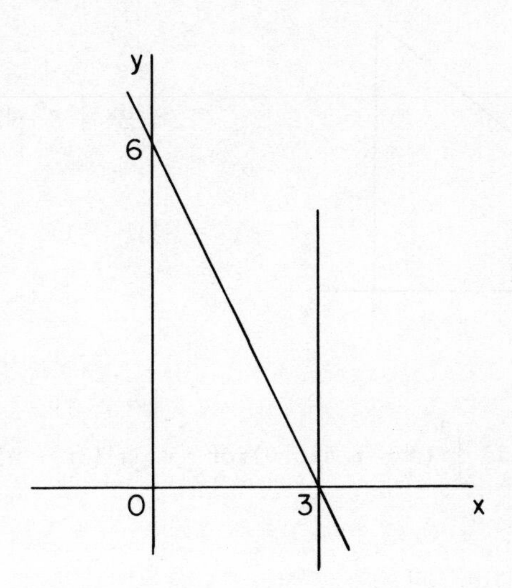

2.

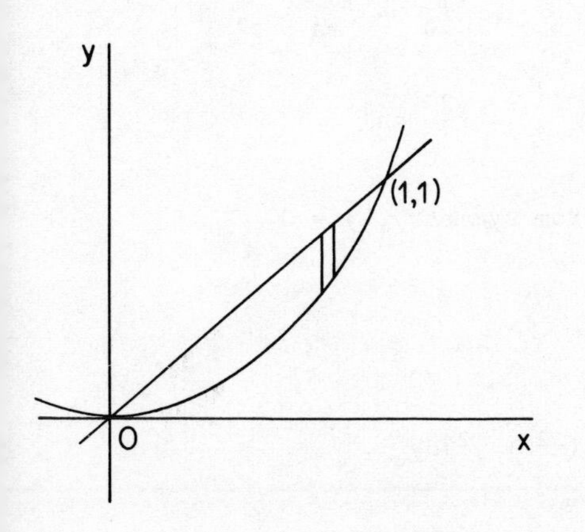

$$\int 2xy(7x + 12y)dS$$

$$= \int_0^1 dx \int_{x^2}^{x} 2xy(7x + 12y)dy \quad \text{if}$$

strips are taken parallel to Oy

$$= \int_0^1 (15x^4 - 7x^6 - 8x^7)dx$$

$$= 1$$

FRAME 34 continued

3.

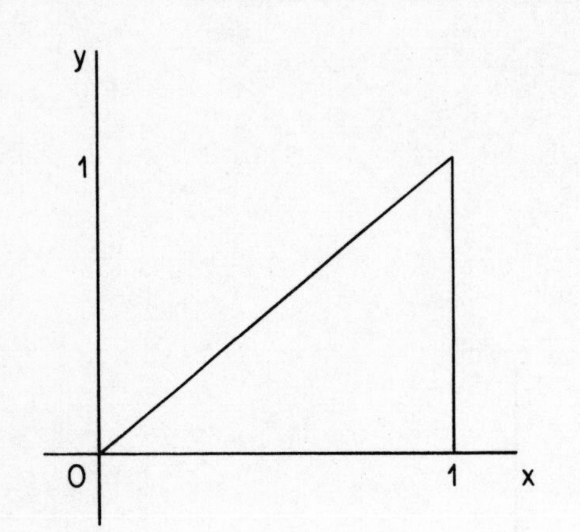

$$\int_0^1 dy \int_y^1 e^{x^2} dx \quad \text{(strips parallel to Ox)}$$

$$= \int_0^1 dx \int_0^x e^{x^2} dy \quad \text{(strips parallel to Oy)}$$

$$= \tfrac{1}{2}(e - 1)$$

4. $\int (a - y)dS = \int_0^{\pi} d\theta \int_0^a (a - r \sin \theta) r dr = a^3(3\pi - 4)/6$

5.

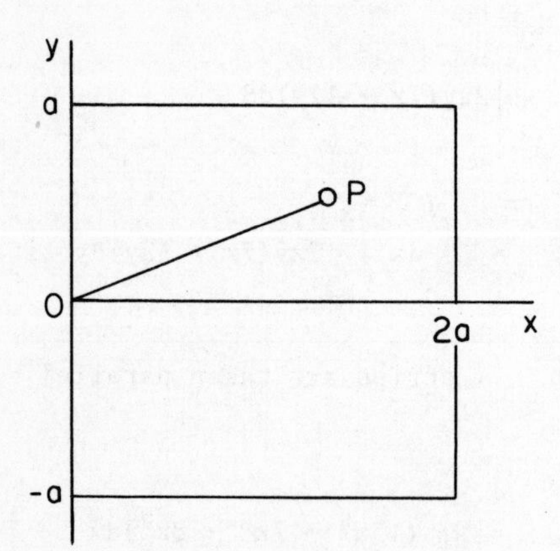

$$\text{Mass} = \int_0^{2a} dx \int_{-a}^{a} k(x^2 + y^2)dy$$

$$= \frac{20}{3} ka^4$$

From symmetry, $\bar{y} = 0$

$$\text{Moment of mass about Oy} = \int_0^{2a} dx \int_{-a}^{a} kx(x^2 + y^2)dy$$

$$= \frac{28}{3} ka^5$$

$$\bar{x} = \frac{7}{5} a$$

FRAME 34 continued

6. Mean value $= \dfrac{1}{\text{Area}} \displaystyle\int e^{-(x^2+y^2)} dS$

$$= \frac{1}{4\pi} \int_0^{2\pi} d\theta \int_0^2 e^{-r^2} r dr$$

$$= \frac{1}{4}(1 - e^{-4})$$

VOLUME INTEGRALS

FRAME 1

Introduction

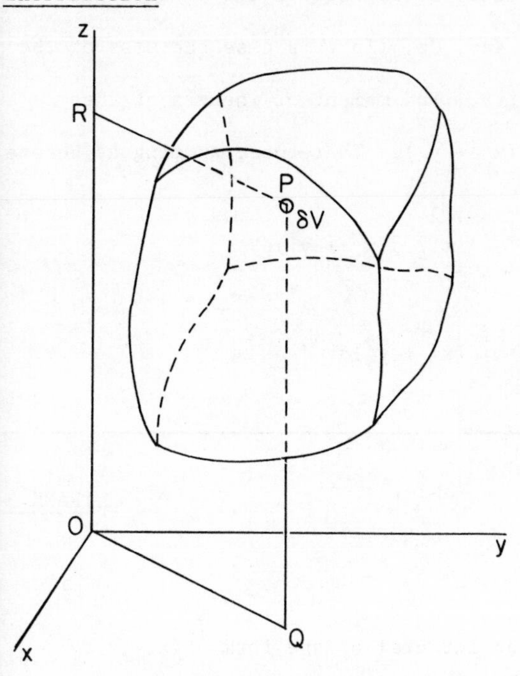

Suppose it is required to find the height above the plane Oxy of the centre of gravity of the solid shown in the diagram.

To find this, it is necessary to obtain the volume of the solid and its first moment of volume about the plane Oxy. A small element of volume δV is taken at a point $P(x,y,z)$ in the solid.

Summing over all the elements δV, and taking the limit as $\delta V \to 0$, gives

$$\text{Total volume } V = \lim_{\delta V \to 0} \sum \delta V$$

$$= \int dV$$

Also the first moment of δV about Oxy is $PQ\delta V$, i.e. $z\delta V$.

$$\therefore \text{ Total first moment } = \lim_{\delta V \to 0} \sum z\delta V$$

$$= \int z dV$$

$\int dV$ and $\int z dV$ are both examples of a VOLUME INTEGRAL, and sometimes written as

$$\int_V dV \quad \text{and} \quad \int_V z dV \quad \text{respectively.}$$

(NOTE: This is a more general approach to finding a volume than that given in FRAME 1 of the programme on Surface Integrals. There we had a special volume in that its cylindrical surface was vertical. In fact, some of the examples which follow can be done without the use of volume integrals, but they serve as simple illustrations of the technique, which comes into its own in more difficult problems.)

FRAME 2

Another example of a volume integral occurs if we require the moment of inertia of the solid in FRAME 1 about, say, Oz. In this case the mass of the element δV is $\sigma\delta V$, where σ is its density. The moment of inertia of the element about Oz is $(\sigma\delta V)PR^2$, i.e. $\sigma\delta V(x^2 + y^2)$. Therefore, summing as before,

$$\text{Total moment of inertia} = \lim_{\delta V \to 0} \sum \sigma(x^2 + y^2)\delta V$$

$$= \int \sigma(x^2 + y^2)dV$$

FRAME 3

Evaluation in Cartesian Coordinates

We shall now consider how to evaluate an integral of the form $\int f(x,y,z)dV$ which, if $f(x,y,z) = \sigma(x^2 + y^2)$, could be the moment of inertia of the solid about Oz.

For δV we start from a point $P(x,y,z)$ and take increments δx, δy and δz in x,y and z respectively, thus forming an element of volume as shown. This gives

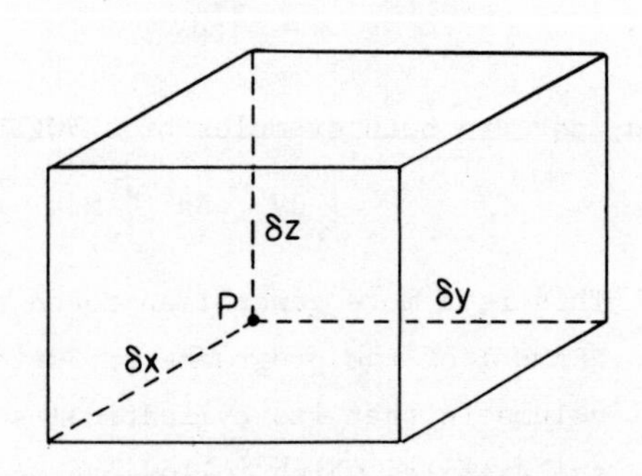

$$\delta V = \delta x\delta y\delta z$$

Summing over all elements δV from Q to R and taking the limit as $\delta z \to 0$, gives approximately the moment of inertia of the column QR about Oz. Thus

FRAME 3 continued

$$\text{M. of I. of column} = \lim_{\delta z \to 0} \sum_{z=z_1}^{z=z_2} \sigma(x^2 + y^2)\delta x \delta y \delta z$$

$$= \left\{ \lim_{\delta z \to 0} \sum_{z=z_1}^{z=z_2} \sigma(x^2 + y^2)\delta z \right\} \delta x \delta y$$

$$= \left\{ \int_{z_1}^{z_2} \sigma(x^2 + y^2)dz \right\} \delta x \delta y$$

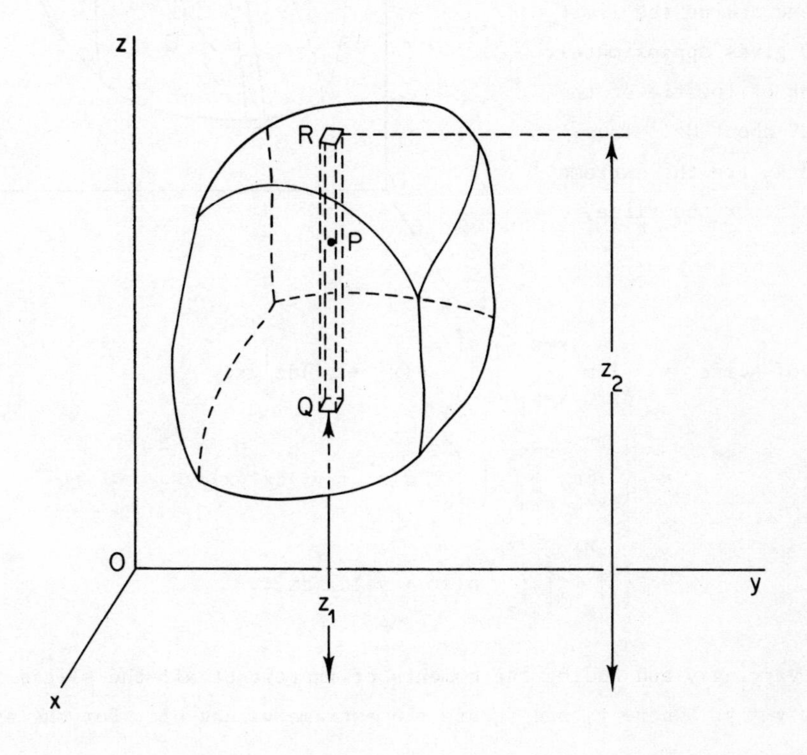

Here, x, y, δx and δy have been kept constant and z allowed to vary. We now allow x to vary but continue to keep y and δy constant (eventually we shall let $\delta x \to 0$).

FRAME 3 continued

The columns thus obtained together make up a slice STUV of thickness δy, perpendicular to Oy, as shown.

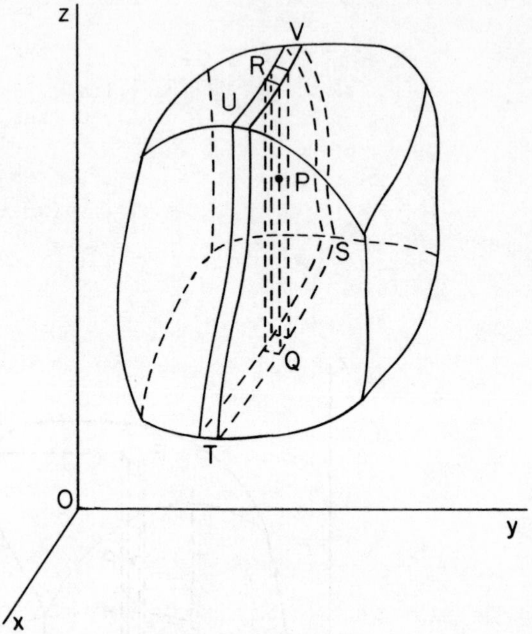

Summing over all these columns and taking the limit as $\delta x \to 0$ gives approximately the moment of inertia of the slice STUV about Oz. Thus, if x_1 and x_2 are the extreme values of x for the slice,

$$\text{M. of I. of slice} = \lim_{\delta x \to 0} \sum_{x=x_1}^{x=x_2} \left(\int_{z_1}^{z_2} \sigma(x^2 + y^2)dz \right) \delta x \delta y$$

$$= \left\{ \lim_{\delta x \to 0} \sum_{x=x_1}^{x=x_2} \left(\int_{z_1}^{z_2} \sigma(x^2 + y^2)dz \right) \delta x \right\} dy$$

$$= \left\{ \int_{x_1}^{x_2} \left(\int_{z_1}^{z_2} \sigma(x^2 + y^2)dz \right) dx \right\} \delta y$$

Finally, varying y and adding the moments of inertia of all the slices from $y = y_1$ to $y = y_2$, where y_1 and y_2 are the extreme values of y for the solid, we have

$$\text{M. of I. of solid} = \lim_{\delta y \to 0} \sum_{y=y_1}^{y=y_2} \left\{ \int_{x_1}^{x_2} \left(\int_{z_1}^{z_2} \sigma(x^2 + y^2)dz \right) dx \right\} \delta y$$

$$= \int_{y_1}^{y_2} \left\{ \int_{x_1}^{x_2} \left(\int_{z_1}^{z_2} \sigma(x^2 + y^2)dz \right) dx \right\} dy \qquad (3.1)$$

FRAME 4

As with double integrals, the notation used for triple integrals in not unique. As three integrations are involved $\int \sigma(x^2 + y^2)dV$ is sometimes written as $\iiint \sigma(x^2 + y^2)dV$. Again, various notations are used for the integral in (3.1), some of which can lead to ambiguity. To avoid this we shall adopt the notation

$$\int_{y_1}^{y_2} dy \int_{x_1}^{x_2} dx \int_{z_1}^{z_2} \sigma(x^2 + y^2)dz$$

which is an extension of the form already used for double integrals.

FRAME 5

We shall now show how to evaluate such an integral, taking as an example $\int_1^4 dy \int_{-3}^{3} dx \int_0^{\frac{1}{2}y} \sigma(x^2 + y^2)dz$.

The first thing to do is to integrate $\sigma(x^2 + y^2)$ with respect to z, treating x and y as constants, and you will then get

$$\int_0^{\frac{1}{2}y} \sigma(x^2 + y^2)dz = \Big[\sigma(x^2 + y^2)z\Big]_0^{\frac{1}{2}y}$$

the limits outside the square brackets being those of z.

Substituting these limits gives $\frac{1}{2}\sigma y(x^2 + y^2)$.

Thus $$\int_1^4 dy \int_{-3}^{3} dx \int_0^{\frac{1}{2}y} \sigma(x^2 + y^2)dz = \int_1^4 dy \int_{-3}^{3} \tfrac{1}{2}\sigma y(x^2 + y^2)dx \qquad (5.1)$$

This is an ordinary double integral which you already know how to evaluate and so we suggest you finish it off.

****

5A

FRAME 6

The integral which has just been evaluated represents the moment of inertia of a certain solid about Oz. To discover what this solid is, it is necessary to examine the limits. These give the boundaries of the solid which in this case are the planes $z = 0$, $z = \frac{1}{2}y$, $x = -3$, $x = 3$, $y = 1$ and $y = 4$. For volume integrals, in general, the boundaries are surfaces, whereas for double integrals you have seen that they are curves. Sketching three-dimensional figures sometimes presents problems, and you will have to use your imagination instead. However in this example the diagram is relatively simple and is shown below.

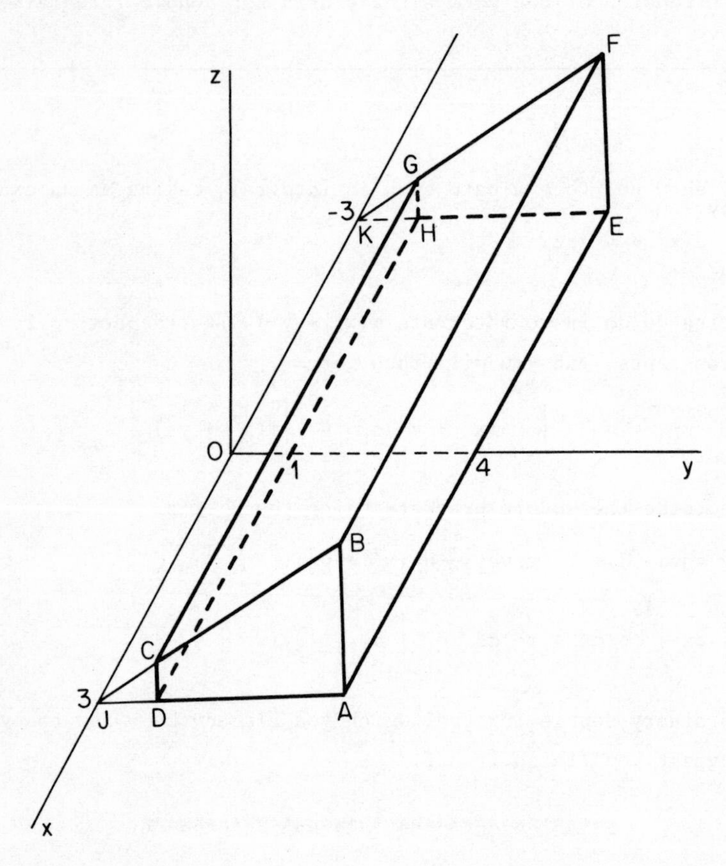

The equation of ADHE is $z = 0$ and that of BCGF is $z = \frac{1}{2}y$.

The equation of HEFG is $x = -3$ and that of DABC is $x = 3$.

The equation of CDHG is $y = 1$ and that of BAEF is $y = 4$.

FRAME 6 continued

A column such as QR in FRAME 3 would be parallel to Oz and extend from face ADHE to face BCGF, and a slice such as STUV would be rectangular and parallel to sides CDHG and BAEF. What would be the form of the integral if the same columns were taken but the slices were parallel to the faces DABC and HEFG?

****

6A

$$\int_{-3}^{3} dx \int_{1}^{4} dy \int_{0}^{\frac{1}{2}y} \sigma(x^2 + y^2)dz$$

FRAME 7

Referring to the integral $\int_{1}^{4} dy \int_{-3}^{3} \frac{1}{2}\sigma y(x^2 + y^2)dx$ in (5.1), what area of integration is defined by the limits?

****

7A

The rectangle bounded by the lines $x = -3$, $x = 3$, $y = 1$ and $y = 4$, i.e. ADHE in the figure in FRAME 6.

It is the area in the plane Oxy over which columns such as QR extend.

FRAME 8

Moments of inertia are usually expressed in terms of mass. In this example a formula is known for the volume, i.e. $6 \times$ Area ABCD $= 6 \times \frac{15}{4} = \frac{45}{2}$.
The moment of inertia is thus $\frac{23}{2}$ M, where M = total mass $= \frac{45}{2}\sigma$.

However, the mass can itself be expressed as a triple integral. Can you give either of the forms it would take in this case if the columns are still parallel to Oz?

****

8A

$$\int_1^4 dy \int_{-3}^3 dx \int_0^{\frac{1}{2}y} \sigma dz \quad or \quad \int_{-3}^3 dx \int_1^4 dy \int_0^{\frac{1}{2}y} \sigma dz$$

FRAME 9

The integral in FRAME 5, i.e. $\int_1^4 dy \int_{-3}^3 dx \int_0^{\frac{1}{2}y} \sigma(x^2 + y^2)dz$, gave the moment of inertia about Oz of the solid shown in FRAME 6. Can you describe the solid whose moment of inertia about Oz is given by

$$\int_0^4 dy \int_{-3}^3 dx \int_0^{\frac{1}{2}y} \sigma(x^2 + y^2)dz?$$

****

9A

The triangular wedge whose ends are the triangles ABJ and EFK in the figure of FRAME 6. Notice that as the lower limit of y decreases from 1 to 0, the face CDHG becomes the line JK. In the latter case three of the bounding planes ($z = 0$, $z = \frac{1}{2}y$ and $y = 0$) meet in the line JK.

FRAME 10

In finding the centre of gravity of a certain solid, the integral

$$\int_0^1 dx \int_0^{1-x} dy \int_0^{1-x-y} ky dz \quad \text{occurred.}$$

(i) Sketch the solid. (HINT: If in difficulty, remember that the limits for x and y define the area in the plane Oxy over which columns extend.)

(ii) Suggest what this integral represented.

****

10A

(i) *The solid is the tetrahedron OABC. The boundaries $z = 0$ and $z = 1 - x - y$ are the planes OAB and ABC respectively. ΔOAB is the area in the plane Oxy over which columns extend.*

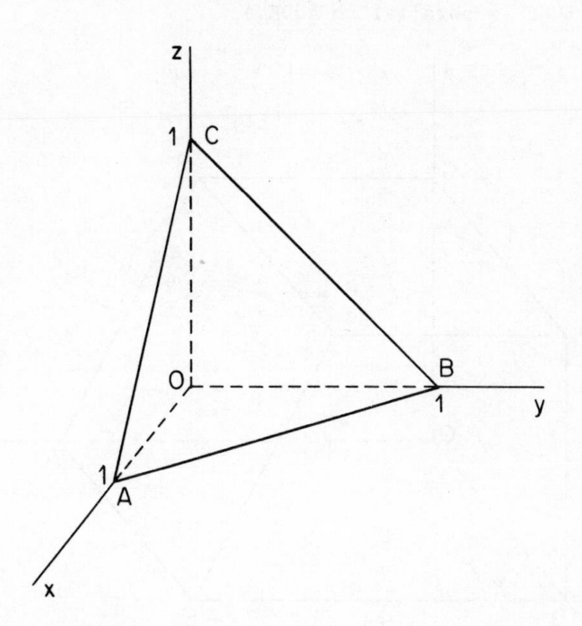

(ii) There are two possibilities:

(a) *For a solid of uniform density k, the integral would have represented the first moment about the plane Oxz, necessary for finding $\bar{y}$.*

(b) *For a solid of variable density, the density at any point in it being equal to ky, the integral would have represented the mass.*

FRAME 11

Having seen how to obtain the shape of the solid from the limits of a triple integral, we shall now look at the reverse problem, i.e. building up the limits from a given solid.

FRAME 11 continued

This will be illustrated by the evaluation of $\int(x^2 + z^2)dV$ throughout the solid shown in the following diagram. (All faces except OABC and DEFG are rectangular, and BCGF is parallel to AODE.)

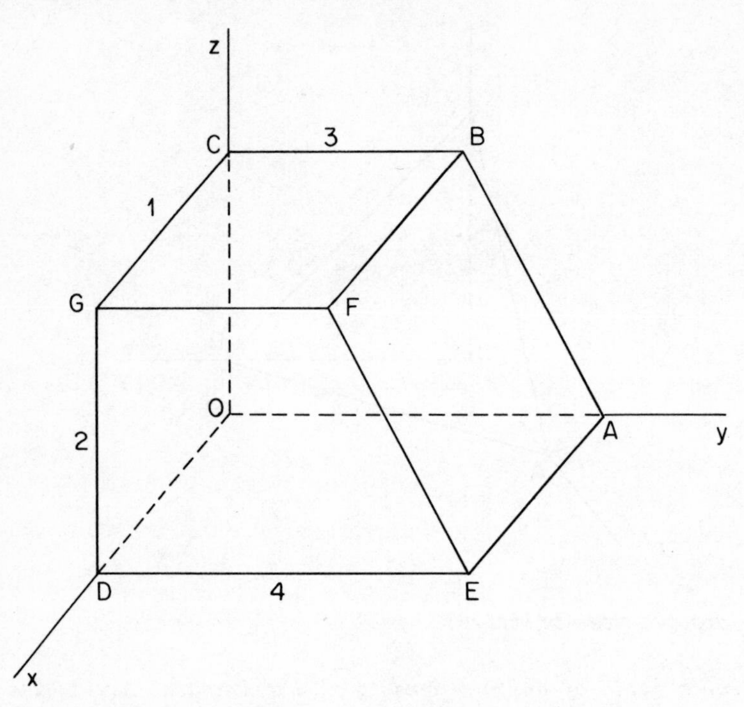

Previously columns were taken parallel to Oz, but in this case we shall take them parallel to Oy. Can you see the disadvantage in taking them parallel to Oz here?

****

11A

Two triple integrals would be required as some columns would extend up to the face BCGF and others to the face ABFE.

This difficulty could be overcome equally well by taking columns parallel to Ox.

FRAME 12

Columns parallel to Oy will have one end in the face OCGD and the other in the face ABFE. The lower limit for y will therefore by $y = 0$. The upper limit $y = 4 - \frac{1}{2}z$ can be obtained from the diagram below, which shows a cross-section of the solid perpendicular to Ox.

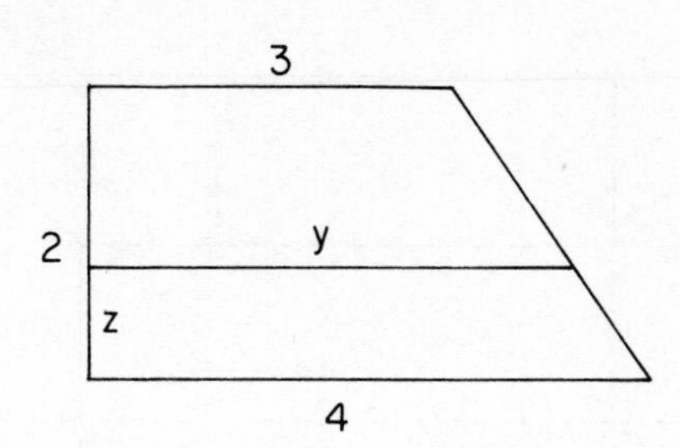

The other limits are obtained by considering the area in the plane Ozx over which the columns extend, i.e. rectangle OCGD. Thus the limits for x are $x = 0$ and $x = 1$, and those for z are $z = 0$ and $z = 2$.

Now write down a triple integral that will give $\int (x^2 + z^2)dV$ throughout the solid.

****

12A

Either $\int_0^1 dx \int_0^2 dz \int_0^{4-\frac{1}{2}z} (x^2 + z^2)dy$ *or* $\int_0^2 dz \int_0^1 dx \int_0^{4-\frac{1}{2}z} (x^2 + z^2)dy$

FRAME 13

Now evaluate whichever of these two you have written down.

****

13A

FRAME 14

Now write down a triple integral which would give $\int \frac{1}{x^2 + y^2 + z^2}\, dV$ throughout the rectangular box shown, taking columns parallel to Ox. (All edges are parallel to the axes.)

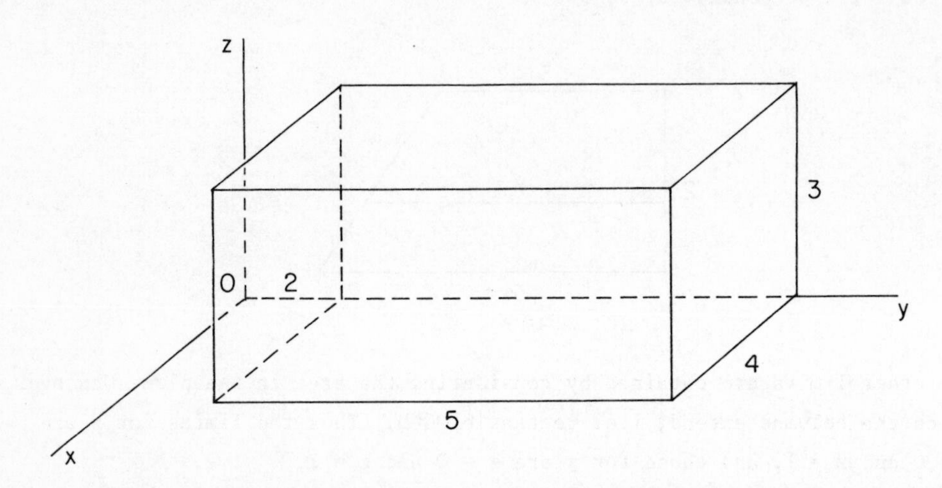

****

14A

Either $\int_2^7 dy \int_0^3 dz \int_0^4 \frac{1}{x^2 + y^2 + z^2}\, dx$ *or* $\int_0^3 dz \int_2^7 dy \int_0^4 \frac{1}{x^2 + y^2 + z^2}\, dx$

It is, of course, just as easy to take columns parallel to Oy or Oz.

FRAME 15

Evaluation in Cylindrical Polar Coordinates

As in two-dimensional problems, Cartesian coordinates are not always the best to use. In three dimensions, two other systems are commonly in use - cylindrical polar coordinates and spherical polar coordinates. We shall first turn our attention to cylindrical coordinates.

FRAME 15 continued

We begin by deciding what element of volume to take for δV. To do this, starting from a point $P(\rho,\phi,z)$, increments $\delta\rho$, $\delta\phi$ and δz are taken. The element of volume δV thus obtained is as shown.

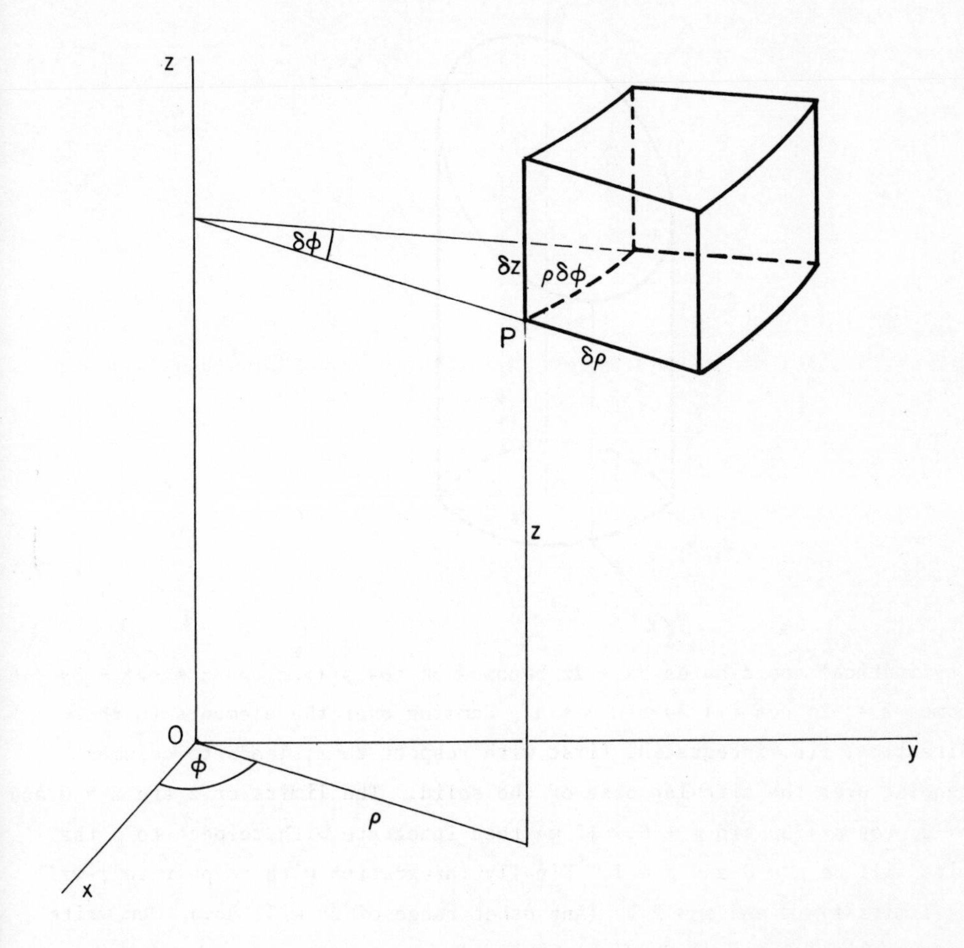

You will notice that the element so formed is approximately a rectangular box whose edges are $\delta\rho$, $\rho\delta\phi$ and δz. Its volume can therefore be taken as $\delta\rho \times \rho\delta\phi \times \delta z$, i.e. δV is replaced by $\rho\delta\rho\delta\phi\delta z$ and $\int f(\rho,\phi,z)dV$ becomes the triple integral $\iiint f(\rho,\phi,z)\,\rho d\rho d\phi dz$.

FRAME 16

To illustrate these ideas, $\int(3x + 2z)dV$ throughout the portion of the cylinder $x^2 + y^2 = 1$ cut off by the planes $z = 0$ and $z = -2x + 3y + 6$ will be formulated in cylindrical coordinates. This solid is illustrated here.

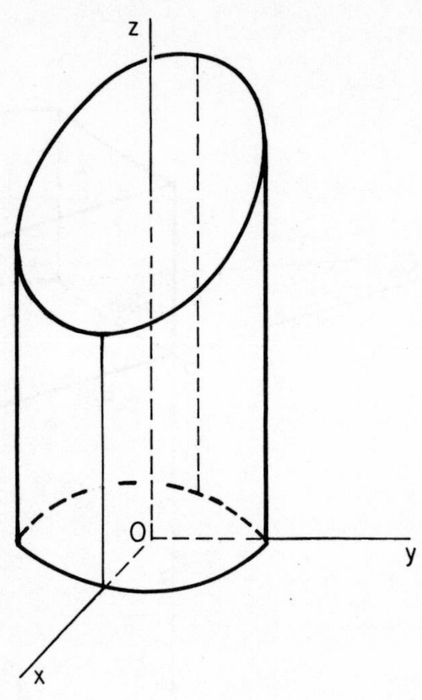

In cylindrical coordinates $3x + 2z$ becomes $3\rho \cos \phi + 2z$ and $z = -2x + 3y + 6$ becomes $z = -2\rho \cos \phi + 3\rho \sin \phi + 6$. Summing over the elements in the z-direction, i.e. integrating first with respect to z, leads to columns extending over the circular base of the solid. The limits of z are $z = 0$ and $z = -2\rho \cos \phi + 3\rho \sin \phi + 6$. If we then integrate with respect to ρ the limits will be $\rho = 0$ and $\rho = 1$. Finally integration with respect to ϕ will give limits $\phi = 0$ and $\phi = 2\pi$. (Any other range of 2π will do.) Now write down the triple integral thus obtained.

****

16A

$$\int_0^{2\pi} d\phi \int_0^1 d\rho \int_0^{-2\rho \cos \phi + 3\rho \sin \phi + 6} (3\rho \cos \phi + 2z)\rho dz$$

FRAME 17

The following diagram represents a quarter of a cylinder of height 3 and radius 2. Express $\int xyz\,dV$ throughout this volume as a triple integral in cylindrical coordinates and evaluate it.

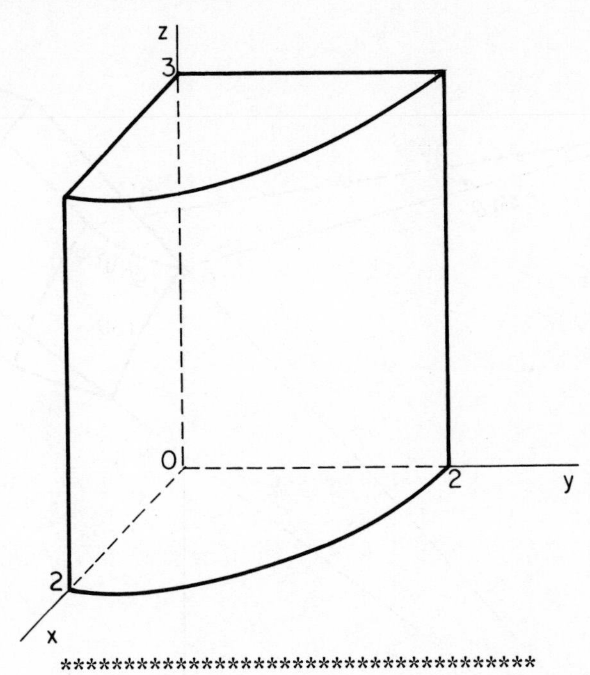

****

17A

One possible integral is $\int_0^{\pi/2} d\phi \int_0^2 d\rho \int_0^3 \rho^3 z \sin\phi \cos\phi \, dz$. *This is for integration in the order* z, ρ, ϕ, *which is usual with cylindrical ccordinates. Other orders of integration could be used in this example, without causing difficulty. Whatever order you have chosen, you should have obtained the answer 9.*

FRAME 18

Evaluation in Spherical Polar Coordinates

Having shown how cylindrical coordinates are used, we shall now have a look at spherical coordinates.

FRAME 18 continued

The first thing to do is to decide what element of volume to take for δV. Starting from a point $P(r,\theta,\phi)$, increments δr, $\delta\theta$ and $\delta\phi$ are taken to give the element of volume shown in the diagram.

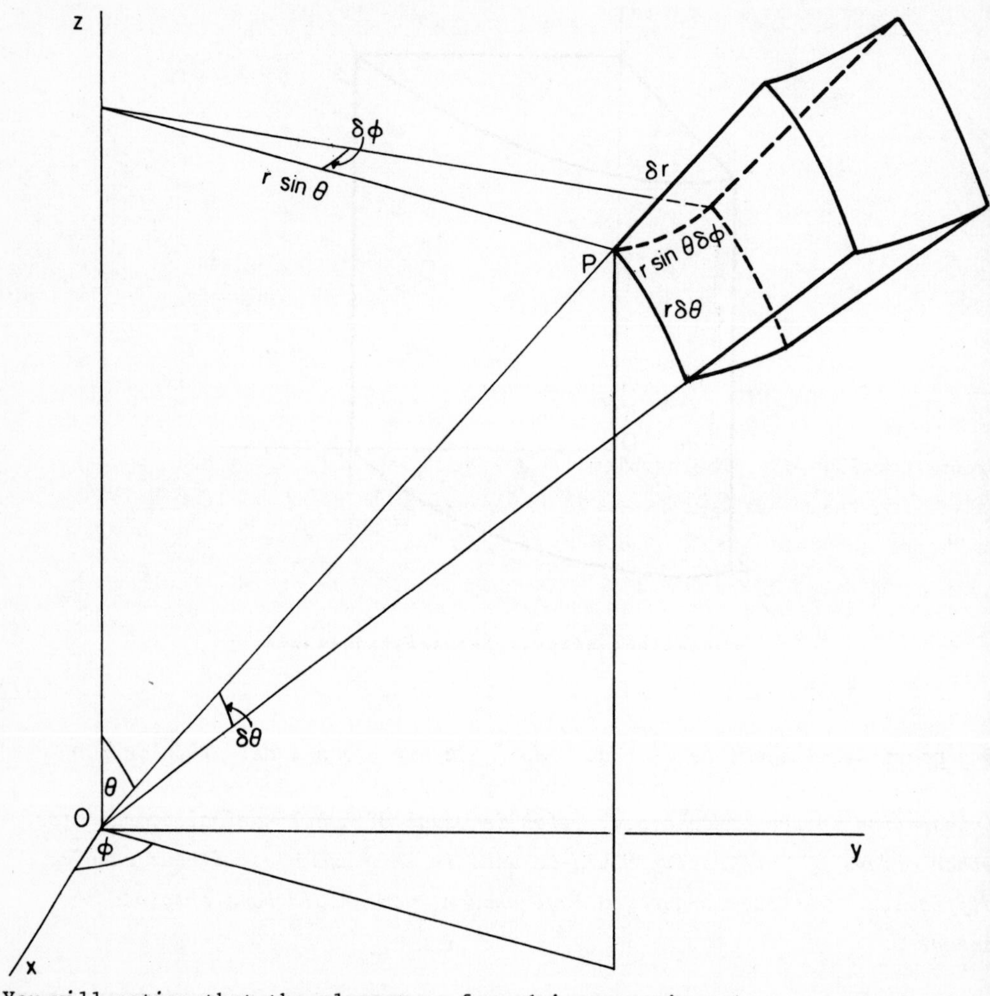

You will notice that the element so formed is approximately a rectangular box whose edges are δr, $r\delta\theta$ and $r \sin \theta \, \delta\phi$. Its volume can therefore be taken as $\delta r \times r\delta\theta \times r \sin \theta \, \delta\phi$, i.e. δV is replaced by $r^2 \sin \theta \, \delta r \delta\theta \delta\phi$ and $\int f(r,\theta,\phi) dV$ becomes the triple integral $\iiint f(r,\theta,\phi) r^2 \sin \theta \, dr d\theta d\phi$.

FRAME 19

The figure represents a cone OABCD surmounted by a spherical cap ABCDE. The axis of the cone is along Oz and its semi-vertical angle is 30°. The sphere, of which the cap is part, has its centre at O and is of radius 2.

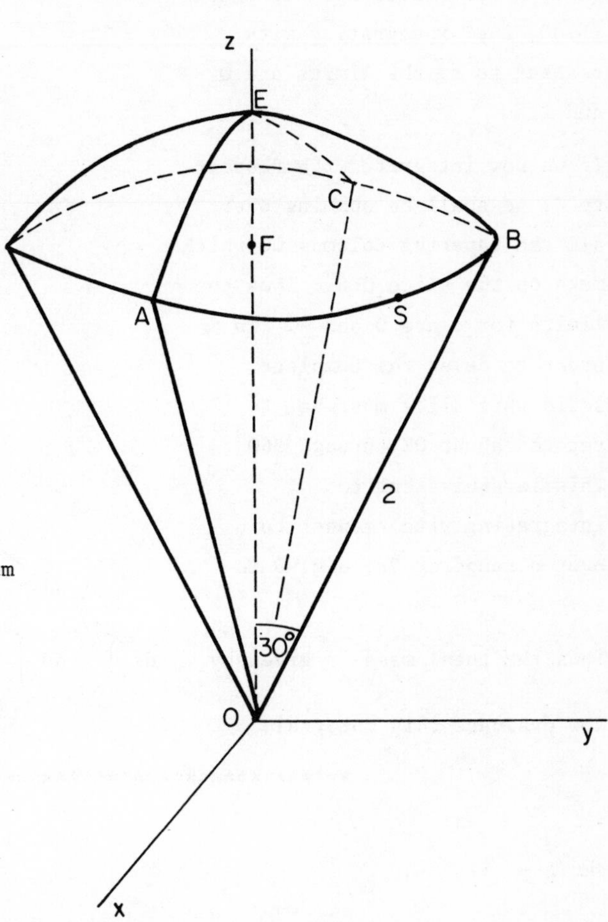

A solid of this shape is of variable density, the density at any point P being proportional to OP. The problem is to find the position of its centre of gravity.

The easiest solution to this problem uses spherical coordinates.

First, it will be necessary to find the mass of the solid. An element of volume δV at P will have mass k OP $\delta V = kr\,\delta V$.

$$\therefore \text{ Total mass } = \int kr\,dV$$

$$= \iiint kr\ r^2 \sin\theta\ dr d\theta d\phi$$

$$= \iiint kr^3 \sin\theta\ dr d\theta d\phi$$

The limits for this integral must now be found.

FRAME 19 continued

Summing first over all elements in OQ, i.e. integrating with respect to r, the limits are 0 and 2.

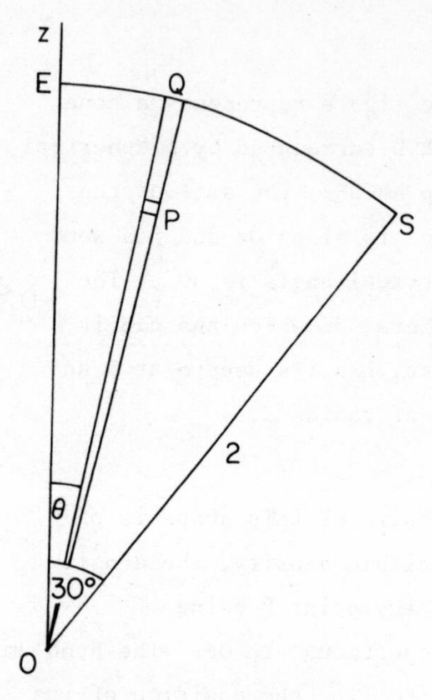

If we now integrate with respect to θ, we shall be summing over all the tapering columns OQ which make up the slice OES. Thus the limits for θ are 0 and $\frac{\pi}{6}$. In order to cover the complete solid this slice must be rotated about OE through 360°. This is equivalent to integrating with respect to ϕ over a range of 2π, e.g. 0 to 2π.

Thus the total mass is given by $\int_0^{2\pi} d\phi \int_0^{\pi/6} d\theta \int_0^2 kr^3 \sin\theta\, dr$

Now evaluate this integral.

****

19A

$4k\pi(2 - \sqrt{3})$

FRAME 20

By symmetry the centre of gravity will be on Oz.
To find $\bar{z}$, we take moments about Oxy.
As the mass of the element δV at P is $kr\delta V$ its moment about Oxy is $(kr\delta V)z$, i.e. $kr^2\cos\theta\ \delta V$ and the total moment for the complete solid is $\int kr^2\cos\theta\ dV$.

Keeping to the same order of integration as before, express this as a triple integral, evaluate it and so find $\bar{z}$.

****

20A

$$\int_0^{2\pi} d\phi \int_0^{\pi/6} d\theta \int_0^2 kr^4 \cos\theta \sin\theta \, dr = \frac{8k\pi}{5}$$

$$\bar{z} = \frac{2}{5(2 - \sqrt{3})}$$

FRAME 21

Now use spherical coordinates to evaluate $\int \frac{1}{x^2 + y^2 + z^2} dV$ throughout the portion of the sphere, centre O and radius a, which lies in the positive octant.

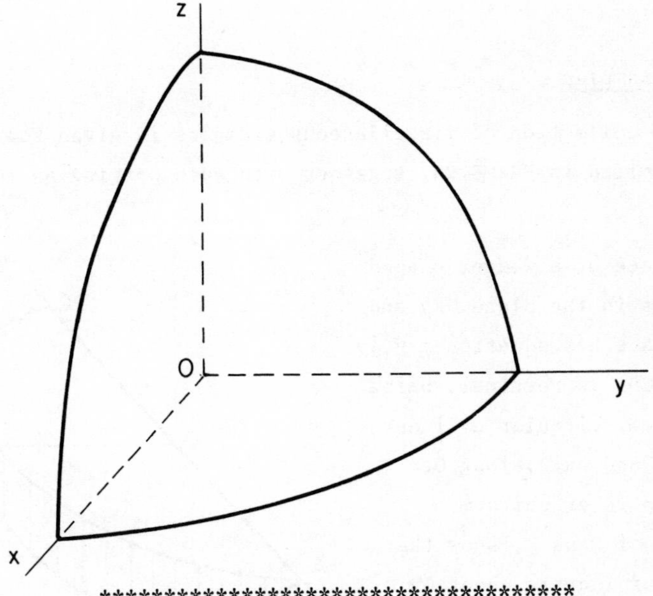

****

21A

$$\int_0^{\pi/2} d\phi \int_0^{\pi/2} d\theta \int_0^a \frac{1}{r^2} r^2 \sin\theta \, dr = \int_0^{\pi/2} d\phi \int_0^{\pi/2} d\theta \int_0^a \sin\theta \, dr$$

$$= \frac{\pi a}{2}$$

Any other order of integration is permissible.

FRAME 22

Choice of system of coordinates

In general, the system of coordinates which it is most convenient to use depends on the shape of the solid throughout which integration is to be taken. If the solid is a cylinder, or part of a cylinder, it is usually best to use cylindrical coordinates.

If the solid is a sphere, or part of a sphere, it is usually best to use spherical coordinates.

In other cases, it is usually best to use Cartesian coordinates.

FRAME 23

Miscellaneous Examples

In this frame a collection of miscellaneous examples is given for you to try. Answers are supplied in FRAME 24, together with such working as is considered helpful.

1. The lower face of a wedge, shaped as shown, is in the plane Oxy and the upper face has equation $z = \frac{1}{2}y$. The side ABCDA is vertical, being part of a semi-circular cylinder of radius a and axis along Oz. If the wedge is of uniform density and of mass M, show that its moment of inertia about Ox is $\frac{13}{30}$ Ma^2.

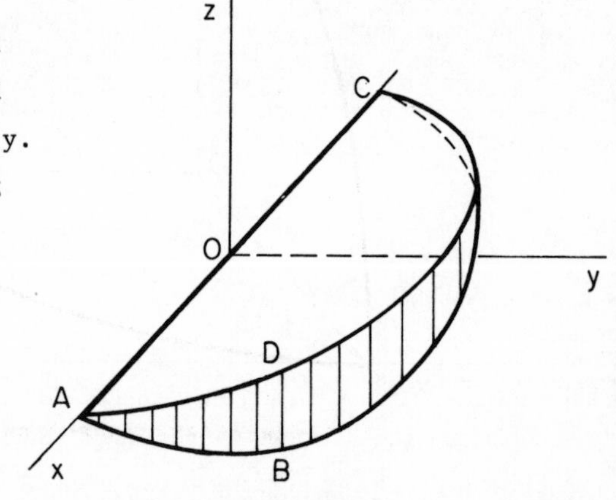

2. A gasholder is in the form of a vertical cylinder of radius a and height h surmounted by a hemispherical top of radius a. The density σ of the gas at a height z above the base of the cylinder is given by $\sigma = ce^{-z}$, where c is a constant. Find the mass of the gas in the hemispherical part. (HINT: Take the pole at the centre of the base of the hemisphere. Look carefully at your integral before evaluating it to see whether a change in the order of integration could help.)

FRAME 23 continued

3. The elastic energy per unit volume of a material is $\frac{q^2}{2E}$ J/m^3 where q N/m^2 is the stress and E N/m^2 the modulus of elasticity (constant for a given material). Find the elastic energy of a circular cylinder radius R m, length ℓ m, in which the stress q varies as the distance from the axis of the cylinder and is equal to q_o at the outer surface.

4.

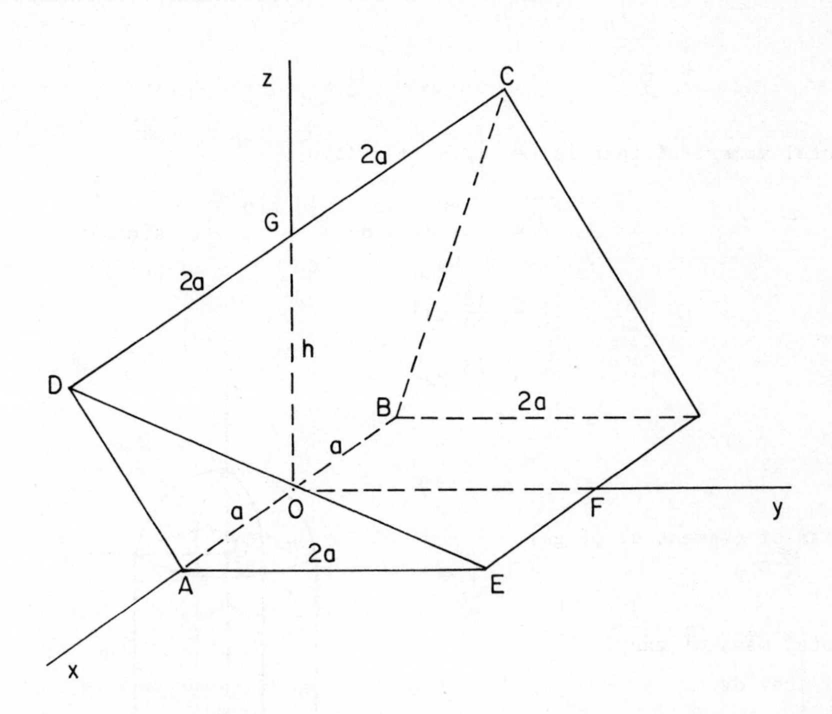

The diagram illustrates a dam. The water face ABCD is vertical and is a trapezium with the dimensions shown. The thickness of the dam increases uniformly from zero at the top to 2a at the bottom. Taking axes as shown, find $\bar{y}$. (It is suggested that you take columns parallel to Ox.)

5. Find the centre of gravity of the conical part of the solid described in FRAME 19.

FRAME 24

Answers to Miscellaneous Examples

1. Mass of element of volume $\delta V = \sigma\delta V$, where σ is its density.
   Moment of inertia of element about Ox $= \sigma(y^2 + z^2)\delta V$
   Total mass $= \int \sigma dV$

$$= \sigma \int_0^{\pi} d\phi \int_0^{a} d\rho \int_0^{\frac{1}{2}\rho \sin\phi} \rho dz \quad \text{(cylindrical coordinates)}$$

$$= \frac{1}{3}\sigma a^3$$

Total moment of inertia $= \int \sigma(y^2 + z^2)dV$

$$= \sigma \int_0^{\pi} d\phi \int_0^{a} d\rho \int_0^{\frac{1}{2}\rho \sin\phi} \{(\rho \sin\phi)^2 + z^2\}\rho dz$$

$$= \frac{13}{90}\sigma a^5$$

$$= \frac{13}{30} Ma^2$$

2.

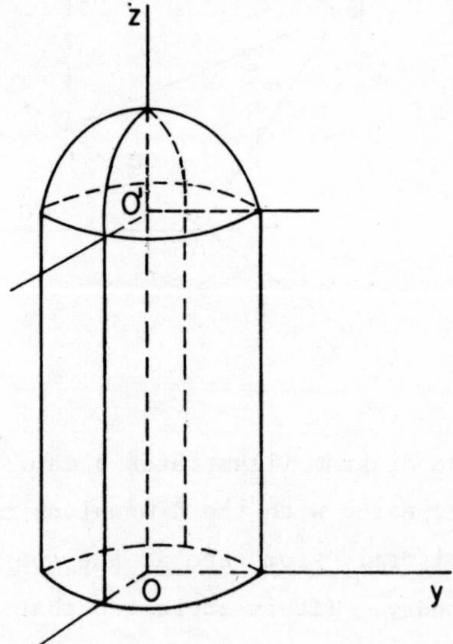

Mass of element δV of gas
$= ce^{-z}\delta V$

Total mass of gas
$= \int ce^{-z}dV$

In this example it is best to work in spherical coordinates with the pole at O'.

Then total mass

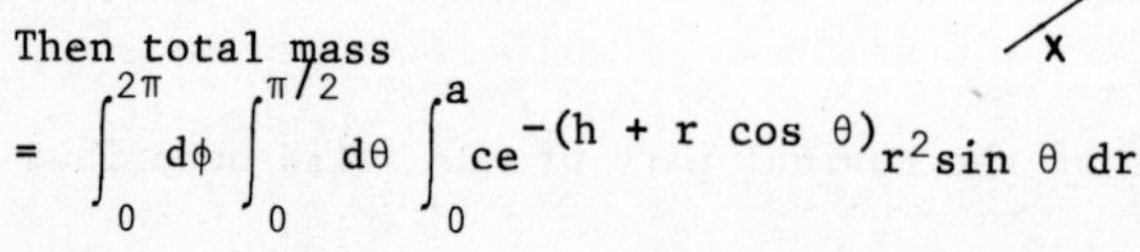

$$= \int_0^{2\pi} d\phi \int_0^{\pi/2} d\theta \int_0^{a} ce^{-(h + r\cos\theta)} r^2 \sin\theta \, dr$$

FRAME 24 continued

It is easier in this case to integrate with respect to θ before integrating with respect to r.

As the limits are all constants,

$$\text{total mass} = \int_0^{2\pi} d\phi \int_0^a dr \int_0^{\pi/2} ce^{-h}e^{-r\cos\theta}\, r^2\sin\theta\, d\theta$$

$$= ce^{-h}\int_0^{2\pi} d\phi \int_0^a r(1 - e^{-r})dr$$

$$= c\pi e^{-h}\{a^2 + 2(a + 1)e^{-a} - 2\}$$

3. $q = \dfrac{q_o}{R}\rho$, using cylindrical coordinates

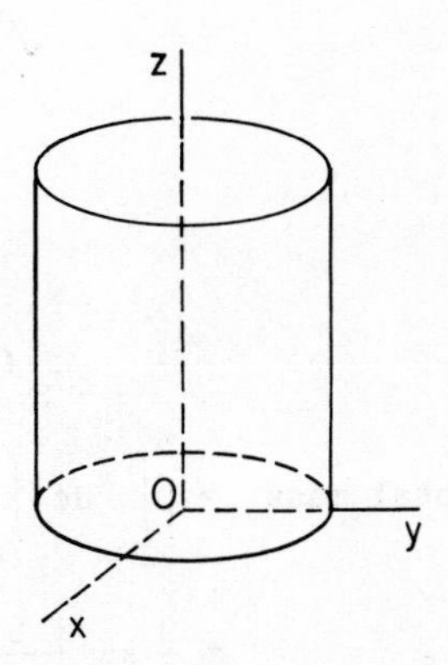

$$\text{Elastic energy} = \int \frac{1}{2E}\left(\frac{q_o}{R}\rho\right)^2 dV$$

$$= \int_0^{2\pi} d\phi \int_0^R d\rho \int_0^{\ell} \frac{q_o^2\rho^2}{2ER^2}\rho dz$$

$$= \frac{1}{4E}\pi q_o^2 R^2\ell \text{ J}$$

4. Coordinates of A are (a,0,0), of D are (2a,0,h) and of E are (a,2a,0).
Equation of plane ADE is $hx - az = ah$.
Mass of dam $= \int \sigma dV$ where σ is its density
First moment of mass of dam about plane Oxz $= \int \sigma y dV$
In two dimensions (plane Oyz), equation of line FG is $hy + 2az = 2ah$
As the dam is symmetrical about the plane Oyz,

$$\text{mass of dam} = 2\sigma\int_0^h dz \int_0^{2a(1-z/h)} dy \int_0^{a(1+z/h)} dx$$

$$= 8\sigma a^2h/3$$

$$\text{Moment of dam about Oxz} = 2\sigma\int_0^h dz \int_0^{2a(1-z/h)} dy \int_0^{a(1+z/h)} y dx$$

$$= 5\sigma a^3h/3$$

$$\bar{y} = 5a/8$$

FRAME 24 continued

5. The diagram is the one for this case corresponding to the second diagram in FRAME 19.

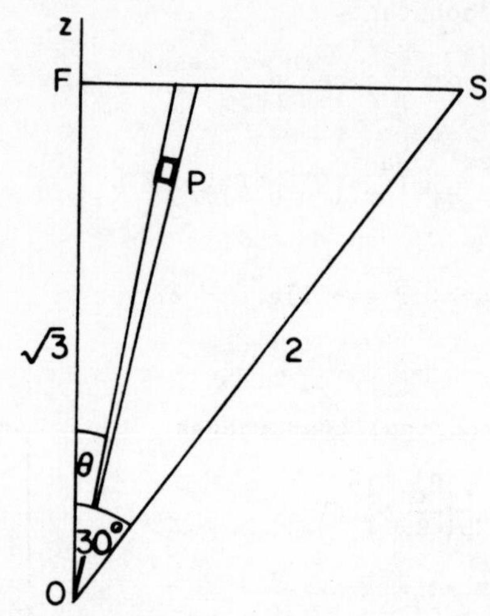

$$\text{Total mass} = \int_0^{2\pi} d\phi \int_0^{\pi/6} d\theta \int_0^{\sqrt{3}\sec\theta} kr^3 \sin\theta \, dr$$

$$= \frac{3}{2} k\pi \left(\frac{8}{3\sqrt{3}} - 1\right)$$

$$\text{Total moment about Oxy} = \int_0^{2\pi} d\phi \int_0^{\pi/6} d\theta \int_0^{\sqrt{3}\sec\theta} kr^4 \cos\theta \sin\theta \, dr$$

$$= \frac{6\sqrt{3}}{5} k\pi \left(\frac{8}{3\sqrt{3}} - 1\right)$$

$$\bar{z} = \frac{4\sqrt{3}}{5}$$

As before, $\bar{x} = 0$, and $\bar{y} = 0$ by symmetry.

UNIT 11

VECTOR ANALYSIS

A.C. Bajpai
I.M. Calus
J.A. Fairley
D. Walker

Loughborough University of Technology

INSTRUCTIONS

This Unit comprises two programmes:

(a) Vector Analysis I
(b) Vector Analysis II

Each programme is divided up into a number of FRAMES which are to be worked *in the order given*. You will be required to participate in many of these frames and in such cases the answers are provided in ANSWER FRAMES, designated by the letter A following the frame number. Steps in the working are given where this is considered helpful. The answer frame is separated from the main frame by a line of asterisks: *********. Keep the answers covered until you have written your own response. If your answer is wrong, go back and try to see why. Do not proceed to the next frame until you have corrected any mistakes in your attempt and are satisfied that you understand the contents up to this point.

Before reading these programmes, it is necessary that you are familiar with the following

Prerequisites

For (a): The contents of the Vector Algebra programme in Unit 3 of Vol I.

For (b): The contents of (a).

Line, surface and volume integrals as covered in Unit 10.

CONTENTS

Instructions

FRAMES PAGE

VECTOR ANALYSIS II

VECTOR ANALYSIS I

FRAME 1

The idea of a Vector Function

In order to obtain an appreciation of what we mean by a "vector function" let us consider a bead moving along a wire AB in space.

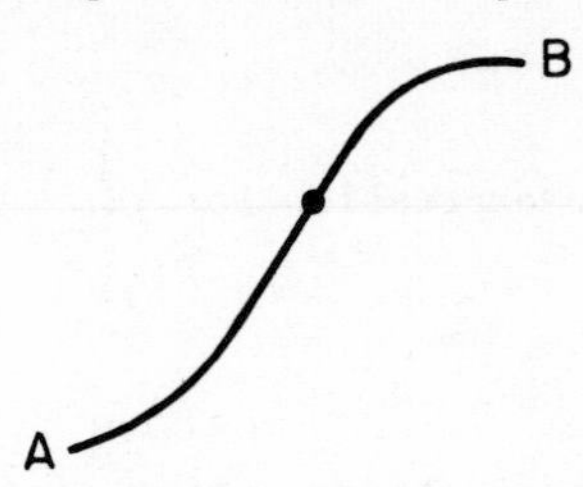

We can specify the position, P, of the bead on AB by taking an origin O and drawing the position vector $\underline{r}$ from O to P.

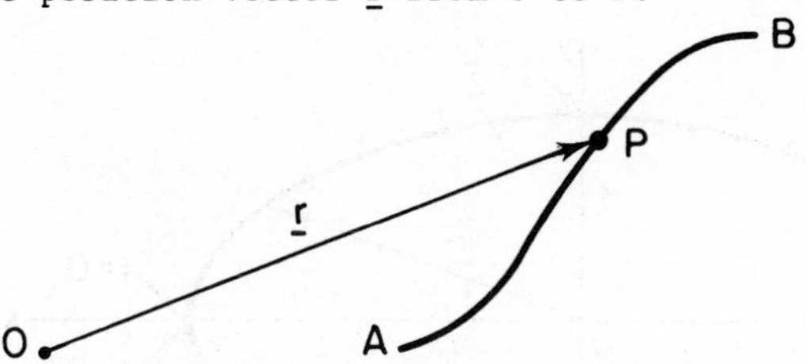

As the bead moves, it will be at different positions on the wire at different times. Thus, its position will be a function of time, and consequently the vector $\underline{r}$ will be a function of time. We denote this by writing

$$\underline{r} = \underline{r}(t)$$

In fact this equation will specify the equation of the curve AB, giving parametric equations for the curve with t as parameter.

As an example, consider a bead moving with constant angular velocity ω on a fixed circular wire, radius a. Take O as the centre of the circle with axes as shown. If A is the position of the bead at time $t = 0$ and P is its position at a subsequent time t, then

$\angle AOP = \omega t$ and $\overrightarrow{OP} = \underline{r} = a \cos \omega t \, \underline{i} + a \sin \omega t \, \underline{j}$

Clearly $\underline{r}$ is a function of t only.

Furthermore we can regard $\underline{r} = a \cos \omega t \, \underline{i} + a \sin \omega t \, \underline{j}$ as being the equation of the circle. Its parametric equations are $x = a \cos \omega t$, $y = a \sin \omega t$.

FRAME 2

We shall now find the form of the curve traversed by the bead when $\underline{r} = 2 \cos \pi t \, \underline{i} + \sin \pi t \, \underline{j}$, sketching the curve and indicating the positions of the bead at $t = 0$, $t = \frac{1}{2}$, $t = 1$, $t = \frac{3}{2}$, $t = 2$.

Now $\quad \underline{r} = x\underline{i} + y\underline{j}$

Comparing this with the given expression for $\underline{r}$,

$$\begin{cases} x = 2 \cos \pi t \\ y = \sin \pi t \end{cases}$$

We can eliminate t from these equations, using $\sin^2\pi t + \cos^2\pi t = 1$, to give

$$\frac{x^2}{4} + y^2 = 1$$

which is an ellipse.

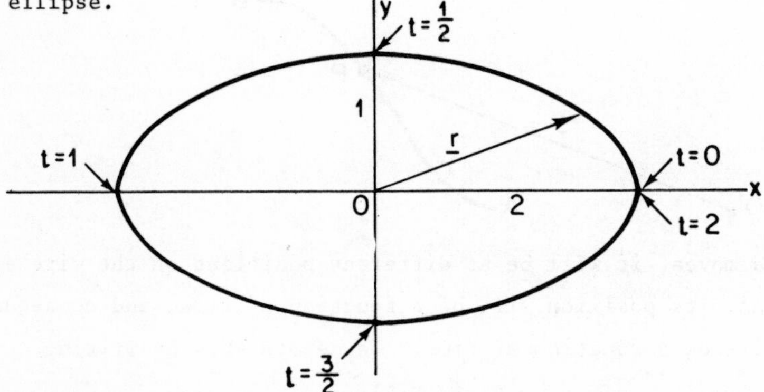

It can be seen that the bead traverses the ellipse between $t = 0$ and $t = 2$.

You should note that we do not always eliminate t (indeed it is sometimes impossible) but we can always plot points on the curve by calculating the x- and y-coordinates for each of a set of values of t.

Now show by a sketch the form of the curve traversed in each of the following cases:

(i) $\underline{r} = 3\underline{i} + t\underline{j}$

(ii) $\underline{r} = \cos \pi t \, \underline{i} + 3 \sin \pi t \, \underline{j}$

(iii) $\underline{r} = \sin t \, \underline{i} + 0\underline{j}$

(iv) $\underline{r} = t\underline{i} + |t|\underline{j}$

****

<u>2A</u>

(i)

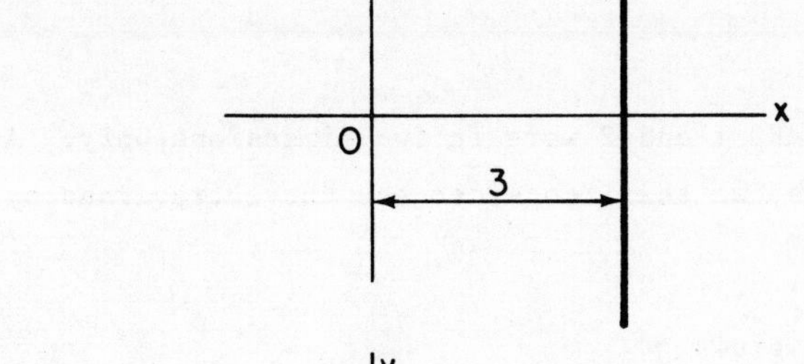

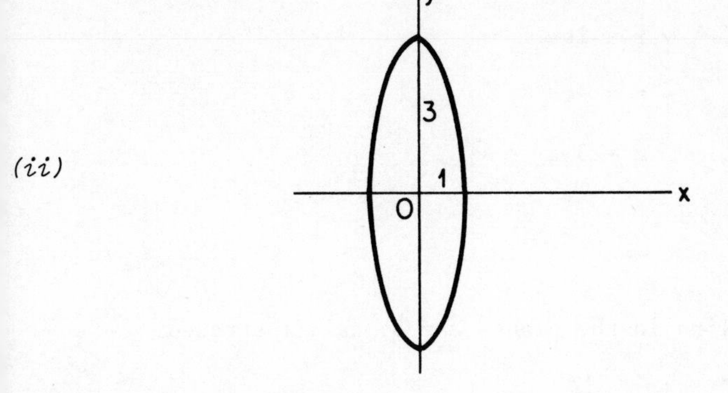

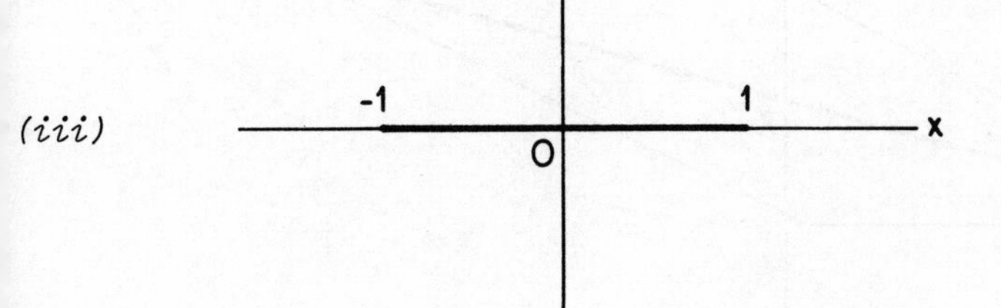

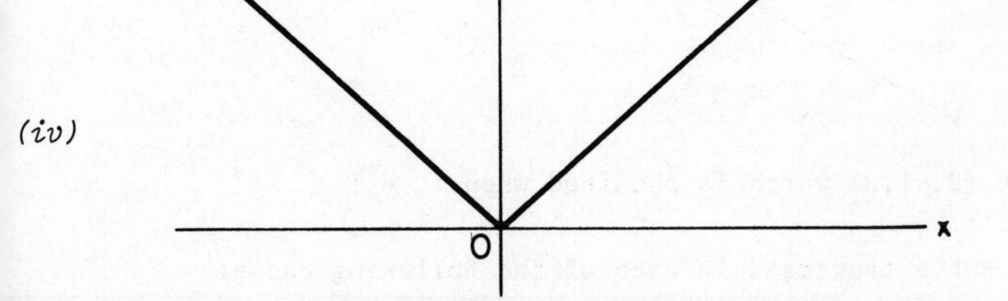

FRAME 3

All of the examples in FRAMES 1 and 2 were in two dimensions only. As an example in three dimensions, we shall consider the curve traversed by the bead when $\underline{r} = 2t\underline{i} - t\underline{j} + 3\underline{k}$.

As $\underline{r} = x\underline{i} + y\underline{j} + z\underline{k}$, this gives

$$\begin{cases} x = 2t \\ y = -t \\ z = 3 \end{cases}$$

and so
$$\begin{cases} x + 2y = 0 \\ z = 3 \end{cases}$$

This represents a straight line in the plane $z = 3$, as illustrated.

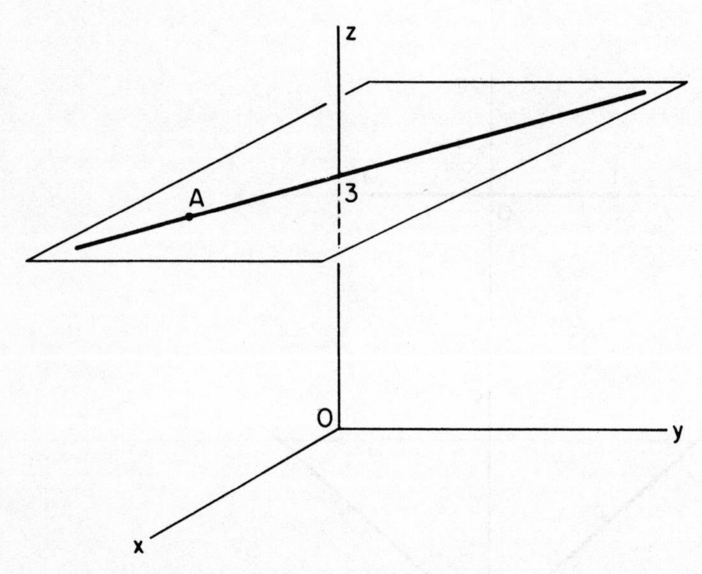

Here A is the point (2,-1,3) which is obtained when $t = 1$

Now illustrate the curve traversed in each of the following cases:

(i) $\underline{r} = 3 \cos t\, \underline{i} + \sin t\, \underline{j} + 2\underline{k}$

(ii) $\underline{r} = 2\underline{i} + t^2\underline{j} + t\underline{k}$

****

(i) $x = 3 \cos t$

$y = \sin t$

$z = 2$

This represents the ellipse $\frac{x^2}{9} + y^2 = 1$ *in the plane* $z = 2$

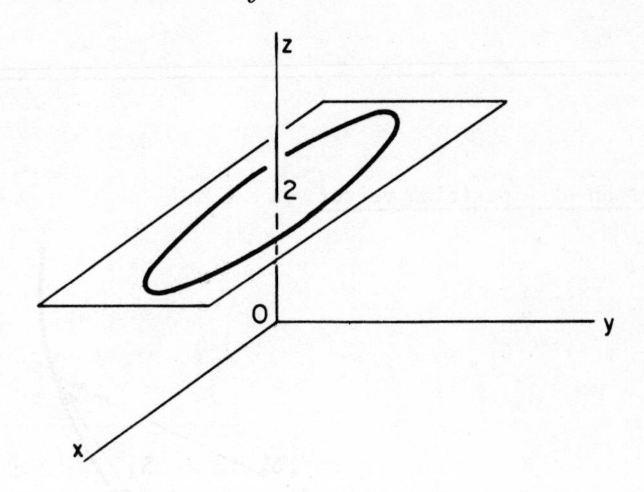

(ii) $x = 2$

$y = t^2$

$z = t$

This represents the parabola $y = z^2$ *in the plane* $x = 2$

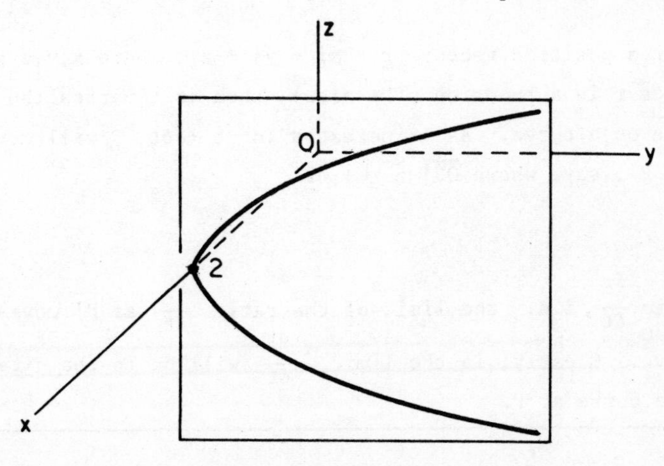

FRAME 4

In the above work we have been considering a vector $\underline{r}$ which is a function of one scalar variable t. We shall find later that we can also have vectors which are functions of more than one scalar variable. For the moment, however, we shall continue with the analysis of vectors which depend on only one scalar quantity.

FRAME 5

Differentiation of a position vector

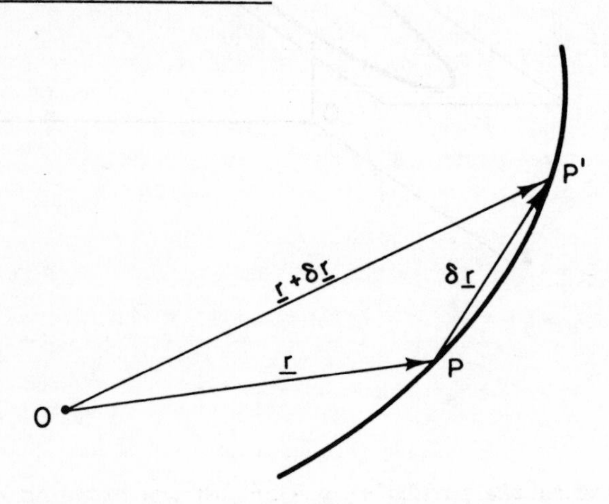

If we consider a position vector $\underline{r} = x\underline{i} + y\underline{j} + z\underline{k}$, where x,y,z are functions of t (and hence $\underline{r}$ is a function of t also), then as t varies the end point P of $\underline{r}$ will move on a curve. As t increases to $t + \delta t$, P will move to a neighbouring point P', say, where $\overrightarrow{OP'} = \underline{r} + \delta\underline{r}$.

Then $\overrightarrow{PP'} = \delta\underline{r}$

Now $\dfrac{d\underline{r}}{dt} = \lim\limits_{\delta t \to 0} \dfrac{\delta\underline{r}}{\delta t}$, i.e. the limit of the ratio $\dfrac{\delta\underline{r}}{\delta t}$ as P' moves towards P along the curve. Clearly, in the limit, $\dfrac{\delta\underline{r}}{\delta t}$ will be in the direction of the tangent to the curve at P.

That is, $\dfrac{d\underline{r}}{dt}$ represents a vector along the tangent at P.

FRAME 5 continued

As $P' \to P$, $|\delta\underline{r}| \to \delta s$, the arc PP'. So $\dfrac{|\delta\underline{r}|}{\delta t} \to \dfrac{\delta s}{\delta t}$, i.e. $\left|\dfrac{\delta\underline{r}}{\delta t}\right| \to \dfrac{\delta s}{\delta t}$.

Therefore, in the limit, $\left|\dfrac{d\underline{r}}{dt}\right| = \dfrac{ds}{dt}$. If t represents time, $\dfrac{ds}{dt}$ represents the speed of P.

Thus $\dfrac{d\underline{r}}{dt}$ represents the velocity of P in both magnitude and direction.

Similarly $\dfrac{d^2\underline{r}}{dt^2}$ is the rate of change of the velocity vector and will represent the acceleration of P in both magnitude and direction.

To illustrate this, we take the example of FRAME 1, since we know what answer to expect for circular motion.

There, $\underline{r} = a \cos \omega t\ \underline{i} + a \sin \omega t\ \underline{j}$

$$\therefore \quad \frac{d\underline{r}}{dt} = -a\omega \sin \omega t\ \underline{i} + a\omega \cos \omega t\ \underline{j}$$

and $$\frac{d^2\underline{r}}{dt^2} = -a\omega^2 \cos \omega t\ \underline{i} - a\omega^2 \sin \omega t\ \underline{j} = -\omega^2\underline{r}$$

Now show that $\left|\dfrac{d\underline{r}}{dt}\right| = \omega a$ and verify that the direction of $\dfrac{d\underline{r}}{dt}$ is along the tangent.

Also interpret physically the result $\dfrac{d^2\underline{r}}{dt^2} = -\omega^2\underline{r}$.

****

5A

$$\left|\frac{d\underline{r}}{dt}\right| = \sqrt{a^2\omega^2 \sin^2\omega t + a^2\omega^2 \cos^2\omega t} = a\omega$$

Drawing the components of $\dfrac{d\underline{r}}{dt}$ *at P gives the diagram shown.*

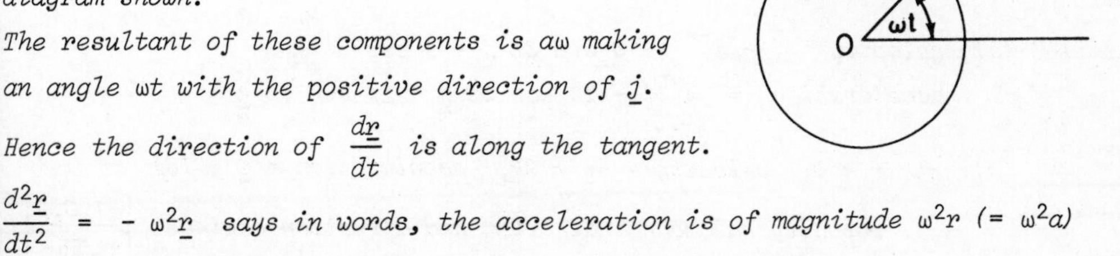

The resultant of these components is $a\omega$ *making an angle* ωt *with the positive direction of* $\underline{j}$.

Hence the direction of $\dfrac{d\underline{r}}{dt}$ *is along the tangent.*

$\dfrac{d^2\underline{r}}{dt^2} = -\omega^2\underline{r}$ *says in words, the acceleration is of magnitude* $\omega^2 r$ $(= \omega^2 a)$ *and is in the negative direction of* $\underline{r}$.

That is, the acceleration is towards the centre and of magnitude $\omega^2 r$.

FRAME 6

Now try these examples:

(i) Given $\underline{r} = \sin t\,\underline{i} + \cos t\,\underline{j} + t\underline{k}$, find

(a) $\dfrac{d\underline{r}}{dt}$ (b) $\dfrac{d^2\underline{r}}{dt^2}$

(ii) If $\underline{r} = 2t\underline{i} - t^2\underline{j} + t^3\underline{k}$, find $\left|\dfrac{d\underline{r}}{dt}\right|$ when $t = 2$.

(iii) A particle moves along a path given by
$\underline{r} = e^{-t}\underline{i} + 2\cos 3t\,\underline{j} + 3\sin 3t\,\underline{k}$ where t is the time.

(a) Determine the particle's velocity and acceleration at any time.

(b) Find the magnitude and direction of the velocity and acceleration at $t = 0$.

(iv) A particle moves so that its position vector is given by
$\underline{r} = \cos \omega t\,\underline{i} + \sin \omega t\,\underline{j}$ where ω = constant.
Show that $\underline{r}\wedge\underline{v}$ is constant, where $\underline{v}$ is the velocity.

****

6A

(i) *(a)* $\dfrac{d\underline{r}}{dt} = \cos t\,\underline{i} - \sin t\,\underline{j} + \underline{k}$

(b) $\dfrac{d^2\underline{r}}{dt^2} = -\sin t\,\underline{i} - \cos t\,\underline{j}$

(ii) $\sqrt{164}$

(iii) *(a)* *velocity* $= -e^{-t}\underline{i} - 6\sin 3t\,\underline{j} + 9\cos 3t\,\underline{k}$
acceleration $= e^{-t}\underline{i} - 18\cos 3t\,\underline{j} - 27\sin 3t\,\underline{k}$

(b) *at* $t = 0$, *velocity* $= -\underline{i} + 9\underline{k}$, *acceleration* $= \underline{i} - 18\underline{j}$

$\therefore$ *velocity has magnitude* $\sqrt{82}$ *and direction cosines* $\left[\dfrac{-1, 0, 9}{\sqrt{82}}\right]$
and acceleration has magnitude $\sqrt{325}$ *and direction cosines*
$\left[\dfrac{1, -18, 0}{\sqrt{325}}\right]$.

6A continued

(iv)

$$\underline{r}\wedge\underline{v} = \begin{vmatrix} \underline{i} & \underline{j} & \underline{k} \\ \cos\omega t & \sin\omega t & 0 \\ -\omega\sin\omega t & \omega\cos\omega t & 0 \end{vmatrix} = \omega\underline{k}$$

FRAME 7

Returning to the diagram in FRAME 5, the position vector $\underline{r}$ could be regarded as a function of s, the arc length measured from some fixed point.

We would then have $$\frac{d\underline{r}}{ds} = \lim_{\delta s \to 0}\left(\frac{\delta \underline{r}}{\delta s}\right) = \lim_{\delta s \to 0}\frac{\overrightarrow{PP'}}{\text{arc } PP'}$$

Clearly the magnitude of this approaches unity as P' moves towards P, and hence $\frac{d\underline{r}}{ds}$ is a unit vector along the tangent at P.

FRAME 8

Scalar Point Function and Scalar Field

To obtain a physical picture of the terms defined here, let us consider the temperature distribution around a hot body at a specific time. To investigate this distribution we would measure the temperature at many points surrounding the body and we could then make a plot of the distribution. Doing this we could find the temperature ϕ at any point and ϕ will necessarily be a function of position. That is, $\phi = \phi(x,y,z)$ and we say that $\phi(x,y,z)$ is a SCALAR POINT FUNCTION since it gives us the value of a scalar ϕ at the point (x,y,z). If we know the value of ϕ at every point of a region we say that we have a SCALAR FIELD.

Now, having plotted the temperature distribution, we could join up points having the same temperature to produce surfaces on which the temperature takes the same value at each point. These are called isothermals in the case of temperature but, in general, would be called LEVEL SURFACES.

That is, $\phi(x,y,z) = C$, where C is constant, is a level surface. These level

FRAME 8 continued

surfaces would take the form of shells, one inside another, surrounding the body and would map out the entire field.

Describe the level surfaces for the scalar point function $\phi = x^2 + z^2$.

****

8A

Cylinders with their axes along Oy.

FRAME 9

Vector Function and Vector Field

Suppose we now consider the case of a magnetic field. At each point (x,y,z) it will have both magnitude and direction and consequently can be represented by a vector, $\underline{H}$, say. If $\underline{H} = H_1\underline{i} + H_2\underline{j} + H_3\underline{k}$, then H_1, H_2 and H_3 will all be functions of x,y and z, and $\underline{H}$ is therefore a vector function.

If the value of $\underline{H}$ is known at every point of a region, we say that we have a VECTOR FIELD.

FRAME 10

Differentiation of a Vector Function

So far, we have only dealt with the differentiation of a position vector. However, a vector function can be differentiated by a similar process. Just as we considered the position vector $\underline{r}$ as being a function of only one variable t, so, at this stage, the vector function $\underline{R}$ will depend on only one variable, u, say. Thus we can write $\underline{R} = \underline{R}(u)$.

If $\underline{R}(u) = R_1(u)\underline{i} + R_2(u)\underline{j} + R_3(u)\underline{k}$, then a change in u will produce changes in both the magnitude and direction of $\underline{R}(u)$. Thus, if u changes to $u + \delta u$, $\underline{R}(u)$ will change to $\underline{R}(u + \delta u)$, and on a diagram this will appear as

FRAME 10 continued

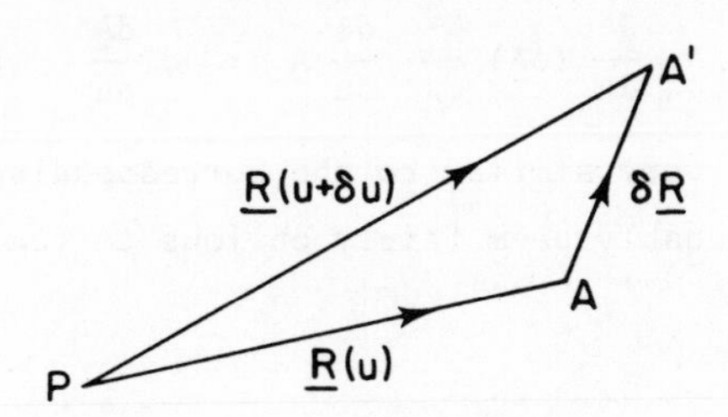

If we write $\underline{R}(u + \delta u) = \underline{R}(u) + \delta\underline{R}$, then

$$\delta\underline{R} = \underline{R}(u + \delta u) - \underline{R}(u) = \overrightarrow{AA'}$$

The ordinary derivative, by definition, is given by

$$\boxed{\frac{d\underline{R}}{du} = \lim_{\delta u \to 0} \frac{\underline{R}(u + \delta u) - \underline{R}(u)}{\delta u}}$$

Note that, from the manner in which it is defined, $\frac{d\underline{R}}{du}$ must be a vector.

In practice, however, as $\underline{R}(u) = R_1(u)\underline{i} + R_2(u)\underline{j} + R_3(u)\underline{k}$, differentiation is performed by differentiating each component separately.
For example, if $\underline{R} = \sin u\ \underline{i} - \sqrt{u} \cos u\ \underline{j} + u\underline{k}$

then $$\frac{d\underline{R}}{du} = \cos u\ \underline{i} + \left(\sqrt{u} \sin u - \frac{1}{2\sqrt{u}} \cos u\right)\underline{j} + \underline{k}$$

Second and higher derivatives, i.e. $\frac{d^2\underline{R}}{du^2}$ etc. can be obtained similarly.

FRAME 11

Differentiation of Sums and Products

Differentiating a position vector $\underline{r}$ in the form $x\underline{i} + y\underline{j} + z\underline{k}$ and a vector function $\underline{R}$ in the form $R_1\underline{i} + R_2\underline{j} + R_3\underline{k}$ involved the differentiation of (a) sums of vectors and (b) scalar multiples of vectors. We were, in fact, using the following general formulae:

$$\frac{d}{du}(\underline{A} + \underline{B}) = \frac{d\underline{A}}{du} + \frac{d\underline{B}}{du} \qquad (11.1)$$

FRAME 11 continued

$$\frac{d}{du}(\phi\underline{A}) = \frac{d\phi}{du}\underline{A} + \phi\frac{d\underline{A}}{du} \qquad (11.2)$$

These two results are very similar to the corresponding formulae in non-vector calculus, and will probably seem fairly obvious to you, so their proofs will not be given here.

Special cases of (11.2) occur when either ϕ or $\underline{A}$ is constant. If ϕ is constant, $\frac{d\phi}{du} = 0$. Thus, for example, $\frac{d}{du}(c\underline{A}) = c\frac{d\underline{A}}{du}$.

Similarly, if $\underline{A}$ is constant, $\frac{d\underline{A}}{du} = 0$. This happens if, for example, $\underline{A}$ is $\underline{i}$, $\underline{j}$ or $\underline{k}$. Thus in the previous frame, $\frac{d}{du}(\sin u\ \underline{i}) = \cos u\ \underline{i}$.

FRAME 12

It will also be necessary to know how to differentiate the product of two vectors, i.e. how to find $\frac{d}{du}(\underline{A}.\underline{B})$ and $\frac{d}{du}(\underline{A}\wedge\underline{B})$. The former is the easier and we shall consider that first.

Taking $\quad \underline{A} = A_1\underline{i} + A_2\underline{j} + A_3\underline{k}$

and $\quad \underline{B} = B_1\underline{i} + B_2\underline{j} + B_3\underline{k}$

then $\quad \underline{A}.\underline{B} = A_1B_1 + A_2B_2 + A_3B_3$

$$\therefore \frac{d}{du}(\underline{A}.\underline{B}) = \left(\frac{dA_1}{du}B_1 + A_1\frac{dB_1}{du}\right) + \left(\frac{dA_2}{du}B_2 + A_2\frac{dB_2}{du}\right) + \left(\frac{dA_3}{du}B_3 + A_3\frac{dB_3}{du}\right)$$

$$= \left(\frac{dA_1}{du}B_1 + \frac{dA_2}{du}B_2 + \frac{dA_3}{du}B_3\right) + \left(A_1\frac{dB_1}{du} + A_2\frac{dB_2}{du} + A_3\frac{dB_3}{du}\right)$$

Now, remembering that $\frac{d\underline{A}}{du} = \frac{dA_1}{du}\underline{i} + \frac{dA_2}{du}\underline{j} + \frac{dA_3}{du}\underline{k}$, express the contents of the first brackets as a scalar product. Then express the contents of the second brackets as a scalar product and hence obtain the formula for $\frac{d}{du}(\underline{A}.\underline{B})$.

****

12A

$$\frac{d\underline{A}}{du} \cdot \underline{B}$$

$$\underline{A} \cdot \frac{d\underline{B}}{du}$$

$$\frac{d}{du}(\underline{A}.\underline{B}) = \frac{d\underline{A}}{du} \cdot \underline{B} + \underline{A} \cdot \frac{d\underline{B}}{du}$$

FRAME 13

The derivation of the formula for $\frac{d}{du}(\underline{A}{}_\wedge\underline{B})$ is also straightforward, but somewhat more tedious. The result, which we shall quote without proof, is

$$\frac{d}{du}(\underline{A}{}_\wedge\underline{B}) = \frac{d\underline{A}}{du} {}_\wedge \underline{B} + \underline{A} {}_\wedge \frac{d\underline{B}}{du}$$

You will probably have noticed the similarity between these last two results and the formula for the differential coefficient of a product in a non-vector situation. However, one important fact must be observed - where vector products are involved, the order of the vectors must be maintained.

By writing $\underline{B}{}_\wedge\underline{C} = \underline{D}$ and using these last two results, obtain the formula for $\frac{d}{du}\{\underline{A}.(\underline{B}{}_\wedge\underline{C})\}$.

****

13A

$$\frac{d}{du}(\underline{A}.\underline{D}) = \frac{d\underline{A}}{du} \cdot \underline{D} + \underline{A} \cdot \frac{d\underline{D}}{du}$$

$$\begin{aligned}\frac{d}{du}\{\underline{A}.(\underline{B}{}_\wedge\underline{C})\} &= \frac{d\underline{A}}{du} \cdot (\underline{B}{}_\wedge\underline{C}) + \underline{A} \cdot \frac{d}{du}(\underline{B}{}_\wedge\underline{C}) \\ &= \frac{d\underline{A}}{du} \cdot (\underline{B}{}_\wedge\underline{C}) + \underline{A} \cdot \left(\frac{d\underline{B}}{du} {}_\wedge \underline{C} + \underline{B} {}_\wedge \frac{d\underline{C}}{du}\right) \\ &= \frac{d\underline{A}}{du} \cdot (\underline{B}{}_\wedge\underline{C}) + \underline{A} \cdot \left(\frac{d\underline{B}}{du} {}_\wedge \underline{C}\right) + \underline{A} \cdot \left(\underline{B} {}_\wedge \frac{d\underline{C}}{du}\right)\end{aligned}$$

FRAME 14

Similarly, it can be proved that

$$\frac{d}{du}\{\underline{A}\wedge(\underline{B}\wedge\underline{C})\} = \frac{d\underline{A}}{du}\wedge(\underline{B}\wedge\underline{C}) + \underline{A}\wedge\left(\frac{d\underline{B}}{du}\wedge\underline{C}\right) + \underline{A}\wedge\left(\underline{B}\wedge\frac{d\underline{C}}{du}\right)$$

The formulae given for the differentiation of sums and products are summarised for your convenience in the next frame.

FRAME 15

$$\frac{d}{du}(\underline{A}+\underline{B}) = \frac{d\underline{A}}{du} + \frac{d\underline{B}}{du} \quad (15.1)$$

$$\frac{d}{du}(\phi\underline{A}) = \frac{d\phi}{du}\underline{A} + \phi\frac{d\underline{A}}{du} \quad (15.2)$$

$$\frac{d}{du}(\underline{A}.\underline{B}) = \frac{d\underline{A}}{du}.\underline{B} + \underline{A}.\frac{d\underline{B}}{du} \quad (15.3)$$

$$\frac{d}{du}(\underline{A}\wedge\underline{B}) = \underline{A}\wedge\frac{d\underline{B}}{du} + \frac{d\underline{A}}{du}\wedge\underline{B} \quad (15.4)$$

$$\frac{d}{du}\{\underline{A}.(\underline{B}\wedge\underline{C})\} = \frac{d\underline{A}}{du}.(\underline{B}\wedge\underline{C}) + \underline{A}.\left(\frac{d\underline{B}}{du}\wedge\underline{C}\right) + \underline{A}.\left(\underline{B}\wedge\frac{d\underline{C}}{du}\right) \quad (15.5)$$

$$\frac{d}{du}\{\underline{A}\wedge(\underline{B}\wedge\underline{C})\} = \frac{d\underline{A}}{du}\wedge(\underline{B}\wedge\underline{C}) + \underline{A}\wedge\left(\frac{d\underline{B}}{du}\wedge\underline{C}\right) + \underline{A}\wedge\left(\underline{B}\wedge\frac{d\underline{C}}{du}\right) \quad (15.6)$$

FRAME 16

Now try the following examples:

(i) Verify formulae (15.3) and (15.4) if $\underline{A} = 5u\underline{i} + u^2\underline{j} - u^3\underline{k}$ and $\underline{B} = \cos u\ \underline{i} - \sin u\ \underline{j}$.

(ii) Use (15.3) to write down a formula for $\frac{d}{du}(\underline{A}.\underline{A})$ and verify your result when $\underline{A} = 2u\underline{i} - u^2\underline{j} + u^3\underline{k}$.

(iii) Verify formulae (15.2), (15.5) and (15.6) where $\phi = u^2$, $\underline{A} = 2u\underline{i} - u^2\underline{j}$, $\underline{B} = u^2\underline{k}$ and $\underline{C} = \cos u\ \underline{j} - 2u\underline{k}$.

****

16A

(i) Each side of (15.3) $= (5 - u^2)\cos u - 7u \sin u$

Each side of (15.4) $= -(u^3\cos u + 3u^2\sin u)\underline{i} + (u^3\sin u - 3u^2\cos u)\underline{j} + (u^2\sin u - 7u\cos u - 5\sin u)\underline{k}$

(ii) $\frac{d}{du}(\underline{A}.\underline{A}) = \frac{d\underline{A}}{du}.\underline{A} + \underline{A}.\frac{d\underline{A}}{du}$

$= 2\underline{A}.\frac{d\underline{A}}{du}$ as with a dot product the order of the factors is immaterial.

Each side $= 8u + 4u^3 + 6u^5$

(iii) Each side of (15.2) $= 2u^2(3\underline{i} - 2u\underline{j})$

Each side of (15.5) $= 2u^2(u\sin u - 3\cos u)$

Each side of (15.6) $= u^3(u\sin u - 4\cos u)\underline{k}$

FRAME 17

Partial Differentiation

In FRAMES 5-7 and 10-16 we considered the ordinary differentiation of vector quantities which were dependent upon only one scalar variable. In physical situations we soon meet the case where a vector is a function of more than one scalar variable, as in the case of a magnetic field which was mentioned in FRAME 9. Another example occurs in hydrodynamics where the velocity of a fluid particle is a function of both position and time. That is, the velocity is a function of four scalar variables x,y,z and t.

For the moment let us consider a vector $\underline{A}$ which is a function of x,y and z.

$$\text{i.e.}\quad \underline{A} = \underline{A}(x,y,z)$$

If y and z are kept constant, while x alone changes, then $\frac{\partial \underline{A}}{\partial x}$ denotes the rate of change of $\underline{A}$ with respect to x. Similarly if we change y or z alone we get $\frac{\partial \underline{A}}{\partial y}$ or $\frac{\partial \underline{A}}{\partial z}$.

Further partial derivatives are possible, just as in non-vector calculus and we may find, for example, $\frac{\partial^2 \underline{A}}{\partial x^2}$, $\frac{\partial^2 \underline{A}}{\partial x \partial y}$ and so on.

FRAME 17 continued

In practice, these differential coefficients are obtained by the partial differentiation of each component of $\underline{A}$. For example, consider

$$\underline{A} = (x^2 + y^2)\underline{i} + (xy + zx)\underline{j} + (z^2 - y^2 - 1)\underline{k},$$

$$\text{then } \frac{\partial \underline{A}}{\partial x} = 2x\underline{i} + (y + z)\underline{j} + 0\underline{k}$$

$$\frac{\partial^2 \underline{A}}{\partial x^2} = 2\underline{i}$$

$$\frac{\partial^2 \underline{A}}{\partial x \partial y} = \underline{j}$$

FRAME 18

If now we assume x, y and z to change simultaneously, in parallel with non-vector calculus we arrive at the total differential of $\underline{A}$ as

$$d\underline{A} = \frac{\partial \underline{A}}{\partial x} dx + \frac{\partial \underline{A}}{\partial y} dy + \frac{\partial \underline{A}}{\partial z} dz$$

This, of course, is a first approximation to the change in $\underline{A}$ when x changes to x + dx, y to y + dy and z to z + dz simultaneously. If, also, x, y and z are all functions of a single variable t, the rate of change of $\underline{A}$ with respect to t is given by the total derivative

$$\frac{d\underline{A}}{dt} = \frac{\partial \underline{A}}{\partial x}\frac{dx}{dt} + \frac{\partial \underline{A}}{\partial y}\frac{dy}{dt} + \frac{\partial \underline{A}}{\partial z}\frac{dz}{dt}$$

As an example, if $\underline{A} = z \sin y\, \underline{i} + x \cos z\, \underline{j} + y \cos x\, \underline{k}$

$$\text{then } d\underline{A} = \frac{\partial \underline{A}}{\partial x} dx + \frac{\partial \underline{A}}{\partial y} dy + \frac{\partial \underline{A}}{\partial z} dz$$

$$= (\cos z\, \underline{j} - y \sin x\, \underline{k})dx + (z \cos y\, \underline{i} + \cos x\, \underline{k})dy + (\sin y\, \underline{i} - x \sin z\, \underline{j})dz$$

If, also, $x = t^2$, $y = 2t$ and $z = 3$,

$$\frac{d\underline{A}}{dt} = \frac{\partial \underline{A}}{\partial x}\frac{dx}{dt} + \frac{\partial \underline{A}}{\partial y}\frac{dy}{dt} + \frac{\partial \underline{A}}{\partial z}\frac{dz}{dt}$$

$$= (\cos z\, \underline{j} - y \sin x\, \underline{k})(2t) + (z \cos y\, \underline{i} + \cos x\, \underline{k})(2) + (\sin y\, \underline{i} - x \sin z\, \underline{j})(0)$$

$$= 2z \cos y\, \underline{i} + 2t \cos z\, \underline{j} + 2(\cos x - ty \sin x)\underline{k}$$

FRAME 18 continued

If the value of this is required when t = 0, then as this gives x = 0, y = 0 and z = 3,

$$\frac{d\underline{A}}{dt} = 6\underline{i} + 2\underline{k}$$

FRAME 19

Now try the following examples:

(i) If $\underline{A} = \cos(x + y)\underline{i} + (x^2 - y^2)\underline{j} + (2x - 3y)\underline{k}$

find $\frac{\partial \underline{A}}{\partial x}$, $\frac{\partial \underline{A}}{\partial y}$, $\frac{\partial^2 \underline{A}}{\partial x^2}$, $\frac{\partial^2 \underline{A}}{\partial y^2}$, $\frac{\partial^2 \underline{A}}{\partial x \partial y}$.

(ii) If $\underline{A} = xyz\underline{i} + yz^2\underline{j} - zx^2\underline{k}$ and $\underline{B} = x\underline{i} + y\underline{j} + z\underline{k}$

find $\underline{A}\wedge\underline{B}$ and hence, $\frac{\partial^2}{\partial x \partial y}(\underline{A}\wedge\underline{B})$ at $(1,0,-1)$

(iii) Prove that $\underline{A} = \underline{\alpha}\,\frac{e^{i\omega(t-(r/c))}}{r}$, where $\underline{\alpha}$ is a constant vector and c,ω are constant scalars,

satisfies $\frac{\partial^2 \underline{A}}{\partial r^2} + \frac{2}{r}\frac{\partial \underline{A}}{\partial r} = \frac{1}{c^2}\frac{\partial^2 \underline{A}}{\partial t^2}$

(This is an important result in electromagnetic theory.)

(iv) If $\underline{A} = x^2y\underline{i} + zx\underline{j} + z^2\underline{k}$, find $d\underline{A}$.

Also, if x = sin t, y = 2t, z = cos t, find $\frac{d\underline{A}}{dt}$ and its value at t = 0.

****

19A

(i) $$\frac{\partial \underline{A}}{\partial x} = -\sin(x + y)\underline{i} + 2x\underline{j} + 2\underline{k}$$

$$\frac{\partial \underline{A}}{\partial y} = -\sin(x + y)\underline{i} - 2y\underline{j} - 3\underline{k}$$

$$\frac{\partial^2 \underline{A}}{\partial x^2} = -\cos(x + y)\underline{i} + 2\underline{j}$$

19A continued

(i) contd. $\dfrac{\partial^2 \underline{A}}{\partial y^2} = -\cos(x+y)\underline{i} - 2\underline{j}$

$$\frac{\partial^2 \underline{A}}{\partial x \partial y} = -\cos(x+y)\underline{i}$$

(ii) $-2\underline{i} - \underline{j} - \underline{k}$

(iv) $d\underline{A} = (2xy\underline{i} + z\underline{j})dx + x^2\underline{i}\,dy + (x\underline{j} + 2z\underline{k})dz$

$$\frac{d\underline{A}}{dt} = (2xy\underline{i} + z\underline{j})\frac{dx}{dt} + x^2\underline{i}\frac{dy}{dt} + (x\underline{j} + 2z\underline{k})\frac{dz}{dt}$$

$$\therefore \text{ at } t = 0, \quad \frac{d\underline{A}}{dt} = \underline{j}$$

FRAME 20

Directional Derivative

The idea of a directional derivative will first be considered in two dimensions and later extended to three. Taking as an example the function $\phi(x,y) = 2x^2y$, AB and CD represent parts of two level curves of this function.

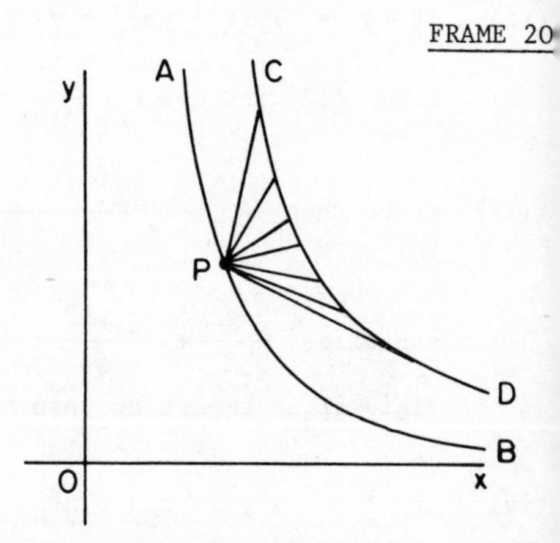

A LEVEL CURVE is a plot of all points at which the function has the same value. It is, of course, the two-dimensional version of a level surface, introduced in FRAME 8. A level curve that you have no doubt met before is a contour on a map, which joins all points at the same height above sea level.

In the present example, the level curves of which AB and CD are parts are $2x^2y = 1$ and $2x^2y = 3/2$ respectively.

Now let P be a fixed point on AB, $(\frac{1}{2},2)$ say, and Q a variable point on CD. The diagram shows a few positions of Q and the corresponding straight lines PQ. We now consider the values of $\dfrac{\delta\phi}{\delta s}$ for various positions of Q, where $\delta\phi$ is the

FRAME 20 continued

increment in ϕ as one goes from P to Q and δs is the distance PQ, i.e.

$$\frac{\delta\phi}{\delta s} = \frac{\phi(Q) - \phi(P)}{PQ}$$

The notation $\phi(P)$ stands for the value of the function ϕ at P.

What will be the value of $\phi(P)$ at the fixed point P and that of $\phi(Q)$ at any point Q on CD?

****

20A

$\phi(P) = 1$ (and would, in fact, be the same for any other position of P on AB).

$\phi(Q) = 3/2$

FRAME 21

Thus in this example $\frac{\delta\phi}{\delta s} = \frac{3/2 - 1}{PQ} = \frac{1}{2PQ}$ and consequently now only depends on the length of PQ. What will happen to the value of $\frac{\delta\phi}{\delta s}$ as Q moves from C to D?

****

21A

It will increase to a maximum when PQ is least and then decrease.

FRAME 22

Now the length of PQ depends on the direction in wnich one moves out from P. Thus $\frac{\delta\phi}{\delta s}$ depends on α, the angle PQ makes with Ox. And we would arrive at exactly the same conclusion whatever level curve Q might lie upon.

If Q is now taken on a level curve very close to AB, can you see in what direction, relative to the curve AB, it is necessary to move out from P in order for $\frac{\delta\phi}{\delta s}$ to be a maximum?

****

22A

Normal to the curve AB at P.

FRAME 23

If we now let $\delta s \to 0$ then Q will tend to P and also $\delta\phi$ will tend to 0. $\frac{\delta\phi}{\delta s}$ will then tend to the derivative $\frac{d\phi}{ds}$ and this will depend on the direction in which Q approaches P. Because it depends upon this direction it is called the DIRECTIONAL DERIVATIVE. Its value at P will be a maximum when Q approaches P along the normal to the level curve at P. This value is often denoted by $\frac{d\phi}{dn}$.

FRAME 24

Looking at this idea from another standpoint, we have from Taylor's Series in two dimensions, which was discussed in Unit 7,

$$\delta\phi = \frac{\partial\phi}{\partial x}\delta x + \frac{\partial\phi}{\partial y}\delta y + \text{terms involving powers and/or products of } \delta x \text{ and } \delta y$$

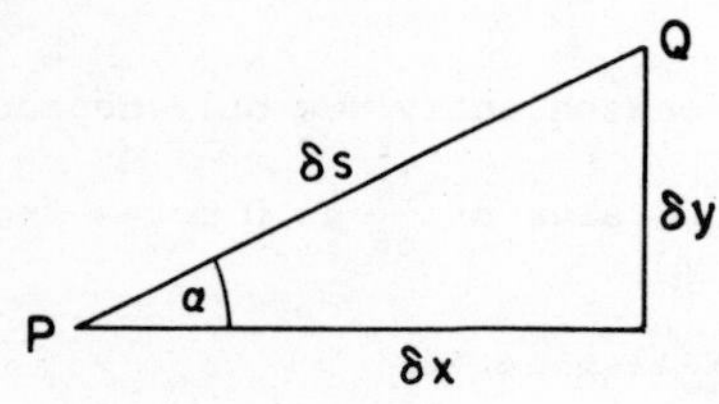

Then, as is seen from the diagram,

$$\delta x = \delta s \cos\alpha$$

$$\delta y = \delta s \sin\alpha$$

$$\delta\phi = \frac{\partial\phi}{\partial x}\delta s\cos\alpha + \frac{\partial\phi}{\partial y}\delta s\sin\alpha + \text{terms involving powers of } \delta s$$

$$\frac{\delta\phi}{\delta s} = \frac{\partial\phi}{\partial x}\cos\alpha + \frac{\partial\phi}{\partial y}\sin\alpha + \text{terms involving } \delta s \text{ and its powers}$$

$\therefore$ In the limit, as $\delta s \to 0$, $\frac{d\phi}{ds} = \frac{\partial\phi}{\partial x}\cos\alpha + \frac{\partial\phi}{\partial y}\sin\alpha$

This, as was to be expected, depends on α.

Returning to $\phi(x,y) = 2x^2y$, find the value of $\frac{d\phi}{ds}$ at $(\frac{1}{2},2)$ when $\alpha = 0$, $\tan^{-1}\frac{1}{8}$, $\frac{\pi}{2}$, $-\pi$.

****

24A

4, $\sqrt{65}/2$, 1/2, -4

$\sqrt{65}/2$ is actually the maximum value of $\frac{d\phi}{ds}$ for this function at $(\frac{1}{2},2)$.

FRAME 25

Gradient of a Function

You will now see that a vector can be associated with the point P. Its direction is that for which $\frac{d\phi}{ds}$ is a maximum and its magnitude is the value of this maximum $\frac{d\phi}{ds}$. This vector is called the GRADIENT of the function ϕ at P. If this is done for all possible positions of P, a vector field results. An abbreviation for the gradient of a function ϕ is grad ϕ.

The vector grad ϕ will be known at any point if its components parallel to the x- and y-axes are known. To find these we proceed as follows:

$\frac{\partial\phi}{\partial x}\cos\alpha + \frac{\partial\phi}{\partial y}\sin\alpha$, i.e. $\frac{d\phi}{ds}$, can be expressed in the form $R\cos(\alpha - \beta)$.

What will then be the expressions for R and β in terms of $\frac{\partial\phi}{\partial x}$ and $\frac{\partial\phi}{\partial y}$?

****

25A

$R = \sqrt{\left(\frac{\partial\phi}{\partial x}\right)^2 + \left(\frac{\partial\phi}{\partial y}\right)^2}$ $\quad$ *β is given by* $\tan\beta = \frac{\partial\phi}{\partial y}\Big/\frac{\partial\phi}{\partial x}$

FRAME 26

Can you now say what value of α will make $\frac{d\phi}{ds}$ a maximum, and what this maximum value will be?

****

26A

$\alpha = \beta$

Maximum value of $\frac{d\phi}{ds} = R = \sqrt{\left(\frac{\partial\phi}{\partial x}\right)^2 + \left(\frac{\partial\phi}{\partial y}\right)^2}$

FRAME 27

The x- and y-components of grad ϕ are thus $R\cos\beta$ and $R\sin\beta$. Express these in terms of $\frac{\partial\phi}{\partial x}$ and $\frac{\partial\phi}{\partial y}$.

****

27A

cos β = $\dfrac{\partial\phi}{\partial x}\Big/R$ *sin* β = $\dfrac{\partial\phi}{\partial y}\Big/R$

Required components are $\dfrac{\partial\phi}{\partial x}$ *and* $\dfrac{\partial\phi}{\partial y}$.

FRAME 28

Thus in the two-dimensional case grad ϕ can be expressed as

$$\text{grad } \phi = \underline{i}\,\frac{\partial\phi}{\partial x} + \underline{j}\,\frac{\partial\phi}{\partial y}$$

For the function $\phi = 2x^2y$, grad $\phi = 4xy\underline{i} + 2x^2\underline{j}$

At $(\frac{1}{2},2)$ this becomes $4\underline{i} + \frac{1}{2}\underline{j}$ the magnitude of which is $\sqrt{65}/2$.
(Although one usually writes $4xy\underline{i}$, rather than $\underline{i}4xy$, it is conventional to put $\underline{i}$ in front of $\frac{\partial\phi}{\partial x}$. You'll see why later.)

Now try the following examples:

(i) If $\phi = x^2 - y^2$ find grad ϕ at the point (2,-3).

(ii) If a cylinder of radius a is placed at the origin in a fluid which is otherwise moving with constant velocity V_o parallel to Ox, the velocity potential is given by $\phi = V_o\left(x + \frac{a^2x}{x^2+y^2}\right)$. Find grad ϕ, which gives the velocity of the fluid outside the cylinder.

****

28A

(i) $4\underline{i} + 6\underline{j}$

(ii) $V_o\left[\left\{1 + \dfrac{a^2(y^2 - x^2)}{(x^2 + y^2)^2}\right\}\underline{i} - \dfrac{2a^2xy}{(x^2 + y^2)^2}\,\underline{j}\right]$

FRAME 29

The concepts considered in the last few frames have corresponding counterparts in three dimensions. Thus, given a function ϕ of x, y and z a directional derivative at any point P is given by

$$\frac{d\phi}{ds} = \frac{\partial\phi}{\partial x}\cos\alpha + \frac{\partial\phi}{\partial y}\cos\beta + \frac{\partial\phi}{\partial z}\cos\gamma$$

FRAME 29 continued

where α, β, γ are the angles made with Ox, Oy, Oz by the direction in which the derivative is being taken. The greatest value of $\frac{d\phi}{ds}$, i.e. $\frac{d\phi}{dn}$, occurs when this direction is normal to the level surface through P.

The directional derivative in any other direction is then given by

$$\frac{d\phi}{ds} = \frac{d\phi}{dn}\cos\theta$$

where θ is the angle between this other direction and the normal. The gradient at P is a vector whose magnitude is $\frac{d\phi}{dn}$ and whose direction is that of the normal. Thus, if $\hat{\underline{n}}$ is a unit vector in this direction,

$$\boxed{\text{grad } \phi = \frac{d\phi}{dn}\hat{\underline{n}}}$$

To emphasise, grad ϕ gives us the greatest rate of change of ϕ and moreover it tells us in which direction this greatest rate of change occurs. This is important in field theory. For example, if ϕ is the electric potential, the electric force at a point will be in a direction normal to the potential surface there and will be of magnitude equal to the rate of change of potential in that direction.

If $\hat{\underline{a}}$ is a unit vector making an angle θ with grad ϕ what quantity will be given by $\hat{\underline{a}}$. grad ϕ?

****

29A

$\hat{\underline{a}}$. *grad* $\phi = \frac{d\phi}{dn}\cos\theta =$ *directional derivative in direction* $\hat{\underline{a}}$

FRAME 30

The gradient at P has components $\frac{\partial\phi}{\partial x}$, $\frac{\partial\phi}{\partial y}$ and $\frac{\partial\phi}{\partial z}$ parallel to the axes and is given by

$$\boxed{\text{grad } \phi = \underline{i}\frac{\partial\phi}{\partial x} + \underline{j}\frac{\partial\phi}{\partial y} + \underline{k}\frac{\partial\phi}{\partial z}}$$

If $\phi = \ell/(x^2 + y^2 + z^2)$, where ℓ is a constant, find grad ϕ.

****

30A

$$\frac{-2\ell(x\underline{i} + y\underline{j} + z\underline{k})}{(x^2 + y^2 + z^2)^2}$$

FRAME 31

We shall now illustrate the use of the gradient by considering the following problem:

Relative to axes Ox, Oy, Oz, the temperature in a certain medium is given by $T = T_o(1 + ax + by)e^{cz}$ where a, b, c, T_o (> 0) are constants.

Find, at the origin, a unit vector in the direction in which the temperature changes most rapidly.

Now grad T gives the maximum rate of change of T in both magnitude and direction. Here grad $T = T_o\{a\underline{i} + b\underline{j} + c(1 + ax + by)\underline{k}\}e^{cz}$.

Hence at the origin, grad $T = T_o(a\underline{i} + b\underline{j} + c\underline{k})$

Then the magnitude of grad T is $T_o\sqrt{a^2 + b^2 + c^2}$,

and a unit vector in the direction of grad T is grad T/|grad T|.

That is, required unit vector is $(a\underline{i} + b\underline{j} + c\underline{k})/\sqrt{a^2 + b^2 + c^2}$.

Now try the following examples:

(i) Suppose the temperature T at (x,y,z) is given by $T = x^2 - y^2 + xyz + 273$, find the vector which gives the magnitude and direction of the maximum rate of change of temperature at the point (-1,2,3).

(ii) Starting from the point (1,1), in what direction does $\phi = x^2 - y^2 + 2xy$ decrease most rapidly?

****

31A

(i) $4\underline{i} - 7\underline{j} - 2\underline{k}$ *(ii)* $-\underline{i}$

FRAME 32

ϕ is sometimes given as a function of r, where $\underline{r} = x\underline{i} + y\underline{j} + z\underline{k}$. To find grad ϕ we make use of the relation $r^2 = x^2 + y^2 + z^2$. Thus, if $\phi = \ln r$

$$\frac{\partial\phi}{\partial x} = \frac{d\phi}{dr}\frac{\partial r}{\partial x} = \frac{1}{r}\frac{x}{r} = \frac{x}{r^2}$$

Similarly $\dfrac{\partial\phi}{\partial y} = \dfrac{y}{r^2}$ and $\dfrac{\partial\phi}{\partial z} = \dfrac{z}{r^2}$

$$\therefore \quad \text{grad}\ \phi = \frac{x}{r^2}\underline{i} + \frac{y}{r^2}\underline{j} + \frac{z}{r^2}\underline{k}$$

$$= \underline{r}/r^2$$

Now try to do these:

(i) If $\phi = f(r)$, show that grad $\phi = \dfrac{f'(r)}{r}\underline{r}$.

(ii) If $f = x^m + y^m + z^m$, show that $\underline{r}$.grad $f = mf$.

(iii) The scalar potential ϕ at (x,y,z) due to a point charge q at the origin is given by $\phi = q/r$. The electrostatic field intensity $\underline{E}$ at this point is given by $\underline{E} = -$ grad ϕ. Find $\underline{E}$.

****

32A

(ii) grad $f = m(x^{m-1}\underline{i} + y^{m-1}\underline{j} + z^{m-1}\underline{k})$

(iii) $q\underline{r}/r^3$ *This is a special case of (i) when* $f(r) = q/r$.

FRAME 33

As grad ϕ is in a direction normal to a level surface ϕ = constant, it can be used to find the direction of the normal to a surface at a point.

For example, if we are considering the spherical surface $x^2 + y^2 + z^2 = a^2$ we think of this as one of the level surfaces $\phi = c$ where $\phi = x^2 + y^2 + z^2$.
Then grad $\phi = 2x\underline{i} + 2y\underline{j} + 2z\underline{k} = 2\underline{r}$.
Thus the direction of the normal is that of the radius vector $\underline{r}$, as, of course, is well-known.

A unit normal is a vector of unit length in the direction of the normal to a surface at a point. Find a unit normal to the surface $x^2y + 2xz = 4$ at the point (2,-2,3).

****

33A

You have probably got $-\frac{1}{3}\underline{i} + \frac{2}{3}\underline{j} + \frac{2}{3}\underline{k}$, *but* $\frac{1}{3}\underline{i} - \frac{2}{3}\underline{j} - \frac{2}{3}\underline{k}$ *is also correct.*

FRAME 34

The Vector Operator ∇

The vector $\underline{i}\frac{\partial\phi}{\partial x} + \underline{j}\frac{\partial\phi}{\partial y} + \underline{k}\frac{\partial\phi}{\partial z}$ can be written as $\left(\underline{i}\frac{\partial}{\partial x} + \underline{j}\frac{\partial}{\partial y} + \underline{k}\frac{\partial}{\partial z}\right)\phi$.

Now you have already met the idea of an operator in non-vector calculus - for example, the operator D which usually means $\frac{d}{dx}$. In a similar way $\underline{i}\frac{\partial}{\partial x} + \underline{j}\frac{\partial}{\partial y} + \underline{k}\frac{\partial}{\partial z}$ can be regarded as a vector operator. It is usually denoted by ∇ (called del or sometimes nabla). Thus

$$\boxed{\nabla \equiv \underline{i}\frac{\partial}{\partial x} + \underline{j}\frac{\partial}{\partial y} + \underline{k}\frac{\partial}{\partial z}}$$

This operator is of great value in simplifying expressions occurring in vector analysis and in simplifying the theories of electromagnetism, hydrodynamics and mechanics. As it is an operator it only has meaning when applied to a function of some kind. Here we have applied it to the scalar function ϕ with the result that

$$\nabla\phi = \text{grad } \phi$$

It can also be applied in various ways to a vector function, as you will see later in the programme.

FRAME 35

Divergence

The idea of divergence will be illustrated by considering the motion of an incompressible fluid. Fig (i) depicts an imaginary rectangular box in the fluid. Its centre is at the point P(x,y,z) and its edges, which are parallel to the axes, are δx, δy, δz.

FRAME 35 continued

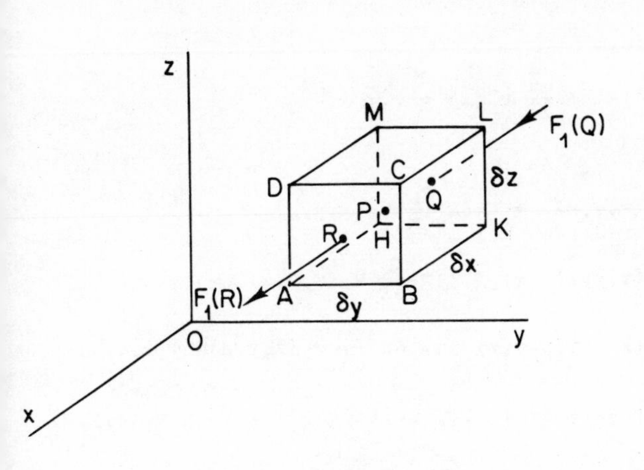

Fig (i)

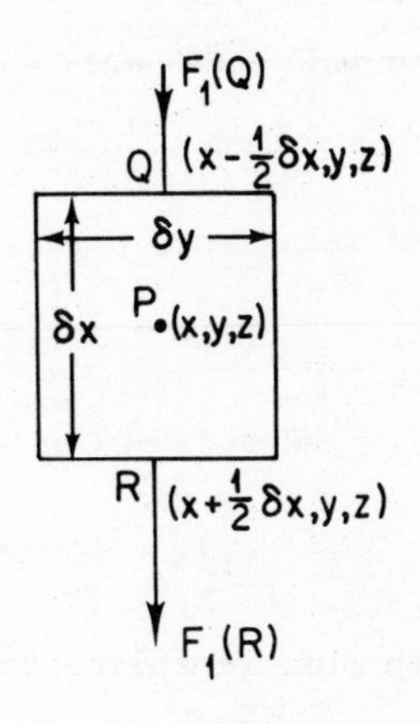

Fig (ii)

If the velocity of the fluid has components v_1, v_2, v_3 parallel to the axes, it can be represented by the vector $\underline{v} = v_1\underline{i} + v_2\underline{j} + v_3\underline{k}$. We start by considering the amount of fluid crossing opposite faces of the box in unit time. The amount of fluid crossing a surface is an example of what is called flux. As the density is constant (the fluid being incompressible) either volume or mass can be used as a measure of amount. For simplicity, volume will be used here.

Consider first the flux across the faces which are perpendicular to Ox (i.e. across the planes ABCD and HKLM). These faces are of area $\delta y\delta z$ and the flux will only depend on v_1, this being the component of $\underline{v}$ perpendicular to these faces. If the box is small we can assume that the value of v_1 on the face ABCD is that attained at its mid-point $(x + \frac{1}{2}\delta x,y,z)$. This will be denoted by $v_1(x + \frac{1}{2}\delta x,y,z)$. Thus the flux through the near face ABCD <u>out</u> of the box

$$= \quad v_1(x + \tfrac{1}{2}\delta x,y,z)\delta y\delta z$$

and similarly the flux through the opposite face <u>into</u> the box

$$= \quad v_1(x - \tfrac{1}{2}\delta x,y,z)\delta y\delta z$$

Hence the net flux out of the box $= \{v_1(x + \frac{1}{2}\delta x,y,z) - v_1(x - \frac{1}{2}\delta x,y,z)\}\delta y\delta z$

In this instance only the x is changing in $v_1(x,y,z)$ so that effectively

FRAME 35 continued

Taylor's series in one dimension can be used.

You will have met this in the form

$$f(x + \alpha) = f(x) + \alpha f'(x) + \frac{\alpha^2}{2!} f''(x) + \ldots$$

and using this here gives

$$v_1(x + \tfrac{1}{2}\delta x, y, z) = v_1(x,y,z) + \tfrac{1}{2}\delta x \frac{\partial}{\partial x} v_1(x,y,z) + \ldots$$

$$\text{and } v_1(x - \tfrac{1}{2}\delta x, y, z) = v_1(x,y,z) + (-\tfrac{1}{2}\delta x)\frac{\partial}{\partial x} v_1(x,y,z) + \ldots$$

$$\therefore v_1(x + \tfrac{1}{2}\delta x, y, z) - v_1(x - \tfrac{1}{2}\delta x, y, z) = \delta x \frac{\partial}{\partial x} v_1(x,y,z) + \ldots$$

Net flux in x-direction $= \left(\frac{\partial v_1}{\partial x}\,\delta x\right)\delta y \delta z + \ldots$, writing $\frac{\partial v_1}{\partial x}$ for $\frac{\partial}{\partial x} v_1(x,y,z)$

Now write down the corresponding expressions for the net flux in the y- and z-directions.

****

35A

Net flux in y-direction $= \left(\frac{\partial v_2}{\partial y}\,\delta y\right)\delta x \delta z + \ldots$

Net flux in z-direction $= \left(\frac{\partial v_3}{\partial z}\,\delta z\right)\delta x \delta y + \ldots$

FRAME 36

Adding these three contributions,

Total flux out of the box $= \left(\frac{\partial v_1}{\partial x} + \frac{\partial v_2}{\partial y} + \frac{\partial v_3}{\partial z}\right)\delta x \delta y \delta z$ + terms involving higher powers of δx, δy, δz, such as $(\delta x)^2 \delta y \delta z$, $\delta x (\delta y)^3 \delta z$ etc.

Now as the volume of the box is $\delta x \delta y \delta z$, we have

$$\text{Flux per unit volume} = \frac{\partial v_1}{\partial x} + \frac{\partial v_2}{\partial y} + \frac{\partial v_3}{\partial z} + \text{terms in } \delta x, \delta y \text{ or } \delta z \text{ and their powers} \qquad (36.1)$$

FRAME 36 continued

Finally, we let δx, δy, δz all tend to zero so that the box shrinks to the point P. The right-hand side of (36.1) then becomes $\frac{\partial v_1}{\partial x} + \frac{\partial v_2}{\partial y} + \frac{\partial v_3}{\partial z}$ and this quantity is defined as the DIVERGENCE of the vector $\underline{v}$. It is customary to abbreviate divergence to div and so write

$$\text{div } \underline{v} = \frac{\partial v_1}{\partial x} + \frac{\partial v_2}{\partial y} + \frac{\partial v_3}{\partial z}$$

Although the idea of divergence has been illustrated here in the context of fluid flow, it has other applications and the divergence of any vector can be found.

For example, if $\underline{A} = x^2z\underline{i} - 2y^3z^2\underline{j} + xyz^2\underline{k}$

$$\text{div } \underline{A} = \frac{\partial A_1}{\partial x} + \frac{\partial A_2}{\partial y} + \frac{\partial A_3}{\partial z}$$

where $A_1 = x^2z$, $A_2 = -2y^3z^2$, $A_3 = xyz^2$

Hence $\text{div } \underline{A} = 2xz - 6y^2z^2 + 2xyz$

Now, if

(i) $\underline{A} = x^2\underline{i} + 2xy\underline{j} + z^2\underline{k}$, find div $\underline{A}$

(ii) $\underline{p} = 3yz\underline{i} + 4xz\underline{j} + 7xy\underline{k}$, find div $\underline{p}$

(iii) $\underline{r} = x\underline{i} + y\underline{j} + z\underline{k}$, find div $\underline{r}$.

****

36A

(i) $2x + 2x + 2z = 2(2x + z)$

(ii) 0

(iii) 3

FRAME 37

It should be noted that divergence is an amount of flux and is therefore essentially scalar. The case of zero divergence has an important physical significance. If div $\underline{v} = 0$ at a point P it follows that there is no flux or net outflow of fluid from P, i.e. that fluid is neither "created" nor "destroyed" at P. This implies that there is no source or sink of fluid at P.

Similarly if $\underline{E}$ is the field intensity vector in an electrostatic field, div $\underline{E} = 0$ implies that there is no charge at P.

In general, at a point where div $\underline{A} = 0$ there is nothing creating flux there. A vector which is such that its divergence vanishes in any given region is said to be SOLENOIDAL in that region.

Now try these examples:

(i) If e is constant, $r^2 = x^2 + y^2 + z^2$, $\phi = e/r$ and $\underline{E} = \text{grad}\ \phi$, prove that $\underline{E}$ is solenoidal.

(ii) If $\underline{F} = r^m \underline{r}$, find the value of m which makes $\underline{F}$ solenoidal.

****

37A

(i) $$\underline{E} = -e\left(\frac{x}{r^3}\underline{i} + \frac{y}{r^3}\underline{j} + \frac{z}{r^3}\underline{k}\right)$$

$$div\ \underline{E} = 0$$

(ii) $m = -3$

FRAME 38

It was seen earlier that the gradient of a scalar function can be expressed in a form involving ∇, and now we shall show you that this can also be done for the divergence of a vector function.

You will remember that

$$\underline{a}.\underline{b} = (a_1\underline{i} + a_2\underline{j} + a_3\underline{k}).(b_1\underline{i} + b_2\underline{j} + b_3\underline{k})$$

$$= a_1b_1 + a_2b_2 + a_3b_3$$

FRAME 38 continued

By analogy, treating ∇ as a vector (although really it is a vector operator)

$$\nabla.\underline{v} = \left(\underline{i}\frac{\partial}{\partial x} + \underline{j}\frac{\partial}{\partial y} + \underline{k}\frac{\partial}{\partial z}\right).\left(v_1\underline{i} + v_2\underline{j} + v_3\underline{k}\right)$$

$$= \frac{\partial}{\partial x}v_1 + \frac{\partial}{\partial y}v_2 + \frac{\partial}{\partial z}v_3$$

$$= \frac{\partial v_1}{\partial x} + \frac{\partial v_2}{\partial y} + \frac{\partial v_3}{\partial z}$$

$$= \text{div } \underline{v}$$

FRAME 39

Curl

Another aspect of the motion of an incompressible fluid will be used to illustrate the idea of curl.

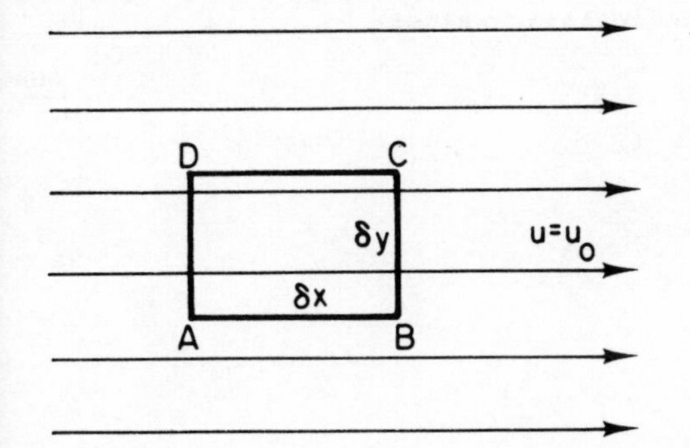

The diagram shows fluid streaming with constant velocity u_o past a small imaginary rectangle ABCD. The stream-lines are all parallel to the plane of the rectangle and also to the sides DC and AB.

Taking first the side AB of the rectangle, if we multiply its length by the velocity of the fluid in the direction AB we get what is called the flow along the path AB.

Thus, flow along AB = $u_o\delta x$.

Taking next the side BC, the fluid has no velocity in that direction and so flow along BC = $0.\delta y = 0$.

Can you now write down the flow along (i) CD, (ii) DA?

****

39A

(i) $(-u_o)\delta x = -u_o\delta x$ *(ii)* 0

FRAME 40

Adding these four results together gives the flow round the complete rectangle ABCDA. The flow in this direction round a closed path is called the circulation. In this example the circulation is zero.

Suppose now we consider the flow of a viscous fluid over a plane. The fact that the fluid is viscous effectively means that there is a sort of friction between adjacent layers so that these flow at different speeds.

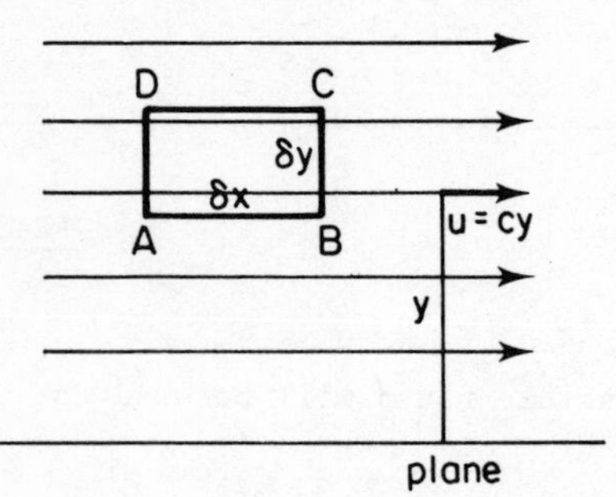

Assuming that the speed of a layer distant y from the plane is $u = cy$, where c is a constant, find the circulation round the rectangular path ABCDA. The rectangle is perpendicular to the plane, AB and DC being parallel to it and AB distant h from it.

****

40A

Flow along AB = $ch\delta x$
Flow along BC = 0
Flow along CD = $-c(h + \delta y)\delta x$
Flow along DA = 0
Circulation = $-c\delta y\delta x$

FRAME 41

Now suppose a small paddle wheel is placed as shown in the fluid such that its axis of rotation, which is fixed, is perpendicular to the rectangle. Can you say what the paddle wheel will do when

(i) $u = u_0$,

(ii) $u = cy$?

****

41A

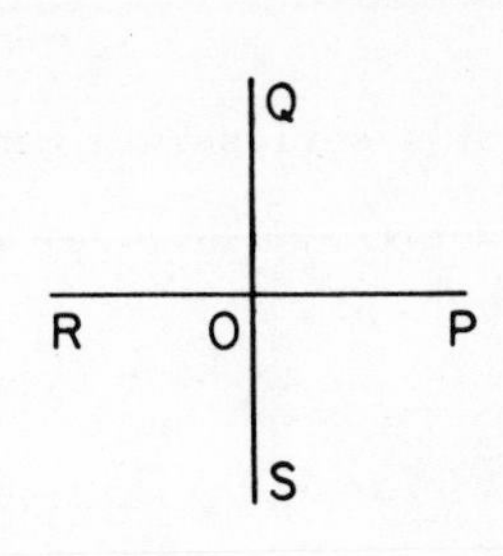

(i) Remain stationary

(ii) Rotate clockwise.

In (i) the same amount of fluid impinges on OQ as on OS, but in (ii) more impinges on OQ than on OS, causing the wheel to turn.

FRAME 42

You will notice that in (i), where the paddle wheel is stationary, the circulation is zero. The circulation is not zero in (ii), where the paddle wheel rotates.

FRAME 43

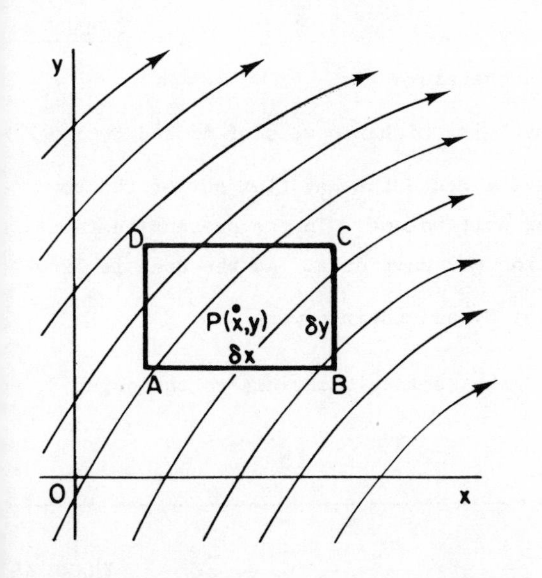

The diagram here shows a slightly more complicated situation. The streamlines are still in planes parallel to the rectangle. v_1, v_2 are the components, parallel to AB and BC, of the velocity of the fluid. P is the centre of the box,

To find the flow along AB we require the velocity along AB. If the rectangle is small, it can be assumed that the value of v_1 along AB is that attained at its mid-point $(x,y-\frac{1}{2}\delta y)$. Denoting this by $v_1(x,y-\frac{1}{2}\delta y)$, the flow along AB = $v_1(x,y-\frac{1}{2}\delta y)\delta x$.

Similarly, the flow along CD = $-v_1(x,y+\frac{1}{2}\delta y)\delta x$.

The sum of these flows is thus $\{v_1(x,y-\frac{1}{2}\delta y) - v_1(x,y+\frac{1}{2}\delta y)\}\delta x$.

FRAME 43 continued

Using Taylor's series in a manner similar to that in FRAME 35 this becomes

$$\left[\{v_1(x,y) + (-\tfrac{1}{2}\delta y)\frac{\partial}{\partial y} v_1(x,y) + \ldots\} - \{v_1(x,y) + \tfrac{1}{2}\delta y \frac{\partial}{\partial y} v_1(x,y) + \ldots\}\right]\delta x$$

$$= -\delta y \frac{\partial}{\partial y} v_1(x,y)\delta x + \ldots$$

$$= -\delta y \frac{\partial v_1}{\partial y}\delta x + \ldots$$

What will be the sum of the flows along BC and DA?

****

43A

$$\{v_2(x + \tfrac{1}{2}\delta x, y) - v_2(x - \tfrac{1}{2}\delta x, y)\}\delta y$$

$$= \delta x \frac{\partial v_2}{\partial x} \delta y + \ldots$$

FRAME 44

The circulation round the rectangle is therefore

$$\left(\frac{\partial v_2}{\partial x} - \frac{\partial v_1}{\partial y}\right)\delta x\delta y + \text{terms involving higher powers of } \delta x \text{ and } \delta y$$

Now when leading up to divergence, having found the net flux out of the box the next step was to find the flux per unit volume. In the present instance the next step is to find the circulation per unit area. As the area is $\delta x\delta y$ this is $\frac{\partial v_2}{\partial x} - \frac{\partial v_1}{\partial y}$ + terms in δx or δy and their powers.

When δx and δy both tend to zero, i.e. the rectangle shrinks to the point P, this becomes $\frac{\partial v_2}{\partial x} - \frac{\partial v_1}{\partial y}$.

FRAME 45

When a rigid body is rotating about a fixed axis a vector is associated with the rotation. This is the 'angular velocity' vector. Its magnitude gives the angular speed and its direction is along the axis of rotation. (See FRAME 22 on page 3:52 in Vol I.)

FRAME 45 continued

You have already seen that when circulation exists a rotational effect is associated with the fluid. This was shown by imagining a paddle wheel to be placed in the fluid, its axis being perpendicular to the rectangle ABCD. By drawing an analogy between the rotational effect of the fluid and the rigid body, a vector is associated with this rotational effect. Its magnitude is $\frac{\partial v_2}{\partial x} - \frac{\partial v_1}{\partial y}$ and its direction is perpendicular to the rectangle ABCD in such a way that Ox, Oy and the vector form a right-handed set. (See FRAME 23 on page 3:53 in Vol I.) This direction is parallel to Oz (not shown in the diagram) and so the vector can be written as $\left(\frac{\partial v_2}{\partial x} - \frac{\partial v_1}{\partial y}\right)\underline{k}$.

FRAME 46

In three-dimensional flow where the streamlines are not, in general, parallel to a plane, the fluid velocity would be given by $\underline{v} = v_1\underline{i} + v_2\underline{j} + v_3\underline{k}$. If the circulation round a small rectangle whose sides are parallel to Ox and Oy is found, the working will be exactly the same as in FRAMES 43 and 44. This is because v_3, being perpendicular to the rectangle, makes no contribution to the circulation round it.

In a similar way the circulation can be calculated round a small rectangle whose sides are parallel to Oy and Oz. This will lead to the vector $\left(\frac{\partial v_3}{\partial y} - \frac{\partial v_2}{\partial z}\right)\underline{i}$.

What vector will be obtained in a similar way if the circulation is found round a small rectangle whose sides are parallel to Oz and Ox?

****

46A

$\left(\frac{\partial v_1}{\partial z} - \frac{\partial v_3}{\partial x}\right)\underline{j}$

FRAME 47

If the circulation is calculated round a small rectangle whose sides are not parallel to any of the axes, this will lead to a vector, perpendicular to the rectangle, whose magnitude is the sum of the components in that direction of the three vectors just found. This is the same as the component of the sum

$$\left(\frac{\partial v_3}{\partial y} - \frac{\partial v_2}{\partial z}\right)\underline{i} + \left(\frac{\partial v_1}{\partial z} - \frac{\partial v_3}{\partial x}\right)\underline{j} + \left(\frac{\partial v_2}{\partial x} - \frac{\partial v_1}{\partial y}\right)\underline{k} \qquad (47.1)$$

in that direction. The vector (47.1) is called the CURL or ROTATION of the vector $\underline{v}$. Rotation is abbreviated to rot but the term curl is more commonly used. Then

$$\boxed{\text{curl } \underline{v} = \left(\frac{\partial v_3}{\partial y} - \frac{\partial v_2}{\partial z}\right)\underline{i} + \left(\frac{\partial v_1}{\partial z} - \frac{\partial v_3}{\partial x}\right)\underline{j} + \left(\frac{\partial v_2}{\partial x} - \frac{\partial v_1}{\partial y}\right)\underline{k}}$$

As curl $\underline{v}$ gives a measure of the rotational effect of a fluid flow, curl $\underline{v} = 0$ means that no such effect is present. In this case the fluid flow is said to be IRROTATIONAL.

As was the case with the divergence it is possible to find the curl of any vector. Thus, for example, if

$$\underline{A} = x^2y^2\underline{i} + 2xz\underline{j} - 2yz\underline{k}$$

$$\begin{aligned}
\text{curl } \underline{A} &= \left(\frac{\partial A_3}{\partial y} - \frac{\partial A_2}{\partial z}\right)\underline{i} + \left(\frac{\partial A_1}{\partial z} - \frac{\partial A_3}{\partial x}\right)\underline{j} + \left(\frac{\partial A_2}{\partial x} - \frac{\partial A_1}{\partial y}\right)\underline{k} \\
&= \left\{\frac{\partial}{\partial y}(-2yz) - \frac{\partial}{\partial z}(2xz)\right\}\underline{i} + \left\{\frac{\partial}{\partial z}(x^2y^2) - \frac{\partial}{\partial x}(-2yz)\right\}\underline{j} \\
&\quad + \left\{\frac{\partial}{\partial x}(2xz) - \frac{\partial}{\partial y}(x^2y^2)\right\}\underline{k} \\
&= -2(z + x)\underline{i} + 0\underline{j} + (2z - 2x^2y)\underline{k}
\end{aligned}$$

Now find curl $\underline{B}$ if $\underline{B} = 2xy\underline{i} - 3yz^2\underline{j} + 4xz^2\underline{k}$

****

47A

$2(3yz\underline{i} - 2z^2\underline{j} - x\underline{k})$

FRAME 48

Just as it has been shown that it is possible to express gradient and divergence in a form involving ∇, so also can the curl of a vector.

You will remember that

$$\underline{a}_\wedge\underline{b} = (a_1\underline{i} + a_2\underline{j} + a_3\underline{k})_\wedge(b_1\underline{i} + b_2\underline{j} + b_3\underline{k})$$

$$= \begin{vmatrix} \underline{i} & \underline{j} & \underline{k} \\ a_1 & a_2 & a_3 \\ b_1 & b_2 & b_3 \end{vmatrix}$$

$$= \underline{i}(a_2b_3 - a_3b_2) + \underline{j}(a_3b_1 - a_1b_3) + \underline{k}(a_1b_2 - a_2b_1)$$

By analogy, treating ∇ as a vector, as in FRAME 38,

$$\nabla_\wedge\underline{v} = \left(\underline{i}\frac{\partial}{\partial x} + \underline{j}\frac{\partial}{\partial y} + \underline{k}\frac{\partial}{\partial z}\right)_\wedge\left(v_1\underline{i} + v_2\underline{j} + v_3\underline{k}\right)$$

$$= \begin{vmatrix} \underline{i} & \underline{j} & \underline{k} \\ \frac{\partial}{\partial x} & \frac{\partial}{\partial y} & \frac{\partial}{\partial z} \\ v_1 & v_2 & v_3 \end{vmatrix}$$

You will see that this gives (47.1) when expanded.

Thus $\nabla_\wedge\underline{v} = \text{curl } \underline{v}$

$$\text{and so curl } \underline{v} = \begin{vmatrix} \underline{i} & \underline{j} & \underline{k} \\ \frac{\partial}{\partial x} & \frac{\partial}{\partial y} & \frac{\partial}{\partial z} \\ v_1 & v_2 & v_3 \end{vmatrix}$$

FRAME 48 continued

You will probably find the determinant form for curl $\underline{v}$ easier to remember than that given in FRAME 47.

As an example, taking the same $\underline{A}$ as in FRAME 47,

$$\text{curl } \underline{A} = \begin{vmatrix} \underline{i} & \underline{j} & \underline{k} \\ \frac{\partial}{\partial x} & \frac{\partial}{\partial y} & \frac{\partial}{\partial z} \\ x^2y^2 & 2xz & -2yz \end{vmatrix}$$

You should now check that this gives the same answer as before. Then find curl $(xy^2\underline{i} + 3yz^2\underline{j} - 5x^3z^3\underline{k})$, using the determinant form.

****

48A

$-6yz\underline{i} + 15x^2z^3\underline{j} - 2xy\underline{k}$

FRAME 49

Now try the following two examples:

1. Find the divergence and curl of the vector field $\underline{F} = xy\underline{i} + yz\underline{j} + 0\underline{k}$ at the point (1,1,1). Also find grad(div $\underline{F}$) at any point.

2. If $\phi = x + y^2 + z^3$, find grad ϕ and hence div(grad ϕ) and curl(grad ϕ).

****

49A

1. *div* $\underline{F} = y + z$ *curl* $\underline{F} = -y\underline{i} - x\underline{k}$

   At (1,1,1) div $\underline{F} = 2$ *and curl* $\underline{F} = -\underline{i} - \underline{k}$

   grad(div $\underline{F}$*)* $= \underline{j} + \underline{k}$

2. *grad* $\phi = \underline{i} + 2y\underline{j} + 3z^2\underline{k}$

   div(grad ϕ*)* $= 2 + 6z$ *curl(grad* ϕ*)* $= \underline{0}$

FRAME 50

Second Order Differential Operators

In the previous frame you have just obtained grad(div $\underline{F}$), div(grad ϕ) and curl(grad ϕ) for certain given $\underline{F}$ and ϕ. These are examples of second order differential operators, and we shall now say a little more about them. Probably the one which you are most likely to meet is the divergence of the gradient, as it can occur in the treatment of problems in heat conduction, electrostatics, etc.

If ϕ represents a scalar function of position, then, as we have seen, grad ϕ is a vector and represents, both in magnitude and direction, the greatest rate of change of ϕ at any point P. It is perpendicular to the level surface, ϕ = constant, which passes through P.

As grad ϕ is a vector whose value can be found at every point, grad ϕ forms a vector field. Thus it can have a divergence, i.e. div(grad ϕ) can be formed.

You are already familiar with the expressions for divergence and gradient, and using these gives

$$\text{div(grad } \phi) = \left(\underline{i}\frac{\partial}{\partial x} + \underline{j}\frac{\partial}{\partial y} + \underline{k}\frac{\partial}{\partial z}\right).\left(\underline{i}\frac{\partial \phi}{\partial x} + \underline{j}\frac{\partial \phi}{\partial y} + \underline{k}\frac{\partial \phi}{\partial z}\right)$$

$$= \frac{\partial^2\phi}{\partial x^2} + \frac{\partial^2\phi}{\partial y^2} + \frac{\partial^2\phi}{\partial z^2}$$

Since div(grad ϕ) can be written $\nabla.(\nabla\phi)$ we also use the notation

$$\text{div grad} \equiv \nabla^2$$

Thus $\nabla^2\phi = \text{div grad } \phi = \dfrac{\partial^2\phi}{\partial x^2} + \dfrac{\partial^2\phi}{\partial y^2} + \dfrac{\partial^2\phi}{\partial z^2}$

and the operator $\nabla^2 \equiv \dfrac{\partial^2}{\partial x^2} + \dfrac{\partial^2}{\partial y^2} + \dfrac{\partial^2}{\partial z^2}$

FRAME 51

As an example, if $\phi = 2x^2 - y^2 - z^2$

$$\frac{\partial^2\phi}{\partial x^2} = 4 \qquad \frac{\partial^2\phi}{\partial y^2} = -2 \qquad \frac{\partial^2\phi}{\partial z^2} = -2$$

$\therefore\ \nabla^2\phi = 4 - 2 - 2 = 0$

Now find $\nabla^2\phi$ when $\phi = x^2y - 2y^3z$

****

51A

2y(1 - 6z)

FRAME 52

In this programme we have not defined grad $\underline{A}$, and consequently cannot give an interpretation to div(grad $\underline{A}$). However, the operator ∇^2 can be applied to a vector.

For a vector $\underline{A}$, $\nabla^2\underline{A} = \nabla^2(A_1\underline{i} + A_2\underline{j} + A_3\underline{k}) = (\nabla^2A_1)\underline{i} + (\nabla^2A_2)\underline{j} + (\nabla^2A_3)\underline{k}$

In other words, $\nabla^2\underline{A}$ means apply ∇^2 to each component of $\underline{A}$.

So if, for example,

$$\begin{aligned} \underline{A} &= (2x - 3y)\underline{i} + 4x^2z\underline{j} + y^3\underline{k} \\ \nabla^2\underline{A} &= 0\underline{i} + 8z\underline{j} + 6y\underline{k} \\ &= 8z\underline{j} + 6y\underline{k} \end{aligned}$$

Now show that, if $\underline{F} = (y^2 - z^2)\underline{i} - (z^2 - x^2)\underline{j} + (x^2 - y^2)\underline{k}$, then $\nabla^2\underline{F} = \underline{0}$.

FRAME 53

Turning to other combinations of div, grad and curl, which of the following could you form?

(i) curl(grad ϕ)

(ii) curl(grad $\underline{A}$)

(iii) grad(div $\underline{A}$)

(iv) curl(curl $\underline{A}$)

(v) div(div $\underline{A}$)

****

53A

(i), (iii) and (iv) can be formed.

In (ii), grad $\underline{A}$ would be required and we have not defined this.

In (v), div $\underline{A}$ can be formed, and is a scalar, but we have not defined the divergence of a scalar.

FRAME 54

We have already considered div grad ϕ and $\nabla^2\underline{A}$ in FRAMES 51 and 52 and you have just seen that curl grad ϕ, grad div $\underline{A}$ and curl curl $\underline{A}$ are possible. The only other combination which you could interpret is div(curl $\underline{A}$). These last four possibilities will not be dealt with more fully here, but, if you are interested, you will find them treated in APPENDIX A. Curl grad ϕ occurs in the study of hydrodynamics and curl curl $\underline{H}$ in Maxwell's equations of electromagnetic theory.

FRAME 55

Div, grad and curl applied to sums and products

As $\underline{A}.\underline{B}$ is a scalar, it is possible to find its gradient. Similarly, $\underline{A}_{\wedge}\underline{B}$ is a vector and so we can find its divergence or curl.

Thus we can have grad($\underline{A}.\underline{B}$), div($\underline{A}_{\wedge}\underline{B}$) and curl($\underline{A}_{\wedge}\underline{B}$).

Which of the following could you form?

(i) div($\underline{A}.\underline{B}$)

(ii) grad($\phi_1\phi_2$)

(iii) curl($\phi\underline{A}$)

(iv) div(ϕ + $\underline{A}$)

(v) grad($\phi\underline{A}$)

****

55A

(ii) and (iii) can be formed.

(i) cannot be formed as $\underline{A}.\underline{B}$ is a scalar.

(iv) cannot be formed as $\phi + \underline{A}$ is not a valid sum.

(v) cannot be formed as $\phi\underline{A}$ is a vector.

FRAME 56

Just as in ordinary calculus there are formulae for the differentiation of sums and products, so in this work there are corresponding formulae for such expressions as $\text{grad}(\phi_1 + \phi_2)$, $\text{div}(\underline{A}_\wedge\underline{B})$ etc.

We shall not pursue this here, but, if you are interested, you will find some of these formulae in APPENDIX B. The theory of hydrodynamics involves the use of such formulae.

FRAME 57

Miscellaneous Examples

In this frame a collection of miscellaneous examples is given for you to try. Answers are supplied in FRAME 58, together with such working as is considered helpful.

1. Find a unit normal to the surface $x^3 + 3xyz + 2y^3 - z^3 - 5 = 0$ at the point (1,1,1).

2. If $\phi = (x^2 + y^2)e^z$, show that curl grad $\phi = 0$.

3. If $\underline{A} = xyz\underline{i} + (x^2 + y^2 + z^2)\underline{j} - (xy + yz + zx)\underline{k}$, show that div curl $\underline{A} = 0$.

4. The rate of change of ϕ with respect to distance in the direction of the vector $2\underline{i} - \underline{j}$ is $7\sqrt{5}$ at the point (1,1,1). If $\phi = cx^2y + 2x^2z + 3y^2z$, find the value of c.

FRAME 57 continued

5. If $\theta = x + y + z$, $\phi = x + y$, $\psi = -2xz - 2yz - z^2$, show that $\nabla\theta.(\nabla\phi_\wedge\nabla\psi) = 0$.

6. If curl $\underline{f} = 0$, where $\underline{f} = (xyz)^m(x^n\underline{i} + y^n\underline{j} + z^n\underline{k})$, show that either $m = 0$ or $n = -1$. (L.U.)

7. Show that the vector $\underline{A} = (4xy - z^3)\underline{i} + 2x^2\underline{j} - 3xz^2\underline{k}$ is irrotational and find a function ϕ such that $\underline{A} = \text{grad }\phi$. (L.U.)

8. Given $\underline{r} = x\underline{i} + y\underline{j} + z\underline{k}$, where $\underline{i}$, $\underline{j}$, $\underline{k}$ are Cartesian unit vectors, and $\underline{p} = a\underline{i} + b\underline{j} + c\underline{k}$ is a constant vector, show that, if $\underline{u} = (\underline{p}.\underline{r})\underline{r}$,

$$\text{div }\underline{u} = 4\underline{p}.\underline{r}$$
$$\text{curl }\underline{u} = \underline{p}_\wedge\underline{r}$$
$$\text{curl}(\underline{p}_\wedge\underline{r}) = 2\underline{p}$$

(L.U.)

9. If r is the distance of a point (x,y,z) from the origin O prove that

(a) $\text{div grad}\left(\frac{1}{r}\right) = 0$

(b) $\text{curl}\left(\underline{k}_\wedge\text{grad }\frac{1}{r}\right) + \text{grad}\left(\underline{k}.\text{grad }\frac{1}{r}\right) = 0$ where $\underline{k}$ is the unit vector in the direction Oz. (L.U.)

FRAME 58

Answers to Miscellaneous Examples

1. Taking $\phi = x^3 + 3xyz + 2y^3 - z^3 - 5$ ($\phi = x^3 + 3xyz + 2y^3 - z^3$ will do just as well)

$$\nabla\phi = (3x^2 + 3yz)\underline{i} + (3xz + 6y^2)\underline{j} + (3xy - 3z^2)\underline{k}$$

At (1,1,1), $\nabla\phi = 6\underline{i} + 9\underline{j}$

Unit normal is $\dfrac{2}{\sqrt{13}}\underline{i} + \dfrac{3}{\sqrt{13}}\underline{j}$

2. $\text{grad }\phi = 2xe^z\underline{i} + 2ye^z\underline{j} + (x^2 + y^2)e^z\underline{k}$

FRAME 58 continued

3. curl $\underline{A} = -(x + 3z)\underline{i} + (xy + y + z)\underline{j} + x(2 - z)\underline{k}$

4. $\nabla\phi = (2cxy + 4xz)\underline{i} + (cx^2 + 6yz)\underline{j} + (2x^2 + 3y^2)\underline{k}$

At (1,1,1), $\nabla\phi = 2(c + 2)\underline{i} + (c + 6)\underline{j} + 5\underline{k}$

$$\frac{2\underline{i} - \underline{j}}{\sqrt{5}} . \{2(c + 2)\underline{i} + (c + 6)\underline{j} + 5\underline{k}\} = 7\sqrt{5}$$

$c = 11$

5. $\nabla\theta = \underline{i} + \underline{j} + \underline{k}$

$\nabla\phi = \underline{i} + \underline{j}$

$\nabla\psi = -2z\underline{i} - 2z\underline{j} - 2(x + y + z)\underline{k}$

$$\nabla\theta.(\nabla\phi_{\wedge}\nabla\psi) = \begin{vmatrix} 1 & 1 & 1 \\ 1 & 1 & 0 \\ -2z & -2z & -2(x+y+z) \end{vmatrix}$$ = 0 using the result in FRAME 37 on page 3:63 in Vol I.

6. $\underline{f} = x^{m+n}y^m z^m \underline{i}$ + corresponding terms in $\underline{j}$ and $\underline{k}$

$$\text{curl } \underline{f} = \begin{vmatrix} \underline{i} & \underline{j} & \underline{k} \\ \frac{\partial}{\partial x} & \frac{\partial}{\partial y} & \frac{\partial}{\partial z} \\ x^{m+n}y^m z^m & x^m y^{m+n} z^m & x^m y^m z^{m+n} \end{vmatrix}$$

$= mx^m y^m z^m(y^{-1}z^n - y^n z^{-1})\underline{i}$ + corresponding terms in $\underline{j}$ and $\underline{k}$

$= 0$ if $m = 0$ or $n = -1$

7.

$$\text{curl } \underline{A} = \begin{vmatrix} \underline{i} & \underline{j} & \underline{k} \\ \frac{\partial}{\partial x} & \frac{\partial}{\partial y} & \frac{\partial}{\partial z} \\ 4xy-z^3 & 2x^2 & -3xz^2 \end{vmatrix} = 0$$

$\therefore$ $\underline{A}$ is irrotational.

We require $\frac{\partial\phi}{\partial x} = 4xy - z^3$, $\frac{\partial\phi}{\partial y} = 2x^2$, $\frac{\partial\phi}{\partial z} = -3xz^2$, from which

$\phi = 2x^2y - xz^3$ + constant.

8. $\underline{u} = (ax + by + cz)(x\underline{i} + y\underline{j} + z\underline{k})$

$\text{div } \underline{u} = 4(ax + by + cz) = 4\underline{p}.\underline{r}$

$\text{curl } \underline{u} = (bz - cy)\underline{i} + (cx - az)\underline{j} + (ay - bx)\underline{k}$

$= \underline{p}\wedge\underline{r}$

$\text{curl}(\underline{p}\wedge\underline{r}) = 2a\underline{i} + 2b\underline{j} + 2c\underline{k} = 2\underline{p}$

Alternatively, if you have read APPENDIX B you could do it this way:

$\underline{u} = (\underline{p}.\underline{r})\underline{r}$

$\text{div } \underline{u} = (\underline{p}.\underline{r})\text{div } \underline{r} + \underline{r}.\text{grad}(\underline{p}.\underline{r})$ using formula (iv) in APPENDIX B

$\text{div } \underline{r} = 3$ and $\text{grad}(\underline{p}.\underline{r}) = \underline{p}$, the latter being more easily worked directly than found by the use of formula (iii).

$\therefore \text{div } \underline{u} = 4\underline{p}.\underline{r}$

$\text{curl } \underline{u} = (\underline{p}.\underline{r})\text{curl } \underline{r} + \text{grad}(\underline{p}.\underline{r})\wedge\underline{r}$ using formula (vi)

$= \underline{p}\wedge\underline{r}$ as $\text{curl } \underline{r} = 0$ and $\text{grad } \underline{p}.\underline{r} = \underline{p}$

$\text{curl}(\underline{p}\wedge\underline{r}) = (\underline{r}.\nabla)\underline{p} - (\underline{p}.\nabla)\underline{r} + \underline{p} \text{ div } \underline{r} - \underline{r} \text{ div } \underline{p}$

The first and last terms are zero as $\underline{p}$ is a constant vector.

$(\underline{p}.\nabla)\underline{r} = \left(a\frac{\partial}{\partial x} + b\frac{\partial}{\partial y} + c\frac{\partial}{\partial z}\right)\underline{r} = \underline{p}$

$\text{div } \underline{r} = 3$

$\text{curl } \underline{p}\wedge\underline{r} = 2\underline{p}$

9. (a) $\text{grad}(1/r) = -\underline{r}/r^3$

$$\text{div grad}\left(\frac{1}{r}\right) = \frac{3x^2 - r^2}{r^4} + \frac{3y^2 - r^2}{r^4} + \frac{3z^2 - r^2}{r^4} = 0$$

Alternatively you could use $\text{div grad}(1/r) = \nabla^2(1/r)$

(b) $\underline{k}\wedge\text{grad}(1/r) = (y\underline{i} - x\underline{j})/r^3$

$$\text{curl}\left(\underline{k}\wedge\text{grad }\frac{1}{r}\right) = \frac{-3xz}{r^5}\underline{i} - \frac{3yz}{r^5}\underline{j} + \left(\frac{3x^2}{r^5} + \frac{3y^2}{r^5} - \frac{2}{r^3}\right)\underline{k}$$

$\underline{k}.\text{grad}(1/r) = -z/r^3$

$$\text{grad}\left(\underline{k}.\text{grad }\frac{1}{r}\right) = \frac{3xz}{r^5}\underline{i} + \frac{3yz}{r^5}\underline{j} + \left(\frac{3z^2}{r^5} - \frac{1}{r^3}\right)\underline{k}$$

Result follows.

FRAME 58 continued

Alternatively, if you have read the appendices, this part can be worked as follows:

$$\text{curl}\left(\underline{k}\wedge\text{grad}\,\frac{1}{r}\right) = \left(\text{grad}\,\frac{1}{r}.\nabla\right)\underline{k} - (\underline{k}.\nabla)\text{grad}\,\frac{1}{r} + \underline{k}\,\text{div grad}\,\frac{1}{r} - \left(\text{grad}\,\frac{1}{r}\right)\text{div}\,\underline{k}$$

from formula (vii) in APPENDIX B

$$= -(\underline{k}.\nabla)\text{grad}(1/r)$$

$$\text{grad}\left(\underline{k}.\text{grad}\,\frac{1}{r}\right) = \left(\text{grad}\,\frac{1}{r}.\nabla\right)\underline{k} + (\underline{k}.\nabla)\text{grad}\,\frac{1}{r} + \text{grad}\,\frac{1}{r}\wedge\text{curl}\,\underline{k} + \underline{k}\wedge\text{curl grad}\,\frac{1}{r}$$

from formula (iii)

$$= (\underline{k}.\nabla)\text{grad}\,\frac{1}{r}$$

APPENDIX A

FRAME A1

Curl grad

As we have seen, if ϕ is a scalar point function, then grad ϕ is the derived vector field. Since it is a vector field it is possible to find its curl, i.e. curl(grad ϕ).

As grad $\phi = \underline{i}\frac{\partial\phi}{\partial x} + \underline{j}\frac{\partial\phi}{\partial y} + \underline{k}\frac{\partial\phi}{\partial z}$

$$\text{curl(grad } \phi) = \begin{vmatrix} \underline{i} & \underline{j} & \underline{k} \\ \frac{\partial}{\partial x} & \frac{\partial}{\partial y} & \frac{\partial}{\partial z} \\ \frac{\partial\phi}{\partial x} & \frac{\partial\phi}{\partial y} & \frac{\partial\phi}{\partial z} \end{vmatrix}$$

Now expand and simplify this determinant.

****

A1A

The $\underline{i}$ component is $\underline{i}\left(\frac{\partial^2\phi}{\partial y\partial z} - \frac{\partial^2\phi}{\partial z\partial y}\right)$, *i.e.* $\underline{0}$

The $\underline{j}$ and $\underline{k}$ components are similarly zero so curl(grad ϕ) = $\underline{0}$

FRAME A2

You have just shown that curl(grad ϕ) is always zero.

This means that if curl $\underline{A} = \underline{0}$, we can find a ϕ such that grad $\phi = \underline{A}$.
This fact is made use of in the theory of hydrodynamics, where the vorticity of the fluid motion is curl $\underline{v}$, $\underline{v}$ being the velocity at any point. For irrotational motion curl $\underline{v} = \underline{0}$ and therefore there exists a ϕ such that $\underline{v}$ = grad ϕ. In this case the function ϕ is called the velocity potential.

FRAME A2 continued

Now try the following example:

If $\underline{A} = 2xz^3\underline{i} + 8y^3\underline{j} + 3x^2z^2\underline{k}$ show that curl $\underline{A} = \underline{0}$ and find the potential function ϕ which is such that grad $\phi = \underline{A}$.

****

A2A

$$\text{curl } \underline{A} = \begin{vmatrix} \underline{i} & \underline{j} & \underline{k} \\ \frac{\partial}{\partial x} & \frac{\partial}{\partial y} & \frac{\partial}{\partial z} \\ 2xz^3 & 8y^3 & 3x^2z^2 \end{vmatrix} = \underline{0}$$

$$\frac{\partial\phi}{\partial x} = 2xz^3 \qquad \frac{\partial\phi}{\partial y} = 8y^3 \qquad \frac{\partial\phi}{\partial z} = 3x^2z^2$$

from which $\phi = x^2z^3 + 2y^4$

FRAME A3

Grad div

If $\underline{A}$ defines a vector field, then the divergence of $\underline{A}$ can be found at any point and will define a scalar function. Hence it is possible to find the gradient of this scalar function giving grad(div $\underline{A}$).

Expanding in cartesians we have

$$\left(\underline{i}\frac{\partial}{\partial x} + \underline{j}\frac{\partial}{\partial y} + \underline{k}\frac{\partial}{\partial z}\right)\left(\frac{\partial A_1}{\partial x} + \frac{\partial A_2}{\partial y} + \frac{\partial A_3}{\partial z}\right)$$

so that $\text{grad div } \underline{A} = \underline{i}\left(\frac{\partial^2 A_1}{\partial x^2} + \frac{\partial^2 A_2}{\partial x\partial y} + \frac{\partial^2 A_3}{\partial x\partial z}\right) + \underline{j}\left(\frac{\partial^2 A_1}{\partial y\partial x} + \frac{\partial^2 A_2}{\partial y^2} + \frac{\partial^2 A_3}{\partial y\partial z}\right)$

$$+ \underline{k}\left(\frac{\partial^2 A_1}{\partial z\partial x} + \frac{\partial^2 A_2}{\partial z\partial y} + \frac{\partial^2 A_3}{\partial z^2}\right)$$

Notice that grad div bears no resemblance to div grad.

Now use this expression for grad div $\underline{A}$ to do the following example:

If $\underline{A} = x^2y\underline{i} + 2x^2z\underline{j} + 5z^2y^4\underline{k}$, find grad div $\underline{A}$.

****

A3A

$2y\underline{i} + (2x + 40zy^3)\underline{j} + 10y^4\underline{k}$

FRAME A4

Div curl

If $\underline{A}$ is a vector field, then curl $\underline{A}$ is itself a vector field and we can find div(curl $\underline{A}$) at every point.

Starting from curl $\underline{A} = \begin{vmatrix} \underline{i} & \underline{j} & \underline{k} \\ \frac{\partial}{\partial x} & \frac{\partial}{\partial y} & \frac{\partial}{\partial z} \\ A_1 & A_2 & A_3 \end{vmatrix}$ expand the determinant and hence find div(curl $\underline{A}$).

****

A4A

$$\begin{aligned} div(curl\ \underline{A}) &= \frac{\partial}{\partial x}\left(\frac{\partial A_3}{\partial y} - \frac{\partial A_2}{\partial z}\right) + \frac{\partial}{\partial y}\left(\frac{\partial A_1}{\partial z} - \frac{\partial A_3}{\partial x}\right) + \frac{\partial}{\partial z}\left(\frac{\partial A_2}{\partial x} - \frac{\partial A_1}{\partial y}\right) \\ &= 0 \end{aligned}$$

FRAME A5

Curl curl

Again, since curl $\underline{A}$ is a vector field we can find curl(curl $\underline{A}$). The general expression for this in terms of the components of $\underline{A}$ is somewhat cumbersome, so we shall not obtain it here. In addition, it can be shown that

$$\text{curl(curl } \underline{A}) = \text{grad(div } \underline{A}) - \nabla^2\underline{A} \qquad (A5.1)$$

but the proof of this is also tedious and will not be given here. However, you may like to verify that this relationship holds when

$$\underline{A} = x^2y\underline{i} - 2xz\underline{j} + 2yz\underline{k}.$$

****

A5A

$curl\ \underline{A} \;=\; 2(z - x)\underline{i} - (2z + x^2)\underline{k}$

$curl(curl\ \underline{A}) \;=\; 2(x + 1)\underline{j}$

$div\ \underline{A} \;=\; 2(x + 1)y$

$grad(div\ \underline{A}) \;=\; 2y\underline{i} + 2(x + 1)\underline{j}$

Alternatively this may be written down using the formula in FRAME A3.

$\nabla^2\underline{A} = 2y\underline{i}$

$\therefore\quad grad(div\ \underline{A}) - \nabla^2\underline{A} \;=\; 2(x + 1)\underline{j}$

FRAME A6

The following equations, due to Maxwell, occur in the theory of electromagnetic waves in a conducting medium:

$$\text{div } \underline{E} = 0 \qquad \text{(A6.1)}$$

$$\text{curl } \underline{E} = -\frac{\mu}{c}\frac{\partial \underline{H}}{\partial t} \qquad \text{(A6.2)}$$

$$\text{curl } \underline{H} = 4\pi\sigma\underline{E} + \frac{k}{c}\frac{\partial \underline{E}}{\partial t} \qquad \text{(A6.3)}$$

μ, c, σ and k are constants for a given medium.

$\underline{H}$ can be eliminated between equations (A6.2) and (A6.3) to give a partial differential equation for $\underline{E}$.

First we differentiate (A6.3) with respect to t giving

$$\frac{\partial}{\partial t}(\text{curl } \underline{H}) = 4\pi\sigma\frac{\partial \underline{E}}{\partial t} + \frac{k}{c}\frac{\partial^2 \underline{E}}{\partial t^2}$$

In partial differentiation the order of differentiation is immaterial

i.e. $\dfrac{\partial^2 z}{\partial x \partial y} = \dfrac{\partial^2 z}{\partial y \partial x}$, so $\dfrac{\partial}{\partial t}(\text{curl } \underline{H}) = \text{curl } \dfrac{\partial \underline{H}}{\partial t}$.

$$\therefore\quad \text{curl } \frac{\partial \underline{H}}{\partial t} = 4\pi\sigma\frac{\partial \underline{E}}{\partial t} + \frac{k}{c}\frac{\partial^2 \underline{E}}{\partial t^2} \qquad \text{(A6.4)}$$

But from (A6.2) $\text{curl}(\text{curl } \underline{E}) = \text{curl}\left(-\dfrac{\mu}{c}\dfrac{\partial \underline{H}}{\partial t}\right)$

$= -\dfrac{\mu}{c}\,\text{curl } \dfrac{\partial \underline{H}}{\partial t}$ as $\dfrac{\mu}{c}$ is a constant

FRAME A6 continued

Substitute this result in (A6.4) and use (A5.1) and (A6.1) to obtain the final result

$$\nabla^2\underline{E} = \frac{k\mu}{c^2}\frac{\partial^2\underline{E}}{\partial t^2} + \frac{4\pi\mu\sigma}{c}\frac{\partial\underline{E}}{\partial t}$$

****

A6A

$$curl\ \frac{\partial\underline{H}}{\partial t} = \frac{-c}{\mu}\ curl(curl\ \underline{E})$$

$$\therefore\ -\frac{c}{\mu}\ curl(curl\ \underline{E}) = 4\pi\sigma\frac{\partial\underline{E}}{\partial t} + \frac{k}{c}\frac{\partial^2\underline{E}}{\partial t^2}$$

$$\nabla^2\underline{E} - grad(div\ \underline{E}) = \frac{4\pi\mu\sigma}{c}\frac{\partial\underline{E}}{\partial t} + \frac{k\mu}{c^2}\frac{\partial^2\underline{E}}{\partial t^2}$$

But div $\underline{E} = 0$ so the result follows.

FRAME A7

The operator $(\underline{A}.\nabla)$

If $\underline{A}$ is $A_1\underline{i} + A_2\underline{j} + A_3\underline{k}$, then

$$(\underline{A}.\nabla) \equiv (A_1\underline{i} + A_2\underline{j} + A_3\underline{k}).(\underline{i}\frac{\partial}{\partial x} + \underline{j}\frac{\partial}{\partial y} + \underline{k}\frac{\partial}{\partial z})$$

$$\equiv (A_1\frac{\partial}{\partial x} + A_2\frac{\partial}{\partial y} + A_3\frac{\partial}{\partial z})$$

You will see that this is a scalar operator. It can therefore be applied to either a scalar or a vector, giving expressions such as

$$(\underline{A}.\nabla)\phi = (A_1\frac{\partial}{\partial x} + A_2\frac{\partial}{\partial y} + A_3\frac{\partial}{\partial z})\phi$$

$$= A_1\frac{\partial\phi}{\partial x} + A_2\frac{\partial\phi}{\partial y} + A_3\frac{\partial\phi}{\partial z}$$

and $(\underline{A}.\nabla)\underline{B} = (A_1\frac{\partial}{\partial x} + A_2\frac{\partial}{\partial y} + A_3\frac{\partial}{\partial z})(B_1\underline{i} + B_2\underline{j} + B_3\underline{k})$

$$= \left(A_1\frac{\partial B_1}{\partial x} + A_2\frac{\partial B_1}{\partial y} + A_3\frac{\partial B_1}{\partial z}\right)\underline{i} + \left(A_1\frac{\partial B_2}{\partial x} + A_2\frac{\partial B_2}{\partial y} + A_3\frac{\partial B_2}{\partial z}\right)\underline{j}$$

$$+ \left(A_1\frac{\partial B_3}{\partial x} + A_2\frac{\partial B_3}{\partial y} + A_3\frac{\partial B_3}{\partial z}\right)\underline{k}$$

FRAME A7 continued

The first of these gives the same result as $\underline{A}.(\nabla\phi)$, so the position of the brackets is not important. However, the position of the brackets is important in $(\underline{A}.\nabla)\underline{B}$, as $\underline{A}.(\nabla\underline{B})$ does not give the same result. You would, in fact, be unable to evaluate $\underline{A}.(\nabla\underline{B})$ as we have not defined $\nabla\underline{B}$.

Now find $(\underline{r}.\nabla)\underline{B}$ when $\underline{r} = x\underline{i} + y\underline{j} + z\underline{k}$ and $\underline{B} = x^2\underline{i} + y^2\underline{j} + z^2\underline{k}$

****

A7A

$2x^2\underline{i} + 2y^2\underline{j} + 2z^2\underline{k} = 2\underline{B}$

APPENDIX B

Div, grad and curl applied to sums and products

The following results may be proved

(i) $\text{grad}(\phi_1 + \phi_2) = \text{grad}\ \phi_1 + \text{grad}\ \phi_2$

(ii) $\text{grad}(\phi_1\phi_2) = \phi_1\ \text{grad}\ \phi_2 + \phi_2\ \text{grad}\ \phi_1$

(iii) $\text{grad}(\underline{A}.\underline{B}) = (\underline{B}.\nabla)\underline{A} + (\underline{A}.\nabla)\underline{B} + \underline{B} \wedge \text{curl}\ \underline{A} + \underline{A} \wedge \text{curl}\ \underline{B}$

(iv) $\text{div}(\phi\underline{A}) = \phi\ \text{div}\ \underline{A} + \underline{A}\ .\ \text{grad}\ \phi$

(v) $\text{div}(\underline{A}\wedge\underline{B}) = \underline{B}\ .\ \text{curl}\ \underline{A} - \underline{A}\ .\ \text{curl}\ \underline{B}$

(vi) $\text{curl}(\phi\underline{A}) = \phi\ \text{curl}\ \underline{A} + (\text{grad}\ \phi)\wedge\underline{A}$

(vii) $\text{curl}(\underline{A}\wedge\underline{B}) = (\underline{B}.\nabla)\underline{A} - (\underline{A}.\nabla)\underline{B} + \underline{A}\ \text{div}\ \underline{B} - \underline{B}\ \text{div}\ \underline{A}$

These results are useful in developing general theories where $\underline{A}$, $\underline{B}$ and ϕ are not known numerically.

As an example on the use of the above formulae we shall find $\text{div}(\underline{r}/r^3)$.

We think of this in the form $\text{div}\left(\frac{1}{r^3}\ \underline{r}\right)$ and use result (iv).

$$\begin{aligned}\text{Thus,}\quad \text{div}(\underline{r}/r^3) &= \frac{1}{r^3}\ \text{div}\ \underline{r} + \underline{r}.\text{grad}\ \frac{1}{r^3}\\ &= \frac{1}{r^3}(3) + \underline{r}.\left(\frac{-3}{r^4}\ \frac{\underline{r}}{r}\right)\\ &= \frac{3}{r^3} - \frac{3}{r^5}(\underline{r}.\underline{r})\\ &= \frac{3}{r^3} - \frac{3}{r^3}\\ &= 0\end{aligned}$$

Now try the following examples:

1. Verify result (iii) in the case where $\underline{A}$ is a constant vector $A_1\underline{i} + A_2\underline{j} + A_3\underline{k}$ and $\underline{r} = x\underline{i} + y\underline{j} + z\underline{k}$.

2. Evaluate $\text{div}(\underline{A}\wedge\underline{r})$ if $\text{curl}\ A = \underline{0}$.

3. Use one of the results listed above to show that

$$(\underline{q}.\nabla)\underline{q} = \text{grad}\ \frac{\underline{q}^2}{2} - \underline{q} \wedge \text{curl}\ \underline{q}$$

****

1. *L.H.S.* = *R.H.S.* = $\underline{A}$

2. *0*

3. *Use (iii) with* $\underline{A} = \underline{B} = \underline{q}$

VECTOR ANALYSIS II

FRAME 1

Vector Integration

(1) Ordinary Integrals

If a vector $\underline{A}$ is a function of a scalar variable u, then we can integrate $\underline{A}$ with respect to u (regarding this as the opposite process to differentiation). In practice, the integral may be found as follows:

$$\int \underline{A}(u)du = \int \{A_1(u)\underline{i} + A_2(u)\underline{j} + A_3(u)\underline{k}\}du$$

$$= \underline{i}\int A_1(u)du + \underline{j}\int A_2(u)du + \underline{k}\int A_3(u)du$$

The following example illustrates this:

The acceleration of a particle at any time t (> 0) is given by

$$\underline{a} = \frac{d\underline{v}}{dt} = 2\cos 2t\,\underline{i} - 8\sin 2t\,\underline{j} + 16t^2\underline{k}.$$

Given that when t = 0, $\underline{v} = 0$ and $\underline{r} = 0$, find $\underline{v}$ and $\underline{r}$ at any time t.

We are given $\frac{d\underline{v}}{dt} = 2\cos 2t\,\underline{i} - 8\sin 2t\,\underline{j} + 16t^2\underline{k}.$

Integrating both sides w.r.t. t, we get

$$\underline{v} = \underline{i}\int 2\cos 2t\,dt - \underline{j}\int 8\sin 2t\,dt + \underline{k}\int 16t^2dt$$

$$\therefore \quad \underline{v} = \underline{i}\sin 2t + \underline{j}\,4\cos 2t + \underline{k}16t^3/3 + \text{constant}$$

Now $\underline{v} = 0$ when t = 0

$$\therefore \quad 0 = 4\underline{j} + \text{constant}$$

yielding, constant = $-4\underline{j}$, which, you will notice, is a vector.

We thus have $\underline{v} = \underline{i}\sin 2t + \underline{j}\,4(\cos 2t - 1) + \underline{k}16t^3/3$

Now use the fact that $\frac{d\underline{r}}{dt} = \underline{v}$ to find $\underline{r}$.

****

1A

$$\underline{r} = \underline{i}\int \sin 2t\,dt + \underline{j}\int 4(\cos 2t - 1)dt + \underline{k}\int 16t^3/3dt$$

$$= -\underline{i}\,\frac{1}{2}\cos 2t + \underline{j}(2\sin 2t - 4t) + \underline{k}\,\frac{4}{3}t^4 + \text{constant}$$

$\underline{r} = 0$ *when* $t = 0$ *gives* $0 = -\frac{1}{2}\underline{i} + \text{constant}$

Hence $\underline{r} = \frac{1}{2}(1 - \cos 2t)\underline{i} + 2(\sin 2t - 2t)\underline{j} + \frac{4}{3}t^4\underline{k}$

FRAME 2

Now try the following examples:

(i) If $\underline{R}(u) = (u^2 - u^3)\underline{i} + 2u^3\underline{j} - 6\underline{k}$, find $\int \underline{R}(u)du$ and $\int_0^2 \underline{R}(u)du$.

(ii) If $\underline{A}(t) = t^2\underline{i} - t^2\underline{j} + (t - 2)\underline{k}$ and $\underline{B}(t) = 2t\underline{i} + 6t^2\underline{k}$,

evaluate $\int_0^2 \underline{A}.\underline{B}\ dt$ and $\int_0^1 \underline{A}\wedge\underline{B}\ dt$.

****

2A

(i) $\underline{i}\left(\frac{u^3}{3} - \frac{u^4}{4}\right) + \underline{j}\,\frac{u^4}{2} - \underline{k}6u + \text{constant}$

$-\frac{4}{3}\underline{i} + 8\underline{j} - 12\underline{k}$

(ii) $0, \quad -\frac{6}{5}\underline{i} - \frac{38}{15}\underline{j} + \frac{1}{2}\underline{k}$

FRAME 3

(2) Line Integrals

Let $\underline{r} = x\underline{i} + y\underline{j} + z\underline{k}$ be the position vector $\overrightarrow{OP}$ and, as $\underline{r}$ varies, suppose P moves along the curve C.

With the usual notation $\overrightarrow{PP'} = \delta\underline{r}$

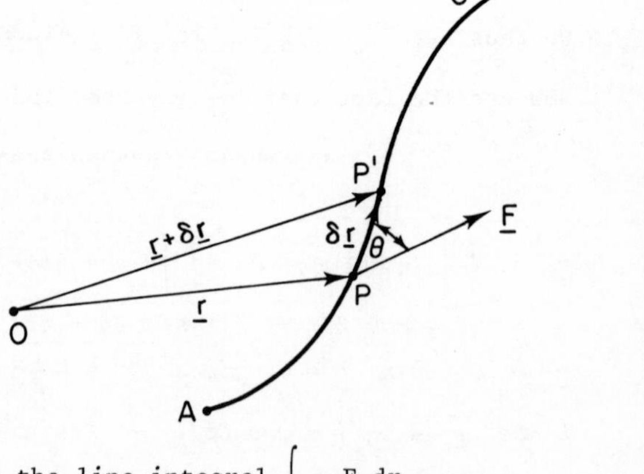

If we know the value of a vector $\underline{F}$ at every point of the region, we can find the scalar product $\underline{F}.\delta\underline{r}$. Adding such products along the curve between A and B and letting $\delta\underline{r} \to 0$, we obtain the line integral $\int_{AB} \underline{F}.d\underline{r}$.

FRAME 3 continued

(To give a physical meaning to the integral, think of $\underline{F}$ as representing the electric field intensity. Then $\underline{F}.\delta\underline{r} = F \cos\theta \, \delta r$ = work done by field in moving a unit charge along the curve over the distance δr.

Hence $\int_{AB} \underline{F}.d\underline{r}$ = total work done between A and B.)

As an example we shall consider $\int_{AB} \underline{F}.d\underline{r}$ where $\underline{F} = \underline{i} + \frac{1}{x}\underline{j} + \underline{k}$ and the path of integration is the semicircle in the x-y plane illustrated in the diagram.

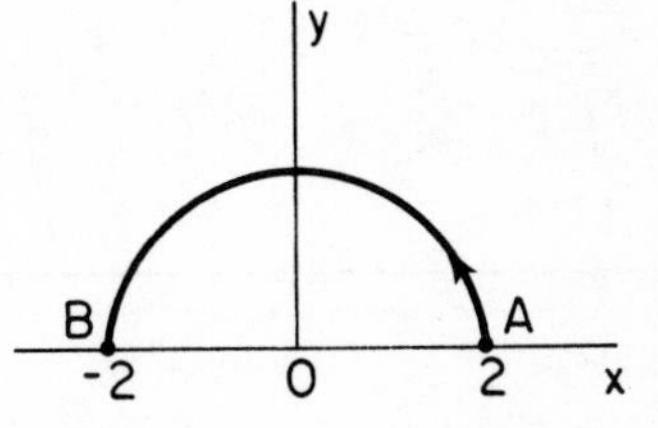

$$\int_{AB} \underline{F}.d\underline{r} = \int_{AB} (\underline{i} + \frac{1}{x}\underline{j} + \underline{k}).(dx\underline{i} + dy\underline{j} + dz\underline{k})$$

$$= \int_{AB} dx + \frac{1}{x}dy + dz$$

Now evaluate this integral along the path specified.

****

3A

The equation of the path is $x^2 + y^2 = 4, \quad z = 0.$

The easiest method uses the parametric equations $x = 2\cos\theta$, $y = 2\sin\theta$. *At A and B,* $\theta = 0$ *and* π *respectively. Also* $dx = -2\sin\theta\, d\theta$, $dy = 2\cos\theta\, d\theta$, $dz = 0$.

$$\therefore \int_{AB} \underline{F}.d\underline{r} = \int_0^{\pi} (-2\sin\theta + 1)d\theta$$

$$= \pi - 4$$

If, however, you have used Cartesian coordinates, then one possible solution is

$$2xdx + 2ydy = 0, \qquad dz = 0$$

and
$$\int_{AB} \underline{F}.d\underline{r} = \int_2^{-2} \left(1 - \frac{1}{\sqrt{4 - x^2}}\right)dx$$

$$= \pi - 4$$

FRAME 4

Now try the following example:

Find $\int_{AB} \underline{F}.d\underline{r}$, where $\underline{F} = (xy + 2y^2)\underline{i} + (3x^2 + y)\underline{j}$,

A is (0,0), B is (1,1) and the path of integration is

(i) $y = x^2$, (ii) $y = x$.

****

4A

(i) $\frac{53}{20}$ *(ii)* $\frac{5}{2}$

FRAME 5

You will remember, from your study of line integrals (in Unit 10 of this Volume) that, in general, if we evaluate a line integral along two different paths, we obtain two different answers even though the two end points are the same. In other words, the value of such an integral depends upon the path of integration. The example in the preceding frame should have reminded you of this.

However, you will also recall that, in certain circumstances, a line integral will give the same answer, whatever the path, and will only depend on the end points. As a reminder, $\int_{AB} Pdx + Qdy + Rdz$ is independent of the path taken if $Pdx + Qdy + Rdz$ is the exact differential of a single-valued function. If a field $\underline{F}$ is such that $\int_{AB} \underline{F}.d\underline{r}$ is independent of the path of integration, then $\underline{F}$ is said to be a CONSERVATIVE FIELD.

If $\underline{F} = \text{grad}\ \phi$, show that $\underline{F}$ is conservative.

****

5A

$$\underline{F} = \underline{i}\frac{\partial\phi}{\partial x} + \underline{j}\frac{\partial\phi}{\partial y} + \underline{k}\frac{\partial\phi}{\partial z}$$

$$\int \underline{F}.d\underline{r} = \int\left(\underline{i}\frac{\partial\phi}{\partial x} + \underline{j}\frac{\partial\phi}{\partial y} + \underline{k}\frac{\partial\phi}{\partial z}\right).(dx\underline{i} + dy\underline{j} + dz\underline{k})$$

$$= \int \frac{\partial\phi}{\partial x}dx + \frac{\partial\phi}{\partial y}dy + \frac{\partial\phi}{\partial z}dz$$

This is the integral of the exact differential $d\phi$, and so $\underline{F}$ is conservative (provided, of course, that ϕ is single-valued).

FRAME 6

Now try this example:

If $\underline{F} = (x + y)\underline{i} + (x - y)\underline{j}$, show, by finding ϕ, that $\underline{F}$ can be expressed in the form grad ϕ.

Hence evaluate $\displaystyle\int_{(0,0,0)}^{(1,1,0)} \underline{F}.d\underline{r}$.

****

6A

ϕ would have to satisfy:

$$\frac{\partial\phi}{\partial x} = x + y \qquad \frac{\partial\phi}{\partial y} = x - y \qquad \frac{\partial\phi}{\partial z} = 0$$

Integrating these gives

$$\phi = \frac{x^2}{2} + xy + f(y,z) \qquad \phi = xy - \frac{y^2}{2} + g(x,z) \qquad \phi = h(x,y)$$

$\phi = \dfrac{x^2}{2} + xy - \dfrac{y^2}{2} + c$ *meets these requirements*

$$\int_{(0,0,0)}^{(1,1,0)} \underline{F}.d\underline{r} = \int_{(0,0,0)}^{(1,1,0)} d\phi = \Big[\phi\Big]_{(0,0,0)}^{(1,1,0)} = 1$$

FRAME 7

We have considered ordinary and line integrals. Perhaps the natural development would be to consider next double integrals, followed by triple integrals. However, since vector double integrals present many new ideas and require a good deal of practice in their evaluation, we shall postpone consideration of these and now discuss triple integrals.

(3) Triple or Volume Integrals

Vector triple integrals are of the form $\iiint\limits_V \underline{A}\, dV$ where the integration is performed throughout a given volume V. This is only a slight extension of the non-vector triple integral with which you are familiar and therefore we shall illustrate their evaluation by considering the following example:

Evaluate $\iiint\limits_V \text{curl } \underline{F}\, dV$ where V is the closed region bounded by the planes $x = 0$, $y = 0$, $z = 0$ and $2x + 2y + z = 4$, and $\underline{F} = 3xy\underline{i} + x^2\underline{j} + zy\underline{k}$.

The region of integration is as shown

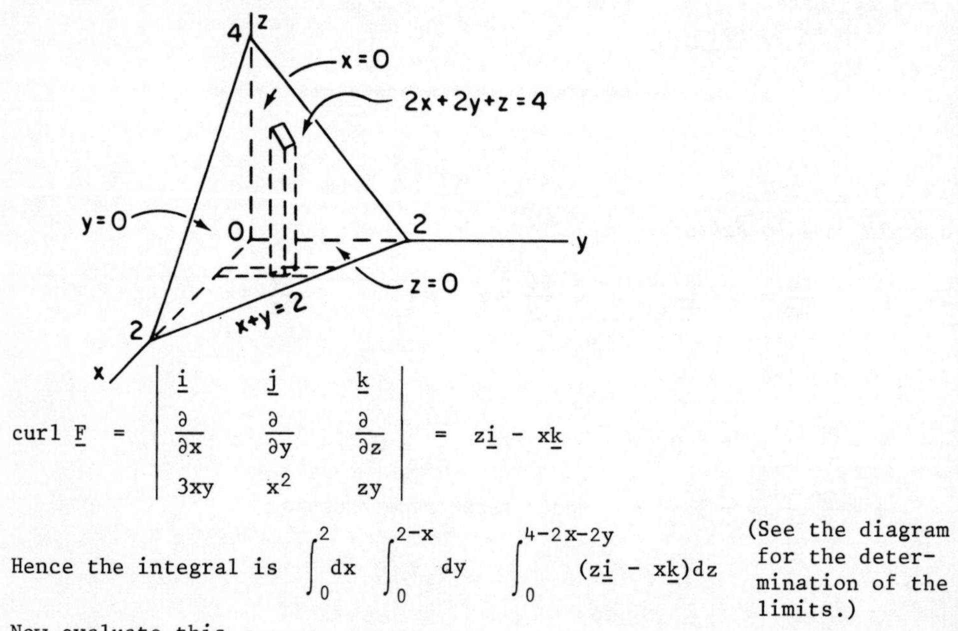

$$\text{curl } \underline{F} = \begin{vmatrix} \underline{i} & \underline{j} & \underline{k} \\ \dfrac{\partial}{\partial x} & \dfrac{\partial}{\partial y} & \dfrac{\partial}{\partial z} \\ 3xy & x^2 & zy \end{vmatrix} = z\underline{i} - x\underline{k}$$

Hence the integral is $\displaystyle\int_0^2 dx \int_0^{2-x} dy \int_0^{4-2x-2y} (z\underline{i} - x\underline{k})dz$ (See the diagram for the determination of the limits.)

Now evaluate this.

****

7A

$$\textit{Integral} = \int_0^2 dx \int_0^{2-x} \left[\frac{z^2}{2}\underline{i} - xz\underline{k}\right]_0^{4-2x-2y} dy$$

$$= \int_0^2 dx \int_0^{2-x} \{2(2 - x - y)^2\underline{i} - 2x(2 - x - y)\underline{k}\}dy$$

$$= \int_0^2 \{\tfrac{2}{3}(2 - x)^3\underline{i} - x(2 - x)^2\underline{k}\}dx$$

$$= \frac{8}{9}\underline{i} - \frac{4}{3}\underline{k}$$

FRAME 8

Now try the following example:

Find $\iiint_V \underline{F}\, dV$ where $\underline{F} = x\underline{i} + y\underline{k}$ and V is the volume bounded by $x = 0$, $y = 0$, $z = 0$, $x = 2$, $y = 2$, $z = 4$.

****

8A

$16(\underline{i} + \underline{k})$

FRAME 9

We shall now give an example of the same type but where the region of integration is more complex.

Find $\int_V \underline{A}\, dV$ where $\underline{A}$ is the vector $x\underline{i} + y\underline{j} + z\underline{k}$ and V is the region bounded by $x = 0$, $y = 0$, $y = 6$, $z = 4$, $z = x^2$.

FRAME 9 continued

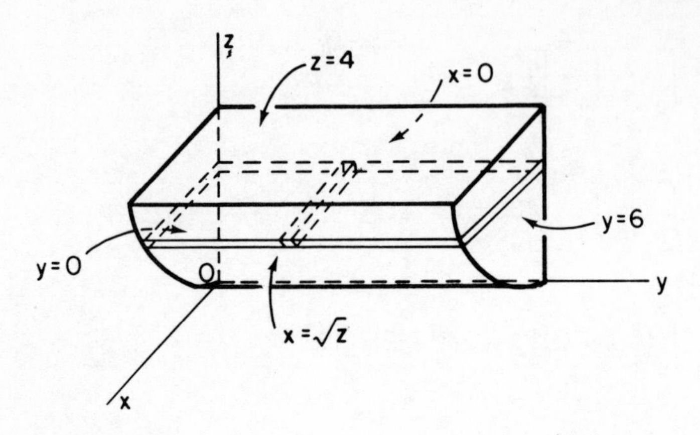

The integration limits are as shown, i.e. $x = 0$ to $x = \sqrt{z}$, followed by $y = 0$ to $y = 6$ and finally $z = 0$ to $z = 4$.

Thus the integral is

$$\int_0^4 dz \int_0^6 dy \int_0^{\sqrt{z}} (x\underline{i} + y\underline{j} + z\underline{k})dx$$

$$= \int_0^4 dz \int_0^6 \left[\frac{x^2}{2}\,\underline{i} + xy\underline{j} + xz\underline{k}\right]_0^{\sqrt{z}} dy$$

$$= \int_0^4 dz \int_0^6 \{\frac{z}{2}\,\underline{i} + \sqrt{z}\; y\underline{j} + z^{3/2}\underline{k}\}dy$$

$$= \int_0^4 \left[\frac{zy}{2}\,\underline{i} + \sqrt{z}\,\frac{y^2}{2}\underline{j} + yz^{3/2}\underline{k}\right]_0^6 dz$$

$$= \int_0^4 (3z\underline{i} + 18\sqrt{z}\;\underline{j} + 6z^{3/2}\underline{k})dz$$

$$= 24\underline{i} + 96\underline{j} + \frac{384}{5}\,\underline{k}$$

FRAME 10

Now try the following examples:

(i) Find $\int_V \underline{A}\,dV$, where $\underline{A} = \underline{i} - y\underline{j} + 2\underline{k}$ and V is the region bounded by the cylinder $x^2 + y^2 = 4$ and the planes $z = 0$ and $z = 3$.
(HINT: It is better to use cylindrical polar co-ordinates here, that is, put $x = \rho\cos\phi$, $y = \rho\sin\phi$.)

(ii) Find $\int_V \underline{A}\,dV$ where $\underline{A} = x\underline{i} - y\underline{j}$ and V is the region in the first octant bounded by the intersecting cylinders $x^2 + y^2 = 1$ and $x^2 + z^2 = 1$ as shown in the diagram.

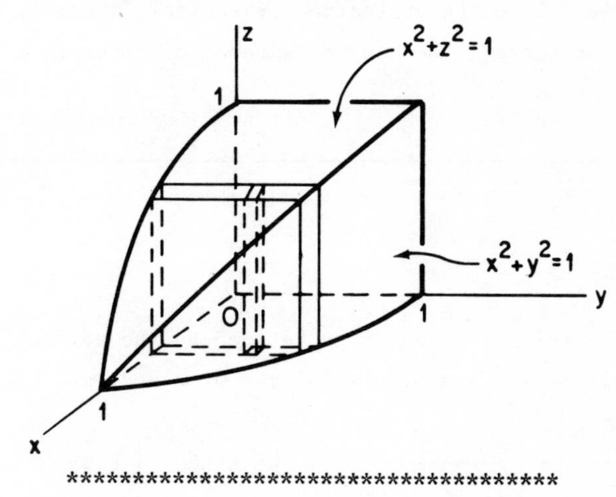

****

10A

(i) $12\pi(\underline{i} + 2\underline{k})$

(ii) $\frac{1}{4}\underline{i} - \frac{3}{32}\pi\underline{j}$

FRAME 11

(4) Surface Integrals

In vector work we often meet surface integrals of the form $\iint_S \underline{F}.d\underline{S}$, where S is the surface over which the integral is taken, and $\underline{F}$ defines a vector

FRAME 11 continued

field. $d\underline{S}$ is a vector whose magnitude is the element of area dS and whose direction is normal to dS and outwards from it. If $\underline{\hat{n}}$ is the unit normal in this direction then $d\underline{S} = dS\underline{\hat{n}}$, and so $\iint_S \underline{F}.d\underline{S} = \iint_S \underline{F}.\underline{\hat{n}}\, dS$.

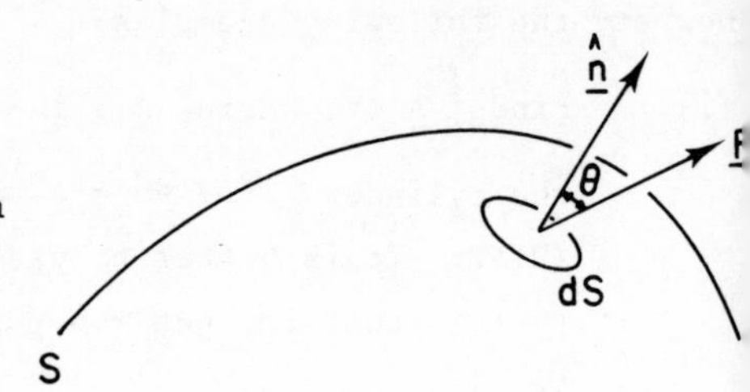

If the angle between $\underline{F}$ and $\underline{\hat{n}}$ is θ, then $\underline{F}.d\underline{S} = F \cos\theta\, dS$.

Hence the integral can also be written as $\iint_S F \cos\theta\, dS$.

In Unit 10 of this volume the evaluation of surface integrals over simple plane areas was considered. We can now extend this process to more complicated surfaces.

FRAME 12

Example 1

Suppose $\underline{F} = x^2\underline{i} + xy\underline{j} + z^2\underline{k}$ and S is defined as the surface of the cube bounded by x = 0, x = 2; y = 0, y = 2; z = 0, z = 2.

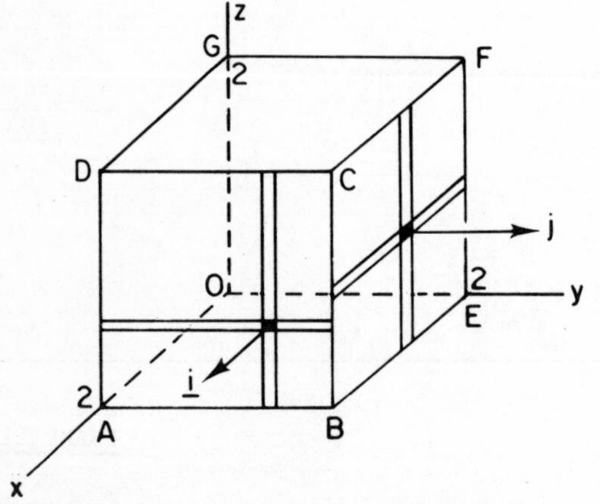

We find $\iint_S \underline{F}.d\underline{S}$ by splitting the integral into integrations over each face of the cube.

Thus the integral can be expressed as the sum

$$\iint_{ABCD} + \iint_{BEFC} + \;.....$$

For the face ABCD, dS = dydz, $\underline{\hat{n}} = \underline{i}$ and x = 2.

$\therefore$ over this face $\iint \underline{F}.d\underline{S} = \iint \underline{F}.(\underline{i}\; dydz)$

FRAME 12 continued

$$= \int_0^2 dz \int_0^2 x^2 dy = \int_0^2 dz \int_0^2 4dy$$

$$= 4 \times \text{area of face} = 16.$$

Similarly for face BEFC, $dS = dxdz$, $\underline{\hat{n}} = \underline{j}$ and $y = 2$.

Hence, the contribution from this face $= \int_0^2 dz \int_0^2 xy\,dx$

$$= \int_0^2 dz \int_0^2 2xdx = \int_0^2 4dz = 8.$$

The contributions from the other faces are determined in a similar way, remembering that $\underline{\hat{n}}$ is always the outward normal. Now find these contributions and the total integral.

****

12A

$$\iint_{CFGD} = 16 \qquad \iint_{OEFG} = 0 \qquad \iint_{OADG} = 0 \qquad \iint_{OABE} = 0 \qquad \iint_S \underline{F}.d\underline{S} = 40$$

FRAME 13

In the example just considered, S was a closed surface and all parts of it were in, or parallel to, the co-ordinate planes. In the next example the surface is not closed and is inclined to these planes.

Example 2

Evaluate $\iint \underline{F}.d\underline{S}$ for $\underline{F} = (x^2 + y^2)\underline{i} - 2\underline{j} + 2yz\underline{k}$ the surface S being that part of the plane $2x + y + 2z = 6$ which lies in the first octant.

Now if $\phi = 2x + y + 2z$, then grad ϕ will be in a direction perpendicular to the surface ϕ = constant, and thus $\frac{\text{grad } \phi}{|\text{grad } \phi|}$ will be a unit normal to the given plane.

FRAME 13 continued

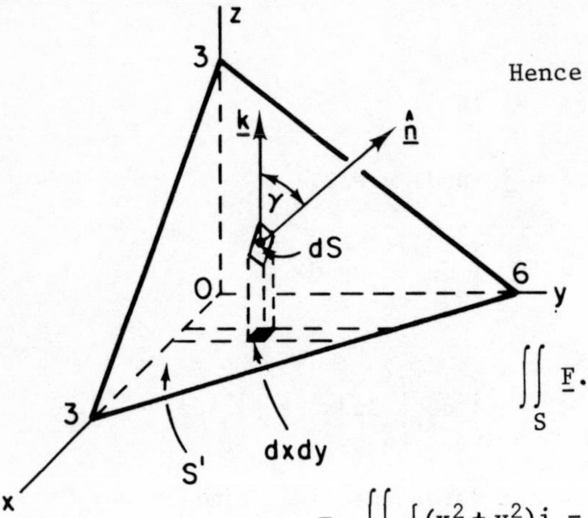

Hence $\hat{\underline{n}} = \dfrac{\text{grad } \phi}{|\text{grad } \phi|}$

$$= \frac{2\underline{i} + \underline{j} + 2\underline{k}}{\sqrt{2^2 + 1^2 + 2^2}}$$

$$= \frac{2}{3}\underline{i} + \frac{1}{3}\underline{j} + \frac{2}{3}\underline{k}$$

$$\iint_S \underline{F}.d\underline{S} = \iint_S \underline{F}.\hat{\underline{n}}dS$$

$$= \iint_S \{(x^2 + y^2)\underline{i} - 2\underline{j} + 2yz\underline{k}\} \cdot \left(\frac{2}{3}\underline{i} + \frac{1}{3}\underline{j} + \frac{2}{3}\underline{k}\right)dS$$

$$= \iint_S \left\{\frac{2}{3}(x^2 + y^2) - \frac{2}{3} + \frac{4}{3}yz\right\}dS$$

It is easier now to project dS onto one of the co-ordinate planes, such as $z = 0$, and integrate over this plane rather than over the slanting plane. The projection of the area dS onto the xy plane is dS cos γ (where γ is the angle between the planes, this being the same as the angle between the normals).

Now $\cos\gamma = \hat{\underline{n}}.\underline{k} = \dfrac{2}{3}$ and $dxdy = dS\cos\gamma$

$\therefore \quad dxdy = dS\left(\dfrac{2}{3}\right)$

or $\quad dS = \dfrac{3}{2}dxdy$

As dS traverses the whole of the given surface, dxdy will cover the whole of the triangular area S' in the xy plane and therefore

$$\iint_S \underline{F}.d\underline{S} = \iint_{S'} \frac{2}{3}\left(x^2 + y^2 - 1 + 2yz\right)\left(\frac{3}{2}dxdy\right)$$

$$= \iint_{S'} (x^2 + y^2 - 1 + 2yz)dxdy$$

FRAME 13 continued

$$= \iint_{S'} \{x^2 + y^2 - 1 + y(6 - y - 2x)\}dxdy,$$ substituting for z from the equation of plane S,

$$= \iint_{S'} (x^2 + 6y - 2xy - 1)dxdy$$

$$= \int_0^3 dx \int_0^{6-2x} (x^2 + 6y - 2xy - 1)dy = \int_0^3 \Big[x^2y + 3y^2 - xy^2 - y\Big]_0^{6-2x} dx$$

$$= \int_0^3 (-6x^3 + 42x^2 - 106x + 102)dx$$

$$= \Big[-\frac{3}{2}x^4 + 14x^3 - 53x^2 + 102x\Big]_0^3$$

$$= 85\tfrac{1}{2}$$

Now try this example:

Example 3

Evaluate $\iint_S \underline{F}.\hat{\underline{n}}\, dS$ where $\underline{F} = -18\underline{i} + 6y\underline{j} + 6z\underline{k}$ and S i[illegible] hat part of the plane $2x + 3y + 6z = 12$ which is located in the first octant.

****

13A

Unit normal $\hat{\underline{n}} = (2\underline{i} + 3\underline{j} + 6\underline{k})/7$

$$\iint_S \underline{F}.\hat{\underline{n}}\, dS = \iint_S \frac{18}{7}(2z - 2 + y)dS$$

$$= \iint_{S'} 3(2z - 2 + y)dxdy$$ *if S is projected onto the plane Oxy*

$$= \int_0^6 dx \int_0^{2(6-x)/3} 2(3 - x)dy$$

$$= 24$$

FRAME 14

In the next example the surface of integration is curved.

Example 4

Find $\iint_S \underline{F}.d\underline{S}$ when

$\underline{F} = (8 + z)\underline{j} + z^2\underline{k}$ and

S is the surface $y^2 = 8x$ in the first octant, bounded by the planes $y = 4$ and $z = 6$.

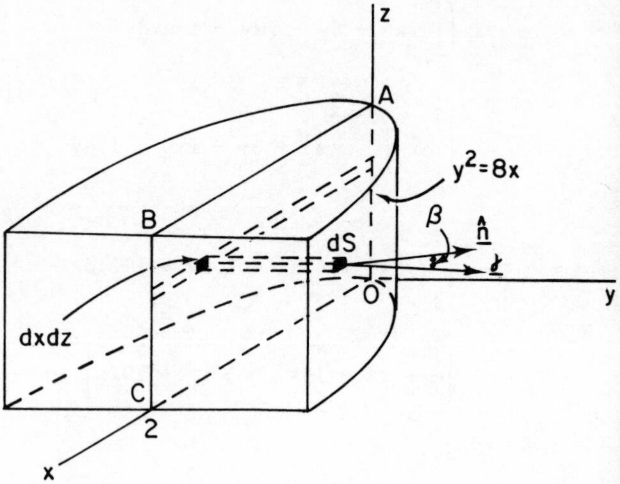

We let $\phi = y^2 - 8x$

so that

$$\hat{\underline{n}} = \frac{-8\underline{i} + 2y\underline{j}}{2\sqrt{16 + y^2}}$$

$$= \frac{-4\underline{i} + y\underline{j}}{\sqrt{16 + y^2}}$$

and $\underline{F}.d\underline{S} = \underline{F}.\hat{\underline{n}}\, dS$

$$= \left[(8 + z)\underline{j} + z^2\underline{k}\right] . \frac{-4\underline{i} + y\underline{j}}{\sqrt{16 + y^2}}\, dS$$

$$= \frac{8y + yz}{\sqrt{16 + y^2}}\, dS$$

Previously the surface has been projected on to the x-y plane, but in the present case this projection is zero as the surface is perpendicular to Oxy. Instead S can be projected on to either the x-z or y-z plane. The x-z plane has been chosen here.

Projected area $dxdz = dS \cos\beta$ where β is the angle between dS and Oxz

$$= dS(\hat{\underline{n}}.\underline{j})$$

$$= dS\, \frac{y}{\sqrt{16 + y^2}}$$

i.e. $dS = \dfrac{\sqrt{16 + y^2}}{y}\, dxdz$

FRAME 14 continued

$$\iint_S \underline{F}.d\underline{S} = \iint_{S'} \frac{(8y + yz)}{\sqrt{16 + y^2}} \frac{\sqrt{16 + y^2}}{y} dxdz,$$ where S' is the projected area in the x-z plane, i.e. OABC

$$= \iint_{S'} (8 + z)dxdz$$

$$= \int_0^6 dz \int_0^2 (8 + z)dx$$

$$= \int_0^6 2(8 + z)dz$$

$$= 2\left[8z + \frac{z^2}{2}\right]_0^6$$

$$= 132$$

FRAME 15

Example 5

Evaluate $\iint_S \underline{F}.\hat{\underline{n}}\, dS$ over the entire surface S of the region in the first octant bounded by the cylinder $x^2 + z^2 = 4$ and the planes $x = 0$, $y = 0$, $z = 0$ and $y = 4$, where $\underline{F} = z\underline{i} + z\underline{j} - y\underline{k}$.

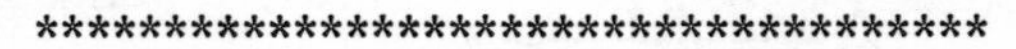

15A

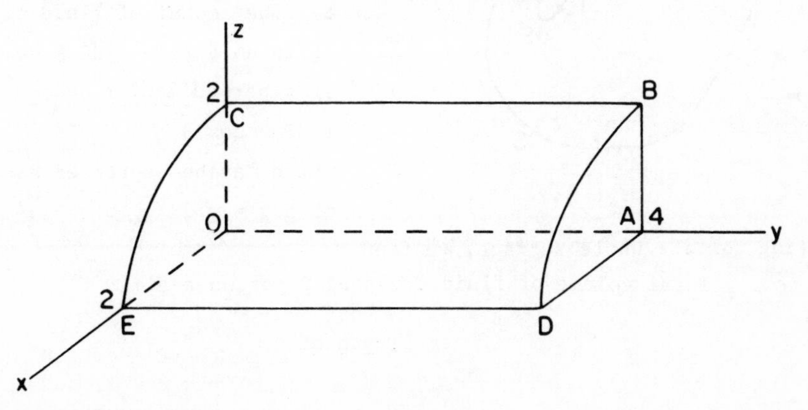

15A continued

$$\iint_{OABC} = -8 \qquad \textit{Note that the outward normal } \underline{\hat{n}} \textit{ is } -\underline{i} \textit{ here.}$$

$$\iint_{OADE} = 16 \qquad \iint_{OEC} = -\frac{8}{3} \qquad \iint_{ABD} = \frac{8}{3} \qquad \iint_{BCED} = -8$$

$$\iint_S \underline{F}.\underline{\hat{n}}\, dS = 0$$

FRAME 16

Flux of a Vector

Vector surface integrals arise from a consideration of the transport, or flux, of a physical quantity across a surface.

To fix ideas, let us consider the quantity of fluid per unit time crossing a surface S in some hydrodynamical flow. This is the generalisation of the simple case which was considered in FRAMES 35-36, pages 11:26-11:29. If you refer back to these frames, you should be able to see that what was done there is a particular case of what follows here. Let $\underline{v}(x,y,z)$ be the velocity of the fluid at some point P of the surface. If dS is a small element of area surrounding P, then $d\underline{S} = dS\ \underline{\hat{n}}$ is the elemental vector area, $\underline{\hat{n}}$ being the outward normal to S. The amount of fluid crossing $d\underline{S}$ in unit time = dS × component of $\underline{v}$ perpendicular to $d\underline{S}$

$= dS\ v \cos\theta$

which may be expressed as

$dS\ \underline{\hat{n}}.\underline{v} = \underline{v}.d\underline{S}$

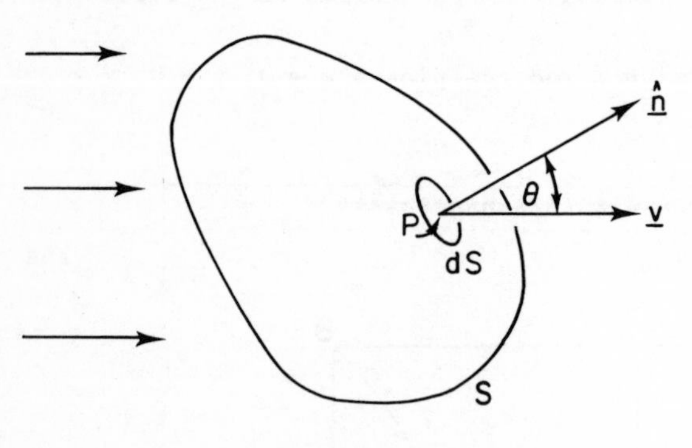

Integrating for the whole surface, we find

Total volume of fluid crossing S per unit time

$$= \int_S \underline{v}.\underline{\hat{n}}\ dS$$

FRAME 16 continued

$$= \int_S \underline{v}.d\underline{S}$$

We define this as the FLUX of the vector $\underline{v}$ through the surface S. Although we have approached the idea of flux from the flow of a fluid across a surface we can have, in a similar way, the flux of an electric field $\underline{E}$ across a surface

$$\text{i.e.} \quad \int_S \underline{E}.d\underline{S}$$

FRAME 17

Divergence of a Vector Field

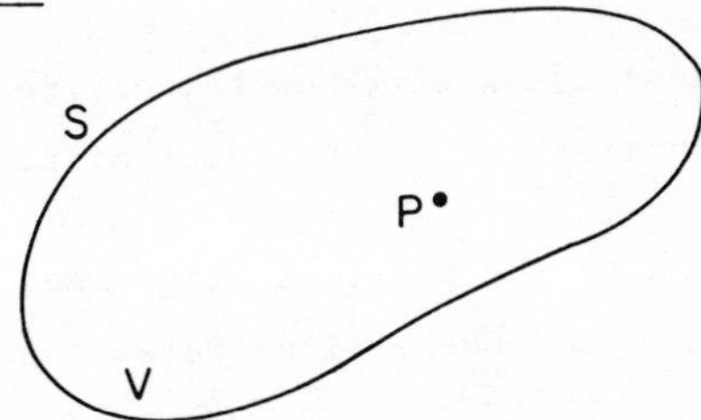

Now let P be any internal point of a volume V bounded by a closed surface S.

The total outward flux of $\underline{v}$ across S $= \int_S \underline{v}.d\underline{S}$

$\therefore$ the average flux per unit volume $= \dfrac{1}{V}\int_S \underline{v}.d\underline{S}$

Now let $V \to 0$, so that $S \to 0$ and the volume shrinks to the point P. The limiting value of this average outward flux per unit volume is then the DIVERGENCE of $\underline{v}$ at P,

$$\text{i.e.} \quad \text{div}\,\underline{v} = \lim_{V\to 0}\left(\frac{1}{V}\int_S \underline{v}.d\underline{S}\right) \tag{17.1}$$

Any volume can be used for V and in FRAME 35, page 11:26, the particular volume chosen was a rectangular box. When you require an expression for the divergence of any vector $\underline{F}$ you can use either

FRAME 17 continued

$$\text{div } \underline{F} = \frac{\partial F_1}{\partial x} + \frac{\partial F_2}{\partial y} + \frac{\partial F_3}{\partial z}$$

or $$\text{div } \underline{F} = \lim_{V \to 0} \left(\frac{1}{V} \int_S \underline{F}.d\underline{S} \right)$$

whichever is the more convenient.

FRAME 18

Flow and Circulation

In FRAMES 39-47, pages 11:31-11:36, we took a simple example illustrating flow and circulation which led on to a definition of curl.

A generalisation of these ideas will now be considered, just as a moment ago we considered the generalisation of the ideas of flux and divergence.

Again, by referring back to the previous programme, you may find it helpful to see how the treatment adopted there illustrates a particular case of what follows here.

You will remember that if $\underline{A}$ represents a vector field of force, then

$\int_{LMP} \underline{A}.d\underline{s}$ represents the work done in moving a unit charge, a unit mass, or unit magnetic pole, along the path LMP.

Generally $\int_{LMP} \underline{A}.d\underline{s}$ is called the FLOW in the path.

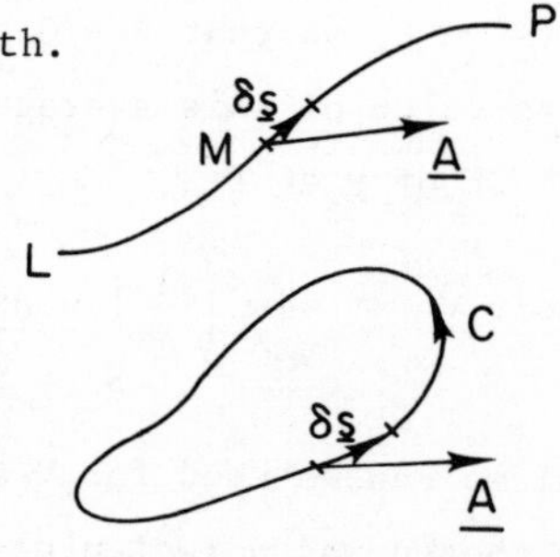

If the line integral is taken round a complete circuit, we then have

$\oint_C \underline{A}.d\underline{s}$ and this is called

the CIRCULATION in the circuit C.

FRAME 18 continued

Consider this example:

Given that $\underline{F} = y^2\underline{i} + x^3\underline{j}$, find the circulation about the circle, radius r, drawn in the x-y plane with centre (1,0,0).

$$\text{Circulation in C} = \oint_C \underline{F}.d\underline{s} = \oint_C (y^2\underline{i} + x^3\underline{j}).(dx\underline{i} + dy\underline{j} + dz\underline{k})$$

$$= \oint_C y^2dx + x^3dy$$

By introducing a parameter, t, we see that the coordinates of any point on the circle may be written

$$x = 1 + r\cos t$$
$$y = r\sin t$$
$$z = 0$$

$$\therefore \text{ Circulation} = \int_0^{2\pi} (r^2\sin^2 t)(-r\sin t)dt + (1 + r\cos t)^3(r\cos t)dt$$

Remembering that $\int_0^{\pi/2} \sin^n x\,dx = \int_0^{\pi/2} \cos^n x\,dx$

$$= \begin{cases} \dfrac{(n-1)\ \ldots\ 3.1}{n\ \ldots\ldots\ 4.2}\dfrac{\pi}{2} & \text{for n even,} \\ \\ \dfrac{(n-1)\ \ldots\ 4.2}{n\ \ldots\ldots\ 5.3}.1 & \text{for n odd,} \end{cases}$$

that $\int_0^{2\pi} \sin^n x\,dx = \begin{cases} 0 & \text{for n odd} \\ 4\int_0^{\pi/2} \sin^n x\,dx & \text{for n even,} \end{cases}$

FRAME 18 continued

$$\text{and} \int_0^{2\pi} \cos^n x \, dx = \begin{cases} 0 \text{ for n odd} \\ 4\displaystyle\int_0^{\pi/2} \cos^n x \, dx \quad \text{for n even,} \end{cases}$$

find the circulation.

****

18A

$$\oint_C \underline{F}.d\underline{s} = \int_0^{2\pi} (-r^3\sin^3 t)dt + (r\cos t + 3r^2\cos^2 t + 3r^3\cos^3 t + r^4\cos^4 t)dt$$

$$= 4\int_0^{\pi/2} (3r^2\cos^2 t + r^4\cos^4 t)dt$$

$$= 3\pi r^2\left(1 + \frac{1}{4}r^2\right)$$

FRAME 19

Curl of a Vector

Now consider a point P in space, and a given direction through P specified by unit vector $\hat{\underline{n}}$.

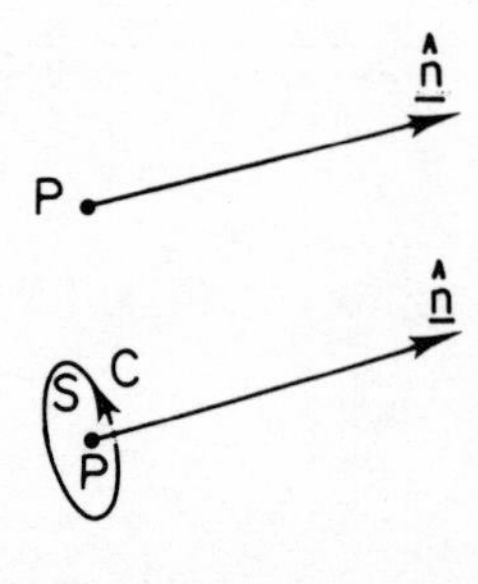

We surround P by a small plane circuit C perpendicular to $\hat{\underline{n}}$ and enclosing area S,

The next step is to calculate the circulation in the small circuit C, i.e. to find

FRAME 19 continued

$\oint_C \underline{A}.d\underline{s}$. By convention, this is determined by proceeding in a clockwise sense around the circuit when the vector $\underline{n}$ is pointing away from you (right-handed screw rule).

The average circulation per unit area is then $\frac{1}{S}\oint_C \underline{A}.d\underline{s}$. Now let the area S shrink to the point P so obtaining $\lim_{S\to 0}\left(\frac{1}{S}\oint_C \underline{A}.d\underline{s}\right)$. Then curl $\underline{A}$ is a vector which is so defined that its component in the direction of $\hat{\underline{n}}$ is this limit. As the component of curl $\underline{A}$ in the direction $\hat{\underline{n}}$ is curl $\underline{A}.\hat{\underline{n}}$, then this result can be written as

$$\text{curl } \underline{A}.\hat{\underline{n}} = \lim_{S\to 0}\left(\frac{1}{S}\oint_C \underline{A}.d\underline{s}\right) \qquad (19.1)$$

The vector curl $\underline{A}$ is specified completely if three such components in mutually perpendicular directions are known. You will recall that in the previous programme the components of curl $\underline{v}$ were found in the three directions Ox, Oy and Oz.

FRAME 20

Conservative Field

In FRAME 19 we defined a component of curl $\underline{F}$ in a particular direction as

$$\lim_{S\to 0}\left(\frac{1}{S}\oint_C \underline{F}.d\underline{s}\right) = \lim_{\text{Area}\to 0}\left(\frac{\text{Circulation}}{\text{Area}}\right)$$

where the circulation is around a suitable path. Now, you will recall that, in a conservative field $\underline{F}$, we know that $\int_{AB} \underline{F}.d\underline{s}$ is independent of the path, and only depends on the positions of the end points. Therefore, if the integration is performed around any closed curve,

$$\oint_C \underline{F}.d\underline{s} = \text{Circulation} = 0 \quad \text{for a conservative field.}$$

FRAME 20 continued

Thus, all the components of curl $\underline{F}$ will be zero. Hence we have the result that if a vector field is conservative, the circulation around any path is zero and curl $\underline{F} = 0$.

Furthermore, when curl $\underline{F} = 0$, we see from FRAME 47, page 11:36, that

$$\frac{\partial F_3}{\partial y} = \frac{\partial F_2}{\partial z}, \quad \frac{\partial F_1}{\partial z} = \frac{\partial F_3}{\partial x}, \quad \frac{\partial F_2}{\partial x} = \frac{\partial F_1}{\partial y}.$$

These are the conditions that $F_1dx + F_2dy + F_3dz$ is a total differential, $d\phi$, say.

Now $d\phi = \frac{\partial \phi}{\partial x}dx + \frac{\partial \phi}{\partial y}dy + \frac{\partial \phi}{\partial z}dz$

and comparing, we get $F_1 = \frac{\partial \phi}{\partial x}$, $F_2 = \frac{\partial \phi}{\partial y}$, $F_3 = \frac{\partial \phi}{\partial z}$

Hence, $\underline{F} = \underline{i}\frac{\partial \phi}{\partial x} + \underline{j}\frac{\partial \phi}{\partial y} + \underline{k}\frac{\partial \phi}{\partial z} = \text{grad } \phi$.

ϕ is called the scalar potential.

To summarise, for a conservative field,

(i) curl $\underline{F} = 0$

(ii) $\underline{F} = \text{grad } \phi$

(iii) circulation $\oint \underline{F}.d\underline{s} = 0$ for any closed path.

Conversely, if curl $\underline{F} = 0$, then $\underline{F} = \text{grad } \phi$ and, as you saw in 5A, the field $\underline{F}$ is then conservative.

Note that (i) and (ii) imply curl(grad ϕ) = 0. This was proved independently in FRAME A1 on page 11:47.

FRAME 21

Now try the following examples:

(i) Look back and note the answer we obtained in FRAME 18 for the circulation about the circle with centre (1,0,0). By using the formula

$$\text{curl } \underline{F} = \lim_{\text{Area}\to 0}\left(\frac{\text{Circulation}}{\text{Area}}\right)$$

show that the component of curl $\underline{F}$ in the positive $\underline{k}$ direction is 3.

FRAME 21 continued

(ii) Verify the above solution by finding the required component of curl $\underline{F}$ when $\underline{F} = y^2\underline{i} + x^3\underline{j}$.

(iii) Prove that curl $\underline{r} = 0$ if $\underline{r} = x\underline{i} + y\underline{j} + z\underline{k}$. Verify, by finding ϕ, that $\underline{r}$ may be expressed as grad ϕ.

(iv) (a) If $\underline{F} = (y^2\cos x + 1)\underline{i} + (2y \sin x - 4z)\underline{j} + (2z - 4y)\underline{k}$, find curl $\underline{F}$ and hence show that $\underline{F}$ is a conservative field of force.

(b) Find the scalar potential for $\underline{F}$.

****

21A

(iii) $\phi = \frac{x^2}{2} + \frac{y^2}{2} + \frac{z^2}{2} + C$

(iv) *(a)* 0

(b) $\phi = y^2\sin x + x - 4yz + z^2 + C$

FRAME 22

Integral Theorems

We now prove two important theorems. They are widely used in the development of the theory of such subjects as electromagnetism and hydrodynamics.

FRAME 23

Gauss' Divergence Theorem

This states that for a closed surface S, bounding a volume V, in a vector field $\underline{F}$,

$$\int_V \text{div } \underline{F}\, dV = \int_S \underline{F}.d\underline{S}$$

FRAME 23 continued

Note that the L.H.S. is a volume (triple) integral, while the R.H.S. is a surface (double) integral.

Proof

Divide the volume V into a system of elementary blocks δV, one line of such blocks being shown in diagram (i). From our previous definition of div $\underline{F}$, given in (17.1), we get approximately, for a small volume δV with closed surface S_i,

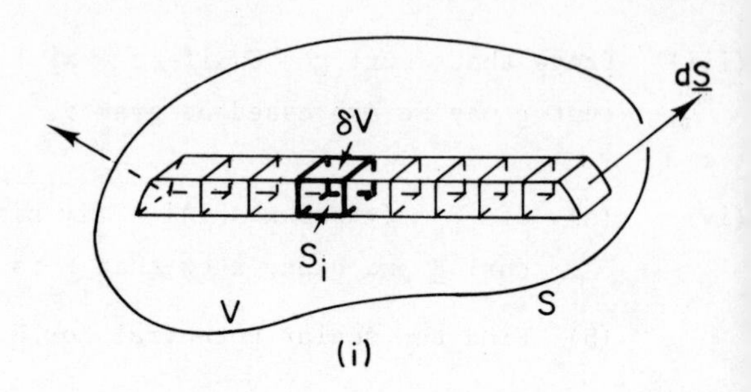

$$\text{div } \underline{F} = \frac{1}{\delta V}\int_{S_i} \underline{F}.d\underline{S} \quad \text{or} \quad \text{div } \underline{F}\ \delta V = \int_{S_i} \underline{F}.d\underline{S}$$

This says in words, div $\underline{F}$ δV = total outward flux of $\underline{F}$ across the surface of δV.

We add now for all the elements and get

$\int_V$ div $\underline{F}$ dV = Total outward flux of $\underline{F}$ across all the surfaces of all such elements δV.

To find an expression for the R.H.S. consider the flux out of any particular element. It is seen that the flux through any face will be equal and opposite to the flux through the face of the adjacent block, unless the face forms part of the surface S.

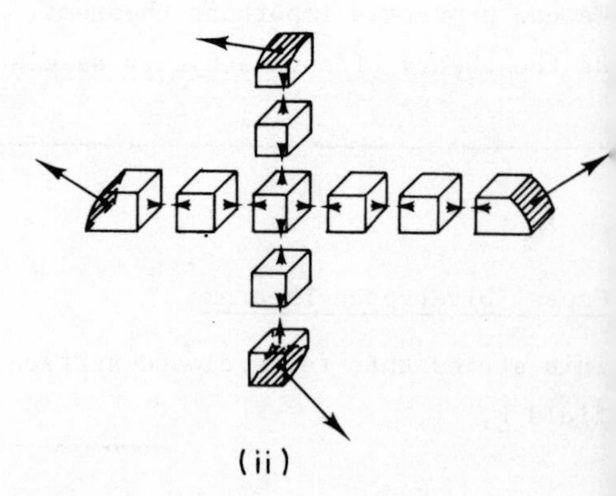

Thus, summing all the fluxes, we must get the total flux through the surface,

i.e. $\int_S \underline{F}.d\underline{S}$

FRAME 23 continued

We find therefore $\int_V \text{div}\, \underline{F}\, dV = \int_S \underline{F}.d\underline{S}$

This is the Divergence Theorem and may be stated in words as:

Total divergence of $\underline{F}$ from a volume V

= Total outward flux of $\underline{F}$ through the surface enclosing V.

FRAME 24

We shall now consider the following example:

Verify the divergence theorem for $\underline{A} = y\underline{i} + x\underline{j} + z^2\underline{k}$ taken over the region bounded by $x^2 + y^2 = 4,\ y \geq 0,\ z = 0,\ z = 3$

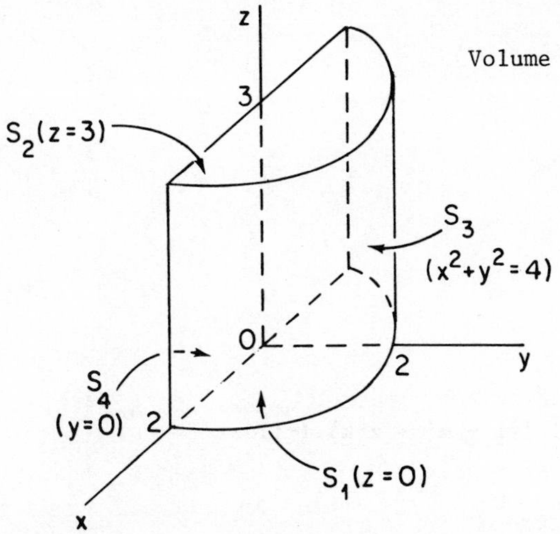

Volume integral $\int_V \nabla.\underline{A}\, dV$

$$= \int_V \left\{\frac{\partial}{\partial x}(y) + \frac{\partial}{\partial y}(x) + \frac{\partial}{\partial z}(z^2)\right\} dV$$

$$= \int_V 2z\, dV$$

$$= \int_{-2}^{2} dx \int_0^{\sqrt{4-x^2}} dy \int_0^3 2z\, dz$$

$$= 18\pi$$

The surface S of the cylinder consists of the base $S_1(z = 0)$, the top $S_2(z = 3)$, the convex portion $S_3(x^2 + y^2 = 4)$ and the face $S_4(y = 0)$.

Then surface integral $\int_S \underline{A}.\hat{\underline{n}}\, dS$

$$= \int_{S_1} \underline{A}.\hat{\underline{n}}\, dS + \int_{S_2} + \int_{S_3} + \int_{S_4}$$

FRAME 24 continued

On $S_1 (z = 0)$, $\underline{\hat{n}} = -\underline{k}$, $\underline{A} = y\underline{i} + x\underline{j}$ $\therefore$ $\underline{A}.\underline{\hat{n}} = 0$ $\therefore$ $\int_{S_1} \underline{A}.\underline{\hat{n}}\, dS = 0$

On $S_2 (z = 3)$, $\underline{\hat{n}} = \underline{k}$, $\underline{A} = y\underline{i} + x\underline{j} + 9\underline{k}$ $\therefore$ $\underline{A}.\underline{\hat{n}} = 9$

$$\therefore \int_{S_2} \underline{A}.\underline{\hat{n}}\, dS = 9 \int_{S_2} dS = 9S_2 = 18\pi$$

On $S_3 (x^2 + y^2 = 4)$ a normal has direction $\nabla(x^2 + y^2) = 2x\underline{i} + 2y\underline{j}$

$\therefore$ a unit normal is $\underline{\hat{n}} = \dfrac{2x\underline{i} + 2y\underline{j}}{\sqrt{4x^2 + 4y^2}} = \dfrac{x\underline{i} + y\underline{j}}{2}$

$$\underline{A}.\underline{\hat{n}} = (y\underline{i} + x\underline{j} + z^2\underline{k}).\left(\frac{x\underline{i} + y\underline{j}}{2}\right)$$

$$= xy$$

$\int_{S_3} xy dS$ can be evaluated by projecting on to the plane Oxz.

Then $(\underline{\hat{n}}.\underline{j})dS = dxdz$

$$\therefore dS = 2\,\frac{dxdz}{y}$$

$$\int_{S_3} \underline{A}.\underline{\hat{n}}\, dS = \int_0^3 dz \int_{-2}^{2} xy\,\frac{2}{y}\, dx$$

$$= 2\int_0^3 dz \int_{-2}^{2} x\, dx$$

$$= 0$$

On $S_4 (y = 0)$, $\displaystyle\int_{S_4} \underline{A}.\underline{\hat{n}}\, dS = \int_0^3 dz \int_{-2}^{2} (y\underline{i} + x\underline{j} + z^2\underline{k}).(-\underline{j})dx$

$$= \int_0^3 dz \int_{-2}^{2} -x\, dx$$

$$= 0$$

$\therefore$ Over the complete surface S, $\int_S \underline{A}.\underline{\hat{n}}\, dS = 18\pi$

$\therefore$ $\int_V \nabla.\underline{A}\, dV = \int_S \underline{A}.\underline{\hat{n}}\, dS$, so verifying the divergence theorem.

FRAME 25

Now try these examples:

(i) Verify the divergence theorem for $\underline{A} = 3xyz\underline{i} + zy^2\underline{j} - 5yz\underline{k}$ taken over the region in the first octant bounded by $x = 2$, $y^2 + z^2 = 9$.

(ii) If S is any surface enclosing a volume V and $\underline{A} = ax\underline{i} + by\underline{j} + cz\underline{k}$, prove $\int_S \underline{A}.\hat{\underline{n}}\, dS = (a + b + c)V$.

****

25A

(i)

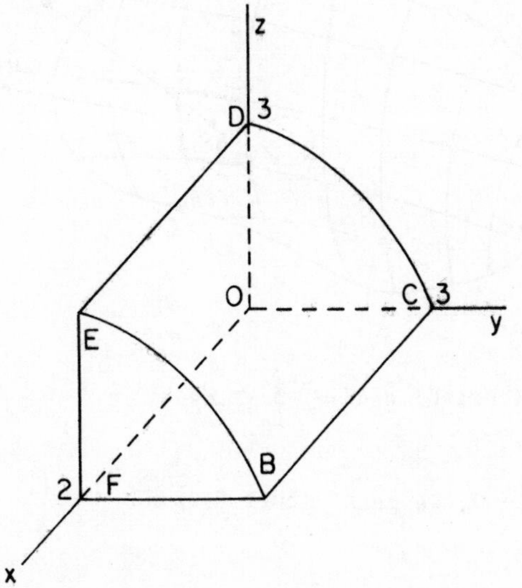

$$\int_V \nabla.\underline{A}\, dV = \frac{45}{4}$$

$$\int_{OFBC} \underline{A}.\hat{\underline{n}}\, dS = 0$$

$$\int_{OFED} \underline{A}.\hat{\underline{n}}\, dS = 0$$

$$\int_{OCD} \underline{A}.\hat{\underline{n}}\, dS = 0$$

$$\int_{FBE} \underline{A}.\hat{\underline{n}}\, dS = \frac{243}{4}$$

$$\int_{BCDE} \underline{A}.\hat{\underline{n}}\, dS = -\frac{99}{2}$$

(ii)

$$\int_S \underline{A}.\hat{\underline{n}}\, dS = \int_V \nabla.\underline{A}\, dV = \int_V (a + b + c)dV = (a + b + c)V$$

FRAME 26

Stokes' Theorem

This states that for any open surface S having for its bounding edge a given closed curve C

$$\int_S \text{curl } \underline{F}.d\underline{S} = \oint_C \underline{F}.d\underline{s}$$

Proof

This follows on similar lines to that of the Divergence Theorem. That is, divide S into small elementary surfaces, and consider one such element $\delta\underline{S} = \hat{\underline{n}}\ \delta S$ with boundary C_i.

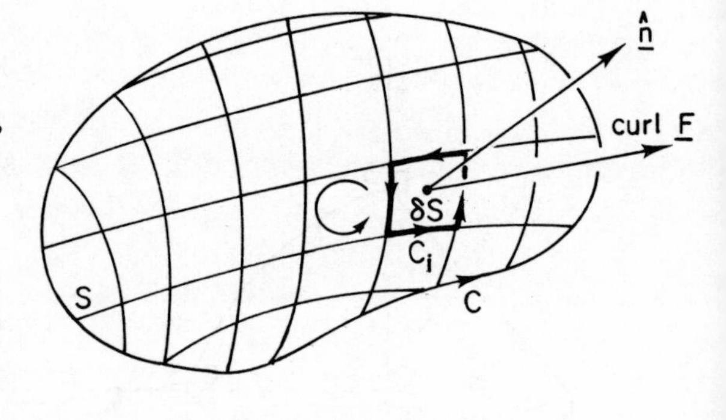

From (19.1) we get approximately, for this small area δS,

$$(\text{curl } \underline{F}).\hat{n} = \frac{1}{\delta S}\oint_{C_i} \underline{F}.d\underline{s}$$

$$\text{i.e.} (\text{curl } \underline{F}).\hat{\underline{n}}\ \delta S \doteq \oint_{C_i} \underline{F}.d\underline{s}$$

Summing over all such elements gives $\sum(\text{curl } \underline{F}).\hat{\underline{n}}\ \delta S \doteq \sum \oint_{C_i} \underline{F}.d\underline{s}$

For the L.H.S., taking the limit as $\delta S \to 0$, we get $\int_S (\text{curl } \underline{F}).\hat{\underline{n}}\ dS$

$$= \int_S \text{curl } \underline{F}.d\underline{S}$$

Now R.H.S. = Sum of the circulations round all the elements.

However, the flow along each edge of C_i will be equal and opposite to the flow along the corresponding edges of the adjacent elements, unless the edge is a part of the boundary C.

FRAME 26 continued

Thus the net flow will be only that obtained from the edges on the boundary, and will equal $\oint_C \underline{F}.d\underline{s}$.

Thus, $\int_S \text{curl } \underline{F}.d\underline{S} = \oint_C \underline{F}.d\underline{s}$

In words this theorem says: the flux of curl F through any open surface S bounded by a circuit C = circulation round the circuit C.

Note that the direction of C must obey the right-handed screw rule given in FRAME 19.

FRAME 27

We shall now consider the following example:

Verify Stokes' theorem for $\underline{A} = -y\underline{i} + 2yz\underline{j} + y^2\underline{k}$ where S is the upper half surface of the sphere $x^2 + y^2 + z^2 = 1$, and C is its boundary.

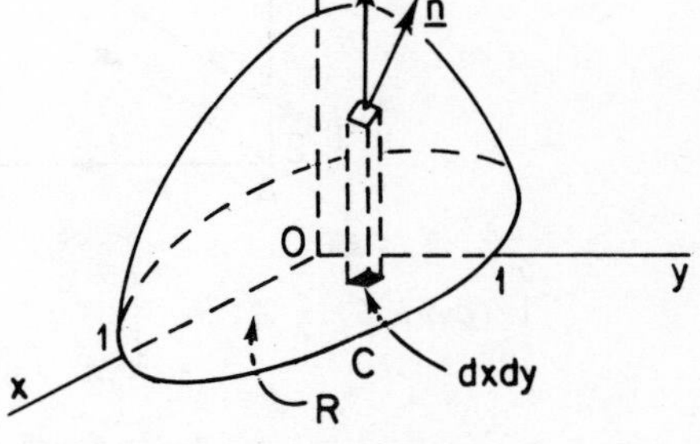

$$\text{Now} \quad \nabla_\wedge \underline{A} = \begin{vmatrix} \underline{i} & \underline{j} & \underline{k} \\ \frac{\partial}{\partial x} & \frac{\partial}{\partial y} & \frac{\partial}{\partial z} \\ -y & 2yz & y^2 \end{vmatrix} = \underline{k}$$

$$\text{Then} \quad \int_S (\nabla_\wedge \underline{A}).\hat{\underline{n}}\, dS = \int_S \underline{k}.\hat{\underline{n}}\, dS = \int_R dxdy$$

since $\hat{\underline{n}}.\underline{k}ds = dxdy$ and R is the projection of S on to the x-y plane.

This last integral equals the area of the circle in the x-y plane = π.

Clearly, the boundary C is a circle of radius 1 and centre (0,0,0) lying in the x-y plane.

Let the parametric equations of C be $x = \cos t$, $y = \sin t$, $z = 0$ $(0 \leq t < 2\pi)$

$$\text{Then} \quad \oint_C \underline{A}.d\underline{s} = \oint_C -ydx + 2yzdy + y^2dz$$

$$= \int_0^{2\pi} \sin^2 t\, dt = \pi$$

FRAME 28

Now try these examples:

(i) Verify Stokes' theorem for $\underline{F} = \left(\frac{z^2}{2} - z\right)\underline{i} - x\underline{j} - \frac{y^2}{2}\underline{k}$ where S is that part of the surface of the cube $x = 0$, $y = 0$, $z = 0$, $x = 2$, $y = 2$, $z = 2$, which lies above the x-y plane.

(ii) If $\underline{H} = \text{curl }\underline{A}$, prove $\int_S \underline{H}.\underline{\hat{n}}\, dS = 0$ for any closed surface S.

****

28A

(i)

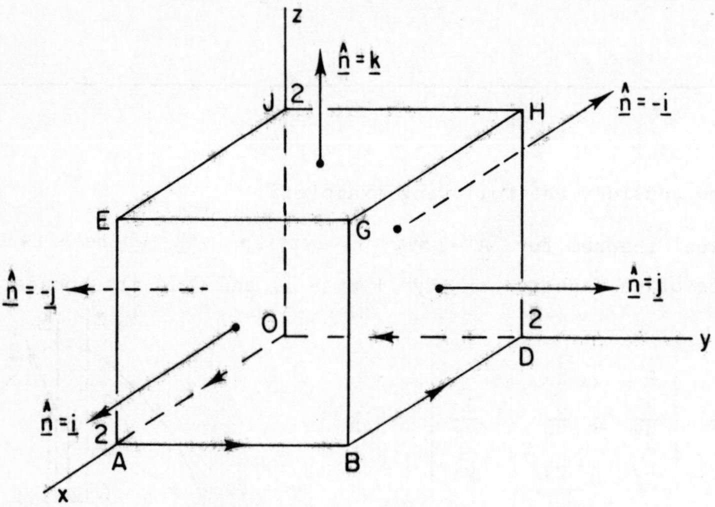

$$\int_S (\text{curl }\underline{F}).\underline{\hat{n}}\, dS = \int_{ABGE} + \int_{BDHG} + \int_{DOJH} + \int_{OAEJ} + \int_{EGHJ}$$

$$= -4 + 0 + 4 + 0 - 4$$

$$= -4$$

$$\oint_C \underline{F}.d\underline{s} = \int_{AB} + \int_{BD} + \int_{DO} + \int_{OA}$$

$$= -4 + 0 + 0 + 0$$

$$= -4$$

(ii) $\int_S \underline{H}.\underline{\hat{n}}\, dS = \int_S (\text{curl }\underline{A}).\underline{\hat{n}}\, dS = \oint_C \underline{A}.d\underline{s} = 0$ *as the bounding curve is of zero length.*

FRAME 29

Green's Theorem in the Plane

This is only a specialisation of Stokes' theorem. The boundary curve C is assumed plane (lying in the x-y plane, say) and the surface S spanning the curve is also assumed to be plane, i.e. the area A in the x-y plane, as shown.

In Stokes' Theorem therefore,

$$d\underline{S} = (dxdy)\underline{k}$$

and $\text{curl } \underline{F} = \left(\frac{\partial F_2}{\partial x} - \frac{\partial F_1}{\partial y}\right)\underline{k}$

$\therefore \int_S \text{curl } \underline{F}.d\underline{S}$ becomes $\int_A \left(\frac{\partial F_2}{\partial x} - \frac{\partial F_1}{\partial y}\right) dxdy$

Also $\oint_C \underline{F}.d\underline{s}$ becomes $\oint_C (F_1\underline{i} + F_2\underline{j}).(dx\underline{i} + dy\underline{j})$

$$= \oint_C F_1 dx + F_2 dy$$

Thus

$$\boxed{\int_A \left(\frac{\partial F_2}{\partial x} - \frac{\partial F_1}{\partial y}\right) dxdy = \oint_C F_1 dx + F_2 dy}$$

FRAME 30

We shall now consider the following example:

Verify Green's Theorem for $\oint_C (x^2 + y^2)dx + (x - y)dy$ where C is the boundary of the region defined by $x = 0,\ y = 0,\ x + y = 1$.

$$\oint_C = \int_{OE} + \int_{EB} + \int_{BO}$$

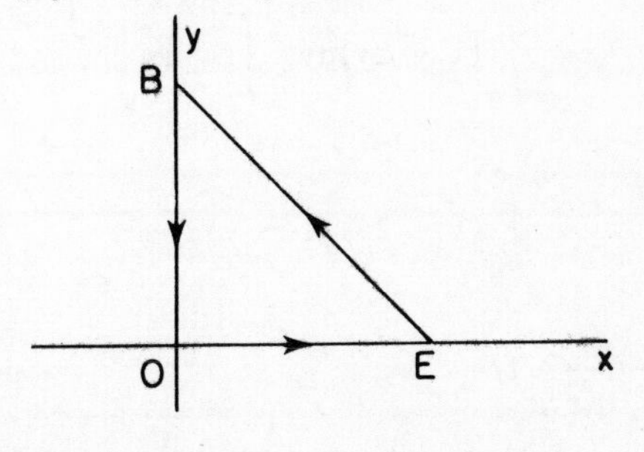

Along OE, $y = 0,\ dy = 0$

$$\int_{OE} = \int_0^1 x^2 dx = \frac{1}{3}$$

FRAME 30 continued

Along EB, $x = 1 - y$, $dx = -dy$

$$\int_{EB} = \int_0^1 \left\{(1 - y)^2 + y^2\right\}(-dy) + (1 - y - y)dy$$

$$= \int_0^1 (-2y^2)dy$$

$$= -2/3$$

Along BO, $x = 0$, $dx = 0$

$$\int_{BO} = \int_1^0 -ydy = \frac{1}{2}$$

$$\therefore \oint_C = \frac{1}{6}$$

Now,

$$\frac{\partial F_2}{\partial x} = 1, \quad \frac{\partial F_1}{\partial y} = 2y$$

$$\therefore \int_A = \int_A (1 - 2y)dxdy$$

$$= \int_0^1 dy \int_0^{1-y} (1 - 2y)dx$$

$$= \int_0^1 (1 - 2y)dy \int_0^{1-y} dx$$

$$= \int_0^1 (1 - 2y)(1 - y)dy$$

$$= 1/6$$

FRAME 31

ow try these examples:

i) Verify Green's Theorem by evaluating the same function as in FRAME 30 in the region defined by $y = \sqrt{x}$, $y = x^2$.

ii) Evaluate $\oint_C (5x - 4y)dx + (2y - 6x)dy$ where C is a circle of radius 2, with centre the origin, traversed positively.

****

31A

i)

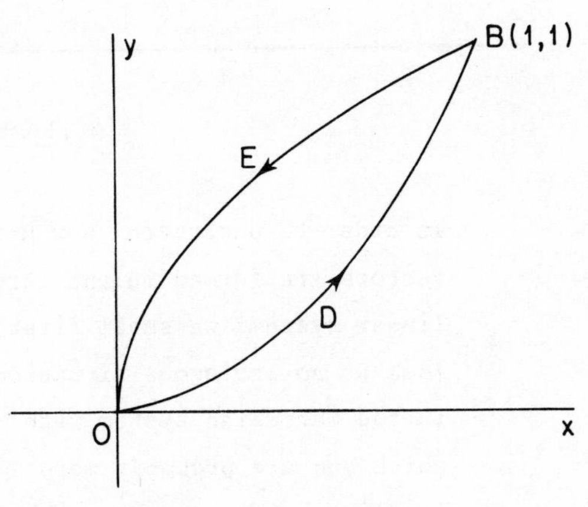

$$\int_{ODB} = \frac{7}{10}$$

$$\int_{BEO} = -\frac{2}{3}$$

$$\oint_C = \frac{1}{30}$$

$$\int_A = \int_0^1 dx \int_{x^2}^{\sqrt{x}} (1 - 2y)dy$$

$$= \frac{1}{30}$$

ii)

$$\oint_C (5x - 4y)dx + (2y - 6x)dy = \int_A (-6 + 4)dxdy$$

$$= -2 \text{ (Area of circle)}$$

$$= -8\pi$$

This could, of course, have been evaluated as an ordinary line integral but the use of Green's Theorem considerably simplifies the working.

FRAME 3

Curvilinear Coordinates

You are already familiar with cylindrical and spherical polar coordinates which were treated in Unit 10 (pages 10:10-10:14) of this volume. You will also know from that programme how to change from Cartesian to polar coordinates and vice versa.

In this section we shall show how div, grad and curl, and also ∇^2, can be expressed in terms of these coordinates. We shall deal first with the case o cylindrical coordinates and then go on to that of spherical coordinates. In each case the question of unit vectors in the system must be considered first

---

FRAME 3

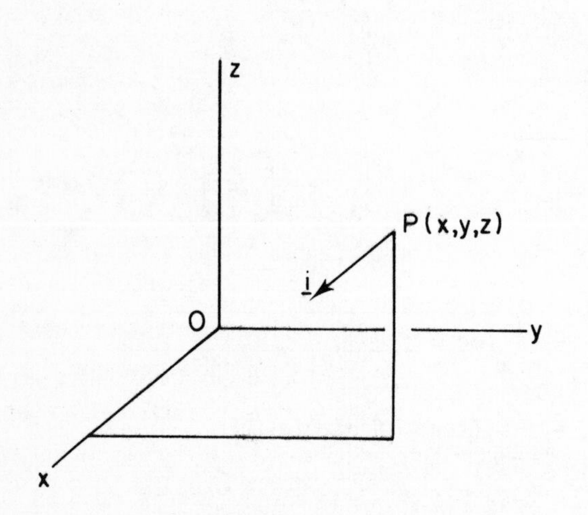

In order to understand how unit vectors are formed in the curvilinear systems we shall first look at an analogous situation in the Cartesian system with which you are probably more familiar.

At P a unit vector can be drawn whose direction is that in which P would move if its y- and z-coordinates were kept fixed and x allowed to increase. Clearly this direction is parallel to Ox and so the unit vector is $\underline{i}$.

If, on the other hand, the x- and z-coordinates of P were kept fixed and y allowed to increase, what unit vector would be formed now?

****

33A

FRAME 34

directions of these vectors $\underline{i}$ and $\underline{j}$ (and, of course, $\underline{k}$) do not vary with position of P. However, when dealing with curvilinear systems we shall d that most of the unit vectors which are formed in a similar manner do y in direction according to the position of P. You have already met a newhat similar situation in the case of the unit normal $\hat{\underline{n}}$ to a curved surface. general the direction of such a unit vector depends on the point at which it drawn.

FRAME 35

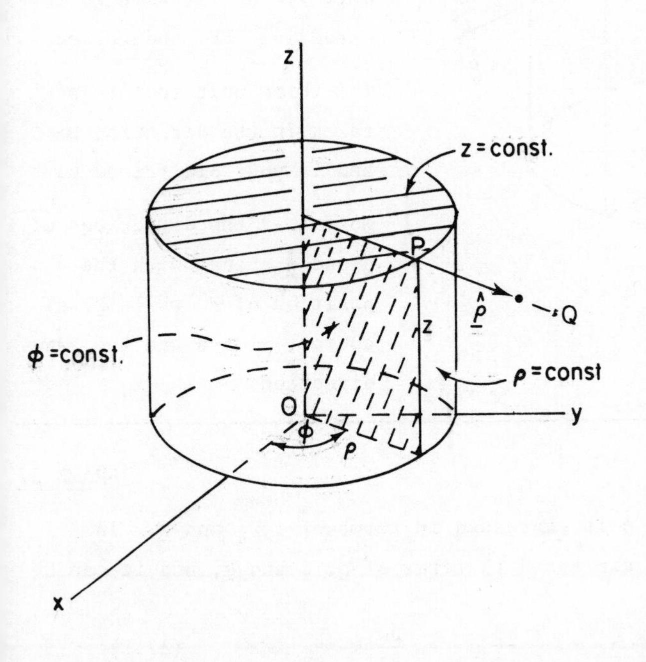

Turning now to cylindrical polar coordinates, let P be any point (ρ,ϕ,z) in the system.

First consider the direction in which P will move if ρ is allowed to increase, while ϕ and z remain fixed. This direction is along PQ, as shown in the diagram (note that when PQ is produced backwards it passes through, and is perpendicular to, Oz). We take our first unit vector at P in this direction and call it $\hat{\underline{\rho}}$.

How will P move if

(i) ρ and z are kept constant and ϕ allowed to increase,

(ii) ρ and ϕ are kept constant and z allowed to increase?

****

35A

(i) *As ϕ increases, P moves round the circle perpendicular to Oz, with radius ρ.*

(ii) *P will move in the direction PS, parallel to Oz. This direction is shown in the diagram in the next frame.*

FRAME 36

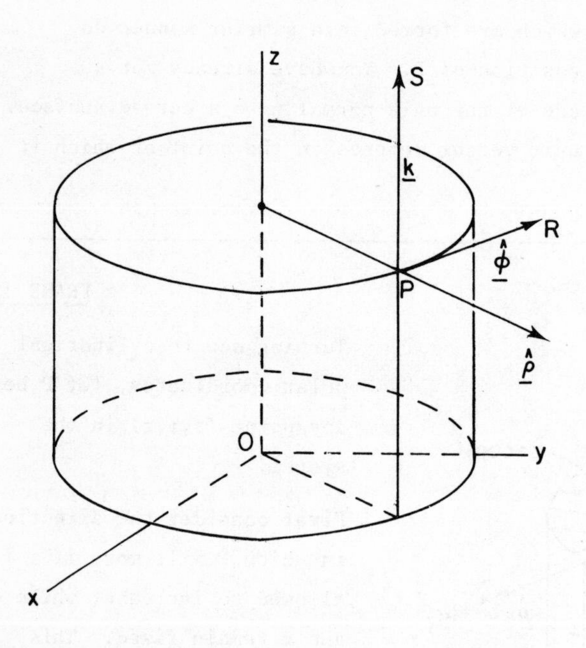

It follows from answer (i) in 35A that, for ϕ increasing, with ρ and z constant, the direction of motion at P is along the tangent there. The second unit vector is taken in this direction, PR, and called $\hat{\underline{\phi}}$.

The third unit vector is taken in the direction PS and is your old friend $\underline{k}$.

Note that the directions of $\hat{\underline{\rho}}$ and $\hat{\underline{\phi}}$ change with the position of P, while $\underline{k}$, of course, always has the same direction.

FRAME 37

In Cartesian coordinates grad ψ is expressed in terms of $\underline{i}$, $\underline{j}$ and $\underline{k}$. In cylindrical coordinates it is expressed in terms of $\hat{\underline{\rho}}$, $\hat{\underline{\phi}}$ and $\underline{k}$, and it can be shown that it is given by

$$\text{grad } \psi = \frac{\partial\psi}{\partial\rho}\hat{\underline{\rho}} + \frac{1}{\rho}\frac{\partial\psi}{\partial\phi}\hat{\underline{\phi}} + \frac{\partial\psi}{\partial z}\underline{k} \qquad (37.1)$$

If a vector $\underline{A}$ is given in cylindrical coordinates it will be of the form

$$A_\rho\hat{\underline{\rho}} + A_\phi\hat{\underline{\phi}} + A_z\underline{k}$$

It can be shown that

$$\text{div } \underline{A} = \frac{1}{\rho}\left\{\frac{\partial}{\partial\rho}(\rho A_\rho) + \frac{\partial}{\partial\phi}(A_\phi) + \frac{\partial}{\partial z}(\rho A_z)\right\} \qquad (37.2)$$

FRAME 37 continued

$$\text{and} \quad \text{curl } \underline{A} = \frac{1}{\rho}\begin{vmatrix} \hat{\underline{\rho}} & \rho\hat{\underline{\phi}} & \underline{k} \\ \frac{\partial}{\partial\rho} & \frac{\partial}{\partial\phi} & \frac{\partial}{\partial z} \\ A_\rho & \rho A_\phi & A_z \end{vmatrix} \qquad (37.3)$$

Using $\nabla^2\psi$ = div grad ψ in conjunction with results (37.1) and (37.2) it immediately follows that

$$\nabla^2\psi = \frac{1}{\rho}\frac{\partial}{\partial\rho}\left(\rho\frac{\partial\psi}{\partial\rho}\right) + \frac{1}{\rho^2}\frac{\partial^2\psi}{\partial\phi^2} + \frac{\partial^2\psi}{\partial z^2} \qquad (37.4)$$

FRAME 38

Now try these examples:

(i) Find grad$(\rho^2 z \sin 2\phi)$ at the point where $\rho = 1$, $\phi = \frac{\pi}{4}$, $z = 2$.

(ii) A certain vector field, in cylindrical coordinates, is $\rho \cos\phi\, \hat{\underline{\rho}} + \rho \sin\phi\, \hat{\underline{\phi}}$. Find its divergence.

(iii) Given that $\underline{q} = q_\rho\hat{\underline{\rho}} + q_\phi\hat{\underline{\phi}} + q_z\underline{k}$ represents the velocity of a fluid, find curl $\underline{q}$ and deduce the conditions for the flow to be irrotational (i.e. curl $\underline{q}$ = 0).

****

38A

(i) $4\hat{\underline{\rho}} + \underline{k}$

(ii) $3\cos\phi$

(iii)

$$\text{curl } \underline{q} = \frac{1}{\rho}\begin{vmatrix} \hat{\underline{\rho}} & \rho\hat{\underline{\phi}} & \underline{k} \\ \frac{\partial}{\partial\rho} & \frac{\partial}{\partial\phi} & \frac{\partial}{\partial z} \\ q_\rho & \rho q_\phi & q_z \end{vmatrix}$$

$$= \frac{1}{\rho}\left\{\left(\frac{\partial q_z}{\partial\phi} - \rho\frac{\partial q_\phi}{\partial z}\right)\hat{\underline{\rho}} + \rho\left(\frac{\partial q_\rho}{\partial z} - \frac{\partial q_z}{\partial\rho}\right)\hat{\underline{\phi}} + \left(q_\phi + \rho\frac{\partial q_\phi}{\partial\rho} - \frac{\partial q_\rho}{\partial\phi}\right)\underline{k}\right\}$$

38A continued

The flow is irrotational if $\rho \neq 0$ *and*

$$\frac{\partial q_z}{\partial \phi} - \rho \frac{\partial q_\phi}{\partial z} = 0$$

$$\frac{\partial q_\rho}{\partial z} - \frac{\partial q_z}{\partial \rho} = 0$$

$$q_\phi + \rho \frac{\partial q_\phi}{\partial \rho} - \frac{\partial q_\rho}{\partial \phi} = 0$$

FRAME 39

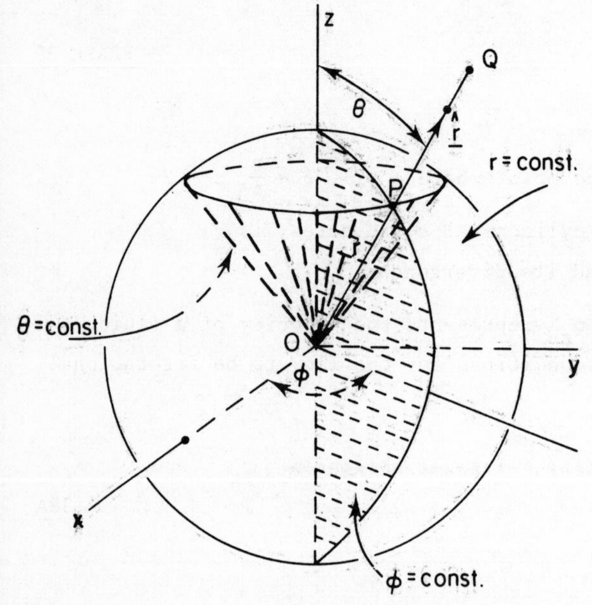

Considering next spherical polar coordinates, let P be any point (r,θ,ϕ) in the system.

If r is allowed to increase, while θ and ϕ remain fixed, P moves in the direction PQ. We take our first unit vector in this direction and call it $\hat{\underline{r}}$.

How will P move if

(i) r and ϕ are kept constant and θ allowed to increase,

(ii) r and θ are kept constant and ϕ allowed to increase?

****

39A

(i) *As* θ *increases, P moves round the circle, centre O, radius r, in the plane* ϕ = *constant.*

(ii) *As* ϕ *increases, P moves round the circle perpendicular to Oz, with radius* $r \sin \theta$.

FRAME 40

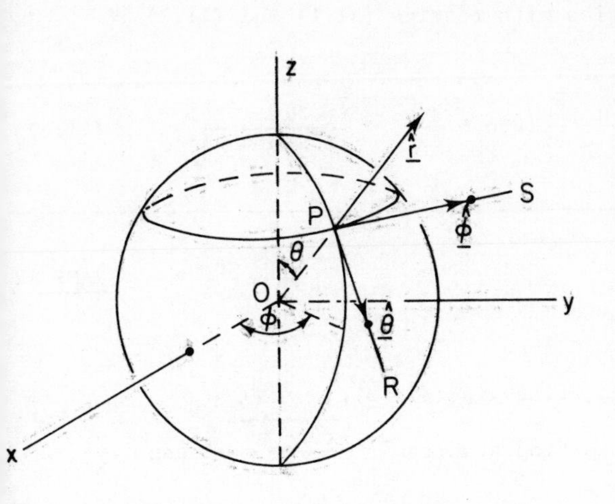

It follows from the two answers in 39A that in each case the direction of motion at P is along the tangent to the appropriate circle.

The second unit vector is taken in the direction PR and called $\hat{\underline{\theta}}$, and the third in the direction PS and called $\hat{\underline{\phi}}$.

Which of these three unit vectors changes direction as P moves?

****

40A

All of them.

FRAME 41

In spherical polar coordinates, it can be shown that

$$\text{grad } \psi = \frac{\partial \psi}{\partial r}\hat{\underline{r}} + \frac{1}{r}\frac{\partial \psi}{\partial \theta}\hat{\underline{\theta}} + \frac{1}{r \sin\theta}\frac{\partial \psi}{\partial \phi}\hat{\underline{\phi}} \tag{41.1}$$

In this system, a vector $\underline{A}$ is of the form

$$A_r\hat{\underline{r}} + A_\theta\hat{\underline{\theta}} + A_\phi\hat{\underline{\phi}}$$

$$\text{Then div } \underline{A} = \frac{1}{r^2}\frac{\partial}{\partial r}(A_r r^2) + \frac{1}{r \sin\theta}\frac{\partial}{\partial \theta}(A_\theta \sin\theta) + \frac{1}{r \sin\theta}\frac{\partial}{\partial \phi}(A_\phi) \tag{41.2}$$

$$\text{and} \quad \text{curl } \underline{A} = \frac{1}{r^2 \sin\theta}\begin{vmatrix} \hat{\underline{r}} & r\hat{\underline{\theta}} & r\sin\theta\,\hat{\underline{\phi}} \\ \frac{\partial}{\partial r} & \frac{\partial}{\partial \theta} & \frac{\partial}{\partial \phi} \\ A_r & rA_\theta & rA_\phi \sin\theta \end{vmatrix} \tag{41.3}$$

FRAME 41 continued

Using $\nabla^2\psi$ = div grad ψ in conjunction with results (41.1) and (41.2) it immediately follows that

$$\nabla^2\psi = \frac{1}{r^2}\frac{\partial}{\partial r}\left(r^2\frac{\partial\psi}{\partial r}\right) + \frac{1}{r^2\sin\theta}\frac{\partial}{\partial\theta}\left(\sin\theta\frac{\partial\psi}{\partial\theta}\right) + \frac{1}{r^2\sin^2\theta}\frac{\partial^2\psi}{\partial\phi^2} \qquad (41.4)$$

FRAME 42

Now try these examples:

(i) Using spherical coordinates, find grad(div $\hat{\underline{\theta}}$).

(ii) Use spherical coordinates to find $\text{grad}\left\{\tan^{-1}\frac{\sqrt{x^2+y^2}}{z} + \tan^{-1}\frac{y}{x}\right\}$.

(iii) If the flow of a fluid is irrotational, there exists a function Φ such that $\underline{q} = -\nabla\Phi$, where $\underline{q}$ is the velocity vector at any point. Φ is then called the velocity potential. If the fluid is of constant density and there are no sources or sinks, the equation of continuity $\nabla.\underline{q} = 0$ is also satisfied. Then as $\underline{q} = -\nabla\Phi$, $\nabla^2\Phi = 0$.

Show, by using spherical polar coordinates, that

$$\nabla^2\left\{U\left(r + \frac{a^3}{2r^2}\right)\cos\theta\right\} = 0$$

Thus $\Phi = U\left(r + \frac{a^3}{2r^2}\right)\cos\theta$ can be the velocity potential of an irrotational flow. It actually represents a uniform flow which is disturbed by inserting a sphere of radius a with centre at the origin. Show that, on the surface of the sphere, the direction of $\underline{q}$ is tangential to the sphere and that $|\underline{q}| \to U$ as $r \to \infty$.

****

42A

(i) $\quad \text{div}\,\hat{\underline{\theta}} = \frac{1}{r}\cot\theta$

$$\text{grad(div}\,\hat{\underline{\theta}}) = \frac{-1}{r^2}\cot\theta\,\hat{\underline{r}} - \frac{1}{r^2}\text{cosec}^2\theta\,\hat{\underline{\theta}}$$

(ii) $\quad \tan^{-1}\frac{\sqrt{x^2+y^2}}{z} + \tan^{-1}\frac{y}{x} = \theta + \phi$

$$\text{grad}(\theta + \phi) = \frac{1}{r}\hat{\underline{\theta}} + \frac{1}{r\sin\theta}\hat{\underline{\phi}}$$

42A continued

(iii) *Using (41.4)*

$$\nabla^2\Phi = \frac{1}{r^2}\frac{\partial}{\partial r}\left\{r^2U\left(1 - \frac{a^3}{r^3}\right)\cos\theta\right\} + \frac{1}{r^2\sin\theta}\frac{\partial}{\partial\theta}\left\{(\sin\theta)U\left(r + \frac{a^3}{2r^2}\right)(-\sin\theta)\right\}$$

$$= \frac{1}{r^2}U\left(2r + \frac{a^3}{r^2}\right)\cos\theta + \frac{1}{r^2\sin\theta}U\left(r + \frac{a^3}{2r^2}\right)(-2\sin\theta\cos\theta)$$

$$= 0$$

Using (41.1)

$$\nabla\Phi = U\left(1 - \frac{a^3}{r^3}\right)(\cos\theta)\hat{\underline{r}} + \frac{1}{r}U\left(r + \frac{a^3}{2r^2}\right)(-\sin\theta)\hat{\underline{\theta}}$$

When $r = a$, $\underline{q} = \frac{3}{2}U\sin\theta\,\hat{\underline{\theta}}$ *which is tangential to the sphere.*

As $r \to \infty$, $\underline{q} \to -U\cos\theta\,\hat{\underline{r}} + U\sin\theta\,\hat{\underline{\theta}}$ *and thus* $|\underline{q}| \to U$.

FRAME 43

Miscellaneous Examples

In this frame a collection of miscellaneous examples is given for you to try. Answers are supplied in FRAME 44, together with such working as is considered helpful.

1. If $\underline{F} = (x^2 + y)\underline{i} + (x + y^2)\underline{j}$, find $\int_{(0,0)}^{(1,1)} \underline{F}.d\underline{r}$ along

   (i) the straight line from (0,0) to (1,0) followed by that from (1,0) to (1,1),

   (ii) the straight line path $y = x$.

   Why do you get the same answer for both (i) and (ii)?

   Find ϕ such that $\underline{F} = \text{grad}\,\phi$.

2. (i) Use the divergence theorem to express $\int_S \underline{r}.d\underline{S}$ in terms of the volume V bounded by the closed surface S.

   (ii) Verify the divergence theorem for $\underline{A} = x^2\underline{i} + y^2\underline{j} + z^2\underline{k}$ taken over the surface of the cube: $0 \leqslant x \leqslant 1$, $0 \leqslant y \leqslant 1$, $0 \leqslant z \leqslant 1$. (C.E.I.)

FRAME 43 continued

3. Use Stokes' Theorem to evaluate $\iint_S \text{curl}\ \underline{A}.d\underline{S}$, where $\underline{A} = (x^2 + y - 4)\underline{i} + 3xy\underline{j} + (2xz + z^2)\underline{k}$ and S is the surface of the paraboloid $z = 4 - (x^2 + y^2)$ above the x-y plane. (C.E.I.)

4. Use Gauss' Divergence Theorem to show that

$$\int_S \frac{1}{r^2}\,\underline{r}\cdot d\underline{S} = \int_V \frac{1}{r^2}\,dV$$

for any closed surface S enclosing a volume V.

5. Use spherical polars to show that $\text{div}\left(\frac{1}{r^3}\,\underline{r}\right) = 0$ and to find $\text{curl}\left(\frac{1}{r^3}\,\underline{r}\right)$. (HINT: Use $\underline{r} = r\,\hat{\underline{r}}$)

6. Show, using Stokes' Theorem, that $\int_C \underline{r}.d\underline{r} = 0$ for any closed curve C.

7. If $\Phi = U\left(\rho + \frac{a^2}{\rho}\right)\cos\phi$, show, by using cylindrical coordinates, that $\nabla^2\Phi = 0$. Thus Φ can represent the velocity potential of an irrotational fluid motion. It actually represents a uniform flow perpendicular to Oz, which is disturbed by placing in it a circular cylinder, of radius a, whose axis is along Oz. Show that on the surface of the cylinder, the direction of $\underline{q}$ ($= -\nabla\Phi$) is tangential to the cylinder and that $|\underline{q}| \to U$ as $\rho \to \infty$.

8. In a certain fluid flow problem $\underline{v}$ is the fluid velocity and the vorticity $\underline{\Omega}$ and vector potential $\underline{u}$ are defined by the equations

$$\underline{\Omega} = \tfrac{1}{2}\,\text{curl}\ \underline{v}, \qquad \underline{v} = \text{curl}\ \underline{u}$$

Show that $\int_V v^2 dV = \oint_S (\underline{u}\wedge\underline{v}).d\underline{S} + 2\int_V \underline{u}.\underline{\Omega}\,dV$ where the closed surface S encloses a volume V of fluid. (You may assume Gauss' Divergence theorem.) (C.E.I.)

(You may also assume formula (v) in APPENDIX B, page 11:53.)

FRAME 43 continued

9. Given that the electric field intensity $\underline{E}$ at P due to a point charge e at S is $e\hat{\underline{r}}/r^2$, where $\hat{\underline{r}}$ is a unit vector along SP, evaluate directly the line integral $\int \underline{E}.d\underline{s}$ taken along the parabola $y^2 = 4ax$ from the origin to infinity, where $\underline{E}$ is due to a point charge e at the focus (a,0) of the parabola. (L.U.)

10. If S is the portion of the surface of the plane $2x + 3y + 6z = 12$ situated in the positive octant, $\underline{n}$ is the unit normal to S in a direction away from the origin and S_1 is the projection of S on to the xOy plane, show that if $\underline{A}$ is a vector field then

$$\int_S \underline{A}.\underline{n}\,dS = \iint_{S_1} \frac{\underline{A}.\underline{n}}{|\underline{n}.\underline{k}|}\,dx\,dy,$$

and that if $\underline{A} = 18z\underline{i} - 12\underline{j} + 3y\underline{k}$ the value of the integral is 24. (L.U.)

11. Prove that if $\underline{F}$ is a vector point function

$$\int_S \underline{n}.\text{curl}\,\underline{F}\,dS = 0,$$

where the integral is taken over a closed surface S, $\underline{n}$ being a unit vector along the outward normal to the element dS.

Evaluate the integral $\int \underline{n}.\text{curl}\,\underline{F}\,dS$ for the vector function

$$\underline{F} = (2y^2 + 3z^2 - x^2)\underline{i} + (2z^2 + 3x^2 - y^2)\underline{j} + (2z^2 + 3y^2 - z^2)\underline{k}$$

over the part of the surface $x^2 + y^2 - 2ax + az = 0$ which lies above the plane $z = 0$. (L.U.)

FRAME 44

Answers to Miscellaneous Examples

1. In both (i) and (ii) the integral equals $1\frac{2}{3}$.

The answers are the same because $(x^2 + y)dx + (x + y^2)dy$ is the exact differential of a single-valued function.

$$\phi = \frac{x^3}{3} + \frac{y^3}{3} + xy + c$$

FRAME 44 continued

2. (i) $\int_S \underline{r}.d\underline{S} = \int_V \nabla.\underline{r}\, dV = \int_V 3dV = 3V$

(ii)

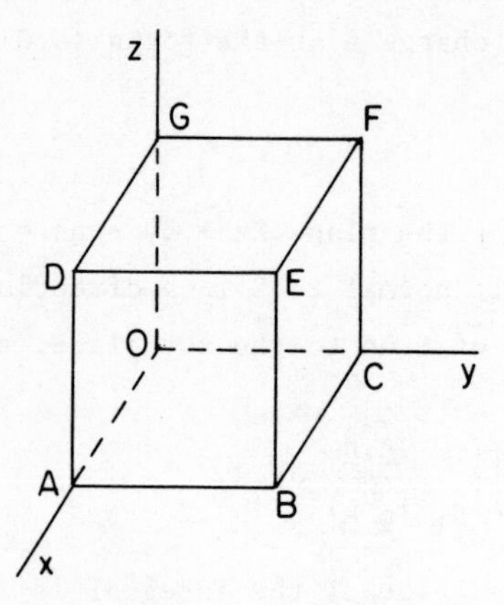

Over OABC, $\int \underline{A}.d\underline{S} = 0$

" OGDA $\int \underline{A}.d\underline{S} = 0$

" ABED $\int \underline{A}.d\underline{S} = 1$

" BCFE $\int \underline{A}.d\underline{S} = 1$

" OCFG $\int \underline{A}.d\underline{S} = 0$

" DEFG $\int \underline{A}.d\underline{S} = 1$

Over complete surface, $\int \underline{A}.d\underline{S} = 3$

$$\int_V \nabla.\underline{A}\, dV = \int_0^1 dx \int_0^1 dy \int_0^1 2(x + y + z)dz = 3$$

3.

y, 2, C, O, 2, x

$$\int_S \text{curl } \underline{A}.d\underline{S} = \int_C \underline{A}.d\underline{r}$$

$$= \int_C (x^2 + y - 4)dx + 3xydy + (2xz + z^2)dz$$

$$= -4\pi$$

4. $$\int_S \frac{1}{r^2}\underline{r}.d\underline{S} = \int_V \text{div}\left(\frac{1}{r^2}\ \underline{r}\right)dV$$

$$= \int_V \text{div}\left(\frac{1}{r}\ \hat{\underline{r}}\right)dV$$

$$= \int_V \frac{1}{r^2}\ \frac{\partial}{\partial r}\left(\frac{1}{r}\ r^2\right)dV \quad \text{using (41.2)}$$

$$= \int_V \frac{1}{r^2}\ dV$$

FRAME 44 continued

5. $\frac{1}{r^3}\,\underline{r} = \frac{1}{r^2}\,\hat{\underline{r}}$

Using (41.2) $\operatorname{div}\left(\frac{1}{r^2}\,\hat{\underline{r}}\right) = \frac{1}{r^2}\frac{\partial}{\partial r}\left(\frac{1}{r^2}\,r^2\right) = 0$

Using (41.3) $$\operatorname{curl}\left(\frac{1}{r^2}\,\hat{\underline{r}}\right) = \frac{1}{r^2\sin\theta}\begin{vmatrix} \hat{\underline{r}} & r\hat{\underline{\theta}} & r\sin\theta\,\hat{\underline{\phi}} \\ \frac{\partial}{\partial r} & \frac{\partial}{\partial\theta} & \frac{\partial}{\partial\phi} \\ \frac{1}{r^2} & 0 & 0 \end{vmatrix} = 0$$

6. $\int_C \underline{r}.d\underline{r} = \int_S \operatorname{curl}\,\underline{r}.d\underline{S} = 0$ as $\operatorname{curl}\,\underline{r} = 0$

7. Using (37.4)

$$\nabla^2\Phi = \frac{1}{\rho}\frac{\partial}{\partial\rho}\left\{\rho U\left(1 - \frac{a^2}{\rho^2}\right)\cos\phi\right\} + \frac{1}{\rho^2}\,U\left(\rho + \frac{a^2}{\rho}\right)(-\cos\phi)$$
$$= 0$$

Using (37.1)

$$\nabla\Phi = U\left(1 - \frac{a^2}{\rho^2}\right)\cos\phi\,\hat{\underline{\rho}} - \frac{1}{\rho}\,U\left(\rho + \frac{a^2}{\rho}\right)\sin\phi\,\hat{\underline{\phi}}$$

When $\rho = a$, $\underline{q} = 2U\sin\phi\,\hat{\underline{\phi}}$ which is tangential to the cylinder.

As $\rho \to \infty$, $\underline{q} \to -U\cos\phi\,\hat{\underline{\rho}} + U\sin\phi\,\hat{\underline{\phi}}$ and thus $|\underline{q}| \to U$.

8. Putting $\underline{A} = \underline{u}$ and $\underline{B} = \underline{v}$ in formula (v) gives

$$\operatorname{div}(\underline{u}\wedge\underline{v}) = \underline{v}.\operatorname{curl}\,\underline{u} - \underline{u}.\operatorname{curl}\,\underline{v}$$

From the divergence theorem

$$\oint_S (\underline{u}\wedge\underline{v}).d\underline{S} = \int_V \operatorname{div}(\underline{u}\wedge\underline{v})dV$$
$$= \int_V (\underline{v}.\operatorname{curl}\,\underline{u} - \underline{u}.\operatorname{curl}\,\underline{v})dV$$
$$= \int_V (v^2 - 2\underline{u}.\underline{\Omega})dV$$

FRAME 44 continued

9.

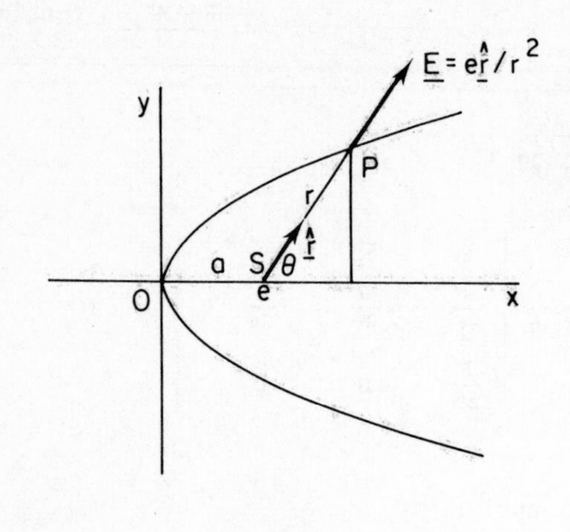

$$\hat{\underline{r}} = \cos\theta\underline{i} + \sin\theta\underline{j}$$

At P, $x = a + r\cos\theta$

$y = r\sin\theta$

$$\underline{E}.d\underline{s} = \frac{e\hat{\underline{r}}}{r^2}.(dx\underline{i} + dy\underline{j})$$

$$= \frac{e}{r^2}(\cos\theta\,dx + \sin\theta\,dy)$$

$$= e\frac{(x - a)dx + ydy}{\{(x-a)^2 + y^2\}^{3/2}}$$

$$\therefore \int \underline{E}.d\underline{s} = e\left[-\left\{(x - a)^2 + y^2\right\}^{-\frac{1}{2}}\right]_{(0,0)}^{(\infty,\infty)}$$

$$= ae$$

10. $dS_1 = dS|\cos\gamma|$ where γ is the angle between $\underline{n}$ and $\underline{k}$

$= dS|\underline{n}.\underline{k}|$

Taking dS_1 as dxdy, result follows.

Letting $\phi = 2x + 3y + 6z - 12$

$\text{grad}\,\phi = 2\underline{i} + 3\underline{j} + 6\underline{k}$

$\underline{n} = (2\underline{i} + 3\underline{j} + 6\underline{k})/7$

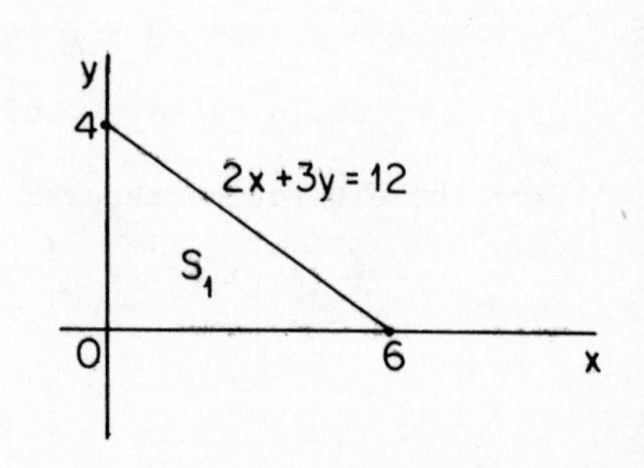

$$\int_S \underline{A}.\underline{n}\,dS = \iint_{S_1} \frac{36z - 36 + 18y}{6}\,dx\,dy$$

$$= \iint_{S_1} (6 - 2x)dx\,dy$$

$$= \int_0^4 dy \int_0^{\frac{1}{2}(12-3y)} 2(3 - x)dx$$

$$= 24$$

FRAME 44 continued

11. $\int_S \underline{n}.\text{curl}\,\underline{F}\,dS = \int_C \underline{F}.d\underline{r} = 0$ as length of bounding curve C is zero.

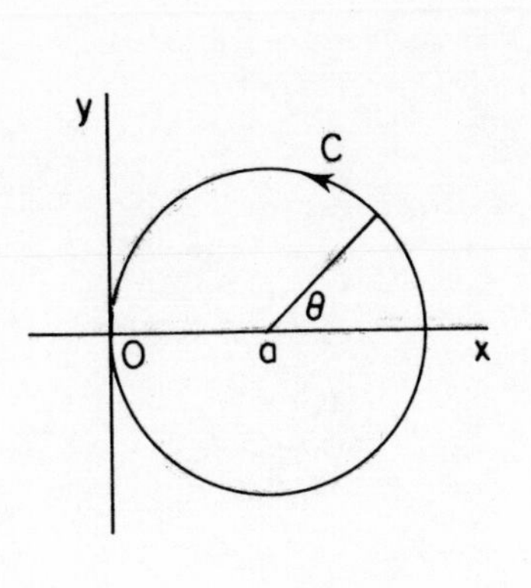

$$\int_S \underline{n}.\text{curl}\,\underline{F}\,dS = \int_C \underline{F}.d\underline{r}$$

$$= \int_C (2y^2 + 3z^2 - x^2)dx + (2z^2 + 3x^2 - y^2)dy + (2x^2 + 3y^2 - z^2)dz$$

$= 6\pi a^3$ by putting $x = a(1 + \cos\theta)$,
$y = a\sin\theta$,
$z = 0$.

UNIT 12

FUNCTIONS OF A COMPLEX VARIABLE

A.C. Bajpai
I.M. Calus
J.A. Fairley

Loughborough University of Technology

INSTRUCTIONS

This Unit comprises four programmes:

(a) Complex Mapping
(b) Theory of Functions
(c) Complex Integration
(d) Conformal Transformations

Each programme is divided up into a number of FRAMES which are to be worked *in the order given*. You will be required to participate in many of these frames and in such cases the answers are provided in ANSWER FRAMES, designated by the letter A following the frame number. Steps in the working are given where this is considered helpful. The answer frame is separated from the main frame by a line of asterisks: *********. Keep the answers covered until you have written your own response. If your answer is wrong, go back and try to see why. Do not proceed to the next frame until you have corrected any mistakes in your attempt and are satisfied that you understand the contents up to this point.

Before reading these programmes, it is necessary that you are familiar with the following

Prerequisites

For (a): The contents of Unit 4 of Vol I.

The contents of the Coordinate Systems programme in Unit 1.

For (b): The contents of Unit 4 of Vol I.

The contents of FRAMES 1-34 of the Functions, Limits and Continuity programme in Unit 1.

The contents of the Infinite Series programme in Unit 1.

Partial Differentiation, as covered in the programme of that name in Unit 1 and FRAMES 1-5 and 21-24 of Unit 7.

For (c): The contents of (b).

The contents of the Line Integrals programme in Unit 10.

The contents of the Partial Fractions supplement in Unit 2.

For (d): The contents of (a) and (b).

CONTENTS

FRAMES		PAGE
	CONFORMAL TRANSFORMATIONS	

COMPLEX MAPPING

FRAME 1

Some Geometrical Results

(FRAMES 1-2 should be simply revision for you.)

In Coordinate Geometry it is possible to define curves and regions by equations and inequalities respectively. According to which is more convenient, either Cartesian or polar coordinates may be used.

Two simple examples in Cartesian coordinates are:

(1) The line AB has the equation

$$\frac{x}{3} + \frac{y}{2} = 1$$

or $2x + 3y = 6$, x and y being the coordinates of any point P on the line.

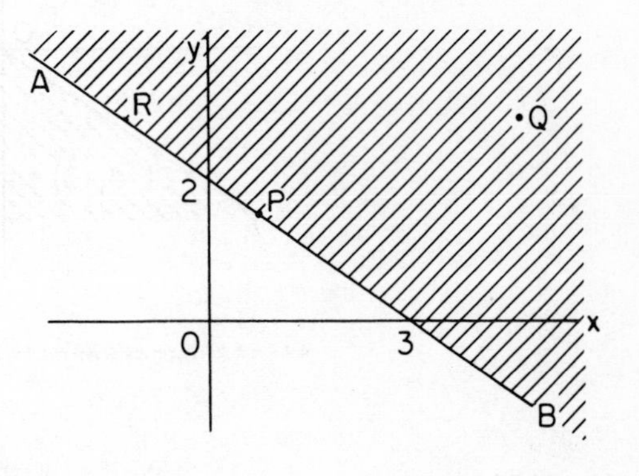

The region shaded is defined by the inequality $2x + 3y > 6$, x and y being the coordinates of any point Q in it.

To see that, for Q, $2x + 3y$ is greater than 6, start from a point R, on AB and at the same height as Q, so that $y_R = y_Q$. For R, $2x + 3y = 6$. But $x_Q > x_R$, so $2x_Q + 3y_Q > 2x_R + 3y_R$ and thus for Q, $2x + 3y > 6$. A similar idea holds for other points not on AB.

(2) The circle centre C(1,2) and radius $1\frac{1}{2}$ has the equation

$$(x - 1)^2 + (y - 2)^2 = \left(\frac{3}{2}\right)^2$$

and the region inside the circle is defined by the inequality

$$(x - 1)^2 + (y - 2)^2 < \left(\frac{3}{2}\right)^2$$

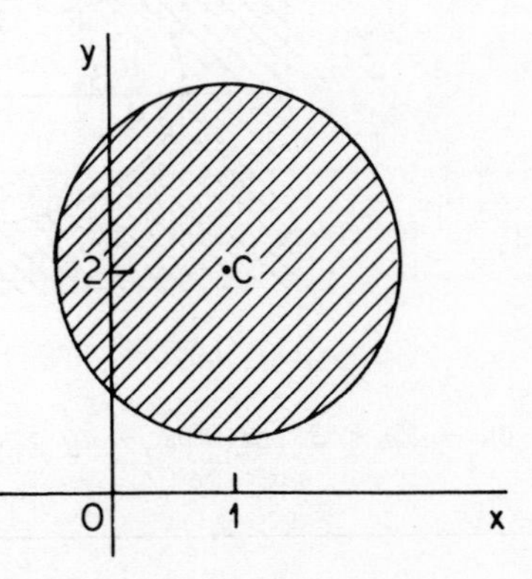

FRAME 1 continued

Now try the following:

(i) What curve is represented by $\frac{(x + 1)^2}{9} + \frac{y^2}{4} = 1$ and what region by $\frac{(x + 1)^2}{9} + \frac{y^2}{4} > 1$?

(ii) Define algebraically (a) the line AB, and (b) the region shaded.

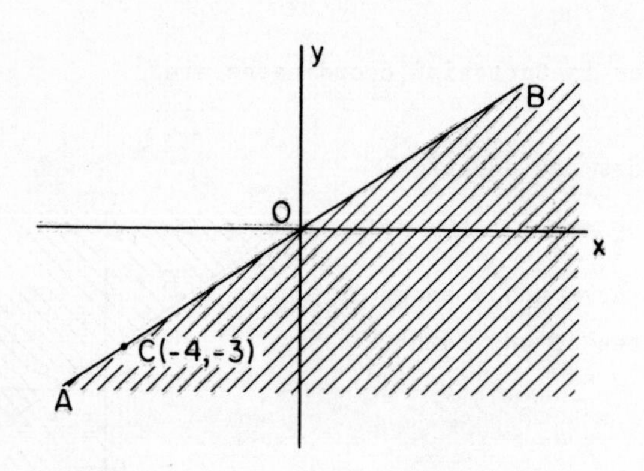

****

1A

(i) Ellipse, centre (-1,0), semi-axes 3,2.
The region outside this ellipse.

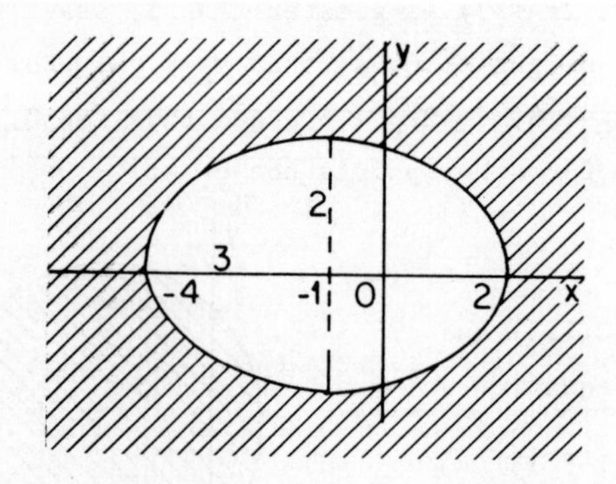

(ii) *(a)* $4y - 3x = 0$

(b) $4y - 3x < 0$ *or* $3x - 4y > 0$

FRAME 2

The following are some examples in polar coordinates. (Here r is assumed to be $\geqslant 0$.)

(1) OA has equation $\theta = \frac{\pi}{6}$.

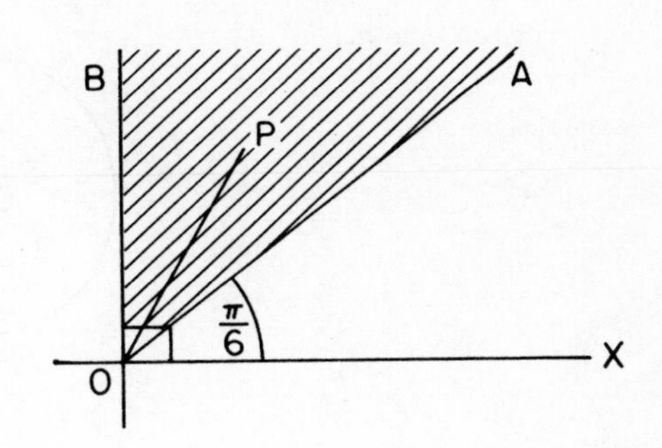

OB has equation $\theta = \frac{\pi}{2}$.

Any point P in the region shown is such that $P\hat{O}X$, i.e. θ if P has coordinates (r,θ), lies between $\frac{\pi}{6}$ and $\frac{\pi}{2}$.

Hence the region is defined by

$\frac{\pi}{6} < \theta < \frac{\pi}{2}$.

(2) The circle, centre the pole and radius 3 has equation $r = 3$ and the region inside it is defined by $r < 3$.

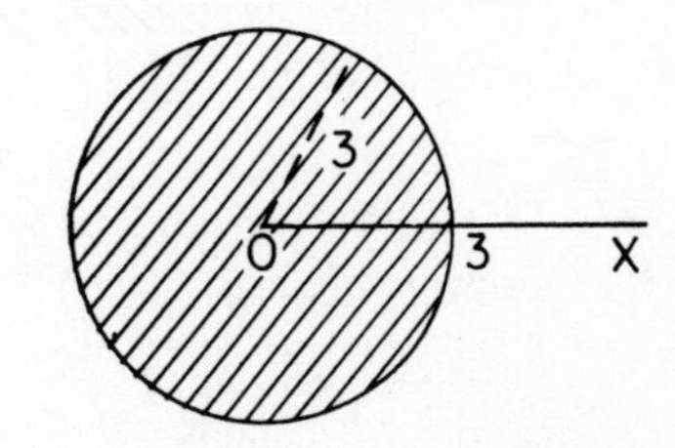

(3) The circle shown has equation $r = 4 \cos \theta$ and the region illustrated is defined by $r > 4 \cos \theta$.

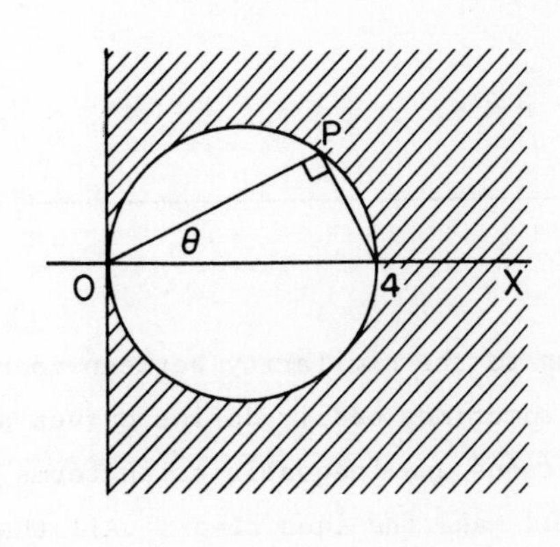

[Negative values of r (which occur when $\frac{\pi}{2} < \theta < \frac{3\pi}{2}$) are ignored.]

FRAME 2 continued

(i) What curve or region is represented by (a) $r = 5$, (b) $3 < r < 5$?

(ii) Give inequalities to define the region shown.

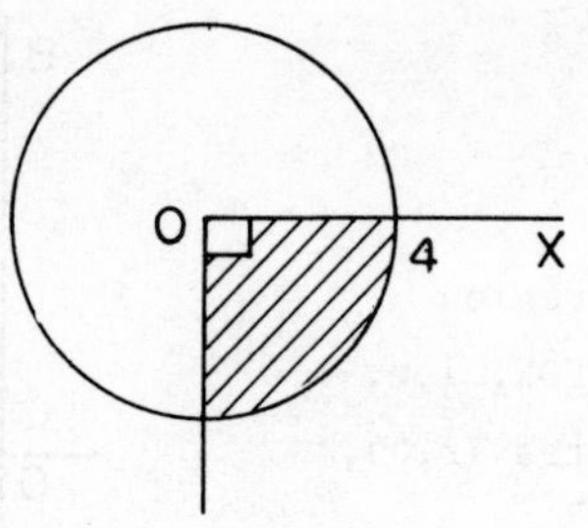

****

2A

(i) (a) Outer circle

(b) Region shaded.

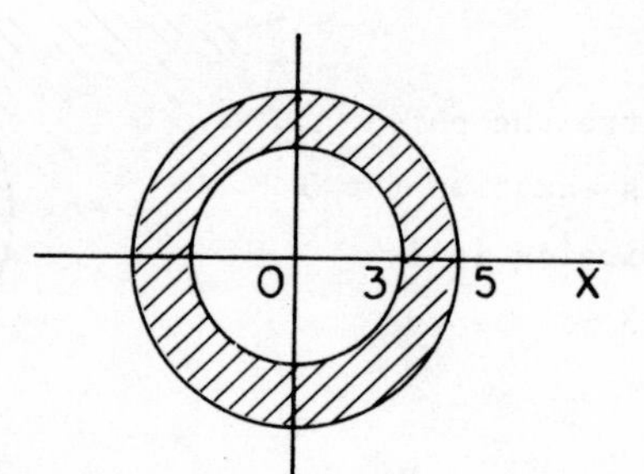

(ii) $r < 4, \quad -\frac{\pi}{2} < \theta < 0$ (or $r < 4, \ \frac{3\pi}{2} < \theta < 2\pi$)

FRAME 3

Owing to the similarity between coordinate geometry and the Argand diagram we can sometimes easily define curves and regions in the Argand diagram by giving equations and inequalities in terms of complex numbers. The following examples should make the idea clear. All the equations and inequalities will be expressed in terms of z ($= x + iy$) and $\bar{z}$ ($= x - iy$), use being made where necessary of their moduli and arguments, also their real and imaginary parts.

FRAME 4

In this frame, we shall interpret the following examples geometrically:

(1) $\text{Re}(z) > 4$ (2) $\arg z = -\dfrac{\pi}{3}$

(3) $|z - (2 + 3i)| \geqslant 4$ (4) $(z - 2)(\bar{z} - 2) = 9$

(1) $\text{Re}(z) > 4$

As $z = x + iy$, $\text{Re}(z) = x$ and so $x > 4$. The region shaded is obtained.

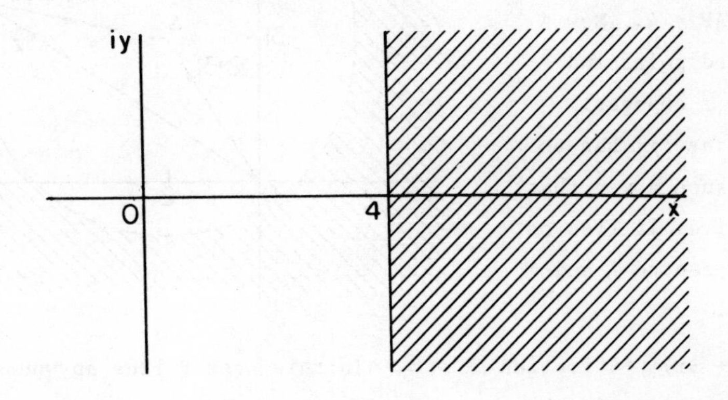

(2) $\arg z = -\pi/3$

If $z = r\underline{/\theta}$, $\arg z = \theta$ and so $\theta = -\pi/3$. This gives the line shown.

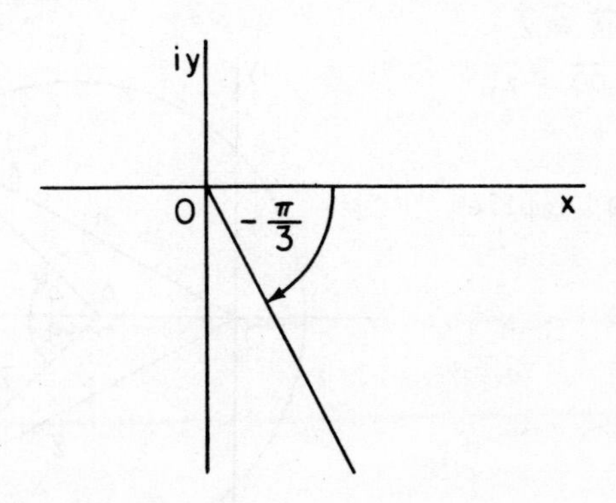

FRAME 4 continued

(3) $|z - (2 + 3i)| \geqslant 4$

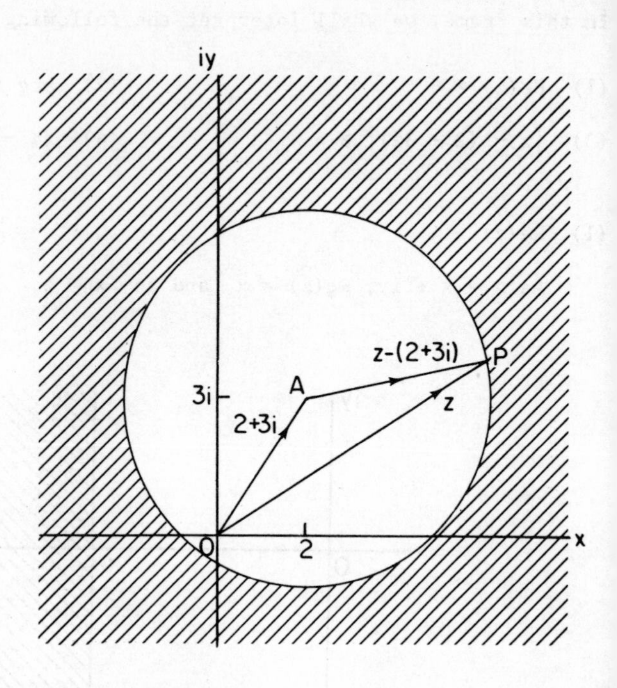

Let $\overrightarrow{OP} = z$ and
$\overrightarrow{OA} = 2 + 3i$,
then as $\overrightarrow{OP} = \overrightarrow{OA} + \overrightarrow{AP}$,
$\overrightarrow{AP} = \overrightarrow{OP} - \overrightarrow{OA}$
$= z - (2 + 3i)$

$|z - (2 + 3i)| = 4$
implies AP = 4. Now A is a fixed point and P moves as z changes. In this case it has to move in such a way that AP = 4, i.e. it lies on a circle centre A and radius 4.

$|z - (2 + 3i)| > 4$ implies AP > 4. In this case P lies anywhere outside the circle, i.e. in the shaded area.

(4) $(z - 2)(\bar{z} - 2) = 9$

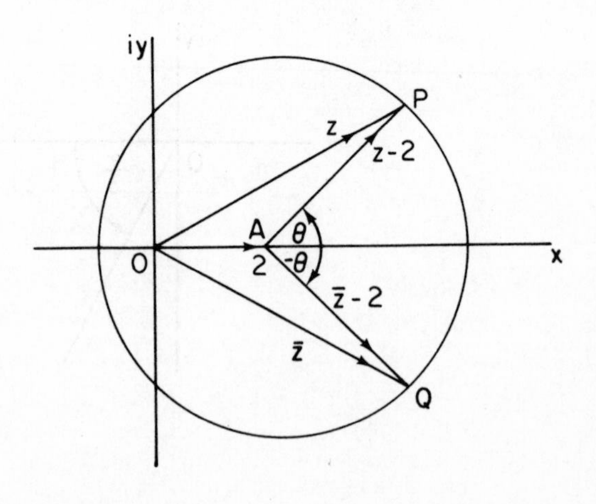

Let $\overrightarrow{OP} = z$ and $\overrightarrow{OA} = 2$
then $\overrightarrow{AP} = z - 2$, $\overrightarrow{OQ} = \bar{z}$,
$\overrightarrow{AQ} = \bar{z} - 2$.

$(z - 2)(\bar{z} - 2) = 9$ implies
$(\overrightarrow{AP})(\overrightarrow{AQ}) = 9$.

Now let $\overrightarrow{AP} = APe^{i\theta}$ then
$\overrightarrow{AQ} = AQe^{-i\theta}$

Therefore,

$(\overrightarrow{AP})(\overrightarrow{AQ}) = (APe^{i\theta})(AQe^{-i\theta})$
$= (AP)(AQ) = 9$

FRAME 4 continued

As AP = AQ, AP = 3, i.e. P lies on a circle, centre A, radius 3. Another way in which we can get this result is to put

$$\begin{aligned}(z - 2)(\bar{z} - 2) &= (x + iy - 2)(x - iy - 2) \\ &= \{(x - 2) + iy\}\{(x - 2) - iy\} \\ &= (x - 2)^2 + y^2\end{aligned}$$

and so $(x - 2)^2 + y^2 = 9$, which is the Cartesian equation of a circle, centre (2,0) and radius 3.

Now interpret the following statements geometrically:

(i) $\text{Im}(z) < -2$ (ii) $2 < |z| < 5$ (iii) $|z - 2| = |z + i|$

****

4A

(i)

Im(z) < −2 gives y < −2.

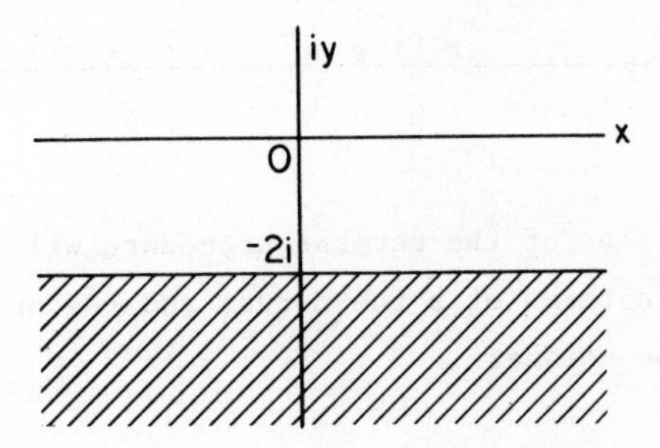

(ii)

$|z| = 2$ is a circle, centre 0, radius 2.

$|z| = 5$ is a circle, centre 0, radius 5.

Here z lies in the annulus enclosed by these two circles.

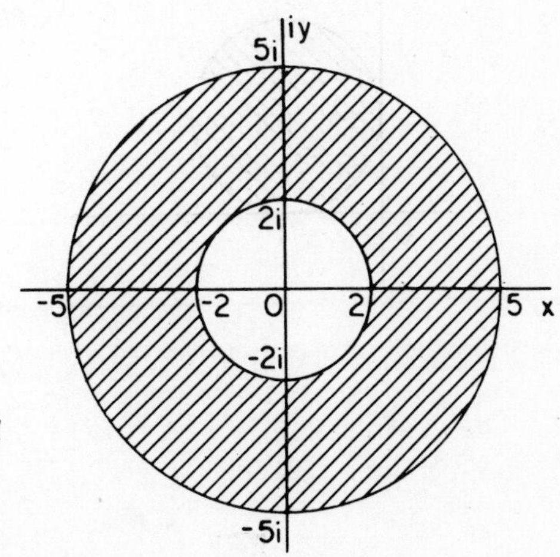

4A continued

(iii)

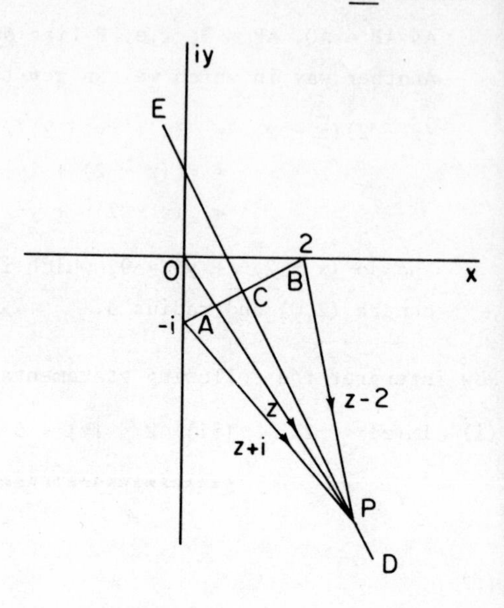

If $\overrightarrow{OP} = z$, *then* $\overrightarrow{BP} = z - 2$,

$\overrightarrow{AP} = z + i$

$|z - 2| = |z + i|$ *gives*

$BP = AP$,

i.e. P lies on ECD, the perpendicular bisector of AB.

FRAME 5

Some examples of the reverse procedure will now be considered. We shall express in terms of z the curves and regions depicted. The representations may not be unique.

(1)

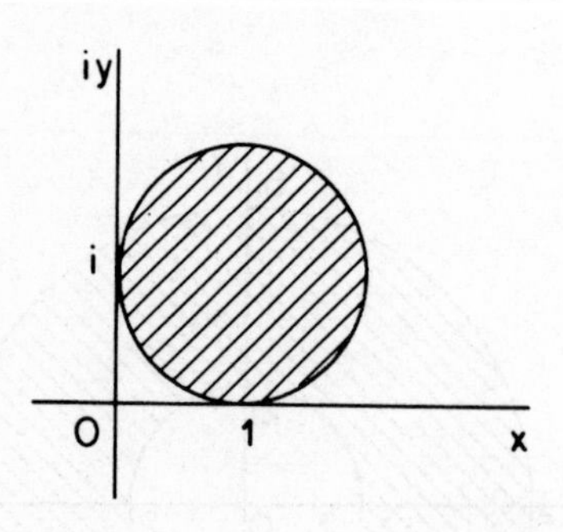

The interior of the circle whose centre is at 1 + i and whose radius is 1.

FRAME 5 continued

Let $\overrightarrow{OP} = z$, then $\overrightarrow{AP} = \overrightarrow{OP} - \overrightarrow{OA}$

$= z - (1 + i)$

The radius of the circle is 1. If P is inside the circle then AP < 1, hence $|z - (1 + i)| < 1$

2)

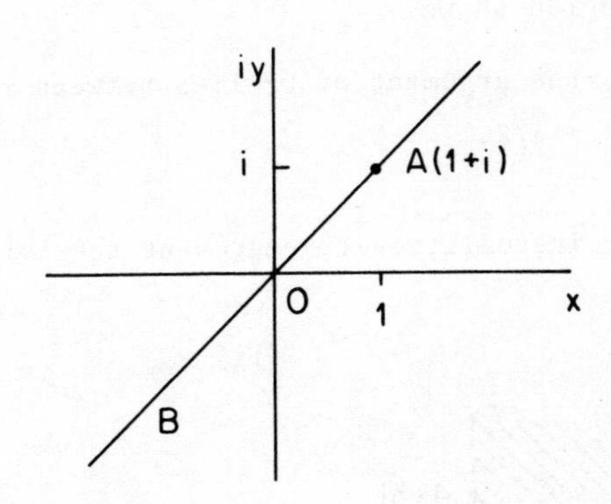

The line AB

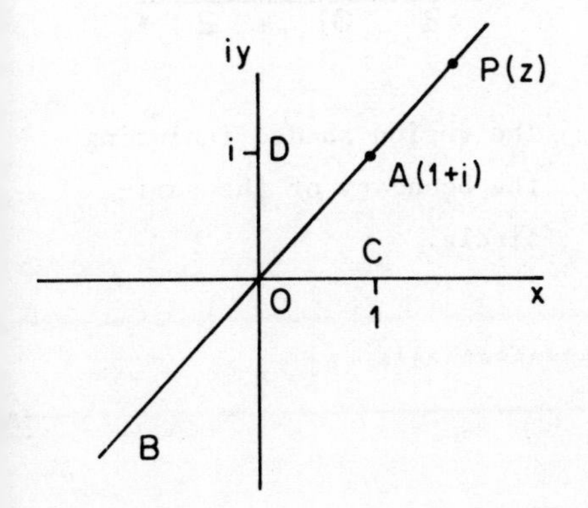

If P is any point on the line, then as the line passes through A and O, CP = DP. But $\overrightarrow{CP} = z - 1$ and $\overrightarrow{DP} = z - i$

$\therefore\ |z - 1| = |z - i|$

NOTE: Any pair of points which are equidistant from the line AB can be used instead of C and D.

FRAME 5 continue

(3)

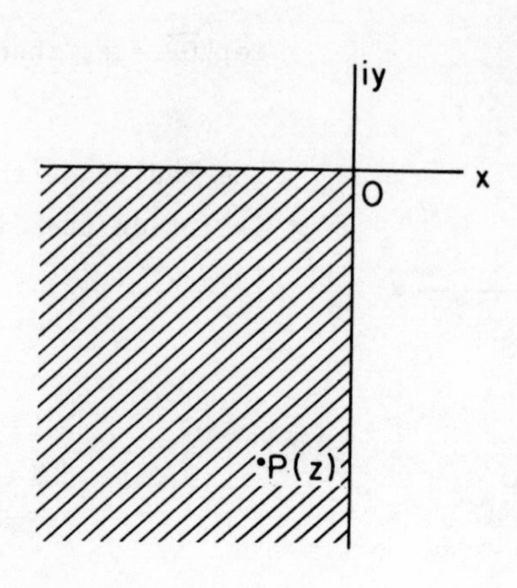

The region shown.

For any point P, the argument of $\overrightarrow{OP}$ lies between $-\pi$ and $-\pi/2$
$\therefore -\pi < \arg z < -\pi/2$.

Now give equations or inequalities to represent the following:

(i)

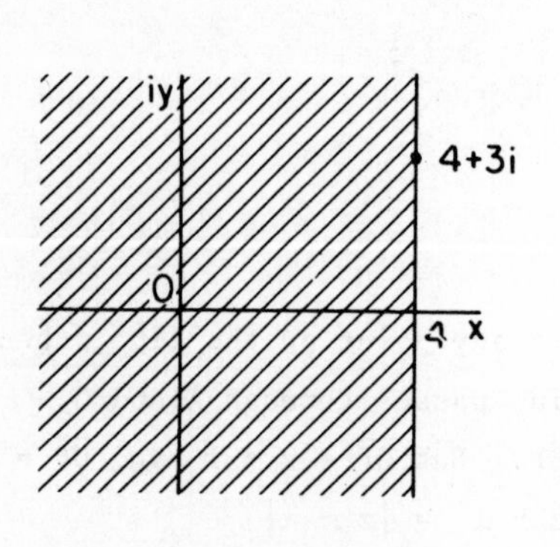

The region shaded

(ii)

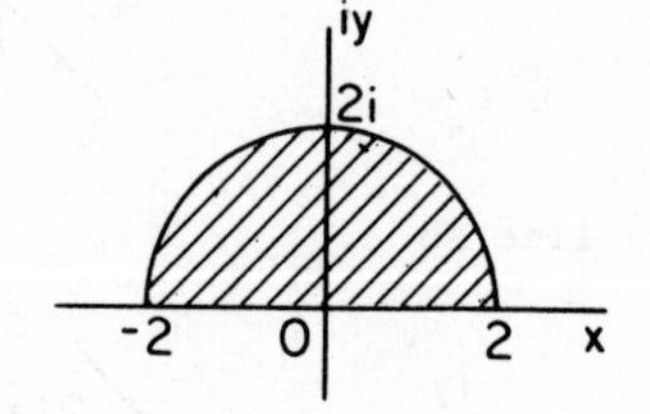

The region shaded including the boundary of the semi-circle.

****

5A

(i) $Re(z) < 4$

(ii) $|z| \leq 2, \quad 0 < \arg z < \pi$; *or* $|z| \leq 2, \ Im(z) > 0$.

FRAME 6

Functions of a Complex Variable

Let us start by considering the function z^2 where, of course, $z = x + iy$. First work out the real and imaginary parts of this.

****

6A

$$(x + iy)^2 = x^2 + 2xiy + i^2y^2$$
$$= (x^2 - y^2) + 2ixy$$
$$\therefore Re(z^2) = x^2 - y^2, \quad Im(z^2) = 2xy$$

FRAME 7

If you have the function sin x, say, in real variable work, then you often denote it by a single letter y, say, and so write $y = \sin x$. More generally, you would write $y = f(x)$ as a functional relationship between the independent variable x and the dependent variable y.

In a similar way, we now extend this idea to complex variables and write a function of z as f(z). It is then found useful to denote this function f(z) by a single letter. Obviously neither x nor y can be chosen for this single letter, as $z = x + iy$, and in practice it is conventional to choose w. This then gives us a functional relationship $w = f(z)$, analogous to what occurs in real variable work.

Thus, for the example in FRAME 6, we would write $w = z^2$.

FRAME 8

Returning now to 6A and using the results there, w can also be written as $(x^2 - y^2) + i2xy$ and so w consists of a real part, $x^2 - y^2$, and an imaginary part, 2xy. Once again, it is useful to use single letters to denote the real and imaginary parts of w. For these, it is conventional to write the real part

FRAME 8 continued

of w as u and the imaginary part as v. Thus, in general, $w = u + iv$, and in our particular example, $u = x^2 - y^2$ and $v = 2xy$. So, as is seen in this example, u and v are real functions of x and y.

Now find u and v if $w = z^3$.

****

8A

$$z^3 = (x + iy)^3$$
$$= x^3 + 3x^2iy + 3xi^2y^2 + i^3y^3$$
$$= (x^3 - 3xy^2) + i(3x^2y - y^3)$$

Thus $u = x^3 - 3xy^2$, $\quad v = 3x^2y - y^3$

FRAME 9

We shall now have a look at one or two other examples where w is given, u and v being required. First, suppose $w = \sin z$. Obviously, $\sin z = \sin(x + iy)$ and the question immediately arises - what can we do with $\sin(x + iy)$? At this stage it will be assumed that this can be expanded by an exactly similar formula to that used for $\sin(A + B)$ in real variable work.

Thus, $\sin(x + iy) = \sin x \cos iy + \cos x \sin iy$

$= \sin x \cosh y + i \cos x \sinh y$ (See FRAME 61, page 4:43 of Vol I if you have forgotten these results.)

Finally, this gives $u = \sin x \cosh y$, $\quad v = \cos x \sinh y$.

Now find u and v if (i) $w = \cosh z$, $\quad$ (ii) $w = \dfrac{z}{z - i}$

****

9A

(i) $\cosh z = \cosh(x + iy)$

$= \cosh x \cosh iy + \sinh x \sinh iy$

$= \cosh x \cos y + i \sinh x \sin y$

$\therefore \quad u = \cosh x \cos y, \qquad v = \sinh x \sin y$

(ii) $\dfrac{z}{z - i} = \dfrac{x + iy}{x + i(y - 1)} \cdot \dfrac{x - i(y - 1)}{x - i(y - 1)}$

$= \dfrac{x^2 + y(y - 1) + ix}{x^2 + (y - 1)^2}$

$\therefore \quad u = \dfrac{x^2 + y(y - 1)}{x^2 + (y - 1)^2}, \qquad v = \dfrac{x}{x^2 + (y - 1)^2}$

FRAME 10

In reverse, given u and v as functions of x and y, it is possible to express w in terms of z and $\bar{z}$. (In the two examples at the end of FRAME 9, w involves z only, but we cannot assume at the start that this is going to happen.)

For example, we might be given $u = x + y$ and $v = xy$, then $w = (x + y) + ixy$.

Now $z = x + iy$ and $\bar{z} = x - iy$. Solve these two equations for x and y in terms of z and $\bar{z}$.

****

10A

$x = \frac{1}{2}(z + \bar{z}), \qquad y = \frac{1}{2i}(z - \bar{z})$

FRAME 11

Substitution of these results gives

$u = \frac{1}{2}(z + \bar{z}) + \frac{1}{2i}(z - \bar{z}) = \frac{1}{2}z(1 - i) + \frac{1}{2}\bar{z}(1 + i)$

$v = \frac{1}{2}(z + \bar{z})\frac{1}{2i}(z - \bar{z}) = -\frac{1}{4}i(z^2 - \bar{z}^2)$

and $w = \frac{1}{2}z(1 - i) + \frac{1}{2}\bar{z}(1 + i) + \frac{1}{4}(z^2 - \bar{z}^2)$

FRAME 11 continued

Thus w has been expressed in terms of z and $\bar{z}$. This can obviously be done for any pair of functions u and v, and hence for any w.

Sometimes, however, it may be possible to short cut the algebra involved. For example, if

$$w = x^2 - y^2 + 2ixy,$$

then it may be spotted that

$$(x + iy)^2 = x^2 + 2ixy + i^2y^2$$
$$= x^2 + 2ixy - y^2$$

and so, immediately

$$w = z^2 \tag{11.1}$$

Now express the following functions w in terms of z and $\bar{z}$.

(i) $w = (x - iy)^2$ (ii) $w = \dfrac{x + iy + 3}{1 - (x + iy)^3}$

(iii) $w = \sqrt{x^2 + y^2}$ (iv) $w = x^2 + iy^2$

****

11A

(i) $w = \bar{z}^2$

(ii) $w = \dfrac{z + 3}{1 - z^3}$

(iii) $w = \sqrt{\frac{1}{4}(z + \bar{z})^2 - \frac{1}{4}(z - \bar{z})^2} = \sqrt{z\,\bar{z}}$ *(You might have spotted that this result follows immediately from* $x^2 + y^2 = z\bar{z}$*)*

(iv) $w = \frac{1}{4}(z + \bar{z})^2 - \frac{1}{4}\,i(z - \bar{z})^2$

FRAME 12

You will notice that w may turn out to be a function of both z and $\bar{z}$ or of either z or $\bar{z}$ alone.

In general, we write w = f(z), where, for the time being at least, it is implied that f(z) may contain $\bar{z}$ as well as z. w is then said to be a

FRAME 12 continued

FUNCTION OF THE COMPLEX VARIABLE z. Actually, as we shall see in the next programme, the only functions that are generally of any use are those which do not contain $\bar{z}$,

e.g. $w = z^2$, $w = \dfrac{z + 3}{1 - z^3}$ [see result (11.1) and example (ii) of FRAME 11].

Because of this, we shall actually, in this programme, restrict ourselves to such functions.

FRAME 13

Complex Mapping - General Ideas

In real variable work, it is possible, given y as a function of x, to draw a graph showing the relation between them. In complex variable theory it is not such a straightforward matter, because, given w as a function of z, there are, as you have seen, really four quantities involved, i.e. u, v, x and y. Of course, on any ordinary graph, only two variables appear and so this topic of COMPLEX MAPPING is much more complicated than ordinary curve sketching.

FRAME 14

To overcome this difficulty, we draw two graphs, i.e. two Argand diagrams, one showing what happens to z and the other what happens to w. In order to do this, however, we require more information than just the relation between w and z. It is necessary to know, as well, what variation in z is taking place. A curve showing this variation can then be plotted and a second curve is drawn showing the corresponding variation in w.

In this programme we shall consider the easier ideas on this topic of mapping. The more difficult work will be left until the last programme - Conformal Transformations - in this Unit.

FRAME 14 continued

It is rather difficult at this stage to give you much idea of the practical use of this work, so here we shall simply say that the shape of a w-curve may make the solution of a problem easier than the shape of the original z-curve. You will find some practical examples involving this work later in this programme and also in the programme on Conformal Transformations.

FRAME 15

As a first example to illustrate the ideas involved, the following problem will be considered:

If $w = 3iz + 1$ and z moves along the straight line from $-1 - i$ to $2 + i$, find what happens to w and illustrate.

First of all, we draw the Argand diagram showing the path followed by z.

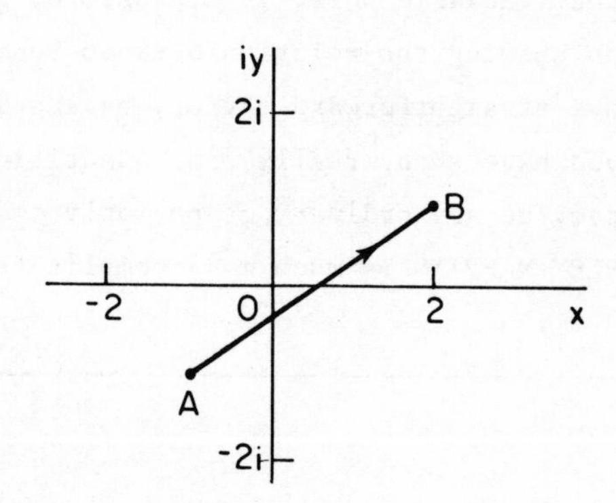

Such a figure as this is usually referred to as the z-plane.

The next thing is to find the positions of w corresponding to the end points A and B of z. These will give us the end points of the path of w. Obtain these by substituting $z = -1 - i$ and $z = 2 + i$ in turn into $w = 3iz + 1$.

****

15A

When $z = -1 - i$, $w = 4 - 3i$

When $z = 2 + i$, $w = -2 + 6i$

FRAME 16

The two points found in 15A are called the CORRESPONDING POINTS to A and B. We shall denote them by A' and B' and so can now begin our sketch of the w-plane.

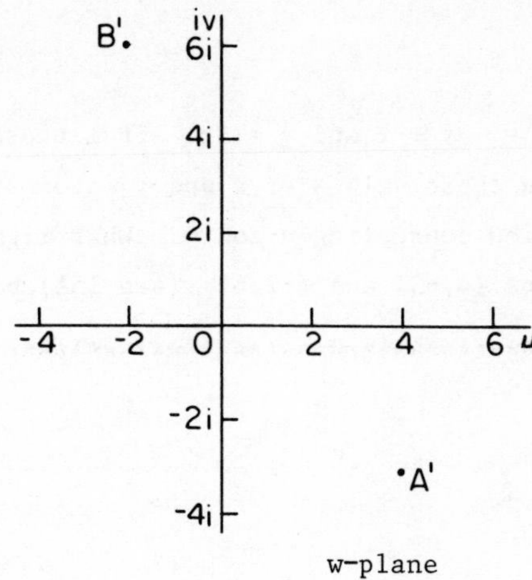

w-plane

The next thing to find is the path of w between A' and B' which we must not assume is a straight line. To do this it is necessary to find u and v in terms of x and y.

Remembering that $w = u + iv$ and $z = x + iy$, find u and v given that (see FRAME 15) $w = 3iz + 1$.

****

16A

$$u + iv = 3i(x + iy) + 1 = 3ix - 3y + 1$$

$$\therefore u = -3y + 1, \qquad v = 3x$$

FRAME 17

At this stage we treat the problem as one in coordinate geometry, and write down the equation of the line joining (-1,-1) to (2,1), i.e. the line AB in the diagram in FRAME 15. (These points in coordinate geometry are, of course, those corresponding to $-1 - i$ and $2 + i$ in complex numbers.)

Write down the equation of the line mentioned above.

****

17A

$3y = 2x - 1$

FRAME 18

In 16A you found that $u = -3y + 1$ and $v = 3x$. From these, $y = \frac{1}{3}(1 - u)$ and $x = \frac{1}{3}v$. Now substitute these values of x and y into $3y = 2x - 1$ and find the corresponding equation connecting u and v. What figure does this equation represent? Do the points (4,-3) and (-2,6) (see 15A) both satisfy it?

****

18A

$1 - u = \frac{2}{3}v - 1$ *or* $3u + 2v = 6$

A straight line

Yes

FRAME 19

So, returning now to complex quantities this means that w actually does follow the straight line joining A' and B'. Our sketch of the w-plane can now be completed.

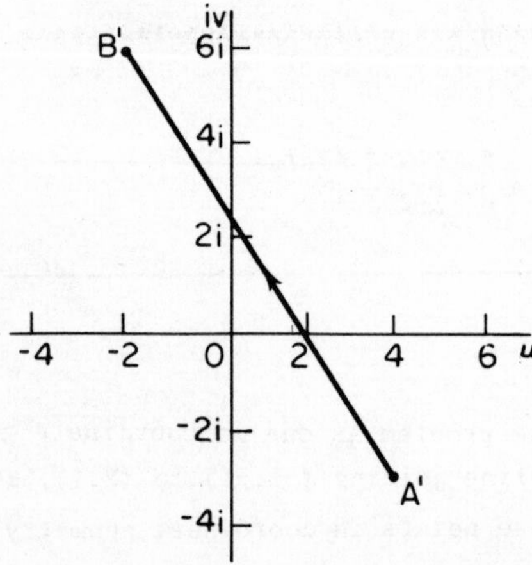

w-plane

FRAME 20

In the example just considered, a straight line in the w-plane did correspond to the given straight line in the z-plane. The next example will show that this is not always so. But whatever their shapes, the two curves (AB and A'B' in the example just considered) are called CORRESPONDING CURVES. It is often useful to insert, on the corresponding curves, arrows to indicate corresponding directions of travel. (If you look back, you will see that this has been done.)

FRAME 21

As a second example, we shall find the w-curve corresponding to the straight line in FRAME 15, but this time for a different functional relation, namely $w = \frac{1}{z}$.

Start by finding the values of w corresponding to $z = -1 - i$ and $z = 2 + i$ in this case.

****

21A

$w = -\frac{1}{2} + \frac{1}{2}i$ and $w = \frac{2}{5} - \frac{1}{5}i$ respectively.

FRAME 22

In 16A you found u and v in terms of x and y when $w = 3iz + 1$. But in order to find the equation of the path of w, these two equations were twisted round in FRAME 18 to give x and y in terms of u and v. These values could then immediately be inserted into the equation $3y = 2x - 1$ of 17A. An alternative way of obtaining the values of x and y in terms of u and v would have been to have solved $w = 3iz + 1$ for z in terms of w and then separated out the real and imaginary parts. This may not always be possible, but if it is, then it is better to proceed in this way. In our present case $w = \frac{1}{z}$ and so $z = \frac{1}{w}$.

Now take this last equation and obtain x and y in terms of u and v.

****

22A

$$x + iy = \frac{1}{u + iv}$$

$$= \frac{u - iv}{u^2 + v^2}$$

$$\therefore \quad x = \frac{u}{u^2 + v^2}, \qquad y = -\frac{v}{u^2 + v^2}$$

FRAME 23

Now substitute these values of x and y into the equation $3y = 2x - 1$ and simplify the result. What curve does your result represent? Does it pass through $\left(-\frac{1}{2}, \frac{1}{2}\right)$ and $\left(\frac{2}{5}, -\frac{1}{5}\right)$ (see 21A)?

****

23A

$$u^2 + v^2 - 2u - 3v = 0 \quad \text{or} \quad (u - 1)^2 + \left(v - \frac{3}{2}\right)^2 = \frac{13}{4}$$

A circle, centre $\left(1, \frac{3}{2}\right)$ *and radius* $\frac{\sqrt{13}}{2}$.

Yes.

FRAME 24

So in this case the path of w corresponding to the straight line path of z is a circle, or rather, it is a part of a circle as we only have to go along it from $-\frac{1}{2} + \frac{1}{2}i$ to $\frac{2}{5} - \frac{1}{5}i$.

Can you spot what new problem is involved here?

****

24A

The problem is: Do we proceed clockwise or anticlockwise round the circle from $-\frac{1}{2} + \frac{1}{2}i$ *to* $\frac{2}{5} - \frac{1}{5}i$*?*

FRAME 25

To find the answer to this problem, we take any other point on the path followed by z and find the corresponding value of w. A simple value for z is $z = \frac{1}{2}$ (this occurs when $x = \frac{1}{2}$, $y = 0$ in the equation $3y = 2x - 1$ of 17A).

What value of w corresponds to this?

****

25A

$w = 2$. The corresponding points $z = \frac{1}{2}$ and $w = 2$ are shown as C and C' respectively in the diagrams for this problem which appear in the next frame.

FRAME 26

So w moves along the arc of the circle that passes through 2. We can now sketch the diagrams in the z- and w-planes.

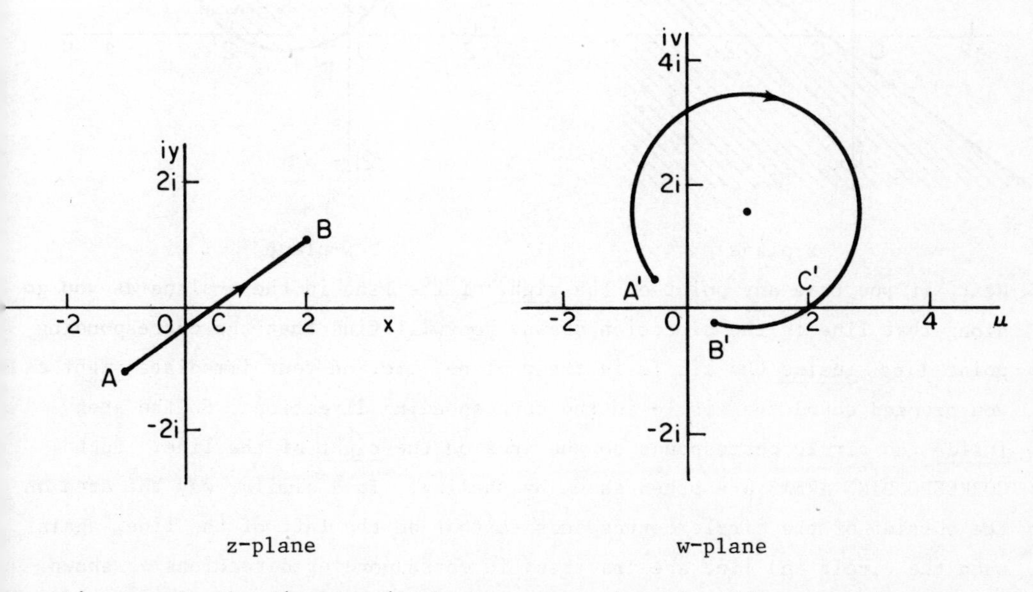

In this example the diagrams in the two planes have been drawn to the same scale, but this is not, of course, necessary.

FRAME 27

Now suppose z is allowed to move along the complete straight line of which AB is part. What do you think will be the corresponding path for w?

****

27A

The whole of the circle $u^2 + v^2 - 2u - 3v = 0$.

FRAME 28

The diagrams in the z- and w-planes are now as shown below.

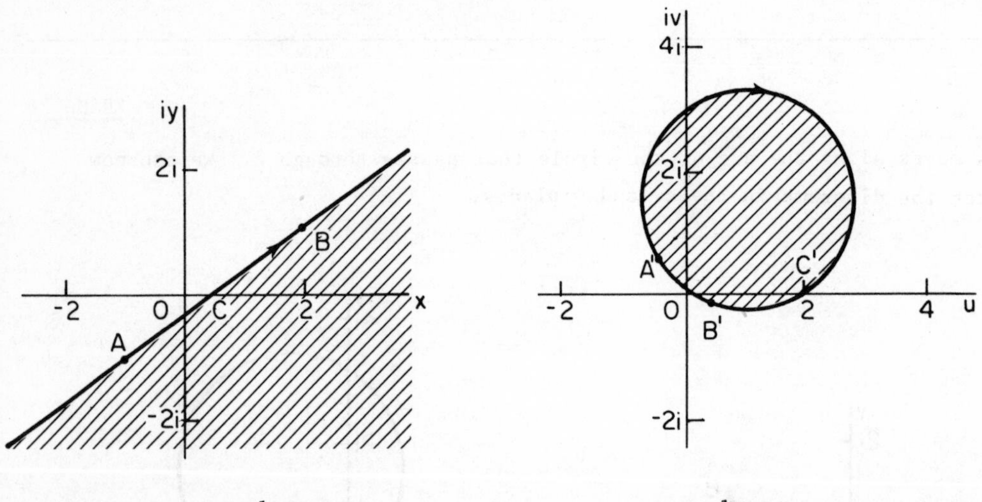

z-plane w-plane

Next, if you take any point on the right of the line in the z-plane as you go along that line in the direction shown, you will find that the corresponding point lies inside the circle in the w-plane, i.e. on your immediate right as you proceed round the circle in the corresponding direction. So the area inside the circle corresponds to the area on the right of the line. Such CORRESPONDING AREAS are often shown by shading. In a similar way the area on the outside of the circle corresponds to that on the left of the line, again when the circle and line are traversed in corresponding directions as shown. You might like to satisfy yourself that the remarks in this frame are reasonable by taking a few widely scattered points in the z-plane and finding the corresponding points in the w-plane.

FRAME 29

Now take the relation $w = \frac{1}{z - 1}$ and find the points in the w-plane corresponding to $z = 2$, $z = -2$, $z = 2i$ and $z = -2i$. Then take the curve $|z| = 2$ and show that the corresponding curve in the w-plane is $3(u^2 + v^2) = 2u + 1$. Illustrate the corresponding points and curves by means of sketches. Finally, by considering corresponding directions of travel, or otherwise, find the region in the w-plane corresponding to $|z| < 2$ in the z-plane.

****

29A

When $z = 2, \quad w = 1$

When $z = -2, \quad w = -\frac{1}{3}$

When $z = 2i, \quad w = \frac{1}{2i - 1} = -\frac{1 + 2i}{5}$

When $z = -2i, \quad w = -\frac{1}{2i + 1} = \frac{-1 + 2i}{5}$

Now $w = \frac{1}{z - 1}$ *gives* $z = \frac{1}{w} + 1$

$$\therefore \; x + iy = \frac{1}{u + iv} + 1$$

$$= \frac{u - iv}{u^2 + v^2} + 1$$

$$\therefore \; x = \frac{u}{u^2 + v^2} + 1 \qquad y = -\frac{v}{u^2 + v^2}$$

$|z| = 2$ *is the equation of the circle whose centre is at the origin and whose radius is 2. As a curve in coordinate geometry, its equation is* $x^2 + y^2 = 4$.

Substitution gives $\left(\frac{u}{u^2 + v^2} + 1\right)^2 + \left(\frac{-v}{u^2 + v^2}\right)^2 = 4$

$$\therefore \quad (u + u^2 + v^2)^2 + v^2 = 4(u^2 + v^2)^2$$

$$\therefore \; u^2 + 2u(u^2 + v^2) + (u^2 + v^2)^2 + v^2 = 4(u^2 + v^2)^2$$

$$\therefore \quad (u^2 + v^2)(1 + 2u + u^2 + v^2) = 4(u^2 + v^2)^2$$

$$\therefore \quad 1 + 2u + u^2 + v^2 = 4(u^2 + v^2)$$

$$\therefore \quad 1 + 2u = 3(u^2 + v^2)$$

29A continued

This reduces further to $u^2 + v^2 - \frac{2}{3}u - \frac{1}{3} = 0,$

i.e. $\left(u - \frac{1}{3}\right)^2 + v^2 = \frac{4}{9}$

which is a circle, centre $\left(\frac{1}{3}, 0\right)$ *and radius* $\frac{2}{3}$.

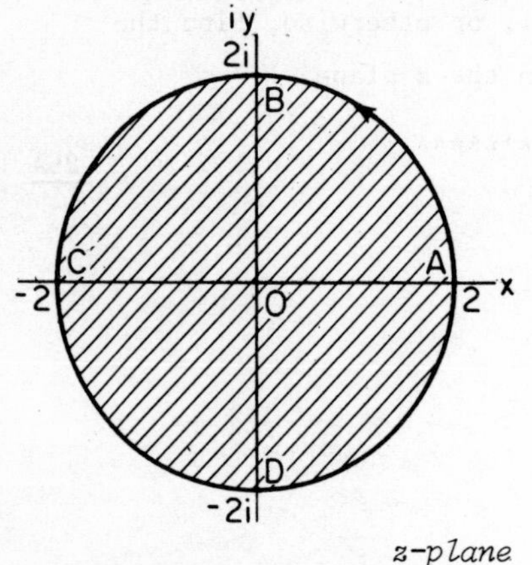

z-plane

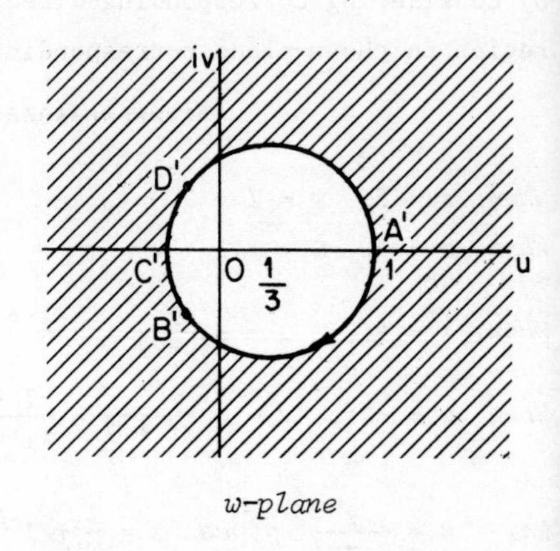

w-plane

The region outside the circle in the w-plane corresponds to that inside the circle in the z-plane.

FRAME 30

The next problem we shall have a look at is this:

If $w = z^2$, find the path traced out by w as z moves round the triangle ABC where A is the origin, B is the point 1 and C the point i.

Start by finding the points in the w-plane corresponding to A, B and C in the z-plane.

****

30A

When $z = 0, \quad w = 0$

When $z = 1, \quad w = 1$

When $z = i, \quad w = -1$

FRAME 31

e are now in a position to draw the Argand diagram for the z-plane and to tart that for the w-plane.

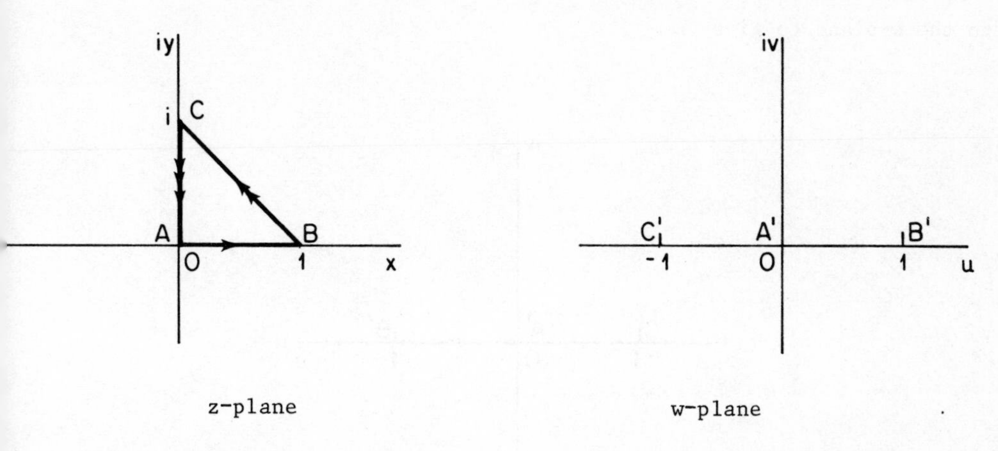

z-plane w-plane

hen you are dealing with problems in this work, it is a good idea to start our sketches early. The one for the z-plane you can complete immediately nd you can add to the other each piece of information as you obtain it.

ow, in the previous examples, it was possible to solve the transformation quation ($w = z^2$ here) for z in terms of w and this simplified the ensuing lgebra. If we do this here, however, we get $z = \pm\sqrt{w}$ and we have to separate ut the real and imaginary parts of a square root. This is more complicated han separating out the real and imaginary parts of a square. So this time $= z^2$ is left as it is and u and v found in terms of x and y. From FRAME 8, $= x^2 - y^2$, $v = 2xy$.

FRAME 32

ach side of the triangle ABC must now be considered in turn, as each side has different equation. Starting with AB, the equation of this line is $y = 0$ so hat $u = x^2$, $v = 0$. Now $v = 0$ represents the real axis in the w-plane just s did $y = 0$ in the z-plane. However, AB is only a section of this axis and o only a section of the u-axis in the w-plane will be involved. Although his might now appear to be obvious, it is as well to check it. For AB,

FRAME 32 continue

$0 \leqslant x \leqslant 1$ and so, as $u = x^2$, $0 \leqslant u \leqslant 1$. Thus, the straight line A'B' in the w-plane corresponds to AB in the z-plane. This information can now be added to the w-plane to give

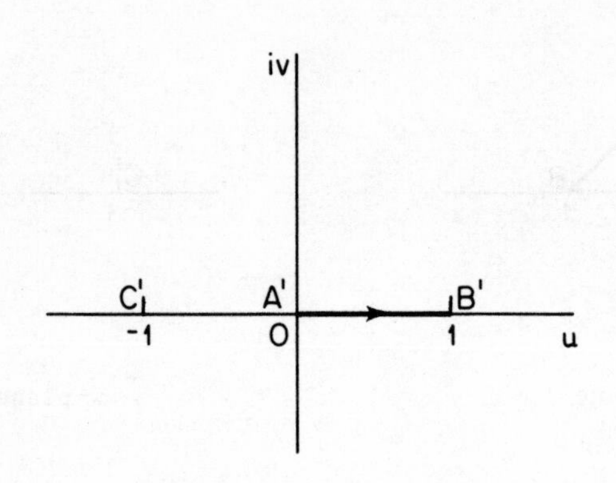

w-plane

Now, for the coordinate geometry situation corresponding to the z-plane, what will be the equation of BC?

****

32

$x + y = 1$ or $y = 1 - x$

FRAME 3

So that, for the section B'C',

$$u = x^2 - (1 - x)^2 \qquad v = 2x(1 - x)$$
$$= 2x - 1$$

Next, between these two equations, eliminate x, so obtaining a relation between u and v. What curve is represented by this equation?

****

33A

$= \frac{1}{2}(u + 1)$

$= (u + 1)\{1 - \frac{1}{2}(u + 1)\}$

$= \frac{1}{2}(1 - u^2)$

r $\quad u^2 = -2(v - \frac{1}{2})$

'his is a parabola in the form of an inverted cup with vertex at (0, ½).

FRAME 34

'he section B'C' corresponding to BC is thus part of this parabola. Check :hat it passes through (1,0) and (-1,0) and add this new information to your -plane diagram.

****

34A

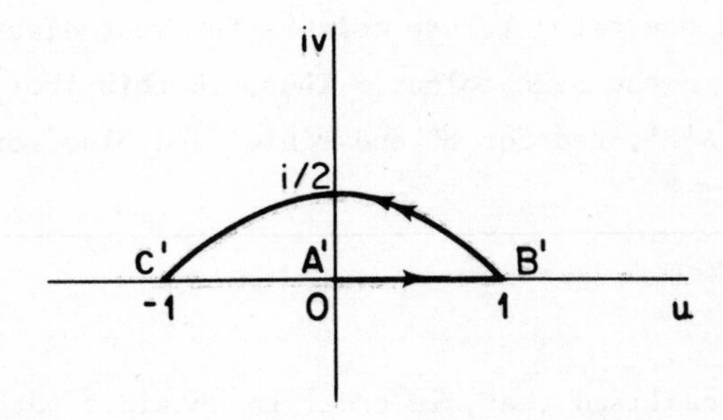

FRAME 35

:an you now decide what happens to w as z moves from C back to A? Having done 30, complete your w-plane diagram.

****

35

Along CA, $x = 0$,

$\therefore \quad u = -y^2 \quad v = 0.$

Also y *decreases from 1 to 0, so* u *increases from -1 to 0.*

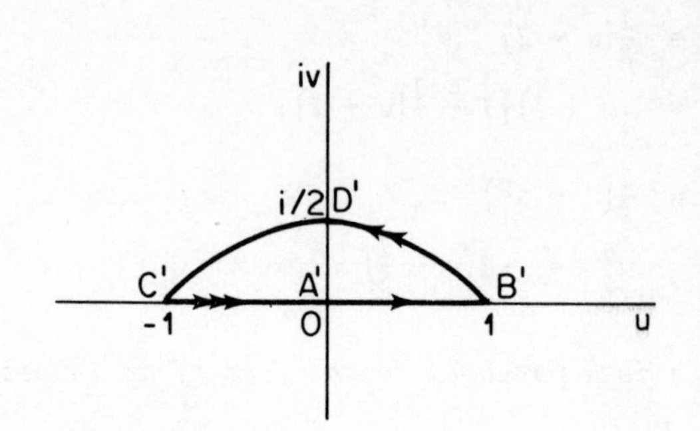

The arrows correspond to those in the diagram of the z-plane in FRAME 31.

FRAME 3(

The example just considered has illustrated how to proceed

(i) when it is inconvenient to solve for z in terms of w,

(ii) when the path followed by z is a series of sections of different curves.

You will find it helpful when more than one path is to be considered (or sections of more than one path) to use colours for your diagrams, drawing corresponding curves in the same colour. Thus, in this last example, you mig use green for AB and A'B', red for BC and B'D'C' and blue for CA and C'A'.

FRAME 37

By now you will have realised that, in order to obtain a path in the w-plane, two pieces of information are necessary, i.e. the equation of the path and the positions of the end points. You will also have probably realised that it is usually easier to find particular points on a path than to get its actual equation.

Sometimes it is possible to determine the shape of the curve in the w-plane without actually finding its equation. If this can be done, then a knowledge of some corresponding points will be sufficient to give us the complete picture. For example, if it can be determined at the start that the w-curve

FRAME 37 continued

is a circle, then it is only necessary to find three points on it for the circle to be determined completely.

We shall now have a look at the effects of certain transformations with this idea in mind.

FRAME 38

Translation, Magnification and Rotation

The very first example we took (see FRAME 15) involved the transformation equation $w = 3iz + 1$. This is an example of the more general equation $w = az + b$, where a and b are constants which may be real, imaginary, or complex. In the particular case we had, $a = 3i$ and $b = 1$. However, in order to build up the ideas involved, we shall start with $a = 1$. Then $w = z + b$. To see the effect of this, let us find what happens to the minor arc of the circle $|z| = 2$ from $z = 2$ to $z = 2i$ under the transformation $w = z + (3 - i)$. The diagram in the z-plane is

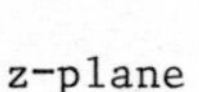

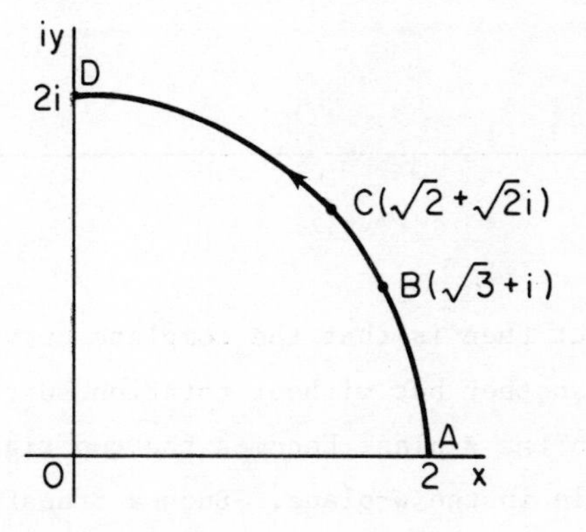

See if you can answer the following questions:

(i) What will be the complex numbers representing the w-points corresponding to A, B, C and D?

(ii) What will be the effect of the transformation on each of these points?

(iii) What will be the effect on any other point on the curve ABCD?

(iv) What will be the effect on the curve itself?

Having answered these questions, sketch the curve in the w-plane.

****

38A

(i) A' *will be* $5 - i$; B', $3 + \sqrt{3}$; C', $(3 + \sqrt{2}) + (\sqrt{2} - 1)i$; D', $3 + i$.

(ii) Each point is moved 3 units to the right and one unit down.

(iii) Any other point will be moved 3 units to the right and one unit down.

(iv) The complete curve will be moved 3 units to the right and one down.

The curve in the w-plane will be

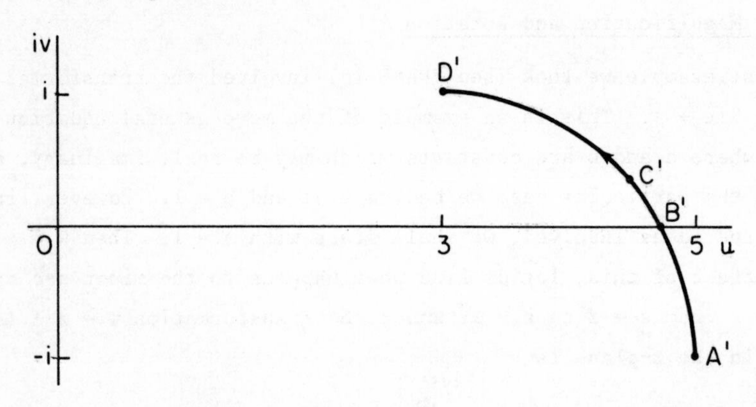

FRAME 39

The net effect then is that the complete curve is moved bodily from one position to another but without rotation so that the top right hand quarter of the circle in the z-plane becomes the top right hand quarter of the corresponding circle in the w-plane. Such a transformation is called a TRANSLATION and takes place whenever the transformation equation is of the form $w = z + b$.

FRAME 40

In the last example, $w = az + b$ was simplified to the form $w = z + b$ by putting $a = 1$. We shall now put $b = 0$ and concentrate on the effect of the value of a. In FRAME 38 it was stated that a could be real, imaginary or complex. The simplest case is when it is real and so first let $a = 4$.

FRAME 40 continued

The diagram shows a particular curve (a straight line in this case) in the z-plane.

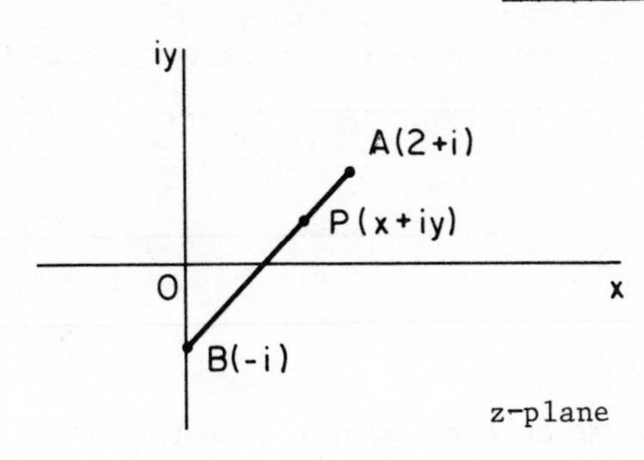

What will be the values of w for A', B' and P'? Also, what are $|\overrightarrow{OA}|$, $|\overrightarrow{OB}|$, $|\overrightarrow{OP}|$ and what will be $|\overrightarrow{OA}'|$, $|\overrightarrow{OB}'|$, $|\overrightarrow{OP}'|$?

****

40A

For A', $w = 8 + 4i$;   for B', $w = -4i$;   for P', $w = 4x + 4iy$.

$|\overrightarrow{OA}| = \sqrt{5}$, $|\overrightarrow{OB}| = 1$, $|\overrightarrow{OP}| = \sqrt{x^2 + y^2}$

$|\overrightarrow{OA}'| = 4\sqrt{5}$, $|\overrightarrow{OB}'| = 4$, $|\overrightarrow{OP}'| = 4\sqrt{x^2 + y^2} = 4|\overrightarrow{OP}|$

FRAME 41

As P is any point on AB and $|\overrightarrow{OP}'| = 4|\overrightarrow{OP}|$ it follows that the figure in the w-plane is a straight line.

The diagram for the w-plane is thus:

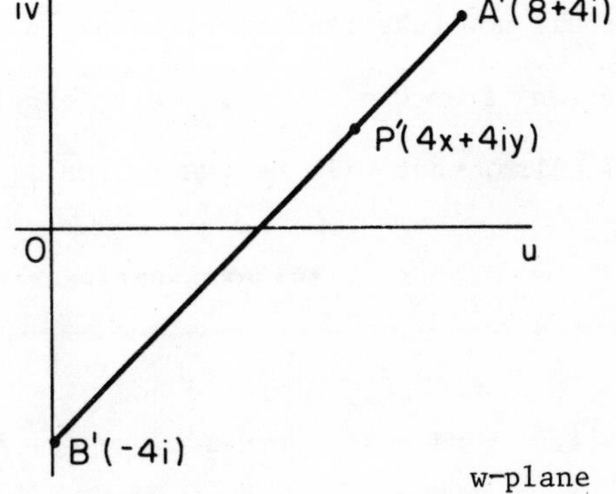

What has been the effect of the transformation $w = 4z$?

****

41A

Every distance on the w-diagram is 4 times the corresponding distance on the z-diagram, i.e. $|\overrightarrow{OA}'| = 4|\overrightarrow{OA}|$, $|\overrightarrow{A'B'}| = 4|\overrightarrow{AB}|$ *etc.*

FRAME 42

This, then, is an example of a transformation, the effect of which is a MAGNIFICATION. If $w = az$, and a is real, every length in the w-plane is a times the corresponding length in the z-plane. What would be the relative sizes of two corresponding areas in this case?

****

42A

That in the w-plane would be a^2 *times that in the z-plane.*

FRAME 43

We shall now take the same line as in FRAME 40 but change the transformation equation from $w = 4z$ to $w = 4iz$. What will be values of w for A', B' and P' now? Also, what will be $|\overrightarrow{OA}'|$, $|\overrightarrow{OB}'|$, $|\overrightarrow{OP}'|$?

****

43A

For A', $w = -4 + 8i$; *for B',* $w = 4$; *for P',* $w = -4y + 4ix$.

$|\overrightarrow{OA}'| = 4\sqrt{5}$, $|\overrightarrow{OB}'| = 4$, $|\overrightarrow{OP}'| = 4\sqrt{x^2 + y^2} = 4|\overrightarrow{OP}|$.

FRAME 44

This time the w-diagram is:

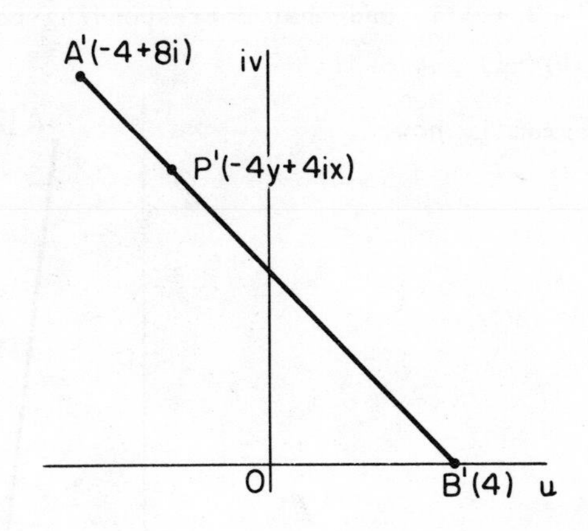

As before, $|\overrightarrow{OA}'| = 4|\overrightarrow{OA}|$, $|\overrightarrow{OB}'| = 4|\overrightarrow{OB}|$ and $|\overrightarrow{OP}'| = 4|\overrightarrow{OP}|$.

Can you say what additional effect has been caused by taking w = 4iz instead of w = 4z?

****

44A

The whole diagram has been rotated anticlockwise about the origin through 90°.

FRAME 45

This transformation therefore gives rise to a MAGNIFICATION and a ROTATION. You will notice that the magnification is 4, i.e. $|4i|$ and the rotation is through 90° anticlockwise, i.e. arg(4i).

If we now go one step further and take w = (3 + 4i)z, what do you think will be the effect of this?

****

45A

Magnification 5 (= $|3 + 4i|$) and rotation anticlockwise through $53^o8'$ [= arg(3 + 4i)].

FRAME 46

In this case the value of w corresponding to $z = 2 + i$ (point A in FRAME 40) is $w = (3 + 4i)(2 + i) = 2 + 11i$, and that corresponding to $z = -i$ (point B in FRAME 40) is $w = (3 + 4i)(-i) = 4 - 3i$.

The corresponding w-diagram is now:

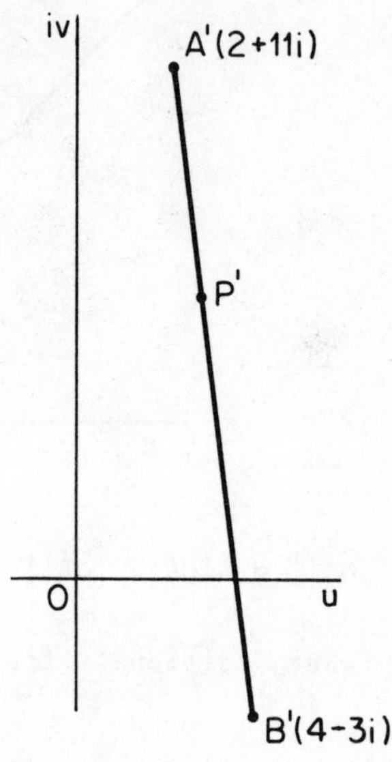

Finally, if we take the complete transformation $w = az + b$, the effect is a magnification $|a|$, a rotation through arg a and then a translation b.

If $|a| < 1$, the effect is actually to diminish the figure and if arg a is negative, the rotation takes place clockwise.

As all distances are magnified by the same amount, $|a|$, as all points are rotated about the origin through the same angle, arg a, and translated by the same amount, b, it follows that the figure in the w-plane is the same shape as that in the z-plane, as none of these operations has any effect shapewise.

FRAME 47

Inversion

The next important transformation is $w = \frac{1}{z}$, called INVERSION. If you turn back to FRAME 21 you will see that we have already looked at one example involving this, and in 22A you obtained $x = \frac{u}{u^2 + v^2}$, $y = -\frac{v}{u^2 + v^2}$. Here we shall

FRAME 47 continued

consider what happens to a circle in the z-plane when this transformation is applied to it. Reverting to our practice of treating a problem in this work by the methods of coordinate geometry, we can say that the equation of any circle in terms of x and y can be written as

$$A(x^2 + y^2) + Cx + Ey + F = 0 \qquad (47.1)$$

(The A is not really necessary here, but inserting it will help us later on.)

Substitute the expressions given above for x and y into this equation and simplify the result as much as possible. Then say what curve this new equation represents.

****

47A

$$A\left\{\frac{u^2}{(u^2+v^2)^2} + \frac{v^2}{(u^2+v^2)^2}\right\} + C\left(\frac{u}{u^2+v^2}\right) + E\left(\frac{-v}{u^2+v^2}\right) + F = 0$$

$$A\left(\frac{1}{u^2+v^2}\right) + C\left(\frac{u}{u^2+v^2}\right) - E\left(\frac{v}{u^2+v^2}\right) + F = 0$$

$$A + Cu - Ev + F(u^2+v^2) = 0 \qquad (47A.1)$$

A circle.

FRAME 48

Thus, a circle in the z-plane transforms into a circle in the w-plane. In practice, a circle is defined if any three points on it are known. For example, suppose we require to find the circle corresponding to $|z| = 2$. This circle passes through 2, 2i and -2. What values of w correspond to these?

****

48A

$\frac{1}{2}$, $-\frac{1}{2}i$ *and* $-\frac{1}{2}$ *respectively.*

FRAME 49

The corresponding curves in the two planes are thus as shown:

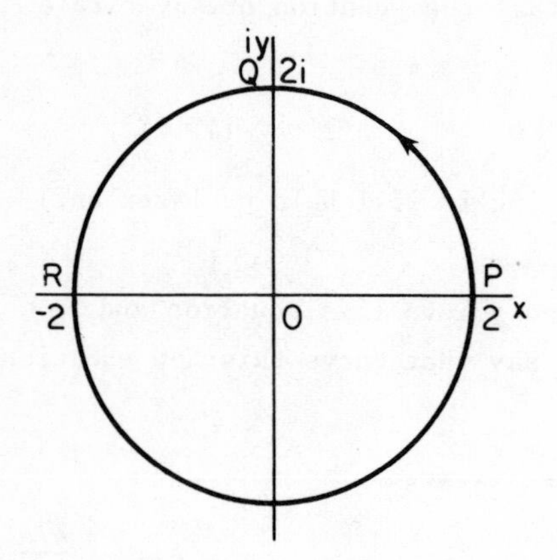

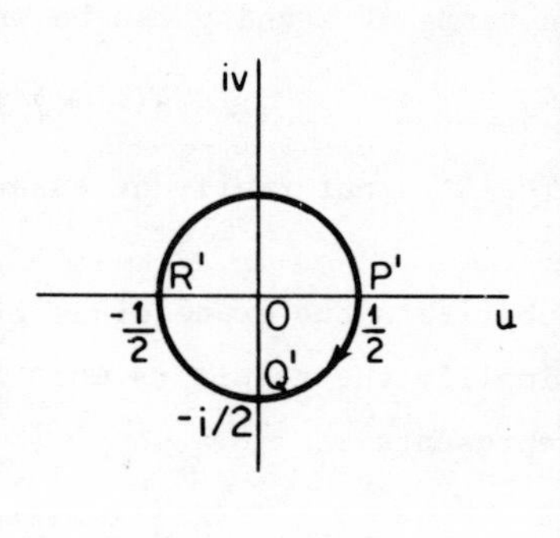

w-plane

z-plane

In this example it would have been just as easy to find the actual equation of the new circle, but later on you will find the three corresponding points idea very useful.

Now what do (47.1) and (47A.1) represent if A = 0?

****

49A

(47.1) represents a straight line.

(47A.1) represents a circle through the origin.

(Look back at result in FRAME 28, which shows an example of this.)

FRAME 50

So we can extend the results of FRAMES 47-48 to include the case of a straight line in the z-plane.

Now what do (47.1) and (47A.1) represent if (a) $F = 0$, (b) $A = 0$ and $F = 0$?

****

50A

(a) If $F = 0$, (47.1) represents a circle through the origin and (47A.1) represents a straight line.

(b) If A and F are both zero, (47.1) and (47A.1) both represent straight lines through the origin.

FRAME 51

So, under inversion, we can make the general statement that a circle in the z-plane becomes a circle in the w-plane where the term 'circle' is extended to include a straight line (i.e. a circle of infinite radius).

Now making use of this general statement so that it is only necessary to find a few corresponding points, find what curves in the w-plane correspond to:

(i) the circle $|z - i| = 1$,

(ii) the real axis,

(iii) the segment of the line in (ii) between $z = 2$ and $z = -2$.

Illustrate by sketches in each case.

****

51A

(i) Three points on $|z - i| = 1$ are 0, $2i$ and $1 + i$. (You may select any three points you like on the circle, but the idea is to choose them so that the algebra involved is as simple as possible.)

The corresponding w-values are ∞, $-\frac{1}{2}i$ and $\frac{1}{1 + i} = \frac{1 - i}{2}$. (A note about ∞ in complex variable work is given in the next frame.)

51A continued

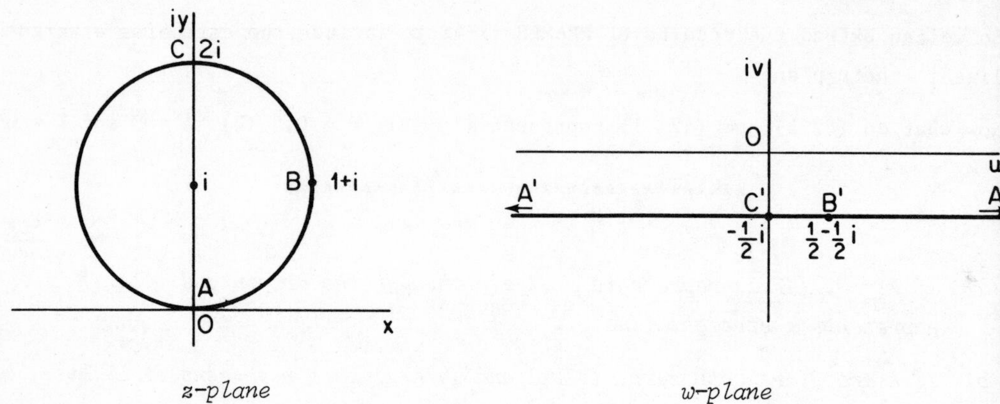

z-plane *w-plane*

(ii) *Three points on the real axis in the z-plane are 2, 0 and -2. For these,* $w = \frac{1}{2}$, ∞ *and* $-\frac{1}{2}$.

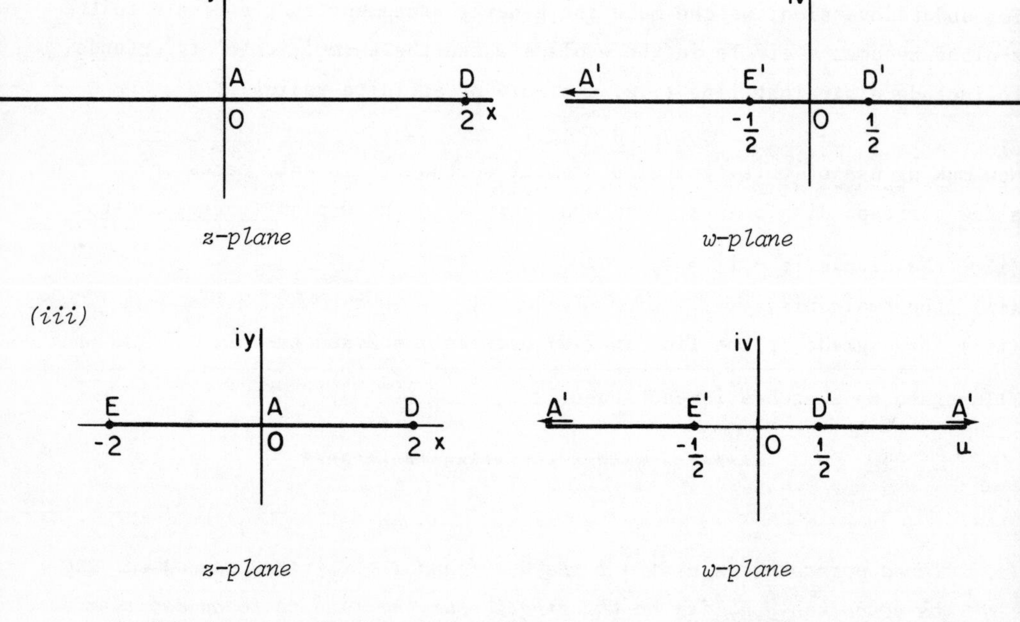

z-plane *w-plane*

If this last result is not immediately obvious, see what happens when you take a few more points between D and E, especially some close to and on both sides of A, and plot the corresponding points in the w-plane.

FRAME 52

A word is necessary here about the idea of infinity in complex variable work.

In real variable work, given an ordinary Ox scale, we think of infinity as being at an infinite distance from O along the line Ox.

In complex variable work we think of it as being anywhere on a circle, centre O, whose radius is infinite. In other words, $z = \infty$ implies $|z| = \infty$, but arg z is not specified.

In our present work, when you get a straight line in either the z- or w-plane and you find, in relation to that line, that $z = \infty$ or $w = \infty$, then this simply means going off an infinite distance along that line. Examples of this occur in FRAME 28 and 51A.

FRAME 53

Impedance and Admittance Loci

There are many practical situations where the mapping of complex quantities is useful. Some topics to which it can be applied are heat flow, fluid flow, electrostatics and certain problems involving impedance and admittance. Most of these require a wider knowledge of complex variable work than has so far been covered and consequently will be left until the last programme in this Unit. However it is possible to consider simple problems on impedance and admittance at this stage. These will be of particular interest to you if your course includes alternating current theory. However, if it doesn't, you should still be able to follow the mathematics as all the necessary electrical formulae will be given.

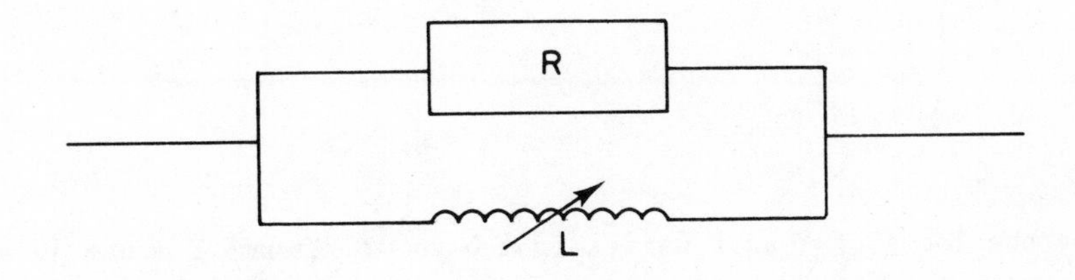

As an example, consider the R, L circuit shown in which it is possible to vary

FRAME 53 continued

L, R remaining fixed. The problem here is to find how the admittance and impedance of the complete circuit vary as L is varied.

For a parallel circuit, the total impedance Z is given by the formula $\frac{1}{Z} = \frac{1}{Z_1} + \frac{1}{Z_2}$ where Z_1 and Z_2 are the impedances of the two branches. Here $Z_1 = R$ and $Z_2 = j\omega L$. (Remember that j is usually used instead of i for $\sqrt{-1}$ in electrical problems.) The admittance Y is given by $\frac{1}{Z}$.

What is Y for this circuit?

****

53A

$$Y = \frac{1}{R} + \frac{1}{j\omega L} = \frac{1}{R} - \frac{j}{\omega L}$$

FRAME 54

It is now possible to draw two diagrams, the L-plane and the Y-plane. In practice, of course, L can only be real and positive, but mathematically it is convenient to think of it as a complex number. If we put $L = x + jy$, then $y = 0$ and $x \geqslant 0$, and the L-plane diagram is

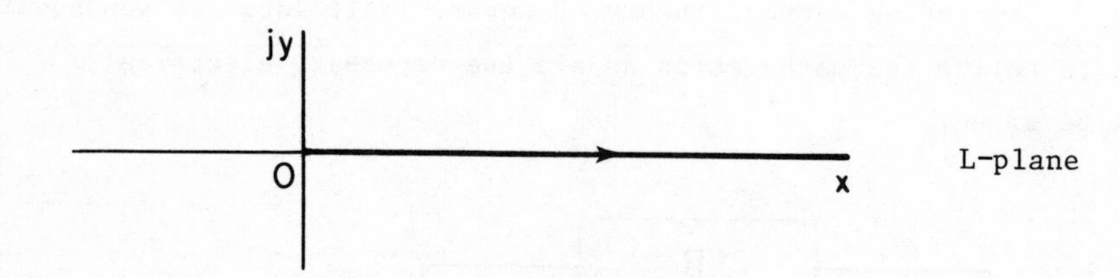

What will be the locus of Y as L varies from 0 to ∞? Assume $Y = u + jv$ and illustrate by a sketch.

****

54A

The straight line shown.

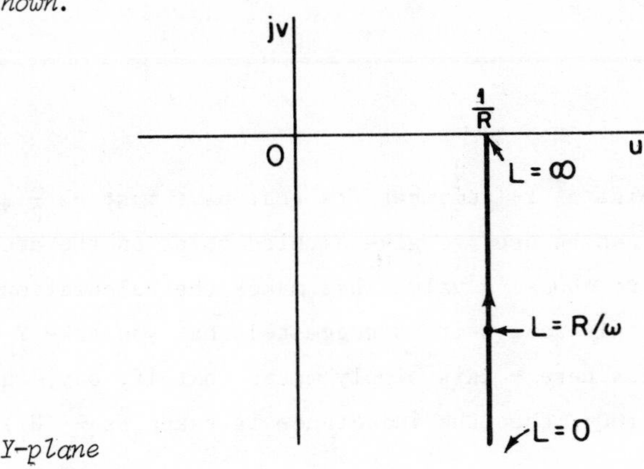

Here $u = \frac{1}{R}$ *and* v *varies from* $-\infty$ *to* 0 *as* L *varies from* 0 *to* ∞.

FRAME 55

Now $Z = \frac{1}{Y}$ and so to obtain the locus of Z (= U + jV, say) it is necessary to apply the process of inversion to the line in the diagram in 54A.

As the locus of Y is part of a straight line, what will be the form of the curve for Z?

****

55A

An arc of a circle.

FRAME 56

As it is part of a circle, it is only necessary to find three points on it to define the circle completely. As we only expect it to be <u>part</u> of the circle, it will be advisable for two of the points to be at the two ends of the particular arc.

What values of Z will correspond to $Y = \frac{1}{R}$ and $Y = \infty$?

****

56A

R and 0 respectively.

FRAME 57

Any suitable value of Y (remember its real part must be $\frac{1}{R}$ and its imaginary part negative) can be used to give a third point on the arc, but it is obviously best to choose a value that makes the calculation as easy as possible. For this reason it is suggested that you take $Y = \frac{1}{R} - \frac{j}{R}$. (Don't worry about units here - this simply means that if, say, the resistance is 1000 Ω and ω = 100π, then the inductance is taken as $\frac{10}{\pi}$ H.)

Now find the corresponding value of Z and sketch the locus.

****

57A

$Z = R(1 + j)/2$

In this particular case it is obvious, from the three points obtained, that the locus is a semicircle, centre ½R + 0j and radius ½R. If you have an example where the result is not obvious, you can proceed alternatively by the method used in 29A.

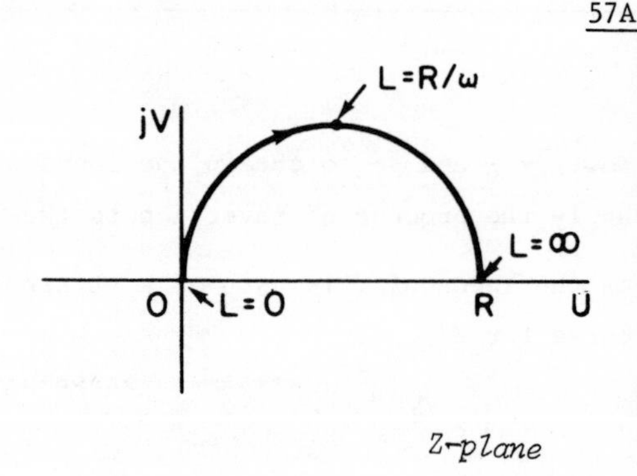

Z-plane

FRAME 58

The second example you should be able to do completely on your own.

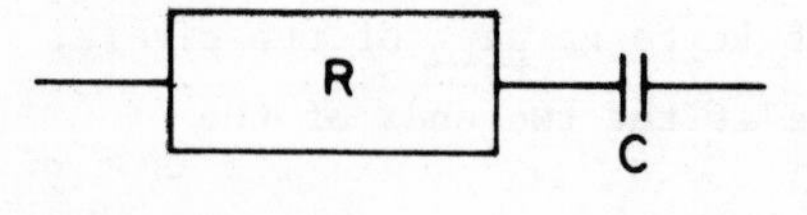

Illustrate the impedance and admittance loci for a circuit consisting of a resistance R and capacitance C in series, R being variable from 0 to ∞, and C being fixed. (The impedance of a capacitance C is $\frac{1}{j\omega C}$ and the impedances for components in series are additive.)

****

58A

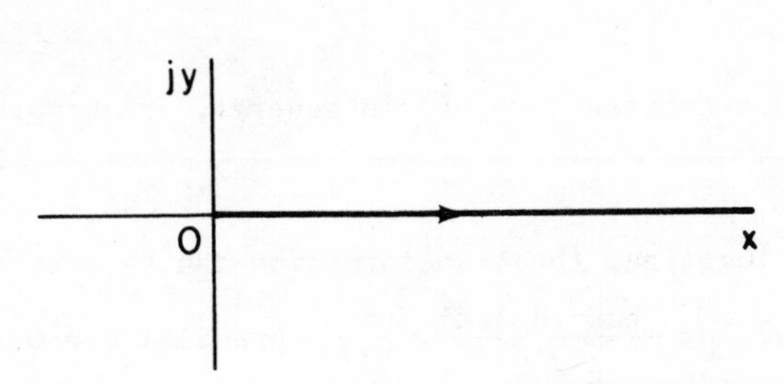

R-plane $(R = x + jy)$

$$Z = R + \frac{1}{j\omega C} = R - \frac{j}{\omega C}$$

$Y = \frac{1}{Z}$. When $Z = -\frac{j}{\omega C}$, $\quad Y = j\omega C$

When $Z = \infty$, $\quad Y = 0$

When $Z = \frac{1}{\omega C} - \frac{j}{\omega C}$, $\quad Y = \omega C\,\frac{1 + j}{2}$

jV
O
U
$-\frac{j}{\omega C}$

Z-plane

jv
$j\omega C$
O
u

Y-plane

FRAME 59

The Bilinear Transformation

In the example you have just considered, had you only been asked about Y, it would still have been necessary to have considered Z first, as $Y = \frac{\omega C}{\omega CR - j}$ and the effect of such a transformation equation has not yet been looked at. For a similar reason, in the first example (FRAME 53) it would have been necessary to have found the locus of Y first even if you had only been asked for that of Z. The BILINEAR TRANSFORMATION enables us to omit this intermediate stage. It is the name given to the transformation

$$w = \frac{az + b}{cz + d} \qquad (59.1)$$

FRAME 59 continued

where a, b, c and d are constants which, in general, are complex and are such that $bc \neq da$.

By a little algebraic juggling, the transformation can be written as

$$w = \frac{(bc - da)/c^2}{z + \frac{d}{c}} + \frac{a}{c}, \quad \text{provided } c \neq 0.$$

{Verify that this form does simplify to (59.1)}

We now proceed from z to w by a succession of steps as follows:

(i) $z_1 = z + \frac{d}{c}$ so that $w = \frac{(bc - da)/c^2}{z_1} + \frac{a}{c}$

(ii) $z_2 = \frac{1}{z_1}$ so that $w = \frac{bc - da}{c^2} z_2 + \frac{a}{c}$

(iii) $z_3 = \frac{bc - da}{c^2} z_2$ so that $w = z_3 + \frac{a}{c}$

(iv) $w = z_3 + \frac{a}{c}$

As the only figure to which inversion has been applied is a circle (including a straight line as the limiting case of a circle), the following remarks will be limited to this shape.

What is transformation (i) and what happens to a circle under it?

****

59A

(i) is a <u>translation</u> and a circle remains a circle (the size of the circle remains the same, but that is not relevant here).

FRAME 60

(ii) is an <u>inversion</u> and once again a circle remains a circle. As the example in FRAMES 48-9 shows, the size of the circle is generally altered.

Now what shape does this circle become when (iii) and then (iv) are applied?

****

60A

(iii) represents a magnification and a rotation. If a circle is magnified and rotated it remains a circle.

(iv) is another translation and does not alter any shape.

FRAME 61

The net effect of all this, then, is that, under the bilinear transformation, a circle remains a circle and so, in order to find the transformed circle, it is only necessary to find three points on it. (Don't forget that a straight line is included in all this as well.)

FRAME 62

As an example of this result we shall find what happens to the circle $|z| = 1$ and the two axes if $w = \dfrac{z}{z - 1}$.

Start by sketching the three curves in the z-plane, putting them all on the same diagram. (Here, different colours will be useful.)

****

62A

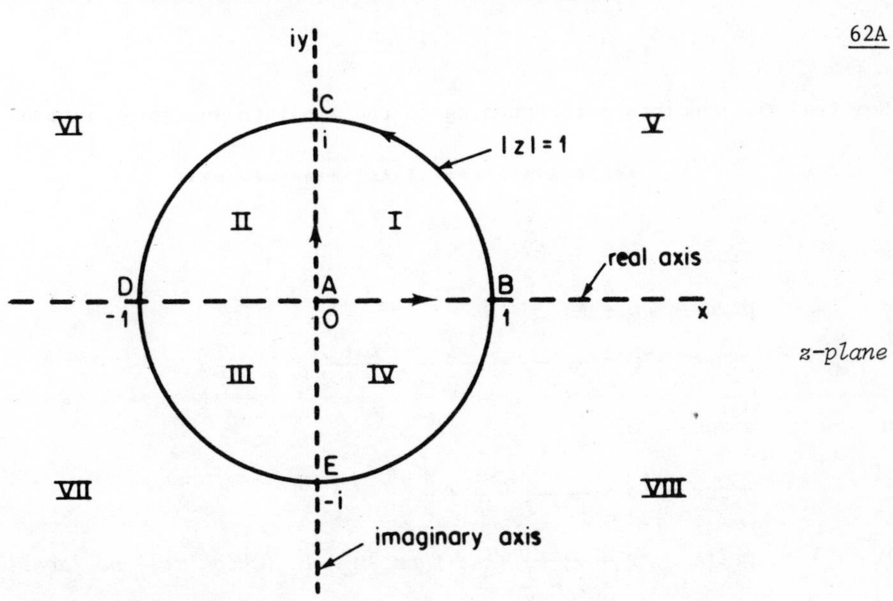

(For the time being, ignore the Roman figures I-VIII.)

62A continued

This diagram, and the one in 65A, will be referred to several times. You might like to keep your own diagrams handy so that you do not have to keep turning back.

FRAME 63

Points in the z-plane are now taken and the corresponding points in the w-plane found. Which points in the z-plane do you suggest should be chosen for this purpose?

****

63A

There are various possibilities, but the best are $z = 0$, $z = 1$, $z = i$, $z = -1$, $z = -i$. *In each case two of the curves pass through the point and so by taking these we shall get information about two of the transformed curves each time. It will also be as well to include* $z = \infty$, *as two of the curves involved (the two axes) are straight lines and so go off to* ∞.

FRAME 64

Now find the w-points corresponding to the z-points suggested in 63A.

****

64A

$z = 0$ *gives* $w = 0$

$z = 1$ *gives* $w = \infty$

$z = i$ *gives* $w = \dfrac{i}{i-1} = \dfrac{i}{i-1} \cdot \dfrac{i+1}{i+1} = \dfrac{1}{2} - \dfrac{1}{2}i$

$z = -1$ *gives* $w = \dfrac{1}{2}$

$z = -i$ *gives* $w = \dfrac{i}{i+1} = \dfrac{1}{2} + \dfrac{1}{2}i$

For $z = \infty$, *write* $w = \dfrac{1}{1 - \dfrac{1}{z}}$ *and then* $w = 1$ *(which will be labelled F').*

FRAME 65

The corresponding points in the w-plane can now be plotted.

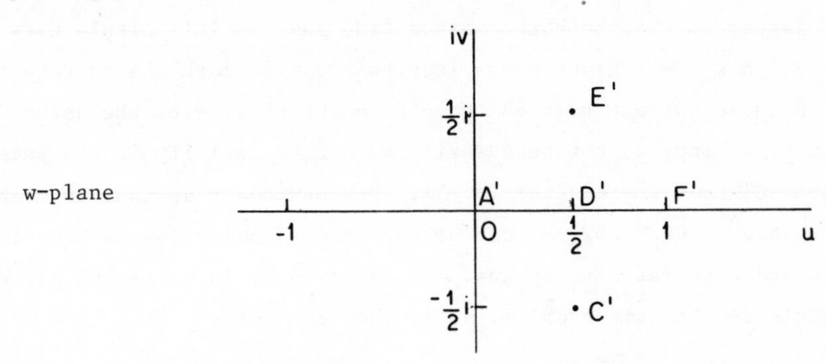

What will now be the curves in the w-plane corresponding to
(i) $|z| = 1$, (ii) $y = 0$, (iii) $x = 0$? Illustrate by adding them to your sketch of the w-plane.

****

65A

(i) *The line* $u = \frac{1}{2}$ *(ii)* *The line* $v = 0$

(iii) The circle $|w - \frac{1}{2}| = \frac{1}{2}$, *i.e.* $A'C'F'E'A'$

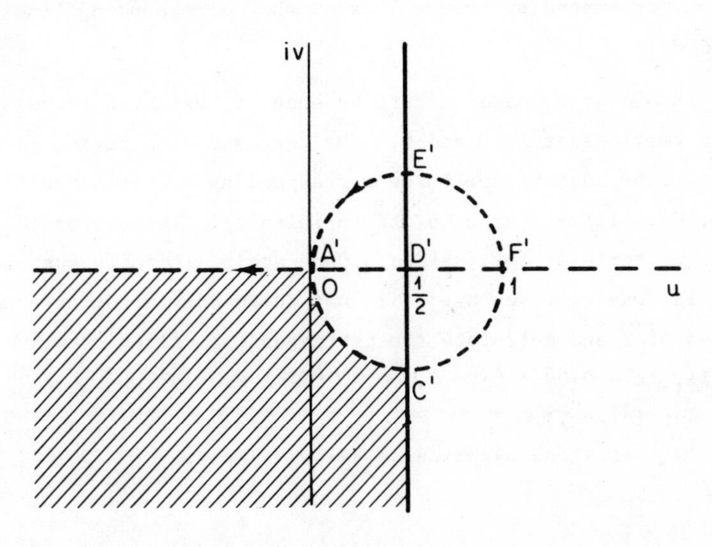

(Ignore the shading for the time being.)

FRAME 66

There is one point here that may require a little more to be said about it. In the diagram in 62A, both the dashed line and the full circle pass through B ($z = 1$). Now $z = 1$ gives $w = \infty$ (see 64A) and so both the corresponding curves to those through B in 62A go off to infinity, i.e. the dashed and full lines in 65A. They do not necessarily go off to infinity in the same direction and from the diagram in 65A, they obviously do not. Remembering what was said in FRAME 52, B' can be regarded as being anywhere on the circle, centre O and with infinite radius. It can even be in different places round this circle in the same problem, as is the case here.

FRAME 67

In some of our previous problems, having obtained corresponding curves, corresponding directions of travel and corresponding regions have been found. Referring back to 62A, three directions of travel are shown and also the eight regions (I - VIII) into which the z-plane is divided by the three curves.

Now the circle in 62A has been traversed in the direction BCDE. Remembering that the w-value corresponding to $z = 1$ is ∞, the corresponding line in 65A will be traversed $\uparrow$.

Also, region I in 62A is bounded in part by each of the three curves in that diagram and has vertices at A, B and C. The corresponding region in 65A will be bounded in part by each of the three corresponding curves and will have vertices at A', C'. It will also go off to infinity. The region in 65A that satisfies these criteria is that shaded. Once again, infinity must be treated in its own special way. If you are still not quite convinced, take the three following values of z and calculate the corresponding values of w:
$z = 0{\cdot}99 + 0{\cdot}01i$, $z = 0{\cdot}99 + 0{\cdot}001i$, $z = 0{\cdot}9997 + 0{\cdot}0099i$. (Approximate values will be sufficient.)

****

67A

The corresponding values of w are $-49 - 50i$, $-98 - 10i$ (approx) and $-2 - 100i$ (approx).

FRAME 68

You will notice that the three z-points chosen in the last frame all lie in region I and are very close to B (see 62A). The three corresponding values of w represent three points in the shaded area in 65A, but very widely scattered throughout it.

Now, on the diagram you drew in answer to the questions in FRAME 65, insert arrows showing directions of travel corresponding to the three directions indicated in 62A and label the regions to correspond to the eight regions depicted in 62A.

****

68A

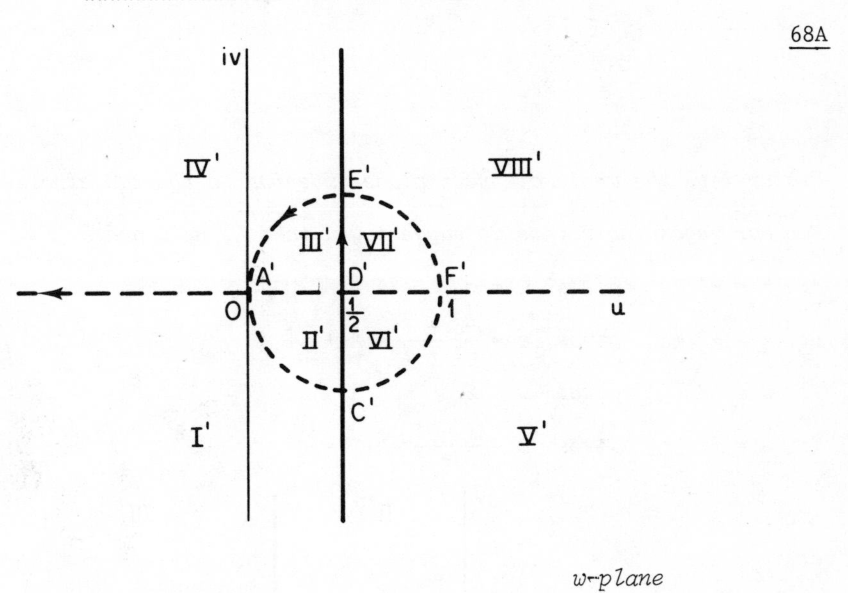

FRAME 69

Now try the following example:

Find the w-curves corresponding to the two circles $|z + 2| = 1$ and $|z + 1 - i| = 1$, given that $w = \dfrac{z - 1}{z + 1}$. Illustrate by diagrams, showing corresponding directions of travel and the four w-regions corresponding to the four z-regions formed by the two given circles.

****

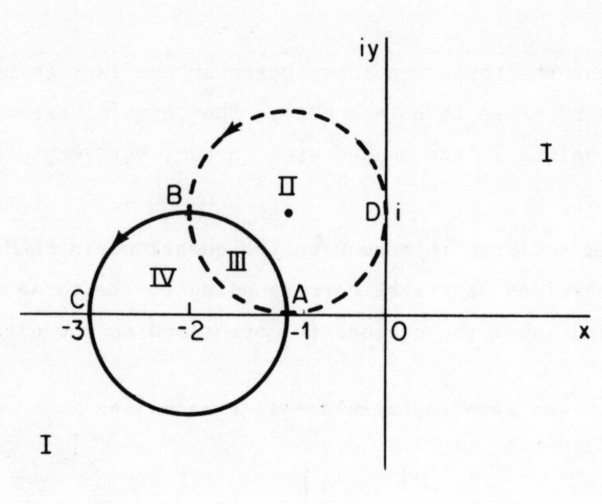

z-plane

Region I is the whole of the z-plane exterior to the two circles.

For corresponding points we suggest you take A, B, C and D.

A: $z = -1$ gives $w = \infty$

B: $z = -2 + i$ gives $w = \dfrac{-3 + i}{-1 + i} = 2 + i$

C: $z = -3$ gives $w = 2$

D: $z = i$ gives $w = \dfrac{i - 1}{i + 1} = i$

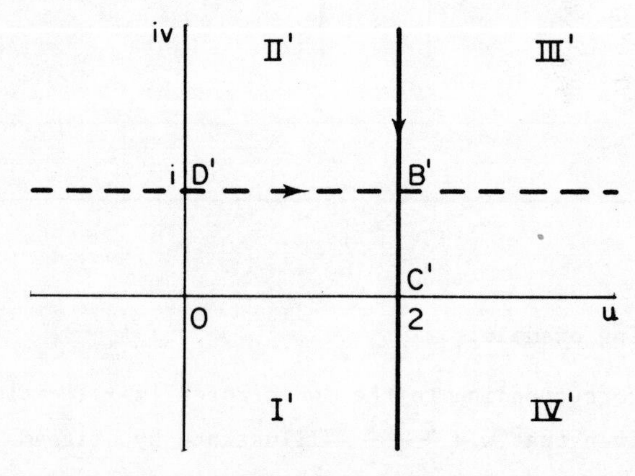

w-plane

The corresponding directions of travel and regions are as indicated.

69A continued

You may find the reasoning for some of the corresponding regions helpful:

Region I is bounded by arc BDA of the dashed circle and arc ACB of the full circle. The corresponding curves are the portion of the dashed line to the left of B' (and going off to infinity) and the portion of the full line below B' (and going off to ∞).

Region II is bounded by the same arc of the dashed circle and the minor arc BA of the full circle. Thus II' is bounded by the same part of the dashed line as I' but the other part of the full line.

FRAME 70

To illustrate the use of the bilinear transformation in a practical situation, the R, C circuit shown will be considered. Here R is kept fixed and C allowed to vary from 0 to ∞.

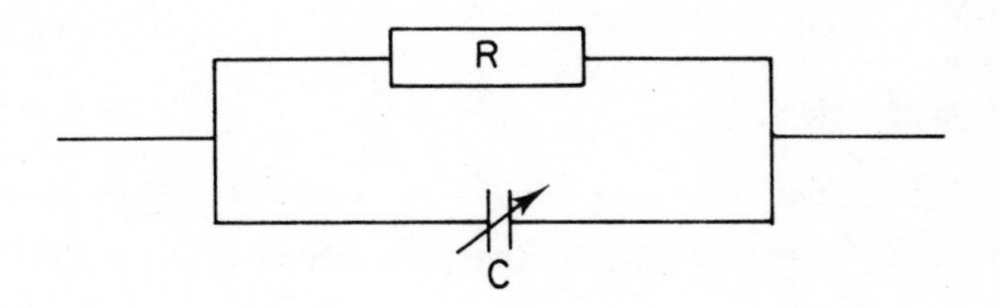

In this case, only the variation in the impedance will be found.

What is Z for this circuit? (This can be left as a fraction with j in the denominator.)

****

70A

$$Z = \frac{R}{1 + j\omega CR}$$

FRAME 71

This can be regarded as a special case of the bilinear transformation

$Z = \dfrac{aC + b}{cC + d}$, in which $a = 0$, $b = R$, $c = j\omega R$ and $d = 1$.

Regarding C as the complex number $x + jy$ in which $y = 0$ and $x \geqslant 0$, the locus of C is part of a straight line. What will be the form of the curve for Z?

****

71A

An arc of a circle, or possibly a part of another straight line.

FRAME 72

Again, it is only necessary to find three points on this arc in order to define it completely. Also, it is again best to choose two of the points to give the end points of the arc. Taking $C = 0$, $C = \infty$, and $C = \dfrac{1}{\omega R}$, find the corresponding values of Z and sketch the locus.

****

72A

$Z = R$, $Z = 0$, $Z = \frac{1}{2}R - \frac{1}{2}Rj$.

The locus of Z is thus the arc of a circle from R to 0 which passes through $\frac{1}{2}R - \frac{1}{2}Rj$. This is as shown in the diagram, in which $Z = U + jV$.

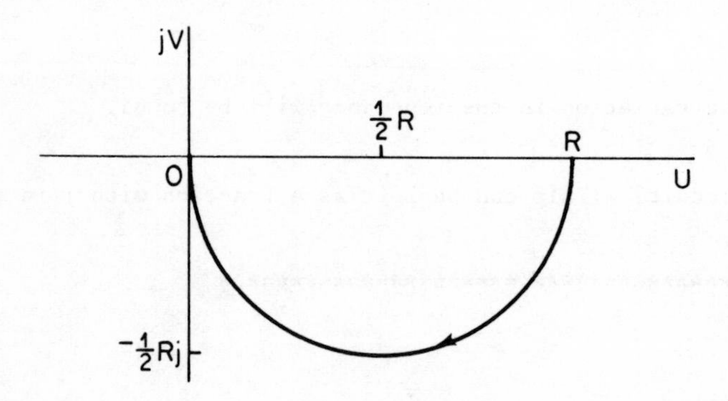

Z-plane

FRAME 73

As a second example, take the R, C, L circuit and consider the variation in the total impedance as C varies.

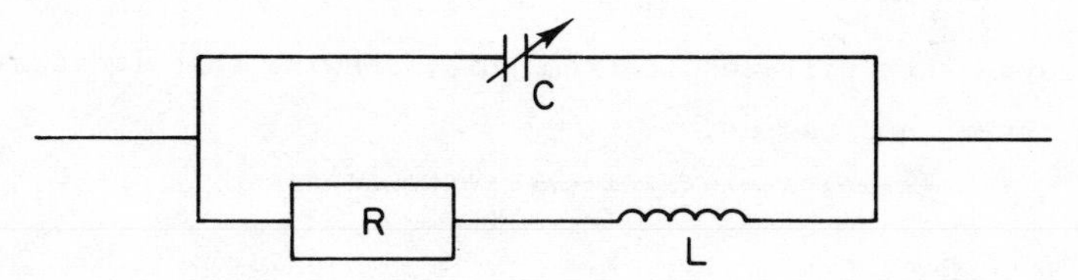

The impedances of the upper and lower branches in this circuit are $\frac{1}{j\omega C}$ and $R + j\omega L$ respectively.

From these values find the total impedance Z and the value of Z when
(i) $C = 0$, (ii) $C = \infty$.

****

73A

$$Y = j\omega C + \frac{1}{R + j\omega L} \qquad (73A.1)$$

$$= \frac{-\omega^2 LC + j\omega RC + 1}{R + j\omega L}$$

$$Z = \frac{R + j\omega L}{(1 - \omega^2 LC) + j\omega RC} \qquad (73A.2)$$

When $C = 0$, $Z = R + j\omega L$

When $C = \infty$, $Z = 0$

FRAME 74

In this example, although a third point can easily be found, it is unlikely to be immediately obvious what the required locus is, although it is possible to proceed in this way. So here we shall adopt the alternative method of solution as was mentioned in FRAME 57.

From (73A.1), $j\omega C = \frac{1}{Z} - \frac{1}{R + j\omega L}$, remembering that $Y = \frac{1}{Z}$.

If, in this, we put $Z = U + jV$, this becomes

$$j\omega C = \frac{1}{U + jV} - \frac{1}{R + j\omega L}$$

$$= \frac{U - jV}{U^2 + V^2} - \frac{R - j\omega L}{R^2 + \omega^2 L^2} \qquad (74.1)$$

FRAME 74 continued

Now C must be completely real and so $j\omega C$ is wholly imaginary. This means that the real part of the R.H.S. of (74.1) is zero.

Find the U-V curve that will result from this, showing that it is a circle and obtaining its centre and radius.

****

74A

$$\frac{U}{U^2 + V^2} - \frac{R}{R^2 + \omega^2 L^2} = 0$$

i.e. $R(U^2 + V^2) = U(R^2 + \omega^2 L^2)$

i.e. $U^2 + V^2 - \frac{U}{R}(R^2 + \omega^2 L^2) = 0$

i.e. $\{U - \frac{1}{2R}(R^2 + \omega^2 L^2)\}^2 + V^2 = \{\frac{1}{2R}(R^2 + \omega^2 L^2)\}^2$

which is a circle, centre $\frac{1}{2R}(R^2 + \omega^2 L^2)$ *and radius* $\frac{1}{2R}(R^2 + \omega^2 L^2)$

FRAME 75

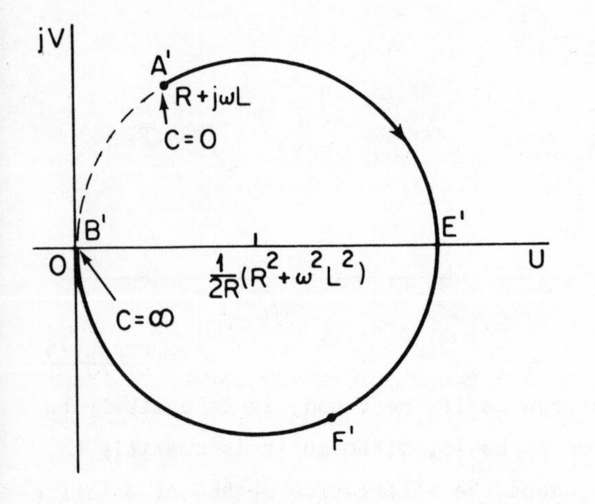

We thus have the circle shown. But this will be the locus for C varying from $-\infty$, through 0 to $+\infty$ (remember C is real) and so it is necessary to decide which portion of the circle must be omitted. The end points of the required arc are A' and B' (see 73A). To decide which of the two possible arcs is required, put $C = \frac{1}{\omega^2 L}$ in (73A.2).

Then $Z = \frac{R + j\omega L}{jR/\omega L}$

$$= \frac{\omega^2 L^2}{R} - j\omega L$$

As the coefficient of j is negative, there is at least one point on the locus which is below the U-axis. Hence the minor arc A'B' is eliminated and so the result is the major arc A'E'F'B'.

FRAME 76

Miscellaneous Examples

In this frame a collection of miscellaneous examples is given for you to try. Answers are supplied in FRAME 77, together with such working as is considered helpful.

1. If $x + iy = \dfrac{2i}{u + iv}$, where x, y, u, v are all real, express x, y in terms of u and v.

   If the point (x,y) moves round a trapezium ABCD whose vertices are the points (1,1), (1,-1), (2,-2), (2,2) in that order, find the equations of the path traced out by the point (u,v) and show it in a separate diagram. (L.U.)

2. If $w = z^2 + 2z - 1$, find the locus of w as z moves round the triangle whose vertices are at z = 0, z = 1, and z = -i.

3. If $w = \dfrac{z}{i - z}$, find the w-curves corresponding to

   (i) $|z| = 2$, (ii) $x = 0$, (iii) $y = 0$.

   Find the regions in the w-plane corresponding to the eight regions into which the z-plane is divided by the three given curves.

4. If z is a complex number and $w = \dfrac{z - 2}{z - i}$

   (a) show that when the point in the Argand diagram represented by w moves along the real axis, z traces a straight line through (2,0) and (0,1)

   (b) determine $|z - 1 - \frac{1}{2}i|$ when w lies on the imaginary axis. What is the locus of z as w moves along the imaginary axis? (C.E.I.)

5.

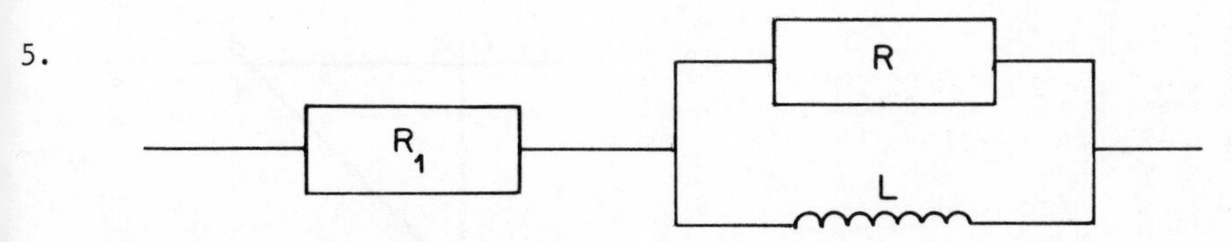

In the above circuit, R_1 and L are kept fixed, while R can take any positive value. Find and sketch the locus of Z, the total impedance of the circuit, as R varies.

FRAME 77

Answers to Miscellaneous Examples

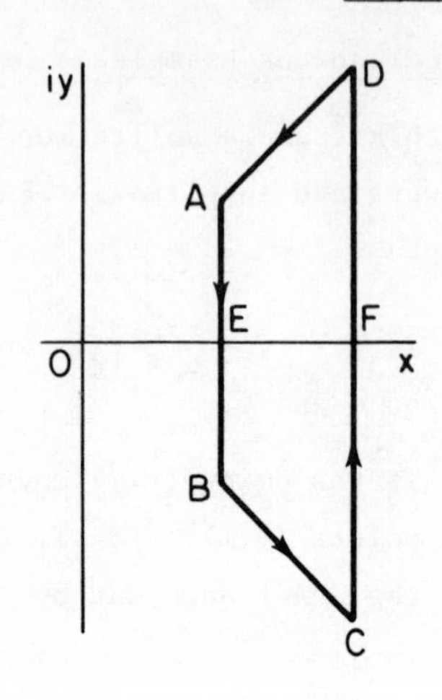

1. $x = \dfrac{2v}{u^2 + v^2}$ $y = \dfrac{2u}{u^2 + v^2}$

Along AB, $x = 1$ $\therefore u^2 + (v - 1)^2 = 1$

Along BC, $x + y = 0$ $\therefore u + v = 0$

Along CD, $x = 2$ $\therefore u^2 + \left(v - \frac{1}{2}\right)^2 = \frac{1}{4}$

Along DA, $y = x$ $\therefore v = u$

Corresponding points are:

$A(z = 1 + i) \rightarrow A'(w = 1 + i)$

$B(z = 1 - i) \rightarrow B'(w = -1 + i)$

$C(z = 2 - 2i) \rightarrow C'(w = -\frac{1}{2} + \frac{1}{2}i)$

$D(z = 2 + 2i) \rightarrow D'(w = \frac{1}{2} + \frac{1}{2}i)$

$E(z = 1) \rightarrow E'(w = 2i)$

$F(z = 2) \rightarrow F'(w = i)$

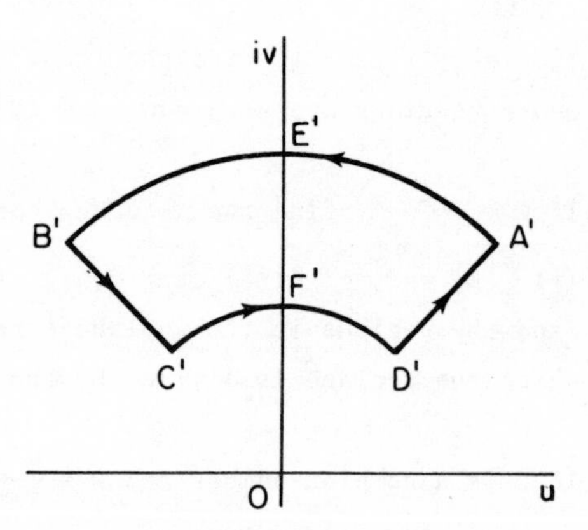

2. $w = z^2 + 2z - 1$

When $z = 0$, $w = -1$

When $z = 1$, $w = 2$

When $z = -i$, $w = -2-2i$

$$u + iv = (x + iy)^2 + 2(x + iy) - 1$$
$$= x^2 + 2ixy - y^2 + 2x + 2iy - 1$$
$$u = x^2 - y^2 + 2x - 1$$
$$v = 2xy + 2y$$

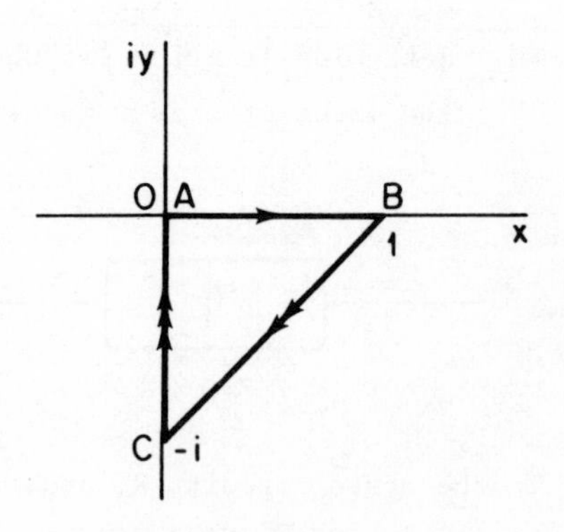

z-plane

FRAME 77 continued

Along AB, $y = 0$

$\therefore u = x^2 + 2x - 1, \quad v = 0$

x increases from 0 to 1

$\therefore$ u increases from -1 to 2.

Along BC, $x - y = 1$ or $y = x - 1$

$\therefore u = x^2 - (x - 1)^2 + 2x - 1$

$= 4x - 2$

$v = 2(x + 1)(x - 1)$

$= 2x^2 - 2$

$\therefore v = 2\left(\frac{u + 2}{4}\right)^2 - 2$

$8(v + 2) = (u + 2)^2$, which is a parabola with vertex at $u = -2$, $v = -2$.

Along CA, $x = 0$

$\therefore u = -y^2 - 1, \quad v = 2y$

$\therefore u = -\left(\frac{v}{2}\right)^2 - 1$

$\therefore v^2 = -4(u + 1)$, which is a parabola with vertex at $u = -1$, $v = 0$.

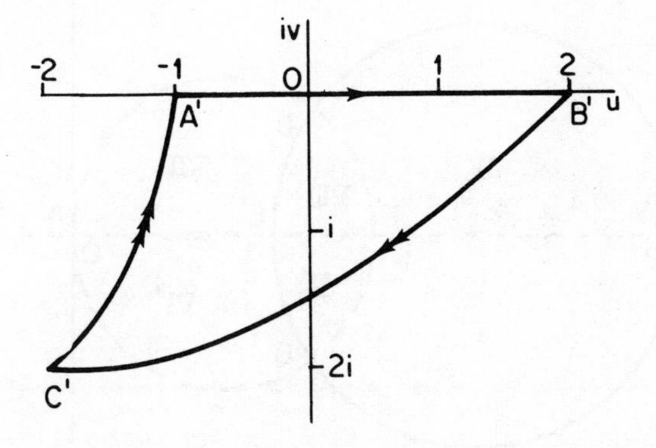

FRAME 77 continued

3.

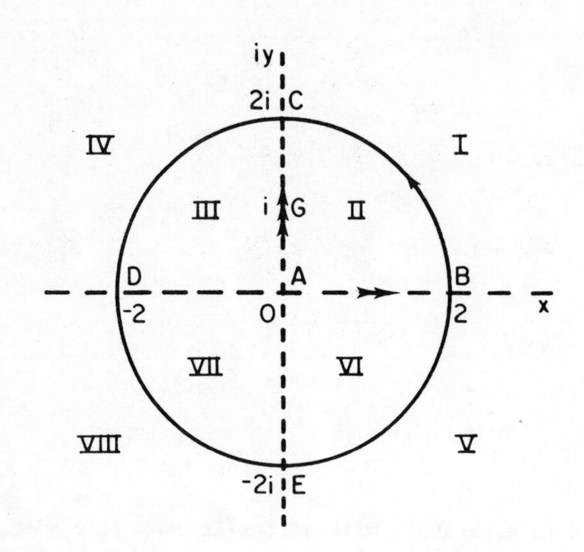

When $z = 0$, $w = 0$

When $z = 2$, $w = \dfrac{2}{i - 2}$

$= -\dfrac{2}{5}(2 + i)$

When $z = 2i$, $w = -2$

When $z = -2$, $w = \dfrac{-2}{i + 2}$

$= \dfrac{2}{5}(-2 + i)$

When $z = -2i$, $w = -\dfrac{2}{3}$

When $z = \infty$, $w = -1$

When $z = i$, $w = \infty$

As the transformation is bilinear, a circle will transform into a circle - the term circle including the limiting case of a straight line. The results are as shown.

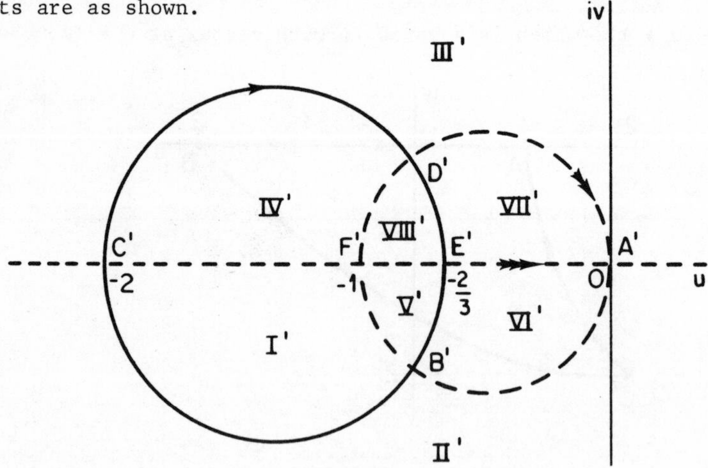

NOTE: The fact that the two circles in the w-plane both pass through B' and D', which are symmetrically placed with respect to the u-axis, means that the u-axis is a diameter of each.

FRAME 77 continued

4. (a) As the transformation is bilinear, the locus of z is a circle or straight line.

From $w = \dfrac{z - 2}{z - i}$, $z = \dfrac{2 - wi}{1 - w}$

Some corresponding points: $A(w = 0) \to A'(z = 2)$

$B(w = 1) \to B'(z = \infty)$

$C(w = \infty) \to C'(z = i)$

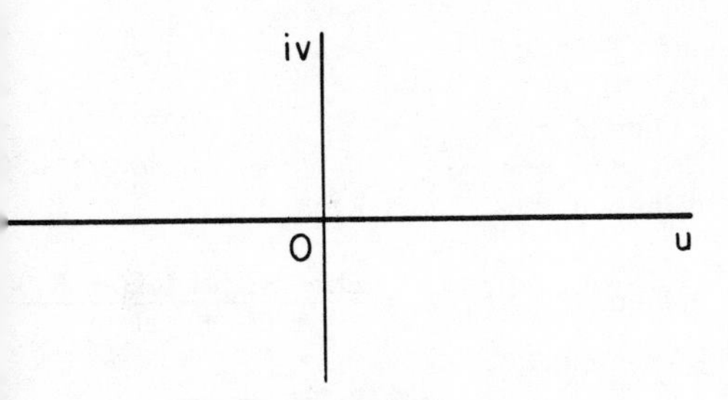

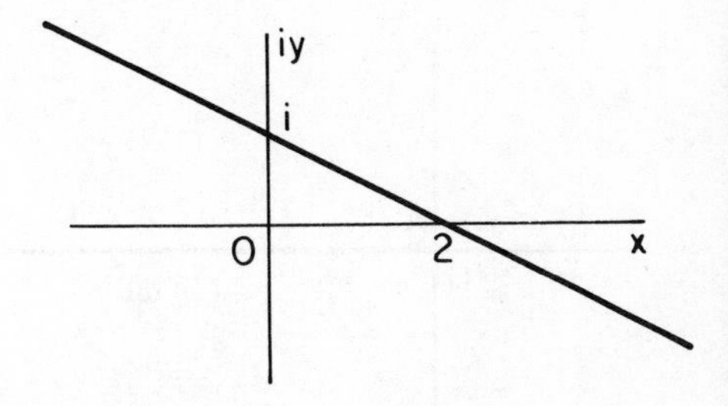

(b) $$z - 1 - \frac{1}{2}i = \frac{2 - wi}{1 - w} - 1 - \frac{1}{2}i$$

$$= \frac{(1 + w)(1 - \frac{1}{2}i)}{1 - w}$$

$$\left|z - 1 - \frac{1}{2}i\right| = \frac{|1 + w|\,|1 - \frac{1}{2}i|}{|1 - w|}$$

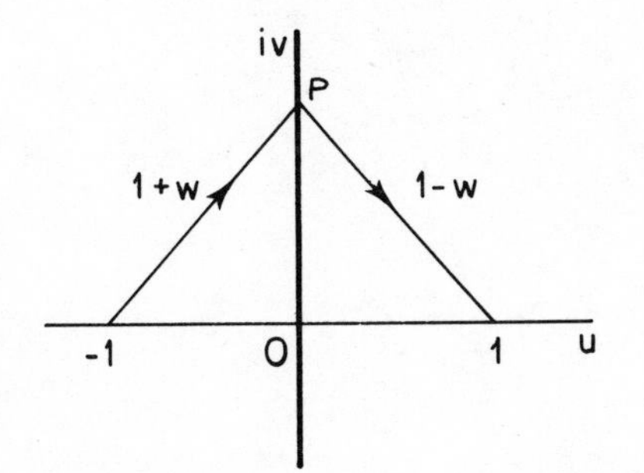

If w is on the imaginary axis, say at P, then as is seen from the diagram, $|1 + w| = |1 - w|$ and $\therefore\ |z - 1 - \frac{1}{2}i| = |1 - \frac{1}{2}i| = \sqrt{5}/2$

Hence the locus of z is the circle, centre $1 + \frac{1}{2}i$ and radius $\sqrt{5}/2$.

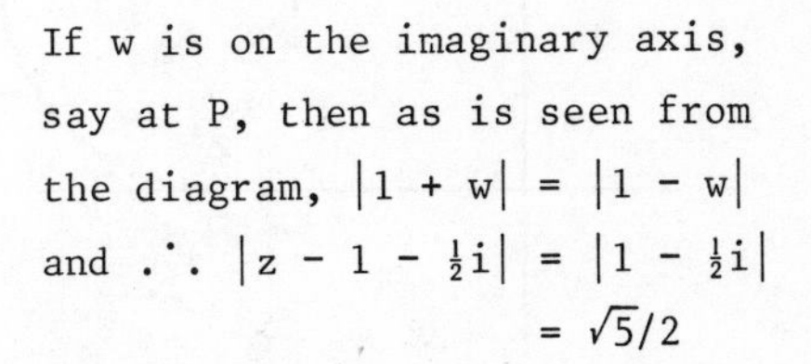

FRAME 77 continued

5. For components in parallel, admittance $= \dfrac{1}{R} + \dfrac{1}{j\omega L} = \dfrac{j\omega L + R}{j\omega RL}$

$$\text{impedance} = \frac{j\omega RL}{R + j\omega L}$$

For complete circuit, $Z = R_1 + \dfrac{j\omega RL}{R + j\omega L}$

$$= \frac{RR_1 + j\omega L(R + R_1)}{R + j\omega L} \qquad (77.1)$$

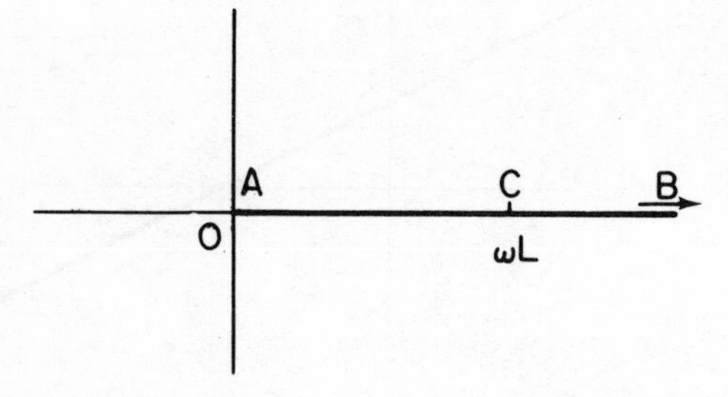

R-plane

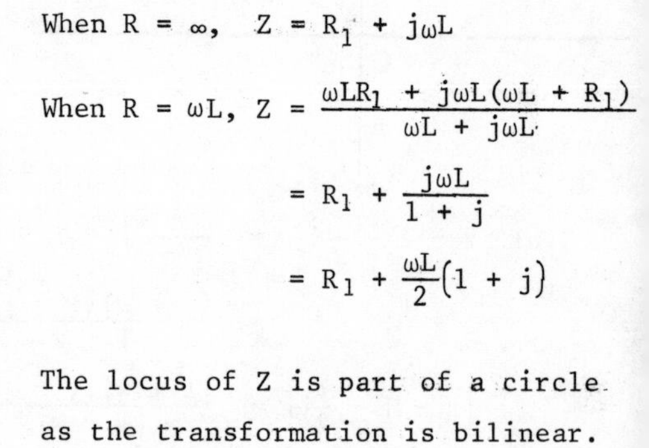

When $R = 0$, $Z = R_1$

When $R = \infty$, $Z = R_1 + j\omega L$

When $R = \omega L$, $Z = \dfrac{\omega LR_1 + j\omega L(\omega L + R_1)}{\omega L + j\omega L}$

$$= R_1 + \frac{j\omega L}{1 + j}$$

$$= R_1 + \frac{\omega L}{2}(1 + j)$$

The locus of Z is part of a circle as the transformation is bilinear.

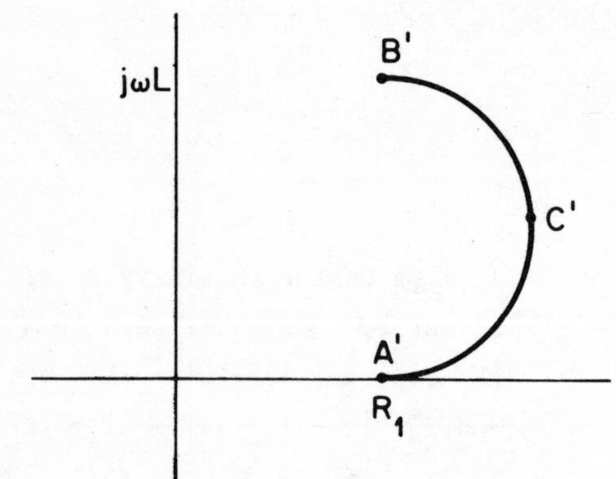

Z-plane

If you did not use $R = \omega L$ to give you C', it is unlikely that the path joining A' and B' is obvious. In this case its equation can be obtained as follows:

From (77.1), $R = \dfrac{j\omega LZ - j\omega LR_1}{R_1 - Z + j\omega L}$

Let $Z = U + jV$, then $R = \dfrac{-\omega LV + j\omega LU - j\omega LR_1}{(R_1 - U) + j(\omega L - V)} \cdot \dfrac{(R_1 - U) - j(\omega L - V)}{(R_1 - U) - j(\omega L - V)}$

FRAME 77 continued

But R is completely real, therefore imaginary part of R is zero

$\therefore\ \omega LV(\omega L - V) + \omega L(U - R_1)(R_1 - U) = 0$

$\therefore\ V^2 - \omega LV + U^2 - 2UR_1 + R_1{}^2 = 0$

$\therefore\ (U - R_1)^2 + \left(V - \frac{\omega L}{2}\right)^2 = \left(\frac{\omega L}{2}\right)^2$

and this is the equation of a circle with centre $\left(R_1, \frac{\omega L}{2}\right)$ and radius $\frac{\omega L}{2}$.

THEORY OF FUNCTIONS

FRAME 1

Introduction

You have already met the idea of a function of a complex variable. Before starting this programme, it will be a good idea for you to remind yourself of the contents of FRAMES 6-12, pages 12:11-12:15 of the previous programme. This present programme will deal with the theory that it is necessary for you to know in order to make further use of functions of a complex variable.

FRAME 2

Single- and Many-Valued Functions

In real variable theory, given an equation between x and y, and also a particular value of x, then it is possible that y may be able to take a different number of values in different cases. For example, if $y = 3x - 7$ and $x = 5$, then y has the single value 8. But if $y^2 = 8x$ and $x = 2$, then y has two possible values +4 and -4.

Similar results can occur when we are dealing with complex quantities. As an example of a SINGLE-VALUED FUNCTION, let us consider $w = z^3$. Here, for each value of z, there is one and only one value of w, e.g., if $z = 2-i$, then $w = 2 - 11i$. But $w = z^{1/4}$ is an example of a MANY-VALUED FUNCTION, as for each value of z, w has more than one value - in this case, four. For example, if $z = -1$, $w = \frac{1}{\sqrt{2}}(1 + i)$, $\frac{1}{\sqrt{2}}(-1 + i)$, $-\frac{1}{\sqrt{2}}(1 + i)$ or $\frac{1}{\sqrt{2}}(1 - i)$.

State whether the following are single- or many-valued functions of z:

(a) $z^2 - 1$ (b) $z^{3/2}$ (c) e^z

****

2A

(a) and (c) are single-valued

(b) is many-valued (two in this case)

FRAME 3

The simplest case of a many-valued function is $w = z^{1/2}$ which has two values of w for each value of z. Expressed in polar form,

$$w = \left[r\{\cos(\theta + 2k\pi) + i\sin(\theta + 2k\pi)\}\right]^{1/2} \qquad (-\pi < \theta \leqslant \pi)$$

$$= \sqrt{r}\{\cos(\tfrac{1}{2}\theta + k\pi) + i\sin(\tfrac{1}{2}\theta + k\pi)\}$$

where $\sqrt{r}$ indicates the ordinary positive arithmetic square root of r, and k needs to take two consecutive values, say 0 and 1. If $k = 0$,

$-\frac{\pi}{2} < \arg w \leqslant \frac{\pi}{2}$ and if $k = 1$, $\frac{\pi}{2} < \arg w \leqslant \frac{3\pi}{2}$.

Corresponding to the two regions in which w may lie, we say that there are two BRANCHES of the function $z^{1/2}$. If a function has branches like this, we often select just one of these branches and then, if we confine ourselves to that one branch, the function can be regarded as single-valued again.

There are other functions, besides roots, which are many-valued and some of these you will meet later.

FRAME 4

Limits

First of all, what is understood by the statement $z \to z_0$, remembering that these are both complex quantities, z being variable and z_0 fixed?

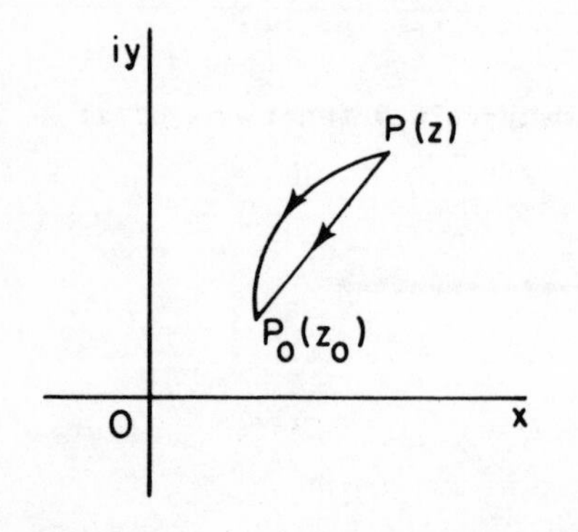

The meaning that is given to this is simply that P must move towards P_0, but that it is free to do so along any path joining P and P_0. Two such paths have been indicated. Thus $|z - z_0|$ must tend to zero, but no restriction is placed on $\arg(z - z_0)$. z_0 is then the LIMIT of z in this case.

FRAME 5

The second statement frequently associated with limits is that $w \to w_0$ as $z \to z_0$. (It is assumed, of course, that w is known in terms of z, so that, given any value of z, the corresponding value of w can be calculated.)

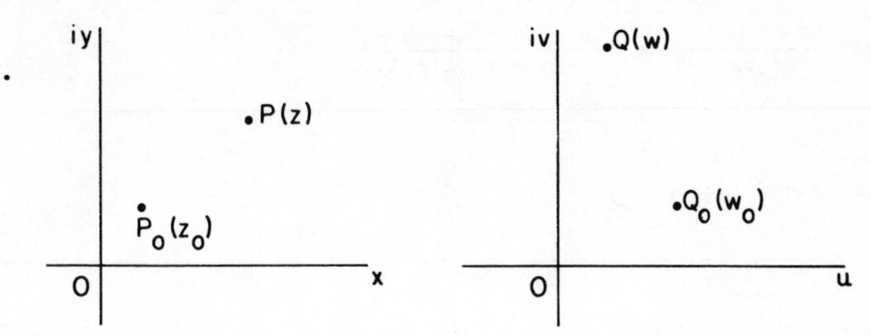

This simply means that Q approaches Q_0 as P approaches P_0, but again there is no restriction on the direction in which it does so. What is important is that $|w - w_0| \to 0$ as $|z - z_0| \to 0$.

Another way of stating that $w \to w_0$ as $z \to z_0$ is to write $\lim_{z \to z_0} w = w_0$.

Some examples of limits are:

(i) $\lim_{z \to 2i} (z^2 - 3) = (2i)^2 - 3 = -7$. If there is no difficulty, as in this case, simple substitution gives the result.

(ii) $\lim_{z \to i} \dfrac{z^2 + 1}{z - i}$. Here, simple substitution gives $\dfrac{0}{0}$. But notice that

$z^2 + 1 = (z + i)(z - i)$ and so $\dfrac{z^2 + 1}{z - i} = z + i$.

If, in this result, $z \to i$, we find the required limit is 2i.

Now find the following limits:

(a) $\lim_{z \to 2+i} \dfrac{z + 1}{z - 1}$

(b) $\lim_{z \to i} \dfrac{z^3 + i}{z^2 + iz + 2}$ (Note that as putting z = i makes both numerator and denominator disappear, it follows that z - i is a factor of each.)

****

5A

(a) $\dfrac{(2 + i) + 1}{(2 + i) - 1} = 2 - i$

(b) Direct substitution gives $\dfrac{0}{0}$

But $\dfrac{z^3 + i}{z^2 + iz + 2} = \dfrac{(z - i)(z^2 + iz - 1)}{(z - i)(z + 2i)}$

$= \dfrac{z^2 + iz - 1}{z + 2i}$

Now $z = i$ gives $\dfrac{-3}{3i} = i.$

FRAME 6

By now, you may have realised that if w is going to have a specific limit as $z \to z_0$, then it must be a single-valued function of z or alternatively on a specified branch of a many-valued function.

FRAME 7

Continuity

In real variable work, the statement that a function is continuous at a point effectively means that its graph does not have a gap there. The function must also be single-valued. For example, the function illustrated in Fig (i) is

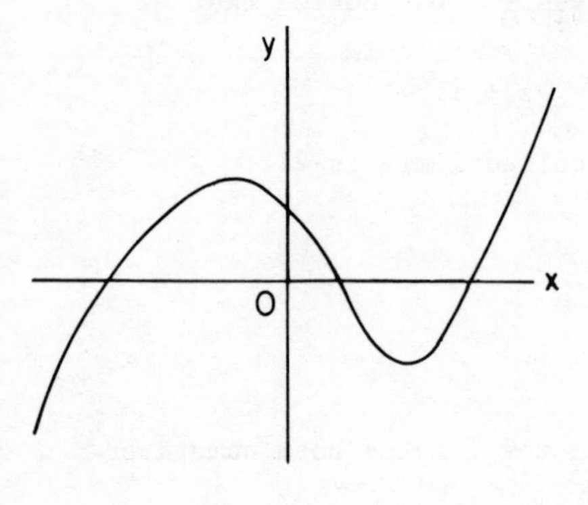

Fig (i)

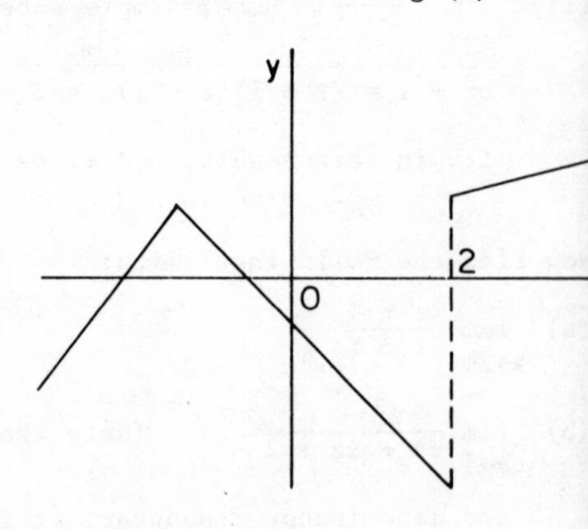

Fig (ii)

FRAME 7 continued

continuous at all points shown but the function illustrated in Fig (ii) is not continuous at $x = 2$.

Expressed more mathematically, $y = f(x)$ is continuous at a point x_0 if

(i) $f(x_0)$ exists and has a definite value,

and (ii) $f(x) \to f(x_0)$ as $x \to x_0$.

In the second figure you will notice that $f(2)$ does not have a definite value, and so the function is not continuous there.

FRAME 8

In a similar way a complex variable function $w = f(z)$ is said to be CONTINUOUS at a point z_0 if

(i) $f(z_0)$ exists and has a definite value,

and (ii) $f(z) \to f(z_0)$ as $z \to z_0$.

Again, w must be single-valued or a specified branch of a many-valued function. As an example, we shall consider whether the two functions

$$\text{(a)} \begin{cases} w = \dfrac{z^2 + 4}{z + 2i} & (z \neq -2i) \\ w = -5i & (z = -2i) \end{cases} \qquad \text{(b)} \quad \begin{matrix} w = \dfrac{z^2 + 4}{z + 2i} & (z \neq -2i) \\ w = -4i & (z = -2i) \end{matrix}$$

are continuous at $z = -2i$.

The function $\dfrac{z^2 + 4}{z + 2i}$ is indeterminate and hence not defined when $z = -2i$ as both numerator and denominator are zero. But $\dfrac{z^2 + 4}{z + 2i} = \dfrac{(z - 2i)(z + 2i)}{(z + 2i)} = z - 2i$ when $z \neq -2i$. As $z \to -2i$ this quantity $\to -4i$.

(a) (i) w exists when $z = -2i$ and has the definite value $-5i$ as it is specially defined for this value of z.

(ii) As $z \to -2i$, w (which is equivalent to $z - 2i$) $\to -4i$. So $w \not\to w_0$ as $z \to z_0$ where $z_0 = -2i$ and so w is not continuous there.

(b) (i) w exists when $z = -2i$ and has the definite value $-4i$.

(ii) as $z \to -2i$, $w \to -4i$ and so here $w \to w_0$ as $z \to z_0$ where $z_0 = -2i$ and so this function is continuous here.

FRAME 8 continued

Now try this example:

$$\text{Is } \left\{ \begin{array}{ll} w = \dfrac{z^3 + i}{z^2 + iz + 2} & (z \neq i \text{ or } -2i) \\ w = i & (z = i) \\ w = 0 & (z = -2i) \end{array} \right\} \text{ continuous at } z = i \text{ and } -2i?$$

****

8A

At i - Yes

At $-2i$ - No. $\because$ $w \to \infty$ as $z \to -2i$ but w is defined as being zero there.

FRAME 9

A function that is continuous at all points in a region is said to be continuous in that region.

The following example will illustrate this.

Are (a) z^2, and (b) $\dfrac{1}{z}$ continuous throughout $|z| < 1$?

(a) (i) $z^2 = (x + iy)^2 = x^2 - y^2 + 2ixy$. For each pair of values of x and y such that $x^2 + y^2 < 1$, this certainly exists and has a definite value.

(ii) $f(z) \to f(z_0)$ as $z \to z_0$ if $|f(z) - f(z_0)| \to 0$ as $|z - z_0| \to 0$ (see FRAME 5). Here $f(z) - f(z_0) = z^2 - z_0{}^2$.
Now $|z^2 - z_0{}^2| = |(z - z_0)(z + z_0)| = |z - z_0||z + z_0| \to 0$ as $|z - z_0| \to 0$. z^2 is therefore continuous everywhere throughout $|z| < 1$.

(b) $\dfrac{1}{z}$ does not have a definite value when $z = 0$, i.e. at the origin. It is therefore not continuous there and consequently is not continuous throughout $|z| < 1$.

Incidentally, if $f(z) = u + iv$ and is continuous, then u and v are also continuous. Conversely, if u and v are continuous, so also is f(z).

Now examine $w = z^3$ to find whether it is continuous throughout $|z| < 2$.

****

9A

$f(z) = (x + iy)^3 = x^3 - 3xy^2 + i(3x^2y - y^3)$ and this exists and has a definite value for each x,y such that $x^2 + y^2 < 4$.

$$|f(z) - f(z_0)| = |z^3 - z_0{}^3| = |(z - z_0)(z^2 + zz_0 + z_0{}^2)|$$

$$= |z - z_0||z^2 + zz_0 + z_0{}^2|$$

$$\to 0 \text{ as } |z - z_0| \to 0$$

$\therefore$ *z^3 is continuous throughout $|z| < 2$.*

FRAME 10

Differentiability

Given y as a real function of x, one does not normally hesitate if it is necessary to find $\frac{dy}{dx}$. Thus if $y = \frac{1}{x + 2}$, you will probably immediately write down $\frac{dy}{dx} = -\frac{1}{(x + 2)^2}$. But be careful! $\frac{1}{x + 2}$ is not defined when $x = -2$ and so neither is the derivative.

Consider also the function illustrated by the graph.

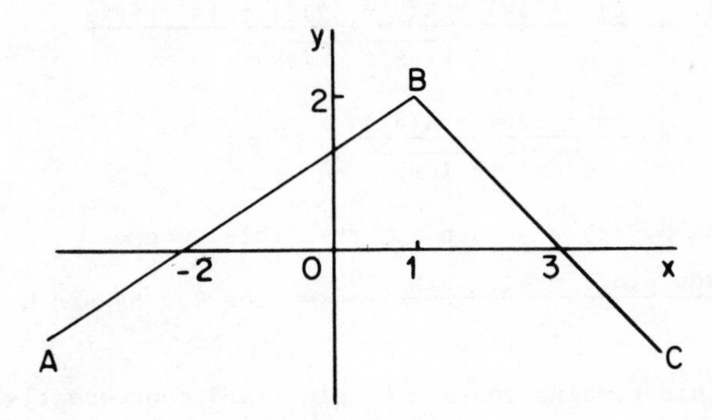

Over the section AB the slope, i.e. $\frac{dy}{dx}$, is $\frac{2}{3}$. Over the section BC it is -1. But what of its value at B? Here it has two different values according to the direction of approach and consequently we can't state "the" value of the derivative there.

FRAME 11

In general, assuming that the derivative of a function f(x) exists for a given value of x, then f'(x) at P is $\lim_{Q \to P} \frac{f(Q) - f(P)}{PQ}$, where Q can approach P from either side.

Now let's have a look at a real function of two variables f(x,y) and the value of a similar expression to that just quoted.

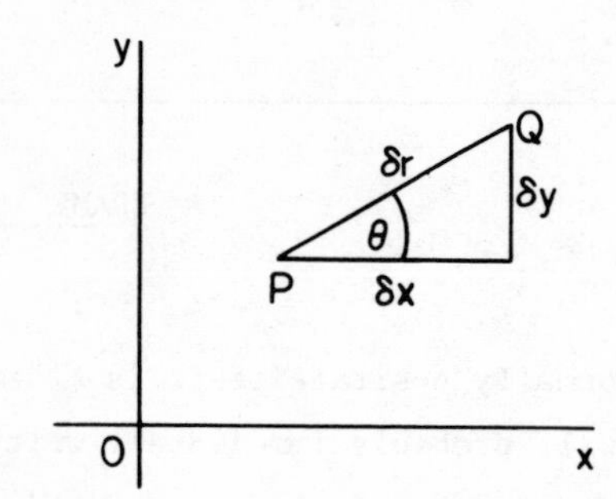

Generally the value of this expression will now depend on the direction of the line PQ. (Remember the ideas of directional derivatives in Vector Analysis if this subject is a part of your course.)

For example, suppose P is the point (x,y), Q the point $(x+\delta x,\ y+\delta y)$ and $f(x,y) = x - 4y^2$.

Then
$$\frac{f(Q) - f(P)}{PQ} = \frac{\{(x + \delta x) - 4(y + \delta y)^2\} - (x - 4y^2)}{\sqrt{(\delta x)^2 + (\delta y)^2}}$$

$$= \frac{\delta x - 8y\delta y - 4(\delta y)^2}{\sqrt{(\delta x)^2 + (\delta y)^2}}$$

Now if $\delta x = \delta r \cos\theta$, $\delta y = \delta r \sin\theta$, then this becomes

$$\frac{\delta r \cos\theta - 8y\delta r \sin\theta - 4(\delta r \sin\theta)^2}{\delta r} = \cos\theta - 8y\sin\theta - 4\delta r \sin^2\theta$$

Then as $PQ \to 0$, this remains $\cos\theta - 8y\sin\theta$ and consequently depends on θ, i.e. on the direction of PQ.

Now find the corresponding result if $f(x,y) = (x - 2)^2 + (y - 3)^2$.

****

11A

$2(x - 2)\cos\theta + 2(y - 3)\sin\theta$

FRAME 12

Again, in general, this depends on the value of θ. But if $x = 2$, $y = 3$, i.e. if P is the point (2,3) this is zero irrespective of θ. So you will see that although $\lim_{PQ\to 0} \frac{f(Q) - f(P)}{PQ}$ generally depends on the direction of PQ when f is a function of x and y, there are certain circumstances in which this direction is immaterial. The number of cases in which this happens is relatively small and if this condition is required the limitations that it places on f and P are very severe.

FRAME 13

Now let us return to complex quantities. P will now represent z $(= x + iy)$ and Q, $z + \delta z$ $\{= x + \delta x + i(y + \delta y)\}$. Thus $\overrightarrow{PQ}$ will represent $\delta x + i\delta y$ and f will be a function of z where we are still assuming (see FRAME 12, page 12:14) that $f(z)$ may contain $\bar{z}$ as well as z. We now consider the corresponding expression

$$\frac{f(Q) - f(P)}{\overrightarrow{PQ}}$$

If, and only if, this tends to a definite limit as $Q \to P$, irrespective of the direction of $\overrightarrow{PQ}$, then $f(z)$ is said to be DIFFERENTIABLE at the point P and this limit is defined as the DERIVATIVE of $f(z)$ at P. The derivative of $f(z)$ can then be written as $f'(z)$ and so

$$f'(z) = \lim_{\delta z\to 0} \frac{f(z + \delta z) - f(z)}{\delta z}$$

the R.H.S. being an alternative form for $\lim_{PQ\to 0} \frac{f(Q) - f(P)}{\overrightarrow{PQ}}$

This requirement of a unique value for this limit, irrespective of the direction of $\overrightarrow{PQ}$, is very restrictive, but in this work we are only really interested in those functions for which it happens.

FRAME 14

In this frame, we shall look at a few examples of functions to see whether they are differentiable.

Example 1: $w = z^2$ [Here w is being used in place of f(z)]

$$w + \delta w = (z + \delta z)^2 = z^2 + 2z\delta z + (\delta z)^2$$

$$\therefore \frac{(w + \delta w) - w}{\delta z} = \frac{z^2 + 2z.\delta z + (\delta z)^2 - z^2}{\delta z}$$

$$= 2z + \delta z$$

Now as $\delta z \to 0$, this tends to the value 2z. It does not depend on the direction of $\overrightarrow{PQ}$ and so the function z^2 is differentiable. Notice that the fact that it is differentiable doesn't even depend upon the position of P. More generally it can be shown that z^n (n a positive integer) can be differentiated and has derivative nz^{n-1}.

Example 2: $w = z\bar{z}$

$w + \delta w = (z + \delta z)(\bar{z} + \overline{\delta z})$ where $\overline{\delta z}$ denotes the conjugate of δz

i.e. $\delta z = \delta x + i\delta y$, $\overline{\delta z} = \delta x - i\delta y$

$$\frac{(w + \delta w) - w}{\delta z} = \frac{(z + \delta z)(\bar{z} + \overline{\delta z}) - z\bar{z}}{\delta z}$$

$$= \frac{z\overline{\delta z} + \bar{z}\delta z + \delta z\overline{\delta z}}{\delta z}$$

$$= z\frac{\overline{\delta z}}{\delta z} + \bar{z} + \overline{\delta z}$$

Now as $\delta z \to 0$, both δx and $\delta y \to 0$, therefore so also does $\overline{\delta z}$. But what happens to $\frac{\overline{\delta z}}{\delta z}$?

FRAME 14 continued

To find out we first put it in terms of δx and δy, thus

$$\frac{\overline{\delta z}}{\delta z} = \frac{\delta x - i\delta y}{\delta x + i\delta y}$$

$$= \frac{\delta r \cos\theta - i\delta r \sin\theta}{\delta r \cos\theta + i\delta r \sin\theta}$$

$$= \frac{\cos\theta - i\sin\theta}{\cos\theta + i\sin\theta}$$

$$= (\cos\theta - i\sin\theta)^2$$

$$= \cos 2\theta - i\sin 2\theta$$

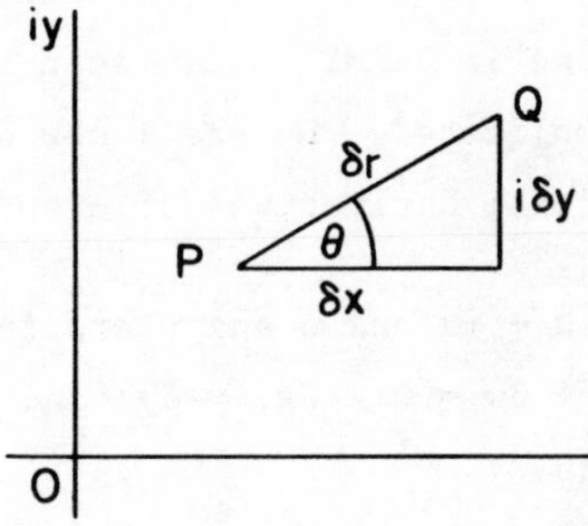

iy
Q
δr
iδy
P
θ
δx
O
x

This is not independent of the direction of $\overrightarrow{PQ}$ as it depends on θ and so the function is not differentiable.

Now test these functions for differentiability:

(a) $w = \dfrac{1}{\bar{z}}$

(b) $w = z^3$ (This is, of course, a particular case of the general statement at the end of Example 1 in this frame.)

****

14A

(a) $\dfrac{(w + \delta w) - w}{\delta z} = -\dfrac{1}{\bar{z}(\bar{z} + \overline{\delta z})} \cdot \dfrac{\overline{\delta z}}{\delta z}$

As $\delta z \to 0$, *this becomes* $-\dfrac{1}{\bar{z}^2}(\cos 2\theta - i\sin 2\theta)$ *and is not independent of* θ. *Hence* $\dfrac{1}{\bar{z}}$ *is not differentiable.*

(b) $\dfrac{(w + \delta w) - w}{\delta z} = 3z^2 + 3z\delta z + (\delta z)^2$

$\to 3z^2$ *as* $\delta z \to 0$.

$\therefore$ z^3 *is differentiable and its derivative is* $3z^2$.

FRAME 15

The Cauchy-Riemann Conditions

Testing a function for differentiability from first principles as has just been done in FRAME 14 can soon become a little tedious if the function is at all complicated. We shall now obtain a test for differentiability in terms of the partial derivatives of u and v w.r.t. x and y.

As $w = u + iv$ and u and v are functions of x and y, we can write $w = w(x,y)$. Then $w + \delta w = w(x+\delta x, y+\delta y)$ and the use of Taylor's Series in two dimensions leads to

$$\delta w = \frac{\partial w}{\partial x}\delta x + \frac{\partial w}{\partial y}\delta y + \text{terms involving powers and products of } \delta x \text{ and } \delta y$$

$$= \frac{\partial w}{\partial x}\delta r\cos\theta + \frac{\partial w}{\partial y}\delta r\sin\theta + \text{terms involving powers of } \delta r$$

(see diagram in FRAME 14)

$$\therefore \frac{\delta w}{\delta z} = \frac{\frac{\partial w}{\partial x}\cos\theta + \frac{\partial w}{\partial y}\sin\theta}{\cos\theta + i\sin\theta} + \text{terms involving } \delta r \text{ and its powers}$$

as $\delta z = \delta x + i\delta y = \delta r(\cos\theta + i\sin\theta)$

Now as $\delta z \to 0$, $\delta r \to 0$ and so $\dfrac{\delta w}{\delta z} \to \dfrac{\frac{\partial w}{\partial x}\cos\theta + \frac{\partial w}{\partial y}\sin\theta}{\cos\theta + i\sin\theta}$

For w to be differentiable, this last expression must be independent of θ. The condition for this is $\dfrac{\partial w}{\partial x}\Big/1 = \dfrac{\partial w}{\partial y}\Big/i$

$$\text{i.e.}\quad \frac{\partial w}{\partial y} = i\frac{\partial w}{\partial x}$$

Replacing $\dfrac{\partial w}{\partial y}$ by $i\dfrac{\partial w}{\partial x}$ in $\dfrac{\frac{\partial w}{\partial x}\cos\theta + \frac{\partial w}{\partial y}\sin\theta}{\cos\theta + i\sin\theta}$, verify that this is now independent of θ.

****

15A

Substitution reduces the expression simply to $\dfrac{\partial w}{\partial x}$.

FRAME 16

We now have the condition for differentiability to be

$$\frac{\partial w}{\partial y} = i\frac{\partial w}{\partial x}$$

Now replace w by u + iv and equate the real and imaginary parts.

****

16A

$$\frac{\partial u}{\partial y} + i\frac{\partial v}{\partial y} = i\left(\frac{\partial u}{\partial x} + i\frac{\partial v}{\partial x}\right)$$

Thus $\begin{cases} \dfrac{\partial u}{\partial x} = \dfrac{\partial v}{\partial y} & \text{(by equating imaginary parts)} \quad (16A.1) \\ \dfrac{\partial u}{\partial y} = -\dfrac{\partial v}{\partial x} & \text{(by equating real parts)} \quad (16A.2) \end{cases}$

FRAME 17

Equations (16A.1) and (16A.2) are called the CAUCHY-RIEMANN EQUATIONS or CAUCHY-RIEMANN CONDITIONS and if a function w is such that these are satisfied, then w is differentiable. Its derivative is $\frac{dw}{dz}$.

FRAME 18

Returning to the examples of FRAME 14 we shall now show that the results obtained there are consistent with those of FRAME 16 and 16A.

Example 1: $w = z^2$

$$u + iv = (x + iy)^2$$
$$= x^2 + 2xiy - y^2$$

Thus $u = x^2 - y^2 \qquad v = 2xy$

$$\therefore \frac{\partial u}{\partial x} = 2x \qquad \frac{\partial v}{\partial x} = 2y$$

$$\frac{\partial u}{\partial y} = -2y \qquad \frac{\partial v}{\partial y} = 2x$$

and the Cauchy-Riemann conditions are satisfied.

(Cauchy-Riemann is often abbreviated to C.R.)

FRAME 18 continued

Example 2: $w = z\bar{z}$

$$u + iv = (x + iy)(x - iy)$$
$$= x^2 + y^2$$

Thus $u = x^2 + y^2 \qquad v = 0$

$$\frac{\partial u}{\partial x} = 2x \qquad \frac{\partial v}{\partial x} = 0$$

$$\frac{\partial u}{\partial y} = 2y \qquad \frac{\partial v}{\partial y} = 0$$

and the C.R. conditions are not satisfied.

Now confirm by this means the results you obtained in 14A.

****

18A

(a) $w = \frac{1}{\bar{z}}$

$$u + iv = \frac{1}{x - iy} = \frac{x + iy}{x^2 + y^2}$$

$$u = \frac{x}{x^2 + y^2} \qquad v = \frac{y}{x^2 + y^2}$$

$$\frac{\partial u}{\partial x} = \frac{y^2 - x^2}{(x^2 + y^2)^2} \qquad \frac{\partial v}{\partial x} = -\frac{2x}{(x^2 + y^2)^2}$$

$$\frac{\partial u}{\partial y} = -\frac{2y}{(x^2 + y^2)^2} \qquad \frac{\partial v}{\partial y} = \frac{x^2 - y^2}{(x^2 + y^2)^2}$$

The C.R. conditions are not satisfied.

(b) $w = z^3$

$$u = x^3 - 3xy^2 \qquad v = 3x^2y - y^3$$

$$\frac{\partial u}{\partial x} = 3x^2 - 3y^2 \qquad \frac{\partial v}{\partial x} = 6xy$$

$$\frac{\partial u}{\partial y} = -6xy \qquad \frac{\partial v}{\partial y} = 3x^2 - 3y^2$$

and the C.R. conditions are satisfied.

FRAME 19

As the derivative $\frac{dw}{dz}$ is independent of the direction of $\overrightarrow{PQ}$ we can use $\theta = 0$ and still get the correct result. In this case $dz = dx$ as $dy = 0$ and $\frac{dw}{dz}$ becomes $\frac{\partial w}{\partial x}$. Notice that w is a function of the single variable z, but when z is replaced by $x + iy$, w is a function of the two variables x and y. Hence $\frac{dw}{dz}$ is an ordinary derivative but $\frac{\partial w}{\partial x}$ is a partial one.

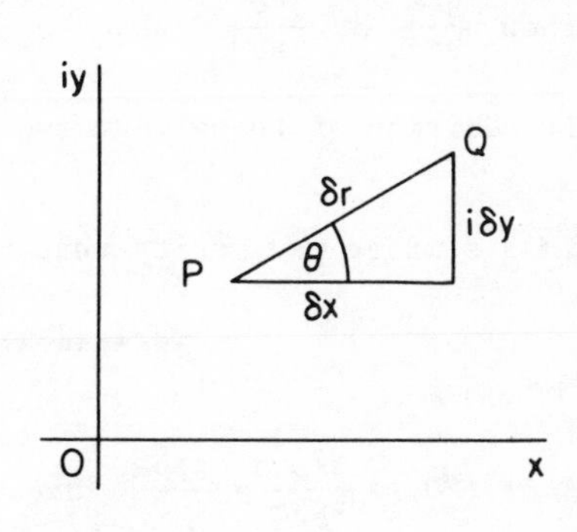

Then as $w = u + iv$,

$$\frac{dw}{dz} = \frac{\partial(u + iv)}{\partial x} = \frac{\partial u}{\partial x} + i\frac{\partial v}{\partial x}$$

By the C.R. conditions this also gives

$$\frac{dw}{dz} = \frac{\partial v}{\partial y} - i\frac{\partial u}{\partial y}$$

If $w = z^2$, verify that $\frac{dw}{dz} = \frac{\partial u}{\partial x} + i\frac{\partial v}{\partial x} = \frac{\partial v}{\partial y} - i\frac{\partial u}{\partial y}$ by obtaining all of these three expressions.

****

19A

$w = z^2$ $\quad\quad \therefore \frac{dw}{dz} = 2z$

Also $u + iv = (x + iy)^2 = x^2 + 2ixy - y^2$

$u = x^2 - y^2$ $\quad\quad v = 2xy$

$$\therefore \frac{\partial u}{\partial x} + i\frac{\partial v}{\partial x} = 2x + i\,2y = 2(x + iy) = 2z$$

$$\frac{\partial v}{\partial y} - i\frac{\partial u}{\partial y} = 2x - i(-2y) = 2z$$

FRAME 20

Conjugate Harmonic Functions

If u and v are two functions that satisfy the C.R. conditions they are said to be CONJUGATE FUNCTIONS. Furthermore as

$$\frac{\partial u}{\partial x} = \frac{\partial v}{\partial y} \qquad \text{and} \qquad \frac{\partial u}{\partial y} = -\frac{\partial v}{\partial x} \qquad (20.1)$$

(differentiate w.r.t. x) (differentiate w.r.t. y)

FRAME 20 continued

then $\frac{\partial^2 u}{\partial x^2} = \frac{\partial^2 v}{\partial x \partial y}$ and $\frac{\partial^2 u}{\partial y^2} = -\frac{\partial^2 v}{\partial y \partial x}$

By addition of these last two results, it follows that $\frac{\partial^2 u}{\partial x^2} + \frac{\partial^2 u}{\partial y^2} = 0$.

In a similar way verify that $\frac{\partial^2 v}{\partial x^2} + \frac{\partial^2 v}{\partial y^2} = 0$.

****

20A

From (20.1) $\frac{\partial^2 u}{\partial y \partial x} = \frac{\partial^2 v}{\partial y^2}$ *and* $\frac{\partial^2 u}{\partial x \partial y} = -\frac{\partial^2 v}{\partial x^2}$

The result follows immediately.

FRAME 21

u and v are therefore both solutions of $\frac{\partial^2 \phi}{\partial x^2} + \frac{\partial^2 \phi}{\partial y^2} = 0$, known as Laplace's equation in two dimensions. A function that satisfies this equation is said to be HARMONIC. u and v are therefore CONJUGATE HARMONIC FUNCTIONS. Given either one of u or v it is possible to find the other. When both u and v are known, w follows immediately. In the next frames we shall consider one or two examples.

FRAME 22

Example: Test whether the following functions are harmonic and find the conjugate harmonic function if it exists. Find also w where possible.

(a) $u = \sin x \cosh y$ (b) $v = 2x - y^2$

(We shall now use the abbreviated notation for partial derivatives e.g. u_x for $\frac{\partial u}{\partial x}$, v_{yy} for $\frac{\partial^2 v}{\partial y^2}$ etc.)

(a) $u = \sin x \cosh y$

First differentiate u to find u_{xx} and u_{yy}.

****

22A

$u_{xx} = -\sin x \cosh y \qquad u_{yy} = \sin x \cosh y$

FRAME 23

From the results in 22A, you will notice that $u_{xx} + u_{yy} = 0$.
Now from the C.R. conditions,

$v_x \; (= -u_y) = -\sin x \sinh y, \qquad v_y \; (= u_x) = \cos x \cosh y.$

Integrate these two equations to find two expressions for v.

****

23A

$v = \cos x \sinh y + f(y) \quad$ *or* $\quad v = \cos x \sinh y + g(x)$

(Hope you didn't forget the arbitrary functions.)

FRAME 24

We now have to choose f(y) and g(x) so that the same expression for v is obtained in each case. This will happen if f(y) = g(x) = c (a constant, which may be zero). Then v = cos x sinh y + c.

Now w = u + iv

$\therefore$ w = sin x cosh y + i(cos x sinh y + c)

As you already know, if w is given in terms of x and y, it can be rewritten in terms of z and $\bar{z}$. In this case, the easiest way of doing this is to remember that cos iy = cosh y and sin iy = i sinh y. (See FRAME 61, page 4:43, Vol I if you have forgotten these results.)

Then w = sin x cos iy + cos x sin iy + ic

= sin (x + iy) + ic

= sin z + ic

FRAME 24 continued

(b) $v = 2x - y^2$ (The second part of the example in FRAME 22.)

$v_x = 2$ $\qquad v_{xx} = 0$

$v_y = -2y$ $\qquad v_{yy} = -2$

$\therefore v_{xx} + v_{yy} = -2$ and v is not harmonic.

Now use the C.R. conditions to obtain u_x and u_y, integrate and show that u does not exist.

****

24A

$u_x = -2y$ $\qquad u_y = -2$

$\therefore u = -2xy + f(y)$ *or* $u = -2y + g(x)$

It is impossible to choose f(y) and g(x) to give the same value of u in each case and therefore u does not exist. This will always happen if v is not harmonic and vice versa.

FRAME 25

Test each of the following functions to find whether it is harmonic and obtain the conjugate function where possible. Find also, where it exists, w in terms of z and/or $\bar{z}$.

(i) $u = x^2 + y^2$ $\qquad$ (ii) $v = 3x^2y - y^3$

****

25A

(i) $u_{xx} + u_{yy} = 4$

v does not exist.

(ii) $v_x = 6xy$ $\qquad v_y = 3x^2 - 3y^2$

$v_{xx} = 6y$ $\qquad v_{yy} = -6y$

$\therefore$ *v is harmonic.*

25A continued

$u_x = 3x^2 - 3y^2 \qquad u_y = -6xy$

$u = x^3 - 3xy^2 + f(y) \qquad u = -3xy^2 + g(x)$

Choose $f(y) = c, \qquad g(x) = x^3 + c$

Then $u = x^3 - 3xy^2 + c$

$$\begin{aligned} w &= x^3 - 3xy^2 + c + i(3x^2y - y^3) \\ &= x^3 + 3x^2iy - 3xy^2 - iy^3 + c \\ &= x^3 + 3x^2iy + 3x(iy)^2 + (iy)^3 + c \\ &= (x + iy)^3 + c \\ &= z^3 + c \end{aligned}$$

[If you don't spot this, put $x = \frac{1}{2}(z + \bar{z}),\ y = \frac{1}{2i}(z - \bar{z})$]

FRAME 26

Several practical situations arise which involve the use of harmonic and conjugate functions.

For example, if heat flows in two dimensions through a body, then, in the steady state, it can be shown that $\frac{\partial^2 T}{\partial x^2} + \frac{\partial^2 T}{\partial y^2} = 0$, i.e. T is harmonic. The diagram illustrates such a case.

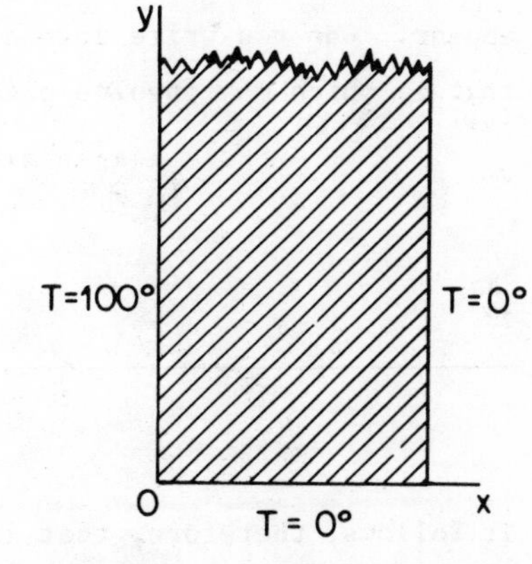

It represents a cross-section of a thick slab of material, the material being so thick that heat flow into and out of the section shown is negligible. The faces of the material are maintained at the temperatures shown and sufficient time is allowed to elapse for the steady state to obtain. The temperature T at any point in this section then satisfies the partial differential equation quoted.

The isothermals are then given by equations of the form T = constant. Furthermore, if T and S are conjugate functions, the lines of heat flow are given by equations of the form S = constant.

Other branches of science and technology where similar results arise are hydrodynamics, diffusion and electrostatics.

FRAME 26 continued

The applications of complex variables to such situations are better considered when you have a greater knowledge of complex mapping and so further discussion will be postponed until the last programme in this Unit.

FRAME 27

You will have noticed that, in the two cases in FRAMES 22-25 where w exists, when w was expressed in terms of z and $\bar{z}$, it only contained z and not $\bar{z}$. This is quite general as we shall now show. It is a direct consequence of the requirement that the function w has a unique derivative at a point, irrespective of the direction of $\overrightarrow{PQ}$ (see FRAME 14). As this requirement gives rise to the C.R. conditions, the non-appearance of $\bar{z}$ must follow directly from these conditions.

FRAME 28

Suppose you have a function ϕ which in general is given in terms of x and y. It may happen that, in some cases, ϕ is a function of x alone, i.e. y does not appear. Can you write down a condition, in terms of a derivative of ϕ, such that ϕ, which may involve both x and y, actually involves x only?

****

28A

$$\frac{\partial \phi}{\partial y} = 0$$

FRAME 29

It follows, therefore, that if we assume w to be a function of z and $\bar{z}$, then $\frac{\partial w}{\partial \bar{z}} = 0$ is the condition that $\bar{z}$ is missing from the relation $w = f(z,\bar{z})$ and so it only contains z.

Now $w = u + iv$ where u and v are functions of x and y and

$$x = \frac{1}{2}(z + \bar{z}), \qquad y = \frac{1}{2i}(z - \bar{z})$$

FRAME 29 continued

Using the chain rule for partial derivatives

$$\frac{\partial u}{\partial \bar{z}} = \frac{\partial u}{\partial x}\frac{\partial x}{\partial \bar{z}} + \frac{\partial u}{\partial y}\frac{\partial y}{\partial \bar{z}}$$

$$= \frac{1}{2}\frac{\partial u}{\partial x} - \frac{1}{2i}\frac{\partial u}{\partial y}$$

Also $\dfrac{\partial v}{\partial \bar{z}} = \dfrac{\partial v}{\partial x}\dfrac{\partial x}{\partial \bar{z}} + \dfrac{\partial v}{\partial y}\dfrac{\partial y}{\partial \bar{z}}$

$$= \frac{1}{2}\frac{\partial v}{\partial x} - \frac{1}{2i}\frac{\partial v}{\partial y}$$

Remembering that as $w = u + iv$, $\dfrac{\partial w}{\partial \bar{z}} = \dfrac{\partial u}{\partial \bar{z}} + i\dfrac{\partial v}{\partial \bar{z}}$, write down the expression for $\dfrac{\partial w}{\partial \bar{z}}$ and show that it is zero.

****

29A

$$\frac{\partial w}{\partial \bar{z}} = \frac{1}{2}\frac{\partial u}{\partial x} + \frac{1}{2}i\frac{\partial u}{\partial y} + i\left(\frac{1}{2}\frac{\partial v}{\partial x} + \frac{1}{2}i\frac{\partial v}{\partial y}\right)$$

$$= \frac{1}{2}\left(\frac{\partial u}{\partial x} - \frac{\partial v}{\partial y}\right) + \frac{1}{2}i\left(\frac{\partial u}{\partial y} + \frac{\partial v}{\partial x}\right)$$

$= 0$ *as by the C.R. conditions,* $\dfrac{\partial u}{\partial x} = \dfrac{\partial v}{\partial y}$ *and* $\dfrac{\partial u}{\partial y} = -\dfrac{\partial v}{\partial x}$

FRAME 30

It therefore follows immediately that w is a function of z alone and does not contain $\bar{z}$. This is the limitation placed on w by the condition that it must have a unique derivative at any point. From now onwards, whenever a function w is mentioned, it will be assumed to contain z only unless otherwise specified.

FRAME 31

Regular or Analytic Functions

A function w is said to be REGULAR (or ANALYTIC) in a certain domain D (region of the Argand diagram) if

(i) for every value of z in D, w is single-valued and is finite (i.e. $|w|$ is finite)

(ii) for every value of z in D, w is differentiable.

FRAME 32

Some functions, which are very important in work involving integration, are regular at all except a finite number of points in a domain. These points are called the SINGULARITIES or SINGULAR POINTS of the function. For example, $w = z^2 - 2z + 3$ is regular for all finite z, but $w = \dfrac{z - 2}{(z + 3)(z + 2i)}$ is not regular at $z = -3$ amd $z = -2i$ as the function is not defined at these points, i.e. becomes infinite.

Where are (i) $\dfrac{1}{z}$ and (ii) $\dfrac{1}{z^2 + 1}$ not regular?

****

32A

(i) $z = 0$ *(ii)* $z = i$ *and* $z = -i$

FRAME 33

Examples

Before considering some particular functions, here are a few general examples on the first part of this programme for you to try. When you have done them, turn to FRAME 34 to check your results.

1. Find $\lim\limits_{z \to i} \dfrac{z^2 + 1}{z^3 - 6z^2 i - 11z + 6i}$

2. Test whether each of the following is harmonic, find the conjugate harmonic function if it exists and where it does, find also w in terms of z:

   (a) $v = e^x \sin y$ (b) $u = x^2 - 2y^2$

FRAME 33 continued

3. Show that if $u = x^2 - y^2$, $v = \frac{-y}{x^2 + y^2}$, then $u + iv$ is not a regular function of z but that both u and v are harmonic. (HINT: If $u + iv$ is regular, it must be differentiable and hence the C.R. conditions must be satisfied.)

FRAME 34

Answers to Examples

1. $-i$

2. (a) Yes $u = e^x \cos y + c$ $w = e^z + c$

   (b) No v and w do not exist.

3. The derivatives of v are best found by writing $v = -\frac{y}{r^2}$ where $r^2 = x^2 + y^2$.

FRAME 35

Power Series

In real variable work you are familiar with the result

$$\frac{1}{1 - x} = 1 + x + x^2 + x^3 + \ldots$$

and know that it is valid for all values of x lying between 1 and -1 but not for 1 and -1 themselves. In complex variable theory the corresponding result would be

$$\frac{1}{1 - z} = 1 + z + z^2 + z^3 + \ldots \qquad (35.1)$$

and the corresponding question would arise that, if this is a valid expansion, for what values of z does it hold?

Such a series as this is called a POWER SERIES and, as the number of terms increases indefinitely, it may CONVERGE to a particular limit or DIVERGE. As is the case with expansions in real power series, divergence spells disaster.

Returning to (35.1), let $z = \frac{1}{2}i$ and show, by considering the real and imaginary parts separately, that the series on the R.H.S. does actually add up to $\frac{4}{5} + \frac{2}{5}i$ which is the value of $1/(1 - \frac{1}{2}i)$.

****

35A

$$R.H.S. = 1 + \frac{1}{2}i + \left(\frac{1}{2}i\right)^2 + \left(\frac{1}{2}i\right)^3 + \ldots$$

$$= 1 + \frac{1}{2}i - \frac{1}{4} - \frac{1}{8}i + \frac{1}{16} + \frac{1}{32}i - \frac{1}{64} - \frac{1}{128}i \ldots$$

$$= \left(1 - \frac{1}{4} + \frac{1}{16} - \frac{1}{64} \ldots\right) + \frac{1}{2}i\left(1 - \frac{1}{4} + \frac{1}{16} - \frac{1}{64} \ldots\right) \qquad (35A.1)$$

Now $1 - \frac{1}{4} + \frac{1}{16} - \frac{1}{64} \ldots = 1 - \frac{1}{4} + \left(\frac{1}{4}\right)^2 - \left(\frac{1}{4}\right)^3 \ldots$

and this is a geometric progression with first term 1 and common ratio $-\frac{1}{4}$.

Its sum to infinity is $\dfrac{1}{1 - \left(-\frac{1}{4}\right)} = \dfrac{4}{5}$. *(35A.1) therefore becomes* $\frac{4}{5} + \frac{2}{5}i$.

FRAME 36

If other values of z are taken instead of $\frac{1}{2}i$ it will be found that (35.1) is valid so long as $|z| < 1$. This means that z can lie anywhere inside the circle whose centre is at the origin and whose radius is 1. This circle is called the CIRCLE OF CONVERGENCE and its radius the RADIUS OF CONVERGENCE. As an example of what happens when $|z| > 1$, try putting $z = 2$ in (35.1).

****

36A

$L.H.S. = \frac{1}{1-2} = -1$. *R.H.S.* $= 1 + 2 + 4 + 8 + \ldots$ *and this certainly does not add up to -1.*

FRAME 37

If you read Unit 1 in Vol I, you will be familiar with the ratio test for real power series. In case you have forgotten it, or if you have not met it before, it states that the series

$$a_0 + a_1x + a_2x^2 + a_3x^3 + \ldots$$

converges if $\lim\limits_{n\to\infty}\left|\dfrac{a_{n+1}x^{n+1}}{a_n x^n}\right|$ i.e. $\lim\limits_{n\to\infty}\left|\dfrac{a_{n+1}}{a_n}x\right| < 1$ and diverges if this

FRAME 37 continued

limit > 1. If the limit = 1, no conclusion can be drawn.

As $\lim_{n\to\infty}\left|\frac{a_{n+1}}{a_n}x\right| = \lim_{n\to\infty}\left|\frac{a_{n+1}}{a_n}\right||x|$, this result can also be expressed slightly differently as follows:

Let $\lim_{n\to\infty}\left|\frac{a_n}{a_{n+1}}\right| = R$. Then if $|x| < R$, the series converges,
$|x| > R$, the series diverges,
$|x| = R$, no conclusion.

Here, you have probably thought of $|x|$ as being the numerical value of x. However, if you think about it more carefully, you will see that this is only a special case of the more general interpretation given to $|z|$.

FRAME 38

In the case of a complex power series, a corresponding result, which we shall not prove, holds as follows:

Given the series

$$a_0 + a_1z + a_2z^2 + a_3z^3 + \ldots$$

let $\lim_{n\to\infty}\left|\frac{a_n}{a_{n+1}}\right| = R$. Then if $|z| < R$, the series converges,
$|z| > R$, the series diverges,
$|z| = R$, no conclusion.

Notice that in this case, the a's may be complex as well as the z's. R is the radius of convergence.

Returning to the series in FRAME 35, all the a's were 1 and hence the series converges so long as $|z| < 1$.

Another result, which is equivalent to this, states that the radius of convergence is the distance from the origin to the nearest singularity of the sum function. In the series already quoted the sum function is $\frac{1}{1 - z}$ which has a singularity at 1. Again we shall not prove this here but it will become more evident in the programme "Complex Integration" in this Unit.

FRAME 38 continued

Example: Expressed as a power series in z,

$$\frac{1}{z+5i} = \frac{1}{5i} - \frac{z}{(5i)^2} + \frac{z^2}{(5i)^3} - \frac{z^3}{(5i)^4} \ldots\ldots + (-1)^n \frac{z^n}{(5i)^{n+1}} + \ldots$$

[NOTE: This series for $\frac{1}{z+5i}$ can be obtained either by long division or by the binomial theorem, first writing it as $\frac{1}{5i\left(1+\frac{z}{5i}\right)}$.]

$$\frac{a_n}{a_{n+1}} = \frac{(-1)^n/(5i)^{n+1}}{(-1)^{n+1}/(5i)^{n+2}} = -5i \quad \therefore R = 5$$

Alternatively $\frac{1}{z+5i}$ has a singularity at $z = -5i$ which gives the same result.

Express each of the following as power series and state the radius of convergence:

(a) $\frac{1}{2z-1}$

(b) $\frac{30}{(z-2)(z+3)}$ (HINT: First put this into partial fractions and leave the result as two separate series.)

****

38A

(a) $-1 - 2z - (2z)^2 - (2z)^3 \ldots\ldots \qquad R = \frac{1}{2}$

(b) $\frac{30}{(z-2)(z+3)} = -3\frac{1}{1-\frac{z}{2}} - 2\frac{1}{1+\frac{z}{3}}$

$$= -3\left(1 + \frac{z}{2} + \frac{z^2}{4} + \frac{z^3}{8} + \ldots\right) - 2\left(1 - \frac{z}{3} + \frac{z^2}{9} - \frac{z^3}{27} + \ldots\right)$$

$$R = 2$$

(The first series gives $R = 2$, i.e. $|z| < 2$ and the second $R = 3$, i.e. $|z| < 3$ To satisfy both these inequalities, we must have $|z| < 2$, i.e. $R = 2$.)

FRAME 39

The Exponential Function

Certain well-known functions exist in real variable work and we shall now have a look at their counterparts in complex quantities.

The exponential function $w = e^z$ or $w = \exp z$ you have already met, having previously used it in the form $e^{x+iy} = e^x(\cos y + i \sin y)$. It is formally defined by the series

$$e^z = 1 + z + \frac{z^2}{2!} + \frac{z^3}{3!} + \dots \qquad (39.1)$$

(Remember $e^x = 1 + x + \frac{x^2}{2!} + \frac{x^3}{3!} + \dots$)

What is the radius of convergence of the series for e^z?

****

39A

∞. *Hence it is valid for all finite z.*

FRAME 40

Most of the rules for the exponential function are similar to those for e^x. Thus

$$e^{z_1} \times e^{z_2} = e^{z_1+z_2}$$

$$e^{-z} = 1/e^z$$

$$\frac{d}{dz} e^z = e^z$$

Verify this last result by

1. differentiating the series for e^z
2. using the result $\frac{dw}{dz} = \frac{\partial u}{\partial x} + i \frac{\partial v}{\partial x}$ (see FRAME 19)

****

40A

1. $\frac{d}{dz}\left(1 + z + \frac{z^2}{2!} + \frac{z^3}{3!} + \ldots\right) = 1 + \frac{2z}{2!} + \frac{3z^2}{3!} + \frac{4z^3}{4!} + \ldots$ (see FRAME 14, Example 1)

$$= 1 + z + \frac{z^2}{2!} + \frac{z^3}{3!} + \ldots$$

2. As $e^z = e^x(\cos y + i \sin y)$

$$u = e^x \cos y \qquad v = e^x \sin y$$

$$\therefore \frac{\partial u}{\partial x} + i\frac{\partial v}{\partial x} = e^x \cos y + ie^x \sin y$$

$$= e^z$$

FRAME 41

There is, however, one property of e^z that does not hold for e^x. This is the fact that z is periodic and has period $2\pi i$.

Show that $e^{z+2\pi i} = e^z$.

****

41A

$$e^{z+2\pi i} = e^z e^{2\pi i}$$

$$= e^z(\cos 2\pi + i \sin 2\pi)$$

$$= e^z$$

FRAME 42

As you know, if a function is periodic, its corresponding inverse function is many-valued, e.g. sin x is periodic and $\sin^{-1} x$ is many-valued. In real variable theory the inverse function corresponding to e^x is ln x and this, together with the fact that e^z is periodic, leads us to suspect that ln z may be many-valued. We shall see later that it is and, as a result, added complications will arise. But before we go on to consider this function, the trigonometric and hyperbolic functions will be mentioned.

FRAME 43

The Trigonometric Functions

When you commenced to study sines and cosines, they were defined in terms of the sides of a right-angled triangle. Later you found that $\sin\theta$, for example, could be expressed as a Maclaurin series, i.e.

$$\sin\theta = \theta - \frac{\theta^3}{3!} + \frac{\theta^5}{5!} - \frac{\theta^7}{7!} \ldots\ldots \quad (43.1)$$

or even in terms of complex numbers, i.e.

$$\sin\theta = \frac{1}{2i}\left(e^{i\theta} - e^{-i\theta}\right) \quad (43.2)$$

(see FRAME 59, page 4:42, Vol I)

Now sin z obviously cannot be expressed in terms of the sides of a triangle for while a triangle can have an angle of 30° or $\frac{\pi}{6}$ at one of its vertices, an angle such as $\frac{\pi}{6} + i$ would be rather difficult to visualise. Consequently, we use one of the two results (43.1) or (43.2) to define sin z. They are really equivalent, as can easily be seen.

Using (43.2), sin z can be defined by the equation

$$\sin z = \frac{1}{2i}\left(e^{iz} - e^{-iz}\right) \quad (43.3)$$

Substitute the series for e^{iz} and e^{-iz} (obtain them from (39.1)) into the R.H.S. of (43.3) and show that the result is $z - \frac{z^3}{3!} + \frac{z^5}{5!} - \frac{z^7}{7!} + \ldots$

****

43A

$$e^{iz} = 1 + (iz) + \frac{(iz)^2}{2!} + \frac{(iz)^3}{3!} + \frac{(iz)^4}{4!} + \frac{(iz)^5}{5!} + \ldots$$

$$= 1 + iz - \frac{z^2}{2!} - \frac{iz^3}{3!} + \frac{z^4}{4!} + \frac{iz^5}{5!} \ldots \quad (43A.1)$$

$$e^{-iz} = 1 - iz - \frac{z^2}{2!} + \frac{iz^3}{3!} + \frac{z^4}{4!} - \frac{iz^5}{5!} \ldots \quad (43A.2)$$

$$e^{iz} - e^{-iz} = 2iz - \frac{2iz^3}{3!} + \frac{2iz^5}{5!} - \frac{2iz^7}{7!} + \ldots$$

$$\frac{1}{2i}\left(e^{iz} - e^{-iz}\right) = z - \frac{z^3}{3!} + \frac{z^5}{5!} - \frac{z^7}{7!} + \ldots$$

FRAME 44

As you have already seen (in FRAME 9, page 12:12) how to find the real and imaginary parts of sin z, there is no need to say anything further about this here.

Now show that $\frac{1}{2}\left(e^{iz} + e^{-iz}\right) = 1 - \frac{z^2}{2!} + \frac{z^4}{4!} - \frac{z^6}{6!} + \ldots$

****

44A

Adding (43A.1) and (43A.2) gives the result immediately.

FRAME 45

cos z is now defined as

$$\cos z = \frac{1}{2}\left(e^{iz} + e^{-iz}\right) \quad \text{or}$$

$$\cos z = 1 - \frac{z^2}{2!} + \frac{z^4}{4!} - \frac{z^6}{6!} + \ldots$$

These are analogous to the results

$$\cos \theta = \frac{1}{2}\left(e^{i\theta} + e^{-i\theta}\right)$$

$$\cos \theta = 1 - \frac{\theta^2}{2!} + \frac{\theta^4}{4!} - \frac{\theta^6}{6!} \ldots\ldots$$

in real variable work.

FRAME 46

The other four trigonometric functions of z are defined by relations similar to their real counterparts:

$$\tan z = \frac{\sin z}{\cos z} \qquad \cot z = \frac{1}{\tan z}$$

$$\sec z = \frac{1}{\cos z} \qquad \operatorname{cosec} z = \frac{1}{\sin z}$$

FRAME 47

You will remember that a function has a singularity at a point if it becomes infinite at that point. Thus tan z has singularities where cos z = 0.

Express cos z in the form u + iv.

****

47A

$\cos z = \cos x \cosh y - i \sin x \sinh y$

FRAME 48

$$\text{If } \cos z = 0, \text{ then } \left.\begin{aligned} \cos x \cosh y &= 0 \\ \sin x \sinh y &= 0 \end{aligned}\right\} \text{ simultaneously.}$$

Find the values of x and y satisfying these two equations.

****

48A

$x = \left(\frac{1}{2} + n\right)\pi$, n an integer. $y = 0$
(To solve these equations note that $\cos x \cosh y = 0$ if $\cos x = 0$ or if $\cosh y = 0$. As y is real the latter is impossible. Thus $\cos x = 0$, $x = \left(\frac{1}{2} + n\right)\pi$. Now as $\cos x = 0$, $\sin x \neq 0$, therefore $\sinh y = 0$ and so $y = 0$.

FRAME 49

tan z therefore is not regular when $z = \left(\frac{1}{2} + n\right)\pi$, (n an integer).
sec z will also not be regular at these points.

Properties of the trigonometric functions are similar to those for real functions. For example:

$$\cos^2 z + \sin^2 z = 1 \qquad (49.1)$$

$$\cos(z_1 + z_2) = \cos z_1 \cos z_2 - \sin z_1 \sin z_2$$

$$\frac{d}{dz} \cos z = - \sin z$$

$$\frac{d}{dz} \sec z = \sec z \tan z \quad \text{etc.}$$

See if you can verify (49.1). (Make use of the exponential definitions of these functions.)

****

49A

$$L.H.S. = \{\tfrac{1}{2}(e^{iz} + e^{-iz})\}^2 + \{\tfrac{1}{2i}(e^{iz} - e^{-iz})\}^2$$
$$= 1$$

FRAME 50

The Hyperbolic Functions

The definitions of the hyperbolic functions follow on in a very similar way as also do their properties. Just a few examples will be sufficient.

sinh z is defined to be $\frac{1}{2}(e^z - e^{-z})$ or $z + \frac{z^3}{3!} + \frac{z^5}{5!} + \ldots$

cosh z is defined to be $\frac{1}{2}(e^z + e^{-z})$ or $1 + \frac{z^2}{2!} + \frac{z^4}{4!} + \ldots$

$\tanh z = \frac{\sinh z}{\cosh z}$ and so on for the other hyperbolic functions.

$$\sinh 2z = 2 \sinh z \cosh z$$
$$\sinh (z_1 + z_2) = \sinh z_1 \cosh z_2 + \cosh z_1 \sinh z_2$$
$$\frac{d}{dz} \sinh z = \cosh z$$
$$\frac{d}{dz} \tanh z = \operatorname{sech}^2 z$$

Show that u and v where $u + iv = \sinh (x + iy)$ satisfy Laplace's equation and the C.R. conditions.

****

50A

$u = \sinh x \cos y$	$v = \cosh x \sin y$
$u_x = \cosh x \cos y$	$v_x = \sinh x \sin y$
$u_y = -\sinh x \sin y$	$v_y = \cosh x \cos y$
$u_{xx} = \sinh x \cos y$	$v_{xx} = \cosh x \sin y$
$u_{yy} = -\sinh x \cos y$	$v_{yy} = -\cosh x \sin y$

Hence results.

FRAME 51

The Logarithmic Function

To find ln z, we first put z into its exponential form.

Suppose $|z| = r$ and $\arg z = \theta + 2k\pi$ ($-\pi < \theta \leq \pi$, k an integer, positive or negative) then $z = re^{i(\theta+2k\pi)}$.

Taking logs of both sides of this equation,

$$\ln z = \ln\{re^{i(\theta+2k\pi)}\}$$
$$= \ln r + i(\theta + 2k\pi) \qquad \text{(NOTE: Remember } \ln e^{\alpha} = \alpha\text{)}$$

or in words, ln c.n. = ln mod + i arg.

For example, to find ln (3 - 4i) we first require the modulus and argument of 3 - 4i. Obtain these for yourself.

****

51A

$|3 - 4i| = 5$

$\arg(3 - 4i) = -53^{o}8' + k360^{o} = -0{\cdot}9273 + 2k\pi$ *radians*

FRAME 52

Then $\ln(3 - 4i) = \ln 5 + i(-0{\cdot}9273 + 2k\pi)$
$= 1{\cdot}6094 + i(-0{\cdot}9273 + 2k\pi)$

This process also enables us to find the log of a negative number.

Try finding ln(-4).

****

52A

$|-4| = 4 \qquad \arg(-4) = \pi + 2k\pi$

$\ln(-4) = \ln 4 + i\pi(1 + 2k)$

FRAME 53

In FRAME 42 we mentioned that we should find that ln z is a many-valued function. This is immediately evident as $\ln z = \ln r + i(\theta + 2k\pi)$ and k can take any integral value.

The principal value of ln z is defined to be that value of ln z for which $k = 0$, i.e. the principal value of ln z is $\ln r + i\theta$, $(-\pi < \theta \leqslant \pi)$.

In FRAME 3 the idea of branches was introduced in the case of many-valued functions. In the case of the log function, each value of k gives one branch. In the case of $w = z^{\frac{1}{2}}$ there were just two branches.

How many branches will ln z have?

****

53A

An infinite number as k can be any integer.

FRAME 54

Now let us take a particular branch of the log function, i.e. a particular value of k, say K, and go on a journey $P_1P_2P_3P_4P_1$, there being two possible paths from P_4 back to P_1, the short path via P_5 and the longer one via P_6 and P_7. Let the argument of P_1, i.e. $P_1\hat{O}x$, be $-\pi + \delta$ and that of $P_4 (P_4\hat{O}x)$ be $\pi - \varepsilon$, the principal value of the argument being taken in each case.

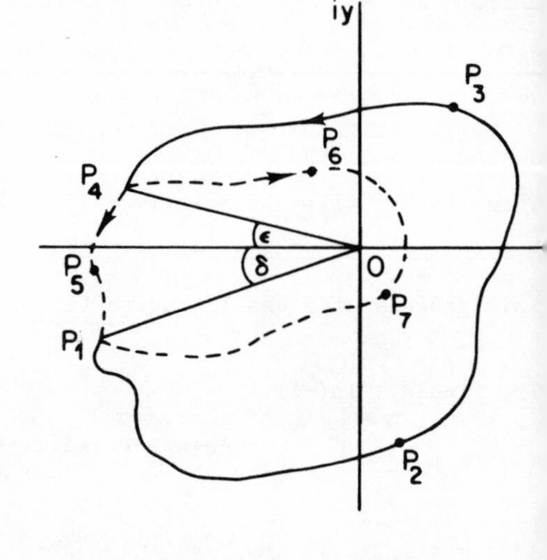

Then $\ln \overrightarrow{OP_1} = \ln OP_1 + i(-\pi + \delta + 2K\pi)$

$\ln \overrightarrow{OP_4} = \ln OP_4 + i(\pi - \varepsilon + 2K\pi)$

As one travels from P_1 to P_4 via P_2 and P_3 the argument has increased from $-\pi + \delta$ to $\pi - \varepsilon$. Now complete the circuit. If the path $P_4P_6P_7P_1$ is followed the argument decreases back to its original value and so the logarithm returns

FRAME 54 continued

to its original value. But if the path $P_4P_5P_1$ is taken then the argument continues to increase and by the time P_1 is reached it is $\pi + \delta$ and so $\ln \overrightarrow{OP}_1 = \ln OP_1 + i(\pi + \delta + 2K\pi)$. This is not the same as the original value. It can be written in the form

$$\ln \overrightarrow{OP}_1 = \ln OP_1 + i(-\pi + \delta + 2\,\overline{K+1}\,\pi)$$

The value of K has effectively been increased by 1 and this means that we are on the next branch.

Draw an Argand diagram to show these two values of $\ln \overrightarrow{OP}_1$.

****

54A

OQ_1 and OQ_1' represent the two values.

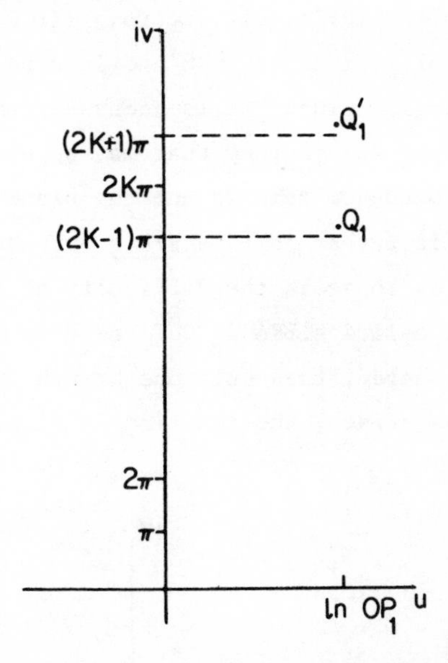

FRAME 55

You will remember that, when we were dealing with limits, continuity, differentiability and regular functions, the requirement was constantly met that the function must be single-valued. Also in the last frame we saw that this condition is satisfied for a return path such as $P_4P_6P_7P_1$ but not for the path $P_4P_5P_1$.

FRAME 55 continued

Can you see the difference between the two paths (in terms of where they cross the real axis) that gives rise to this distinction?

****

55A

The path $P_4P_6P_7P_1$ crosses the positive real axis.

The path $P_4P_5P_1$ crosses the negative real axis.

FRAME 56

If you select any other path from P_4 back to P_1 (except one that passes through 0) you will find that a similar result occurs. (If 0 is a point on the path there is another difficulty, i.e. $|z| = 0$ and ln z becomes infinite.) Thus, if we wish to keep ln z single-valued, it is necessary that the negative real axis be not crossed during any journey that may be undertaken in the Argand diagram depicting z. To ensure this we cut the plane right along the negative real axis and regard this cut as an impassable barrier. The point 0 is also included in the cut so as to avoid the difficulty of ln z becoming infinite. The cut that is made is called a BRANCH CUT and 0 is called the BRANCH POINT. If this restriction is obeyed, then only one branch of the log function is involved and on that one branch, the function is single-valued.

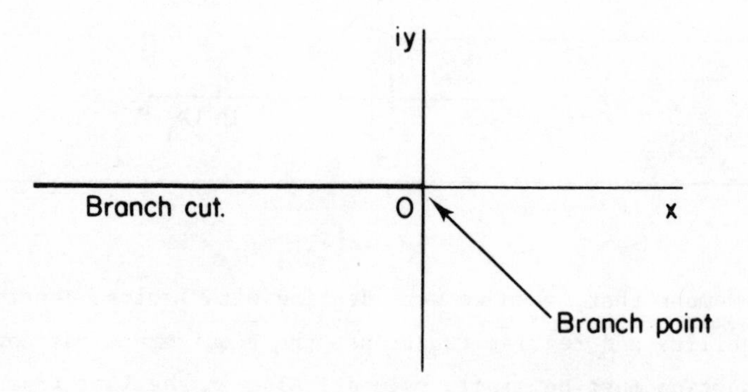

FRAME 57

Selecting a specified branch of ln z, say, for simplicity, the principal branch $\ln z = \ln r + i\theta$, we can now proceed to find its derivative. Now $r = |z| = \sqrt{x^2 + y^2}$ and $\theta = \arg z = \tan^{-1}\frac{y}{x}$.

So $\ln z = \frac{1}{2}\ln(x^2 + y^2) + i\tan^{-1}\frac{y}{x}$, or $u = \frac{1}{2}\ln(x^2 + y^2)$, $v = \tan^{-1}\frac{y}{x}$.

Apply the result $\frac{dw}{dz} = \frac{\partial u}{\partial x} + i\frac{\partial v}{\partial x}$ to obtain the derivative in this case and simplify the result.

****

57A

$$\frac{dw}{dz} = \frac{x}{x^2 + y^2} + i\frac{-y}{x^2 + y^2}$$

$$= \frac{x - iy}{x^2 + y^2}$$

$$= \frac{1}{x + iy}$$

$$= \frac{1}{z}$$

FRAME 58

Thus, so long as we remain on one branch, $\frac{d}{dz}(\ln z) = \frac{1}{z}$.

Now take the two complex numbers i and $-1 + i$ and work out the following:

(a) $\ln i$

(b) $\ln(-1 + i)$

(c) $\ln i + \ln(-1 + i)$

(d) $\ln(-1 - i)$

Take the principal value of each logarithm, i.e. remain on the same branch.

****

58A

(a) $i\frac{\pi}{2}$

(b) $\frac{1}{2}\ln 2 + i\frac{3\pi}{4}$

(c) $\frac{1}{2}\ln 2 + i\frac{5\pi}{4}$

(d) $\frac{1}{2}\ln 2 - i\frac{3\pi}{4}$

FRAME 59

Now $-1 - i = i(-1 + i)$ but you will notice that the result in (d) is not the sum of the results in (a) and (b).

Now see whether $\ln(-1 + i)$, which is $\ln i(1 + i)$, is the same as $\ln i + \ln(1 + i)$.

****

59A

(a) $i\frac{\pi}{2}$ *(b)* $\frac{1}{2}\ln 2 + i\frac{\pi}{4}$

(c) $\frac{1}{2}\ln 2 + i\frac{3\pi}{4}$ *(d)* $\frac{1}{2}\ln 2 + i\frac{3\pi}{4}$

FRAME 60

Here you will notice that the log of the product is the sum of the logs.

Can you, by considering the arguments involved, spot the reason for the difference in the two cases?

****

60A

In the first case the sum of the arguments is greater than π *and so we are on the next branch of the log. This has not happened in the second case.*

FRAME 61

From this you will see that care is necessary about making the statement $\ln z_1 + \ln z_2 = \ln z_1 z_2$. If a particular branch is required for the logs then the result is only true if $-\pi < \arg z_1 + \arg z_2 \leqslant \pi$. If any branch is permissible then it is true to say that

$$\text{a value of } \ln z_1 + \text{a value of } \ln z_2 = \text{a value of } \ln z_1 z_2 .$$

A similar result holds in the case of division, but now the same branch is only maintained if the difference of the arguments lies between $-\pi$ and π (including π but not $-\pi$).

FRAME 62

The Generalised Power $w = a^z$ (a complex)

In real variable theory, if $y = a^x$, then

$$\ln y = x \ln a$$

$$\text{and so} \quad y = e^{x \ln a}$$

In complex variable theory, we define $w = a^z$ as $w = e^{z \ln a}$. As ln a is many-valued so also is w and its principal value is obtained by taking the principal value of ln a.

As an example let us consider $(2 - i)^i$.

By the definition $(2 - i)^i = e^{i \ln (2-i)}$

First find the general value of ln (2 - i).

****

62A

$0{\cdot}8047 + i(-0{\cdot}4634 + 2k\pi)$

FRAME 63

Then $i \ln (2 - i) = (0{\cdot}4634 - 2k\pi) + 0{\cdot}8047i$

$$\text{and} \quad e^{i \ln (2-i)} = e^{(0\cdot4634-2k\pi)+0\cdot8047i}$$

$$= e^{0\cdot4634-2k\pi}(\cos 46^{o}6' + i \sin 46^{o}6')$$

The principal value is $e^{0\cdot4634}(\cos 46^{o}6' + i \sin 46^{o}6')$ which can easily be evaluated.

Now find the principal value of i^{1-i}.

****

63A

$ie^{-\pi/2}$

FRAME 64

L'Hopital's Rule

If you read Unit 1 in Vol I you will be familiar with this result in real variable theory. To remind you, it states that if $f(a) = 0$ and $g(a) = 0$, then

$$\lim_{x\to a} \frac{f(x)}{g(x)} = \lim_{x\to a} \frac{f'(x)}{g'(x)}$$

As an example,

$$\lim_{x\to 0} \frac{\sin x}{x} = \lim_{x\to 0} \frac{\cos x}{1} = 1$$

It is, of course, necessary to check that both numerator and denominator are zero before applying the rule.

FRAME 65

An exactly similar result holds in the case of complex variables. We shall simply state it here without giving a proof.

Given two regular functions of z, i.e. $f(z)$ and $g(z)$, then, if both $f(a) = 0$ and $g(a) = 0$,

$$\lim_{z\to a} \frac{f(z)}{g(z)} = \lim_{z\to a} \frac{f'(z)}{g'(z)}$$

a, in general, will be complex.

Example 1: $$\lim_{z\to 0} \frac{\sin z}{z} = \lim_{z\to 0} \frac{\cos z}{1} = 1$$

Example 2: $$\lim_{z\to i} \frac{z^2 - 2iz - 1}{z^4 + 2z^2 + 1} = \lim_{z\to i} \frac{2z - 2i}{4z^3 + 4z} = \lim_{z\to i} \frac{2}{12z^2 + 4} = -\frac{1}{4}$$

Note that, in this case, both $2z - 2i$ and $4z^3 + 4z$ are zero when $z = i$ and so, just as in real variable work, the process is repeated.

Evaluate (a) $\displaystyle\lim_{z\to 1+i} \frac{z^2 - (1 + 3i)z - 2(1 - i)}{z^2 - (1 + i)z}$ (b) $\displaystyle\lim_{z\to 0} \frac{z - \sin z}{z^3}$

65A

(a) $-i$ *(b)* $\frac{1}{6}$ *(after applying the rule three times)*

FRAME 66

Miscellaneous Examples

In this frame a collection of miscellaneous examples is given for you to try. Answers are supplied in FRAME 67 together with such working as is considered helpful.

1. Show that $u = 2 \cos x \sinh y + 4xy$ is harmonic. Find the conjugate harmonic function v and also w as a function of z.

2. By expressing sin z and cos z in terms of exponential functions, find all the solutions of $\sin z = 2i \cos z$.

3. Express i^i in the form $a + ib$.

4. Show that the points representing the roots of the equation $(z + i)^6 = 64(z - i)^6$ lie on a circle and determine this circle completely. (HINT: Work in terms of moduli.)

5. If $w = u + iv$ is a regular function, show that the two families of curves in the x,y plane whose equations are $u = \lambda$ and $v = \mu$ are orthogonal, λ and μ being constants. (HINT: In each case, find an expression for dy/dx, remembering that u and v are both functions of x and y.)

6. Is the function

$$\begin{cases} f(z) = \dfrac{z + i}{z^2 + 1} & (z \neq i \text{ or } -i) \\ f(z) = 0 & (z = i) \\ f(z) = \dfrac{1}{2}i & (z = -i) \end{cases}$$

continuous at the points i and -i?

FRAME 66 continued

7. Evaluate $\lim_{z \to 0} \dfrac{1 - \cos z}{\sin z^2}$.

8. An identity which occurs in the quantum theory of photoionisation is

$$\left(\frac{ia - 1}{ia + 1}\right)^{ib} = e^{-2b \cot^{-1} a}$$

where a and b are real. Verify this result.

FRAME 67

Answers to Miscellaneous Examples

1. $v = -2 \sin x \cosh y + 2y^2 - 2x^2 + c$

   $w = -2i \sin z - 2iz^2 + ic$

2. $$\frac{1}{2i}\left(e^{iz} - e^{-iz}\right) = i\left(e^{iz} + e^{-iz}\right)$$

$$3e^{iz} = -e^{-iz}$$

$$e^{2iz} = -\frac{1}{3}$$

$$2iz = \ln \frac{1}{3} + i(\pi + 2k\pi)$$

$$z = \frac{1}{2} i \ln 3 + \left(k + \frac{1}{2}\right)\pi$$

3. $i^i = e^{i \ln i}$

   $\ln i = i\left(\frac{\pi}{2} + 2k\pi\right)$

   $i \ln i = -\left(\frac{1}{2} + 2k\right)\pi$

   $i^i = e^{-(\frac{1}{2}+2k)\pi}$

FRAME 67 continued

4. $|z + i|^6 = 64|z - i|^6$

$|x + i(y+1)| = 2|x + i(y - 1)|$

$x^2 + (y + 1)^2 = 4\{x^2 + (y - 1)^2\}$

$\therefore\ 3x^2 + 3y^2 - 10y + 3 = 0$

5. $u = \lambda$ $\qquad$ $v = \mu$

$du = \dfrac{\partial u}{\partial x}dx + \dfrac{\partial u}{\partial y}dy = 0$ $\qquad$ $dv = \dfrac{\partial v}{\partial x}dx + \dfrac{\partial v}{\partial y}dy = 0$

$\dfrac{dy}{dx} = -\dfrac{u_x}{u_y}$ $\qquad$ $\dfrac{dy}{dx} = -\dfrac{v_x}{v_y}$

Product of slopes $= \dfrac{u_x v_x}{u_y v_y}$

$= -1$ (by C.R. conditions).

6. $f(z) = \dfrac{1}{z - i}$ $\quad (z \neq -i)$

When $z = i$ this is infinite. But $f(i) = 0$ by definition

$\therefore$ Not continuous at i.

$\lim_{z\to -i} \dfrac{z + i}{z^2 + 1} = \dfrac{1}{2}i$ and this is the defined value at $z = -i$

$\therefore$ Continuous at $-i$.

7. $$\lim_{z\to 0} \frac{1 - \cos z}{\sin z^2} = \lim_{z\to 0} \frac{\sin z}{2z \cos z^2}$$

$$= \lim_{z\to 0} \frac{\cos z}{2 \cos z^2 - 4z^2 \sin z^2}$$

$$= \frac{1}{2}$$

FRAME 67 continued

8. $\left(\frac{ia - 1}{ia + 1}\right)^{ib} = \exp\left(ib \ln \frac{ia - 1}{ia + 1}\right)$

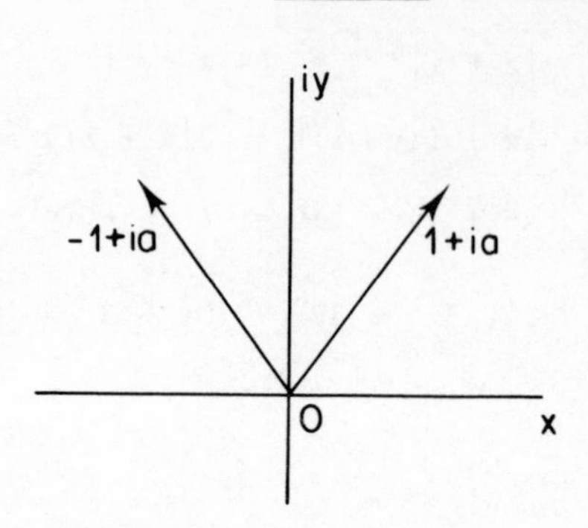

$$\ln \frac{ia - 1}{ia + 1} = \ln (ia - 1) - \ln (ia + 1)$$

$$= i(\pi - \tan^{-1} a) - i \tan^{-1} a$$

$$= 2i\left(\frac{\pi}{2} - \tan^{-1} a\right)$$

$$= 2i \cot^{-1} a$$

$$\therefore \left(\frac{ia - 1}{ia + 1}\right)^{ib} = \exp(-2b \cot^{-1} a)$$

As was mentioned in FRAME 61, similar remarks apply to the log of a quotient as do to the log of a product. In this particular case, no complication arises.

COMPLEX INTEGRATION

FRAME 1

Introduction

If you are given a function of x in ordinary integration, say $x^2 - \frac{2}{x}$, you know how to find the value of an expression such as $\int_1^4 \left(x^2 - \frac{2}{x}\right)dx$. More generally, you know how to deal with $\int_a^b f(x)dx$, assuming that f(x) is a function whose integral can be found. Other types of integral which occur when dealing with real variables are $\int_C f\ ds$, $\int_S f\ dS$ and $\int_V f\ dV$ where f is a function which may contain any or all of x, y and z in each case. The first of these three is a line integral, the second a surface integral and the third a volume integral. Incidentally, $\int_1^4 \left(x^2 - \frac{2}{x}\right)dx$ is a very simple example of a line integral, the line just being part of the x-axis.

In complex variable work we often come across an integral of the form $\int f(z)dz$, where here z = x + iy. Now as z varies, i.e. as x and y vary, the point representing z will trace out a curve. Because of this, which of the types of integral just mentioned do you think $\int f(z)dz$ will be?

****

1A

$\int (z)dz$ is a line integral, as it is this type of integral which arises when the point at which the function is evaluated moves along a curve.

FRAME 2

As $\int f(z)dz$ is a line integral, it is necessary to specify the curve along which z moves before it can be evaluated. If this curve is denoted by C, then we have an integral of the form $\int_C f(z)dz$ in any given case.

Furthermore, as $\int_C f(z)dz$ is a line integral, then, if C comprises two parts, C_1 and C_2, say,

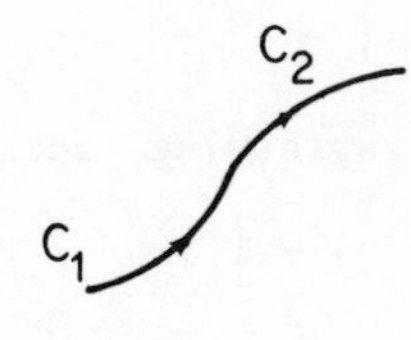

$$\int_C f(z)dz = \int_{C_1} f(z)dz + \int_{C_2} f(z)dz.$$

FRAME 3

The forms of line integral in two dimensions with which you are familiar are $\int_C f(x,y)ds$ and $\int_C Pdx + Qdy$. As you know, these forms are really equivalent. Two dimensions have been specified here because, as the complex quantity z varies, P, the point in the Argand diagram representing z, moves only in two dimensions.

It is a relatively simple matter to express $\int_C f(z)dz$ in a form of line integral with which you are familiar. In the previous programme in this Unit, the form u + iv was frequently used for f(z), u and v being real functions of x and y.

Using this form for f(z) and dx + idy for dz, write down f(z)dz and separate it into its real and imaginary parts.

****

3A

$$\begin{aligned} f(z)dz &= (u + iv)(dx + idy) \\ &= (udx - vdy) + i(udy + vdx) \end{aligned}$$

FRAME 4

$$\begin{aligned} \text{Thus} \int_C f(z)dz &= \int_C \{(udx - vdy) + i(udy + vdx)\} \\ &= \int_C (udx - vdy) + i\int_C (udy + vdx) \end{aligned}$$

The integral $\int_C f(z)dz$ can therefore be expressed as the sum of two real line integrals, each of which is of the form $\int_C Pdx + Qdy$. Assuming that these can be evaluated, it is possible to find $\int_C f(z)dz$.

As an illustration, we shall consider $\int_C zdz$ where C is the straight line path joining A(1 - i) to B(4 + 5i).

Start by expressing $\int zdz$ in the form $\int (udx - vdy) + i\int (udy + vdx)$ and by finding the Cartesian equation of the line joining the points (1,-1) and (4,5).

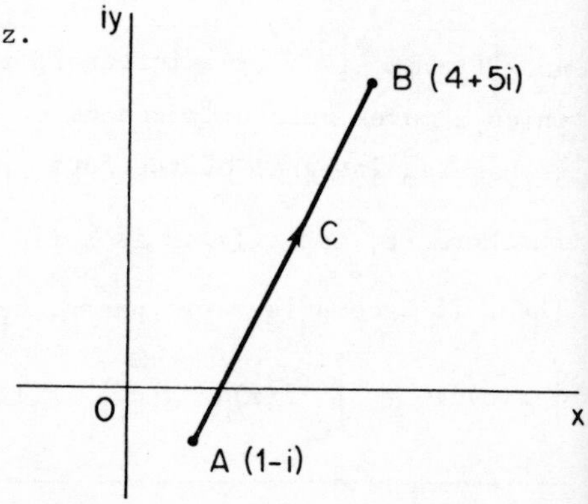

****

4A

$$\int z dz = \int (x dx - y dy) + i\int (x dy + y dx) \qquad (4A.1)$$

Equation of line is $y = 2x - 3$.

FRAME 5

Substitution gives

$$\int_C z\,dz = \int_1^4 \{x dx - (2x - 3)2dx\} + i\int_1^4 \{x2dx + (2x - 3)dx\}$$

Now simplify and complete this example.

****

5A

$$\int_C z dz = \int_1^4 (-3x + 6)dx + i\int_1^4 (4x - 3)dx$$

$$= -\frac{9}{2} + 21i$$

FRAME 6

Now find $\int_C z^2 dz$ along the path C from $-1 + i$ to $5 + 3i$ composed of two straight lines, the first from $-1 + i$ to $5 + i$ and the second thence to $5 + 3i$.

****

6A

$$\int z^2 dz = \int \{(x^2 - y^2)dx - 2xy dy\} + i\int \{(x^2 - y^2)dy + 2xy dx\} \qquad (6A.1)$$

From $-1 + i$ *to* $5 + i$, $y = 1$, $dy = 0$ *and integral becomes*

$$\int_{-1}^5 (x^2 - 1)dx + i\int_{-1}^5 2x\,dx = 36 + 24i$$

From $5 + i$ *to* $5 + 3i$, $x = 5$, $dx = 0$ *and integral becomes*

$$\int_1^3 (-10y)dy + i\int_1^3 (25 - y^2)dy = -40 + \frac{124}{3}i$$

Total result $= (36 + 24i) + (-40 + \frac{124}{3}i) = -4 + \frac{196}{3}i$

FRAME 7

Have you noticed anything about each of the two integrals on the R.H.S.'s of both (4A.1) and (6A.1)? If you haven't, go back to each of them and see if you can spot any property common to them all.

****

7A

They are all integrals of exact differentials of single-valued functions. If you didn't notice this, go back and check that it is so.

FRAME 8

As a result of this, they are independent of the paths joining the end points and can thus be evaluated more simply. (Congratulations if you spotted this earlier and made use of it.)

Thus the R.H.S. of (4A.1) becomes

$$\left[\left(\frac{1}{2}x^2 - \frac{1}{2}y^2\right) + i(xy)\right]_{(1,-1)}^{(4,5)} = -\frac{9}{2} + 21i$$

What does $\frac{1}{2}x^2 - \frac{1}{2}y^2 + ixy$ become when expressed as a function of z?

****

8A

$\frac{1}{2}z^2$

FRAME 9

If you evaluate $\frac{1}{2}z^2$ for $z = 4 + 5i$ and $z = 1 - i$ and subtract the results, you will obtain $-\frac{9}{2} + 21i$. Thus $\int_C z\,dz$ where C is the straight line from $1 - i$ to $4 + 5i$ can be evaluated as $\left[\frac{1}{2}z^2\right]_{1-i}^{4+5i}$ and it doesn't matter what path is followed between $1 - i$ and $4 + 5i$. Note that $\frac{d}{dz}(\frac{1}{2}z^2)$ is z.

Now verify that $\left[\frac{1}{3}z^3\right]_{-1+i}^{5+3i} = -4 + \frac{196}{3}i$ (Compare this result with FRAME 6 and 6A.)

FRAME 10

A very important property of z and z^2 (the quantities which have just been integrated) is that they are both regular. The main part of this programme will be devoted to the integration of functions such as these and ones like $\frac{z^2}{(z-1)(z+2)}$. This last function is regular except at z = 1 and z = −2 where there are singularities. However, before continuing with the methods and results used for such integrals, we shall have a look at some cases where complex integrals can arise in practice.

FRAME 11

Some Instances where Complex Integrals arise

You may be familiar with the method of solving differential equations by the use of Laplace Transforms. If so, you will know that F(s), the Laplace transform of f(t), is given by

$$F(s) = \int_0^\infty e^{-st} f(t)dt$$

From this, a table can be built up from which the transforms of common functions can be obtained. This table can then, of course, be used backwards to find the inverse transforms of certain functions of s. However, it may happen that the inverse transform of a function that does not appear in the table is required. If G(s) is the function whose inverse transform is needed, it can be shown that the required inverse is given by

$$\frac{1}{2\pi i}\int_C e^{st} G(s)ds$$

where C is the straight line path shown. In Unit 5 (in Vol I), s was thought of as being real but in Unit 13 the wider concept of s as a complex quantity will be introduced. In this formula, s, like z, will thus be of the form x + iy. The line C extends an infinite distance in both directions. γ is to some extent arbitrary, but must be so chosen that all the singularities of G(s) lie to its left.

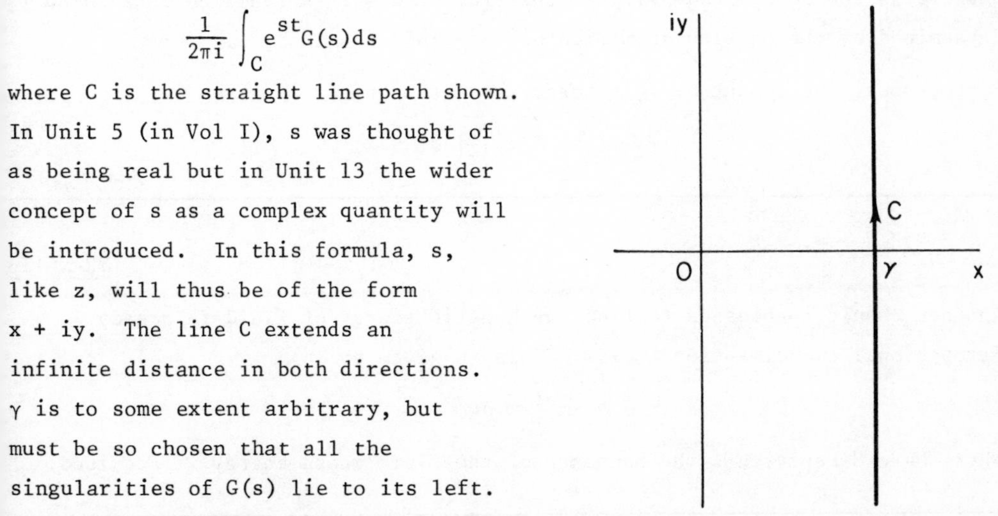

FRAME 12

Another instance of a complex integral occurs in the study of a fluid in steady two-dimensional irrotational motion with a fixed cylinder placed in its path. The term "two-dimensional motion" means that the fluid can be regarded as streaming over a plane in such a way that its motion in all other planes parallel to the first is the same. "Steady" means that the motion has settled down so that it does not change with time. The flow is irrotational if a small paddle wheel placed in the fluid does not rotate. The diagram shows the fluid streaming past the cylinder which is placed so that its generators are perpendicular to the z-plane.

Under such conditions, it is only natural to expect that the fluid exerts a force on the cylinder. If the components of this force per unit length of cylinder are X and Y parallel to Ox and Oy respectively, then it can be shown that

$$X - iY = \frac{1}{2}\, i\rho \int_C \left(\frac{dw}{dz}\right)^2 dz$$

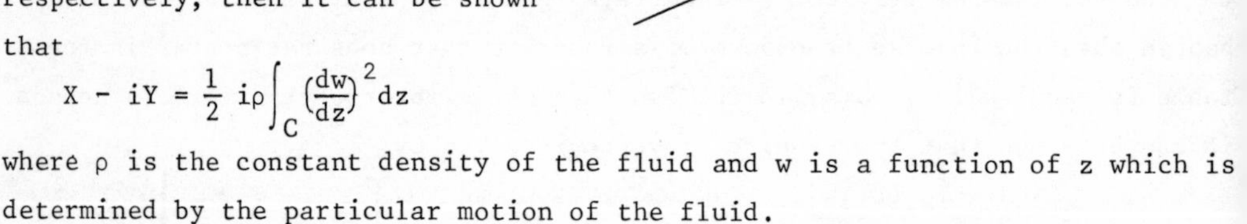

where ρ is the constant density of the fluid and w is a function of z which is determined by the particular motion of the fluid.

Furthermore, the moment of this force about the origin is given by

$$-\frac{1}{2}\,\rho\ \mathrm{Re} \int_C z\left(\frac{dw}{dz}\right)^2 dz$$

FRAME 13

Another result enables us to find the kinetic energy of fluid in steady irrotational two-dimensional flow. This is given by

$$-\frac{1}{4}\,\rho\ \mathrm{Im} \int_C \bar{w}\ dw$$

where here C represents the boundary of the fluid whose energy is required.

FRAME 14

The equation $\frac{\partial^2\theta}{\partial x^2} - \frac{\partial\theta}{\partial t} = 0$ occurs in problems involving heat and moisture diffusion through rectangular slabs and in the transmission of electrical energy in a uniform unloaded cable. For certain boundary conditions, the solution for θ is given by

$$\frac{\theta_0}{2\pi i}\int_C \frac{e^{zt-x\sqrt{z}}}{z}\,dz$$

where C is the same path as was specified in FRAME 11.

FRAME 15

When a gun is fired from an aeroplane, the recoil has an effect on the motion of the plane. In a particular case, it was found that in order to calculate the pitching angle, it was necessary to evaluate

$$\frac{1/38\cdot6}{2\pi i}\int_C \frac{9\cdot56e^{zt}}{z^3}\,dz$$

where C is the same path as was specified in FRAME 11. Here t represents time and this was an approximate formula for small values of t.

FRAME 16

Cauchy's Theorem

We shall now return to the evaluation of complex integrals and consider a very important result.

You will remember that in FRAMES 4 and 6, $\int z\,dz$ and $\int z^2 dz$ were evaluated along certain paths and it was found that, in each case, the value of the integral was independent of the path followed between the end points. In FRAME 10, your attention was drawn to the fact that z and z^2 are both regular. You will also remember, from your work on line integrals, that, if a line integral is independent of the path of integration, then its value taken round a closed curve is zero.

FRAME 16 continued

In FRAME 4, $\int_C f(z)dz$ was expressed as

$$\int_C (udx - vdy) + i\int_C (udy + vdx) \qquad (16.1)$$

Now, if you have read the programmes on Vector Analysis, you will already have met Green's Theorem (see FRAME 29, page 11:87). If Vector Analysis is not part of your course, you will find this theorem given in APPENDIX A at the end of this programme (see page 12:165.)

Taking the first of the integrals in (16.1) and using Green's Theorem,

$$\int_C (udx - vdy) = \iint_S \left(-\frac{\partial v}{\partial x} - \frac{\partial u}{\partial y}\right) dxdy$$

where S is the area enclosed by the curve C. Now, if z is regular everywhere within and on C, then $\frac{\partial u}{\partial y} = -\frac{\partial v}{\partial x}$ and so $-\frac{\partial v}{\partial x} - \frac{\partial u}{\partial y} = 0$.

$$\therefore \int_C (udx - vdy) = 0$$

Now apply Green's Theorem to the second integral in (16.1). Can you deduce anything about its value?

****

16A

$$\int_C (udy + vdx) = \iint_S \left(\frac{\partial u}{\partial x} - \frac{\partial v}{\partial y}\right) dxdy$$

$= 0$ *if z is regular everywhere within and on C.*

FRAME 17

Thus, if f(z) is regular everywhere within and on C, a closed curve, then

$$\int_C f(z)dz = 0$$

This is CAUCHY'S THEOREM.

FRAME 18

Deformation of Contour

Very often, the integrals which have to be evaluated involve functions such as $\frac{1}{z-2}$ and $\frac{z}{(z-1)(z+5i)^2}$ and C encloses at least one singularity. To start with, we shall assume that a function f(z) has only one singularity (at the point A for which z = a) and that C encloses it. By convention, the direction of C is taken as shown.

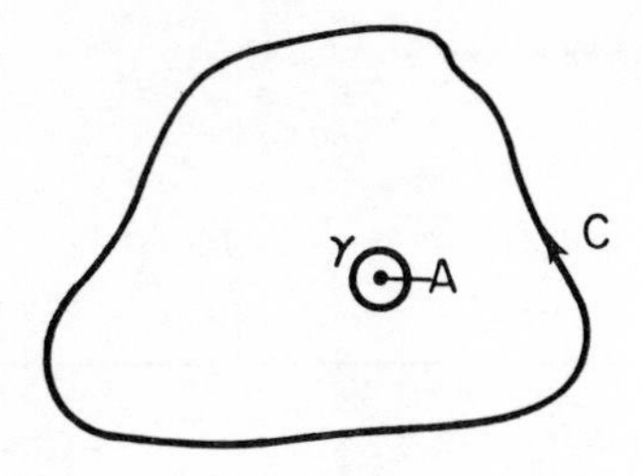

Fig (i)

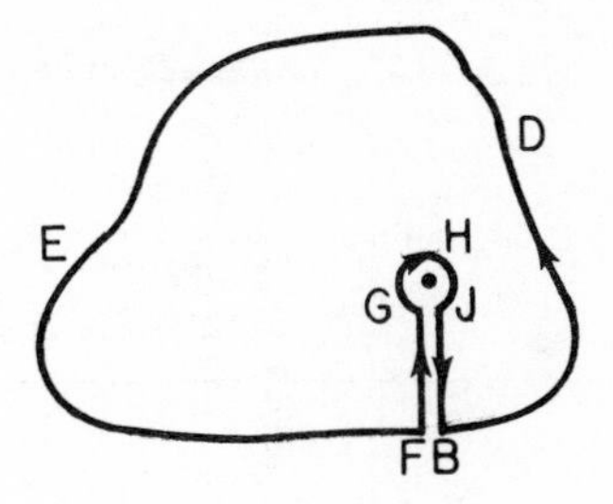

Fig (ii)

Fig (i) shows the path C of integration, usually referred to as the CONTOUR, enclosing the singularity. γ is a small circle surrounding A. It has its centre at A and its radius is ρ. The only restriction placed on ρ is that it is sufficiently small for γ to lie entirely within C.

Fig (ii) shows the contour BDEFGHJB. It comprises most of C, most of γ and the two lines FG and JB. (The small gaps FB and GJ can be taken in any convenient positions.)

What will be the value of $\int_{BDEFGHJB} f(z)dz$?

****

18A

0, since f(z) is regular at all points within and on this new contour.

FRAME 19

$$\therefore \int_{BDEF} f(z)dz + \int_{FG} f(z)dz + \int_{GHJ} f(z)dz + \int_{JB} f(z)dz = 0 \qquad (19.1)$$

as the L.H.S. of this equation is $\int_{BDEFGHJB} f(z)dz$ from the properties of line integrals.

Now let B tend to F and J to G so that the lines FG and BJ coincide. What can you say about $\int_{FG} f(z)dz$ and $\int_{JB} f(z)dz$ now?

****

19A

$\int_{FG} f(z)dz = -\int_{JB} f(z)dz$, *again from the properties of line integrals.*

FRAME 20

Now as $B \to F$, $\int_{BDEF} f(z)dz \to \int_C f(z)dz$, and as $J \to G$, $\int_{GHJ} f(z)dz \to \int_{GHJG} f(z)dz$.

Thus (19.1) becomes

$$\int_C f(z)dz + \int_{GHJG} f(z)dz = 0$$

$$\text{i.e. } \int_C f(z)dz = -\int_{GHJG} f(z)dz$$

$$= \int_{GJHG} f(z)dz$$

If this direction, i.e. GJHG is now assigned to γ, then

$$\int_C f(z)dz = \int_\gamma f(z)dz.$$

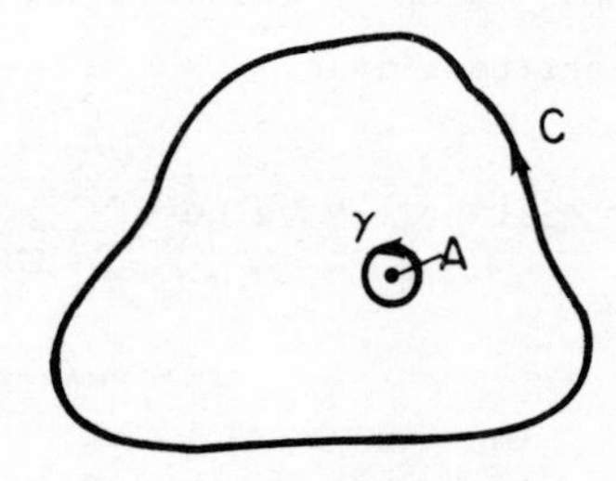

FRAME 21

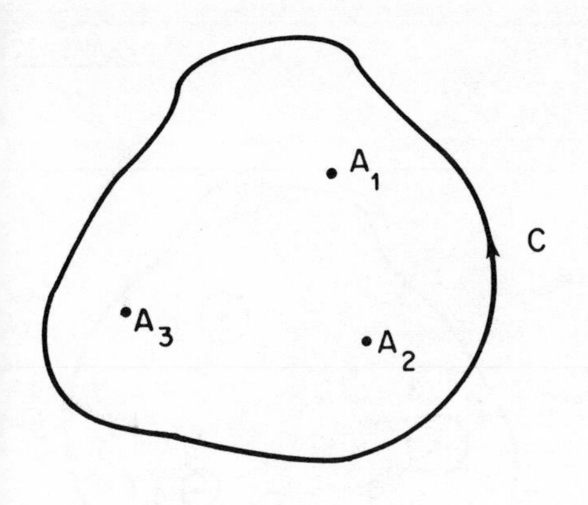

This result can easily be extended to the case of a function that has more than one singularity inside C. For example, suppose that there are three singularities at A_1, A_2 and A_3 for which $z = a_1$, a_2 and a_3 respectively.

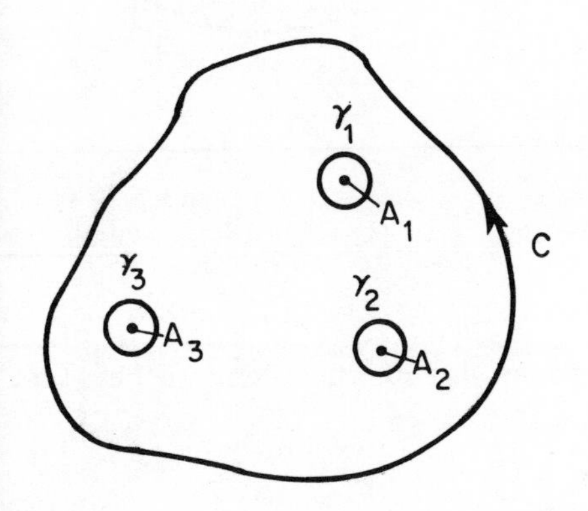

A_1, A_2 and A_3 are now surrounded by three small circles, γ_1, γ_2 and γ_3 having their centres at A_1, A_2 and A_3 and radii ρ_1, ρ_2 and ρ_3. Each ρ is sufficiently small that the corresponding γ does not extend over the boundary C or to any one of the other singularities.

Can you suggest a contour for this diagram which is such that for it, $\int f(z)dz = 0$ and which will contain most of C, γ_1, γ_2 and γ_3?

****

21A

There are many possibilities, of which the following are two:

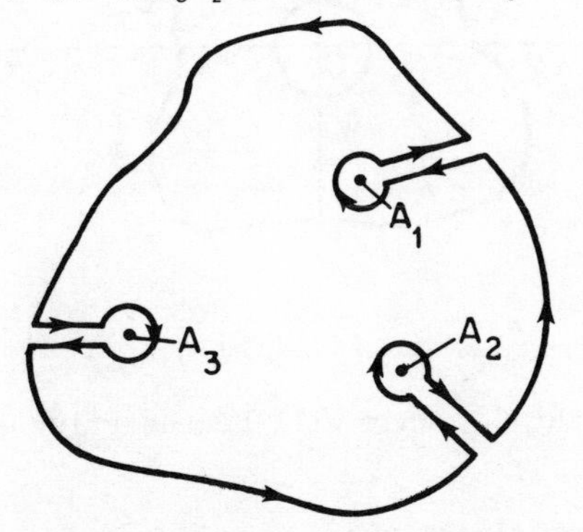

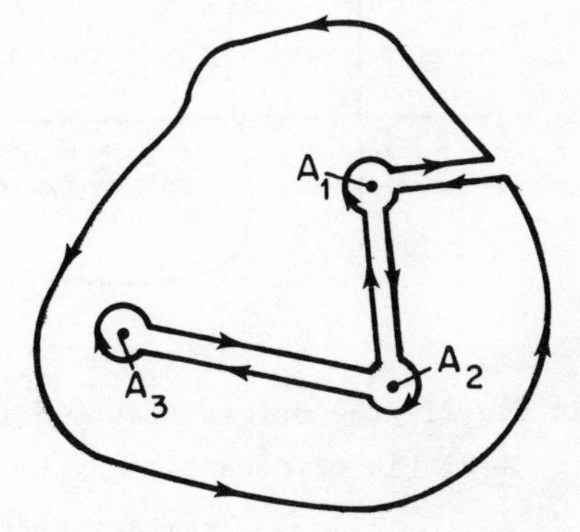

FRAME 22

Taking either of the contours shown in 21A and proceeding in a similar manner to that adopted in FRAMES 18-20, the result

$$\int_C f(z)dz = \int_{\gamma_1} f(z)dz + \int_{\gamma_2} f(z)dz + \int_{\gamma_3} f(z)dz$$

is obtained, where directions are now associated with the small circles as shown.

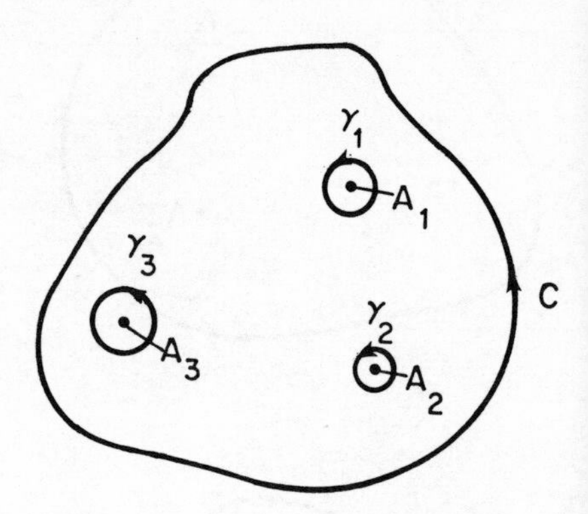

This process can be extended to cover any finite number of singularities within C. Note that no case has been considered where there is a singularity actually on C.

FRAME 23

Some Standard Integrals

The simplest integral to be considered that makes use of the ideas in the last few frames is

$$\int_C \frac{1}{z}\, dz$$

$\frac{1}{z}$ has a singularity at the origin and so two cases are taken as shown:

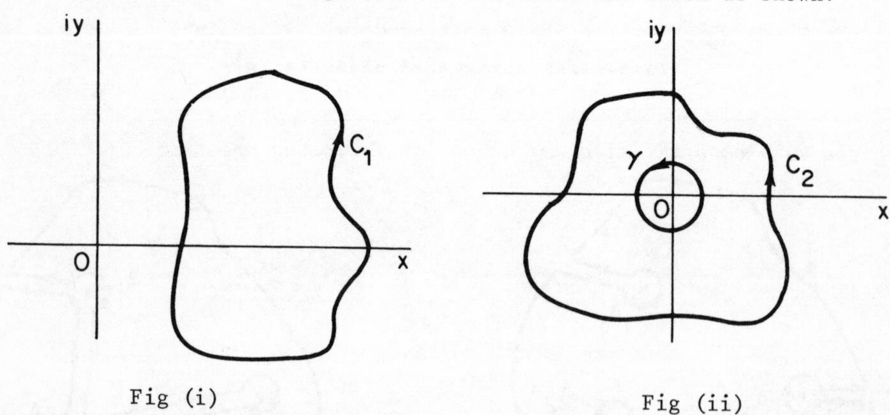

Fig (i) Fig (ii)

In Fig (i) the origin lies outside the contour C_1. What will be the value of $\int_{C_1} \frac{1}{z}\, dz$ in this case?

****

23A

0. Cauchy's Theorem applies.

FRAME 24

In Fig (ii), γ is a circle of radius ρ, with its centre at the origin, lying entirely within C_2. We thus have

$$\int_{C_2} \frac{1}{z}\, dz = \int_{\gamma} \frac{1}{z}\, dz$$

Now you are familiar with the exponential form of a complex quantity (see FRAMES 52-53, pages 4:38-4:39, Vol I, if you have forgotten). We now make use of this, and z for any point on γ can be written as $\rho e^{i\theta}$.

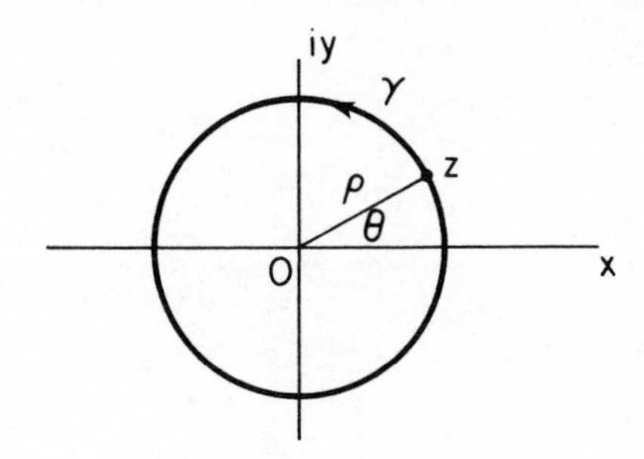

The diagram is an enlargement of the circle γ in Fig (ii), FRAME 23. As z moves completely round γ once, θ will vary from $-\pi$ to π (or from 0 to 2π).

Substituting into $\int_{\gamma} \frac{1}{z}\, dz$ gives

$$\int_{-\pi}^{\pi} \frac{1}{\rho e^{i\theta}}\, \rho i e^{i\theta} d\theta \quad \text{as if } z = e^{i\theta}, \text{ then } dz = \rho i e^{i\theta} d\theta$$

Simplify and evaluate this integral.

****

24A

$$\int_{-\pi}^{\pi} i d\theta = 2\pi i$$

FRAME 25

Thus $\int_C \frac{1}{z}\,dz = 0$ if C <u>does not</u> enclose the origin

but $\int_C \frac{1}{z}\,dz = 2\pi i$ if C <u>does</u> enclose the origin.

Working along similar lines, now try and find $\int_C \frac{1}{z^2}\,dz$ in each of the two cases, i.e. (i) C does not, (ii) C does, enclose the origin.

****

25A

(i) If C does not enclose the origin, $\int_C \frac{1}{z^2}\,dz = 0$ *by Cauchy's Theorem.*

(ii) If C does enclose the origin, then with the notation of FRAME 24,

$$\int_C \frac{1}{z^2}\,dz = \int_\gamma \frac{1}{z^2}\,dz$$

$$= \int_{-\pi}^{\pi} \frac{1}{(\rho e^{i\theta})^2}\,\rho i e^{i\theta}\,d\theta$$

$$= \int_{-\pi}^{\pi} \frac{i}{\rho}\,e^{-i\theta}\,d\theta$$

$$= \frac{i}{\rho}\Big[-\frac{1}{i}\,e^{-i\theta}\Big]_{-\pi}^{\pi}$$

$$= \frac{1}{\rho}\left(-e^{-i\pi} + e^{i\pi}\right)$$

$$= \frac{1}{\rho}(1 - 1)$$

$$= 0$$

FRAME 26

Thus $\int_C \frac{1}{z^2}\,dz = 0$ whether C encloses the origin or not.

In a similar way,

$$\int_C \frac{1}{z^n}\,dz = 0$$

for n any positive integer except 1 and for C any closed contour.

FRAME 27

These results enable us to obtain $\int_C \frac{1}{(z-a)^n} dz$ for n a positive integer. In this case the distinction has to be drawn between C not enclosing the point z = a and C enclosing it as shown:

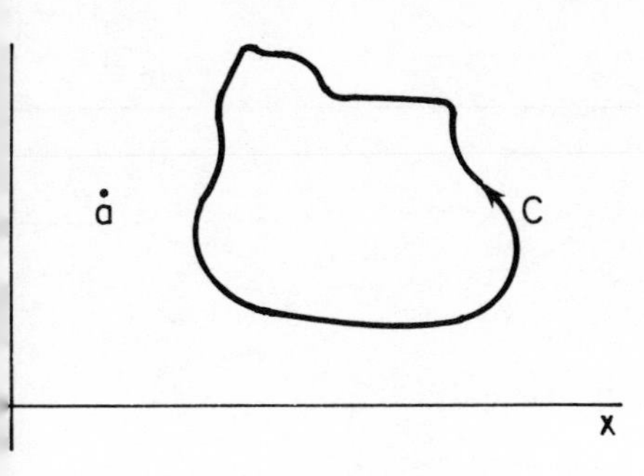

Fig (i)

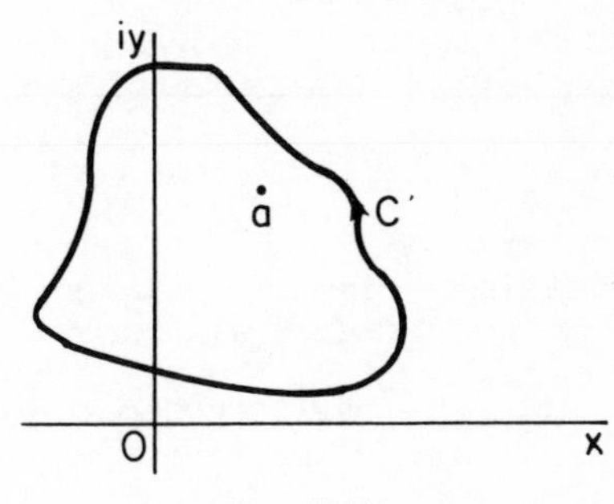

Fig (ii)

If the substitution w = z - a is made, $\int \frac{1}{(z-a)^n} dz$ becomes $\int \frac{1}{w^n} dw$ and the contours shown in Fig (i) and Fig (ii) become those in Fig (iii) and Fig (iv) respectively.

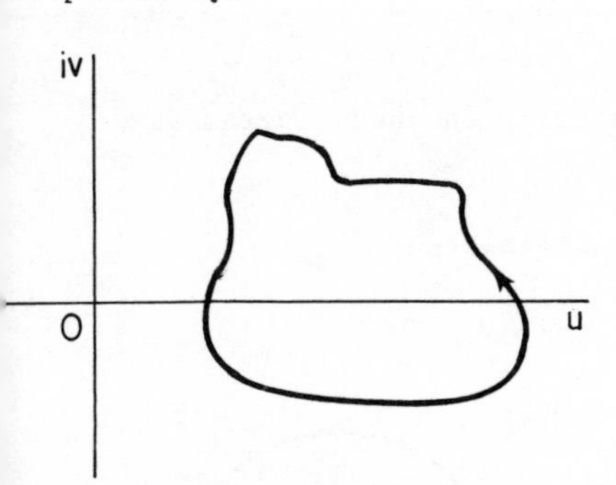

Fig (iii)

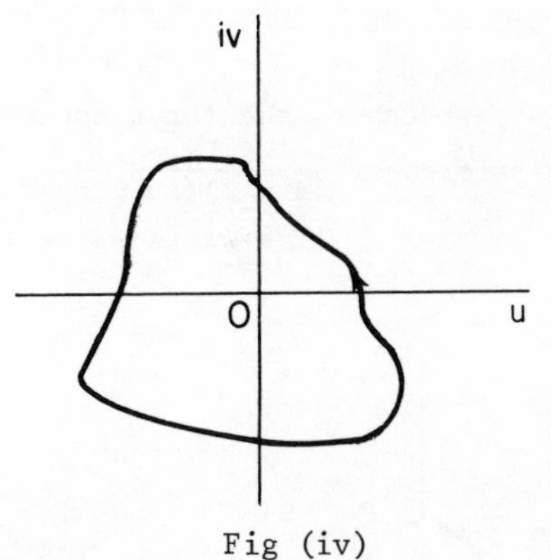

Fig (iv)

Can you now give the value of $\int_C \frac{1}{(z-a)^n} dz$ in the four cases

(i) n = 1, C does not enclose a

(ii) n = 1, C encloses a

(iii) n > 1, C does not enclose a

(iv) n > 1, C encloses a ?

****

27A

In cases (i), (iii) and (iv), $\int_C \frac{1}{(z-a)^n}\,dz = 0$

In case (ii) $\int_C \frac{1}{z-a}\,dz = 2\pi i$

FRAME 28

Summarising, $\int_C \frac{1}{(z-a)^n}\,dz = 0 \quad (n \neq 1)$

$\int_C \frac{1}{z-a}\,dz = 2\pi i$ if C encloses a, but is 0 otherwise.

With the aid of partial fractions, these results enable us to obtain many other integrals. As an example $\int_C \frac{z}{(z-1)(z+2i)}\,dz$ will be considered for three different contours C, i.e. C_1, C_2 and C_3 where C_1 is $|z| = \frac{1}{2}$, C_2 is $|z| = \frac{3}{2}$ and C_3 is the rectangle whose vertices are 2 + i, -1 + i, -1 - 3i, 2 - 3i.

Start by sketching the three contours and inserting on each diagram the singularities of $\frac{z}{(z-1)(z+2i)}$.

****

28A

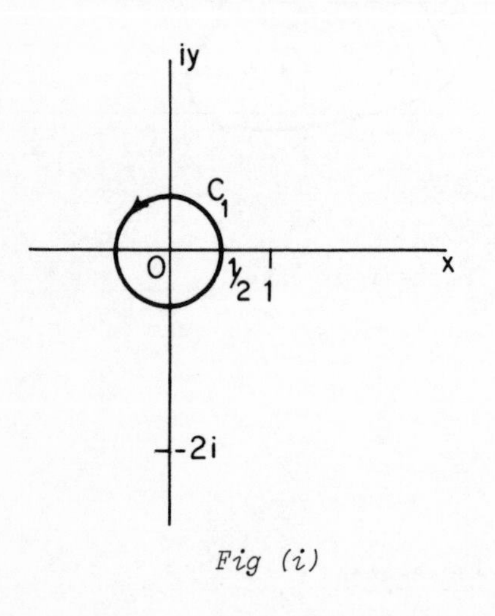

Fig (i)

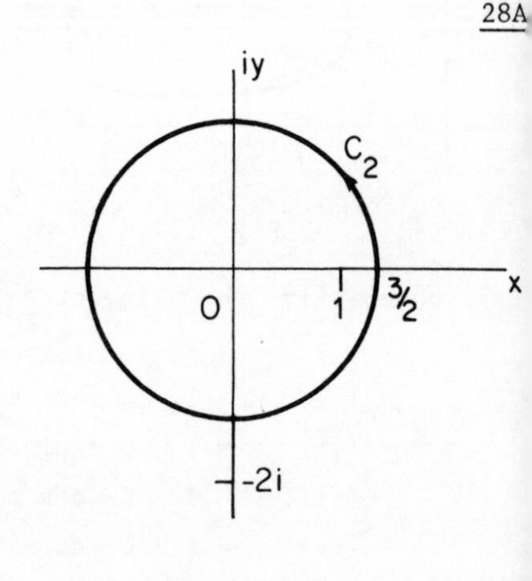

Fig (ii)

28A continued

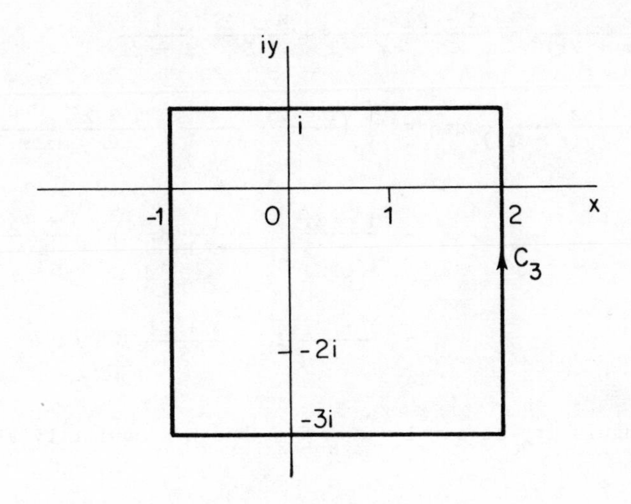

Fig (iii)

The singularities of $\dfrac{z}{(z - 1)(z + 2i)}$ *are at 1 and -2i.*

FRAME 29

The next thing to do is to express $\dfrac{z}{(z - 1)(z + 2i)}$ in partial fractions.

The fact that z is complex makes no difference to the method so let

$$\frac{z}{(z - 1)(z + 2i)} \equiv \frac{A}{z - 1} + \frac{B}{z + 2i}$$

Now find A and B.

****

29A

$$A = \frac{1}{1 + 2i} = \frac{1 - 2i}{5}$$

$$B = \frac{2i}{1 + 2i} = \frac{4 + 2i}{5}$$

FRAME 30

Thus $\dfrac{z}{(z-1)(z+2i)} = \dfrac{1-2i}{5}\dfrac{1}{z-1} + \dfrac{4+2i}{5}\dfrac{1}{z+2i}$

and so
$$\int_C \frac{z}{(z-1)(z+2i)}\,dz = \int_C\left(\frac{1-2i}{5}\frac{1}{z-1} + \frac{4+2i}{5}\frac{1}{z+2i}\right)dz$$
$$= \frac{1-2i}{5}\int_C \frac{1}{z-1}\,dz + \frac{4+2i}{5}\int_C \frac{1}{z+2i}\,dz$$
$$= \frac{1-2i}{5} I_1 + \frac{4+2i}{5} I_2 \quad \text{say.}$$

$\dfrac{1}{z-1}$ has a singularity at $z = 1$ and $\dfrac{1}{z+2i}$ has a singularity at $z = -2i$.

In the case of contour C_1, both of these singularities lie outside it and so $I_1 = 0$ and $I_2 = 0$.

$$\therefore \int_{C_1} \frac{z}{(z-1)(z+2i)}\,dz = 0$$

However, contour C_2 encloses 1 but not 2i, so in this case $I_1 = 2\pi i$ and $I_2 = 0$

$$\therefore \int_{C_2} \frac{z}{(z-1)(z+2i)}\,dz = \frac{1-2i}{5} \times 2\pi i = \frac{2\pi(2+i)}{5}$$

Now find the value of $\displaystyle\int_{C_3} \frac{z}{(z-1)(z+2i)}\,dz$.

****

30A

C_3 encloses both singularities, $\therefore$ $I_1 = 2\pi i$ and $I_2 = 2\pi i$.

$$\therefore \int_{C_3} \frac{z\,dz}{(z-1)(z+2i)} = \frac{1-2i}{5} \times 2\pi i + \frac{4+2i}{5} \times 2\pi i$$
$$= 2\pi i$$

FRAME 31

Now work the following example completely.

Find $I = \int_C \frac{2z - 1}{z(z - 1)}\, dz$ in the cases where C is (i) C_1, $|z - 3| = 1$,

(ii) C_2, $|z| = \frac{1}{2}$, (iii) C_3, $|z| = 2$.

****

31A

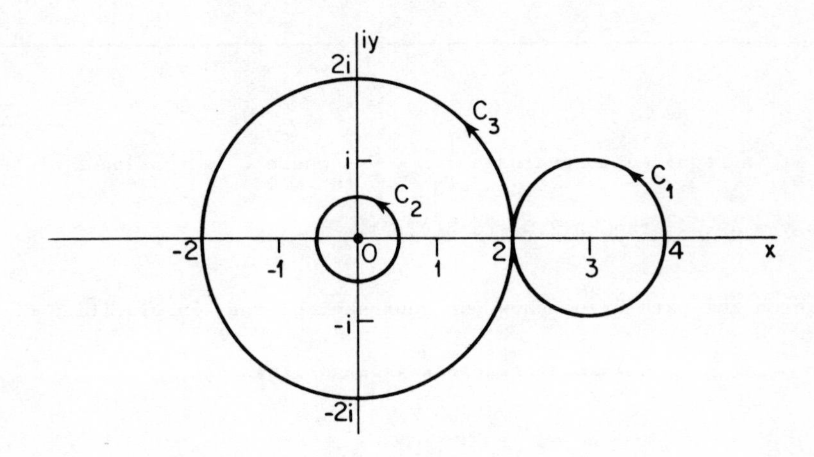

Singularities occur at $z = 0$ and $z = 1$

$$\frac{2z - 1}{z(z - 1)} = \frac{1}{z} + \frac{1}{z - 1}$$

$$\int_C \frac{2z - 1}{z(z - 1)}\, dz = \int_C \frac{1}{z}\, dz + \int_C \frac{1}{z - 1}\, dz$$

or $$I = I_1 + I_2$$

(i) C_1 *encloses no singularity.* $I_1 = 0$, $I_2 = 0$.

$$I = 0$$

(ii) C_2 *encloses singularity at 0.* $I_1 = 2\pi i$, $I_2 = 0$

$$I = 2\pi i$$

(iii) C_3 *encloses singularities at 0 and 1.* $I_1 = 2\pi i$, $I_2 = 2\pi i$

$$I = 4\pi i$$

FRAME 32

In all the cases just considered, integration has been round a closed curve so that Cauchy's Theorem and the results in FRAMES 23-27 have been sufficient to give the answers directly. However it may happen that the value of an integral along a path that is not closed is required. Such an integral can, of course, be found by the method used in FRAMES 4-6 but this may sometimes become tedious. The example in the next frame illustrates an alternative approach. The idea involved will be found useful later on.

FRAME 33

Suppose it is required to evaluate $\int_C \frac{dz}{z^2 + 1}$ where C is that part of the parabola $y = 4 - x^2$ from A(2,0) to B(-2,0).

First sketch the path C and show, on your sketch, the singularities of $\frac{1}{z^2 + 1}$.

****

33A

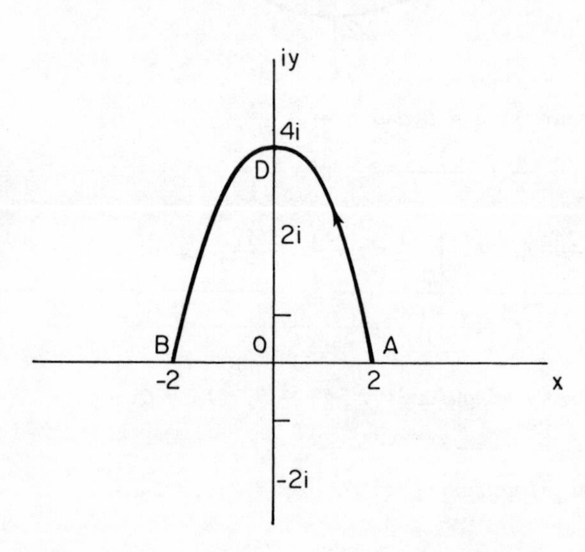

Singularities occur at i and -i and C is the path ADB.

FRAME 34

We now choose a closed path, consisting in part of the path ADB required and in part of a curve (or straight line) from B back to A. This latter is chosen in such a way that the integral of the given function along it is relatively easy to obtain. As will be seen shortly the straight line path BOA is one for which this happens. Its virtue lies in the fact that everywhere along it $y = 0$.

Now find the value of $\int_\Gamma \frac{1}{z^2 + 1}\,dz$ where Γ is the path ADBOA.

[HINT: Note that $z^2 + 1$ factorises into $(z + i)(z - i)$.]

****

34A

$$\frac{1}{(z + i)(z - i)} = \frac{i}{2}\frac{1}{z + i} - \frac{i}{2}\frac{1}{z - i}$$

The only singularity inside the contour is at $z = i$

$$\therefore \int_\Gamma \frac{1}{z^2 + 1}\,dz = \frac{i}{2}\int_\Gamma \frac{1}{z + i}\,dz - \frac{i}{2}\int_\Gamma \frac{1}{z - i}\,dz$$

$$= \frac{i}{2} \times 0 - \frac{i}{2} \times 2\pi i$$

$$= \pi$$

FRAME 35

$$\text{Now } \int_\Gamma \frac{1}{z^2 + 1}\,dz = \int_{ADB} \frac{1}{z^2 + 1}\,dz + \int_{BOA} \frac{1}{z^2 + 1}\,dz$$

$$= \pi$$

$$\therefore \int_{ADB} \frac{1}{z^2 + 1}\,dz = \pi - \int_{BOA} \frac{1}{z^2 + 1}\,dz$$

Along BOA, $z = x$ and so $dz = dx$

$$\therefore \int_{BOA} = \int_{-2}^{2} \frac{1}{x^2 + 1}\,dx$$

$$= \left[\tan^{-1} x\right]_{-2}^{2}$$

$$= 2 \tan^{-1} 2$$

$$\therefore \int_{ADB} \frac{1}{z^2 + 1}\,dz = \pi - 2 \tan^{-1} 2$$

FRAME 36

Examples

Now try these examples to see whether you have followed the work covered so far in this programme. Then turn to FRAME 37 to check your answers.

In questions 1-2, evaluate the integral shown along each of the paths indicated.

1. $\int_C \frac{1}{(2z - 3i)(z + i)}\, dz$

   (i) $|z| = 2$ (ii) $|z - 1| = 1$

   (iii) Rectangle with vertices at ± 1, $\pm 1 - 3i$.

2. $\int_C \frac{1}{z^2(z - 1)^2}\, dz$

   (i) $|z| = \frac{1}{2}$ (ii) $|z + 1 - i| = 2$

   (iii) Triangle with vertices at 3, $-1 + i$, $-1 - i$.

3. By integrating round a suitable closed path, find $\int_{AB} \frac{1}{9 + z^2}\, dz$ where AB is that part of the parabola $y^2 = 4(1 - x)$ from $(0,2)$ to $(0,-2)$ in the complex z-plane.

FRAME 37

Answers to Examples

1. (i) 0 (ii) 0 (iii) $-2\pi/5$

2. (i) $4\pi i$ (ii) $4\pi i$ (iii) 0

3. $-\frac{i}{3}\ln 5$

FRAME 38

Taylor's and Laurent's Series

In the previous programme (Theory of Functions) the idea of power series was introduced and some examples given (see FRAMES 35-50, pages 12:87-12:96). You have, of course, already met Taylor's series for real variables in the form

$$f(x) = A_0 + A_1(x - a) + A_2(x - a)^2 + \dots$$

where $A_0 = f(a)$, $A_1 = f'(a)$, $A_2 = f''(a)/2!$ etc.

We shall now consider some more examples of series.

1. $$\frac{1}{z-2} = \frac{1}{-2\left(1 - \frac{z}{2}\right)}$$

$$= -\frac{1}{2}\left(1 - \frac{z}{2}\right)^{-1}$$

$$= -\frac{1}{2}\left\{1 + \left(\frac{z}{2}\right) + \left(\frac{z}{2}\right)^2 + \left(\frac{z}{2}\right)^3 + \dots\right\} \quad \text{provided } |z| < 2$$

$$= -\frac{1}{2} - \frac{z}{4} - \frac{z^2}{8} - \frac{z^3}{16} - \dots \qquad (38.1)$$

and this is an example of Taylor's series in which a = 0 and x is replaced by z.

2. Dividing (38.1) by z^2 gives, provided $|z| < 2$,

$$\frac{1}{z^2(z-2)} = -\frac{1}{2z^2} - \frac{1}{4z} - \frac{1}{8} - \frac{z}{16} - \dots \qquad (38.2)$$

Here z = 0 must also be excluded if it is desired to keep the function finite.

3. $$\frac{1}{z-2} = \frac{1}{z\left(1 - \frac{2}{z}\right)}$$

$$= \frac{1}{z}\left(1 - \frac{2}{z}\right)^{-1}$$

$$= \frac{1}{z}\left\{1 + \frac{2}{z} + \left(\frac{2}{z}\right)^2 + \left(\frac{2}{z}\right)^3 + \dots\right\} \qquad \text{if } |z| > 2$$

$$= \frac{1}{z} + \frac{2}{z^2} + \frac{4}{z^3} + \frac{8}{z^4} + \dots \qquad (38.3)$$

FRAME 38 continued

In all of these expansions the binomial theorem has been used. As the index in each case is -1, it is necessary to arrange the fraction so that the first term in the bracket is 1 and the modulus of the second term is <1.

By making use of what you know about Taylor's series in real variable work and assuming that similar properties hold for functions of a complex variable, can you say why (38.2) and (38.3) are not examples of Taylor's series in z?

****

38A

They both contain negative powers of z.

FRAME 39

It is also possible to express $\frac{1}{z-2}$ in terms of powers of, say, $z - 1$, $z + 1$, etc. and, more generally, $z - a$ where a is a complex constant.

For example,

$$\frac{1}{z-2} = \frac{1}{(z-1)-1}$$

$$= \frac{1}{w-1} \quad \text{if we put } w = z - 1$$

Then, if $|w| < 1$, this gives

$$\frac{1}{-(1-w)} = -(1 + w + w^2 + w^3 + \ldots)$$

and so

$$\frac{1}{z-2} = -\{1 + (z-1) + (z-1)^2 + (z-1)^3 + \ldots\} \qquad (39.1)$$

But if $|w| > 1$, we have

$$\frac{1}{w-1} = \frac{1}{w\left(1 - \frac{1}{w}\right)}$$

$$= \frac{1}{w}\left\{1 + \left(\frac{1}{w}\right) + \left(\frac{1}{w}\right)^2 + \left(\frac{1}{w}\right)^3 + \ldots\right\}$$

$$= \frac{1}{w} + \frac{1}{w^2} + \frac{1}{w^3} + \frac{1}{w^4} + \ldots$$

and so

$$\frac{1}{z-2} = \frac{1}{z-1} + \frac{1}{(z-1)^2} + \frac{1}{(z-1)^3} + \frac{1}{(z-1)^4} + \ldots \qquad (39.2)$$

FRAME 39 continued

Can you write down the two regions in the z-plane in which (39.1) and (39.2) are valid respectively? Illustrate each of them by a sketch.

****

39A

(39.1) $|w| < 1$ *gives* $|z - 1| < 1$, *i.e.*

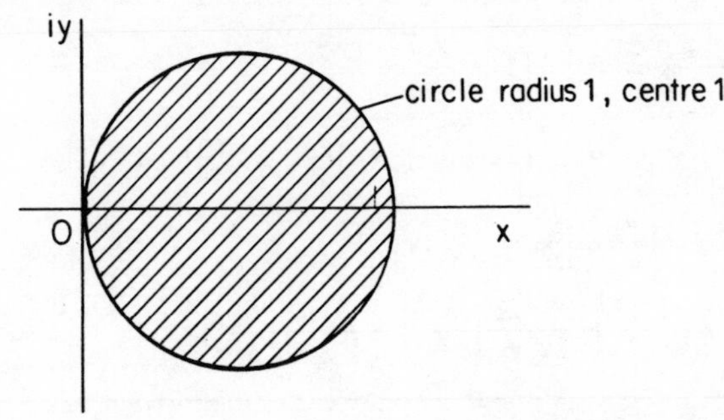

(39.2) $|w| > 1$ *gives* $|z - 1| > 1$, *i.e. the region outside the circle shown above.*

FRAME 40

As you already know, a series such as (38.1) or (39.1) which contains only positive powers is called a Taylor series.

A series which contains negative powers is called a LAURENT SERIES. (38.2), (38.3) and (39.2) are examples of Laurent series. It may consist only of negative powers, for example (38.3) and (39.2) or it may contain both negative and positive powers such as, for example, (38.2).

In either a Taylor or a Laurent series, the powers may be just of z or of a linear expression such as $z - 1$, $z + 1$, $z - 2i$, etc.

Now obtain Taylor or Laurent series for each of the following, stating in each case which it is:

(i) $\dfrac{z}{z + 3}$ in powers of z and valid for $|z| < 3$

(ii) $\dfrac{z - 1}{z + 3}$ in powers of $z + 1$ and valid for $|z + 1| > 2$.

(HINT: In (ii) put $w = z + 1$ so that $\dfrac{z - 1}{z + 3} = \dfrac{w - 2}{w + 2} = 1 - \dfrac{4}{w + 2}$)

****

40A

(i) $\dfrac{1}{z+3} = \dfrac{1}{3\left(1+\frac{z}{3}\right)} = \dfrac{1}{3}\left(1+\frac{z}{3}\right)^{-1}$

$= \dfrac{1}{3}\left\{1 - \left(\frac{z}{3}\right) + \left(\frac{z}{3}\right)^2 - \left(\frac{z}{3}\right)^3 \dots\right\}$

$\dfrac{z}{z+3} = \dfrac{z}{3} - \dfrac{z^2}{9} + \dfrac{z^3}{27} \dots\dots$ *a Taylor series*

(ii) $\dfrac{1}{w+2} = \dfrac{1}{w\left(1+\frac{2}{w}\right)} = \dfrac{1}{w}\left(1+\frac{2}{w}\right)^{-1}$

$= \dfrac{1}{w}\left\{1 - \left(\frac{2}{w}\right) + \left(\frac{2}{w}\right)^2 - \left(\frac{2}{w}\right)^3 \dots\right\}$

$1 - \dfrac{4}{w+2} = 1 - \dfrac{4}{w} + \dfrac{8}{w^2} - \dfrac{16}{w^3} \dots$

$\dfrac{z-1}{z+3} = 1 - \dfrac{4}{z+1} + \dfrac{8}{(z+1)^2} - \dfrac{16}{(z+1)^3} \dots\dots$ *a Laurent series*

FRAME 41

It is a simple matter to extend this process to more complicated expressions.

For example, suppose it is required to expand $\dfrac{2z-1}{(z+1)(z-4)}$ in powers of z.

First of all $\dfrac{2z-1}{(z+1)(z-4)}$ can be split into partial fractions and gives $\dfrac{3}{5}\dfrac{1}{z+1} + \dfrac{7}{5}\dfrac{1}{z-4}$.

Now, if $|z| < 1$, $\dfrac{1}{z+1} = (1+z)^{-1}$

$= 1 - z + z^2 - z^3 \dots$

but, if $|z| > 1$, $\dfrac{1}{z+1} = \dfrac{1}{z}\dfrac{1}{1+\frac{1}{z}}$

$= \dfrac{1}{z}\left(1+\dfrac{1}{z}\right)^{-1}$

$= \dfrac{1}{z}\left(1 - \dfrac{1}{z} + \dfrac{1}{z^2} - \dfrac{1}{z^3} \dots\right)$

$= \dfrac{1}{z} - \dfrac{1}{z^2} + \dfrac{1}{z^3} - \dfrac{1}{z^4} \dots$

Now express $\dfrac{1}{z-4}$ in series form in the two cases

(i) $|z| < 4$, (ii) $|z| > 4$

****

41A

(i) $$\frac{1}{z-4} = -\frac{1}{4}\frac{1}{1-\frac{z}{4}}$$
$$= -\frac{1}{4}\{1 + \frac{z}{4} + \left(\frac{z}{4}\right)^2 + \left(\frac{z}{4}\right)^3 + \ldots\}$$
$$= -\frac{1}{4} - \frac{z}{16} - \frac{z^2}{64} - \frac{z^3}{256} - \ldots$$

(ii) $$\frac{1}{z-4} = \frac{1}{z}\frac{1}{1-\frac{4}{z}}$$
$$= \frac{1}{z}\{1 + \frac{4}{z} + \left(\frac{4}{z}\right)^2 + \left(\frac{4}{z}\right)^3 + \ldots\}$$
$$= \frac{1}{z} + \frac{4}{z^2} + \frac{16}{z^3} + \frac{64}{z^4} + \ldots$$

FRAME 42

Thus: if $|z| < 1$,

$$\frac{2z-1}{(z+1)(z-4)} = \frac{3}{5}\left(1 - z + z^2 - z^3 \ldots\right) - \frac{7}{5}\left(\frac{1}{4} + \frac{z}{16} + \frac{z^2}{64} + \frac{z^3}{256} + \ldots\right) \quad (42.1)$$

if $1 < |z| < 4$,

$$\frac{2z-1}{(z+1)(z-4)} = \frac{3}{5}\left(\frac{1}{z} - \frac{1}{z^2} + \frac{1}{z^3} - \frac{1}{z^4} \ldots\right) - \frac{7}{5}\left(\frac{1}{4} + \frac{z}{16} + \frac{z^2}{64} + \frac{z^3}{256} + \ldots\right) \quad (42.2)$$

if $|z| > 4$,

$$\frac{2z-1}{(z+1)(z-4)} = \frac{3}{5}\left(\frac{1}{z} - \frac{1}{z^2} + \frac{1}{z^3} - \frac{1}{z^4} \ldots\right) + \frac{7}{5}\left(\frac{1}{z} + \frac{4}{z^2} + \frac{16}{z^3} + \frac{64}{z^4} + \ldots\right) \quad (42.3)$$

If $|z| < 1$, the result is a Taylor series, but if $|z| > 1$, it is a Laurent series.

FRAME 43

Functions other than purely algebraic ones can also be expressed in series form. As has already been mentioned in the previous programme (Theory of Functions)

$$e^z = 1 + z + \frac{z^2}{2!} + \frac{z^3}{3!} + \ldots$$
$$\sin z = z - \frac{z^3}{3!} + \frac{z^5}{5!} - \frac{z^7}{7!} \ldots.$$

and so on for other functions such as cos z, cosh z, etc.

FRAME 43 continued

If you are familiar with the $\sum$ notation, you will realise that the results can often be expressed more compactly by its use. For example,

$$e^z = \sum_0^\infty \frac{z^n}{n!}$$

$$\sin z = \sum_1^\infty (-1)^{n+1} \frac{z^{2n-1}}{(2n-1)!}$$

FRAME 44

In general, the Laurent series for a function f(z) can be written in terms of z - a, as

$$f(z) = A_0 + A_1(z-a) + A_2(z-a)^2 + A_3(z-a)^3 + A_4(z-a)^4 + \dots$$
$$+ \frac{B_1}{z-a} + \frac{B_2}{(z-a)^2} + \frac{B_3}{(z-a)^3} + \frac{B_4}{(z-a)^4} + \dots \qquad (44.1)$$

The results in FRAME 42 give a clue to the region in which such an expansion is valid. In that example, the regions of validity are those shown below, and also on the diagram are marked the singularities (A and B) of $\frac{2z-1}{(z+1)(z-4)}$.

(42.1) is valid in the small circle, shaded ////

(42.2) is valid in the annulus between the circles, shaded ═══

(42.3) is valid outside the large circle, shaded \\\\\\\\

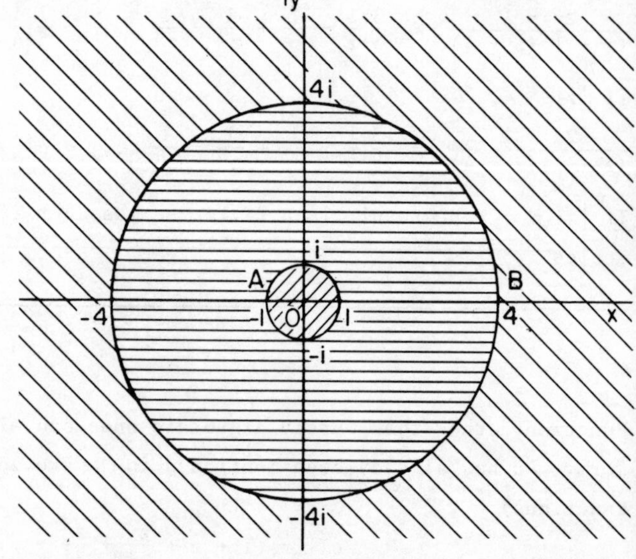

Do you notice anything about the positions of the singularities with respect to the various regions indicated?

****

44A

Each singularity lies on the circle which is the boundary between two of the regions.

FRAME 45

In general, for the series in (44.1), to find the regions of validity, we draw circles centre a and radii such that one circle passes through each singularity. Then a series such as (44.1) is valid for all points in one of the regions so formed. If we wish to pass outside of this region, a different series must be found which is valid in the next region, and so on.

Now go back to the example in FRAME 39 and check that the singularity of $\frac{1}{z-2}$ lies on the boundary of the circle shown in 39A.

****

45A

The singularity is at $z = 2$ which lies on the circle whose centre is 1 and whose radius is 1.

FRAME 46

Poles and Zeros

Certain terminology is used in connection with a function whose Laurent series is as given in (44.1). For convenience we shall repeat this series here.

$$\begin{aligned} f(z) &= A_0 + A_1(z-a) + A_2(z-a)^2 + A_3(z-a)^3 + A_4(z-a)^4 + \dots \\ &\quad + \frac{B_1}{z-a} + \frac{B_2}{(z-a)^2} + \frac{B_3}{(z-a)^3} + \frac{B_4}{(z-a)^4} + \dots \end{aligned} \tag{46.1}$$

You know that the Laurent series for a function takes different forms according to the region in which we are interested. In what follows the form is that applicable to the interior of the smallest circle of convergence.

1. If all the B's are zero, the Laurent series reduces to a Taylor series. If, in addition, $A_0 = A_1 = A_2 = \dots = A_{n-1} = 0$, so that $f(z) = A_n(z-a)^n + A_{n+1}(z-a)^{n+1} + A_{n+2}(z-a)^{n+2} + \dots,$ $f(z)$ is said to have a ZERO of order n at a.

FRAME 46 continued

2. The B section, i.e. $\frac{B_1}{z - a} + \frac{B_2}{(z - a)^2} + \frac{B_3}{(z - a)^3} + \ldots$ is called the PRINCIPAL PART of the Laurent series. If a function is such that $B_{m+1} = B_{m+2} = B_{m+3} = \ldots = 0$ and $B_m \neq 0$, i.e. the principal part reduces to $\frac{B_1}{z - a} + \frac{B_2}{(z - a)^2} + \frac{B_3}{(z - a)^3} + \ldots + \frac{B_{m-1}}{(z - a)^{m-1}} + \frac{B_m}{(z - a)^m}$, then $f(z)$ is said to have a POLE of order m at a. If m = 1, then the pole is said to be SIMPLE.

   For example, (38.2) indicates that $\frac{1}{z^2(z - 2)}$ has a pole of order 2 at z = 0.

Locate and state the order of the pole or zero indicated by the following series:

(i) $(z - 1)^4 + (z - 1)^5 + (z - 1)^6 + \ldots$

(ii) $\frac{1}{z + 2} - 1 + (z + 2) - (z + 2)^2 + (z + 2)^3 - \ldots$

(iii) $\frac{1}{(z + i)^3} + \frac{2}{(z + i)^2} + \frac{3}{z + i} + 4 + 5(z + i) + 6(z + i)^2 + \ldots$

****

46A

(i) Zero of order 4 at z = 1

(ii) Simple pole at z = -2

(iii) Pole of order 3 at z = -i.

FRAME 47

The poles and zeros of a function can also be located by examining the function itself. Thus the fact that $\frac{1}{z^2(z - 2)}$ has a pole of order 2 at z = 0 is indicated by the presence of the z^2 term in the denominator.

Now expand this function in powers of z - 2 for $|z - 2| < 2$ and show that there is a simple pole at z = 2. This is also indicated by the presence of z - 2 in the denominator.

****

47A

If $z - 2 = w$, *then* $\dfrac{1}{z^2(z - 2)} = \dfrac{1}{(w + 2)^2 w}$

$$\frac{1}{(w + 2)^2} = \frac{1}{2^2\left(1 + \frac{w}{2}\right)^2}$$

$$= \frac{1}{4}\left(1 + \frac{w}{2}\right)^{-2}$$

$$= \frac{1}{4}\left\{1 - 2\left(\frac{w}{2}\right) + 3\left(\frac{w}{2}\right)^2 - 4\left(\frac{w}{2}\right)^3 \ \dots \right\}$$

i.e. $\dfrac{1}{z^2} = \dfrac{1}{4}\left\{1 - 2\,\dfrac{z - 2}{2} + 3\,\dfrac{(z - 2)^2}{4} - 4\,\dfrac{(z - 2)^3}{8} \dots\right\}$

$$\therefore \frac{1}{z^2(z - 2)} = \frac{1}{4(z - 2)} - \frac{1}{4} + \frac{3}{16}(z - 2) - \frac{1}{8}(z - 2)^2 \ \dots \qquad (47A.1)$$

and so there is a simple pole at $z = 2$.

FRAME 48

In practice the poles and zeros of a function f(z) are located by examining f(z) rather than by finding its Laurent expansion.

For example, $\dfrac{(z - 3i)^2}{z(z + i)^3(z - 1)^4}$ has a zero of order 2 at $z = 3i$, a simple pole at $z = 0$, a pole of order 3 at $z = -i$ and a pole of order 4 at $z = 1$.

As another example, $\tan z = \dfrac{\sin z}{\cos z}$ and has zeros when $\sin z = 0$, i.e. when $z = k\pi$, k an integer and simple poles when $\cos z = 0$, i.e. when $z = (k + \frac{1}{2})\pi$.

Locate the poles and zeros of (i) $\dfrac{z}{z^2 - 3z + 2}$ and (ii) cosec z.

****

48A

(i) $z^2 - 3z + 2 = (z - 1)(z - 2)$

Thus $\dfrac{z}{z^2 - 3z + 2}$ *has a zero at* $z = 0$ *and simple poles at* $z = 1$ *and* $z = 2$.

(ii) $\operatorname{cosec} z = \dfrac{1}{\sin z}$. *It has simple poles when* $\sin z = 0$, *i.e. when* $z = k\pi$ *(k an integer).*

FRAME 49

The effect of a pole at a point can easily be illustrated pictorially. For simplicity a simple pole at $z = 0$ will be taken.

First sketch the real function $y = |1/x|$.

****

49A

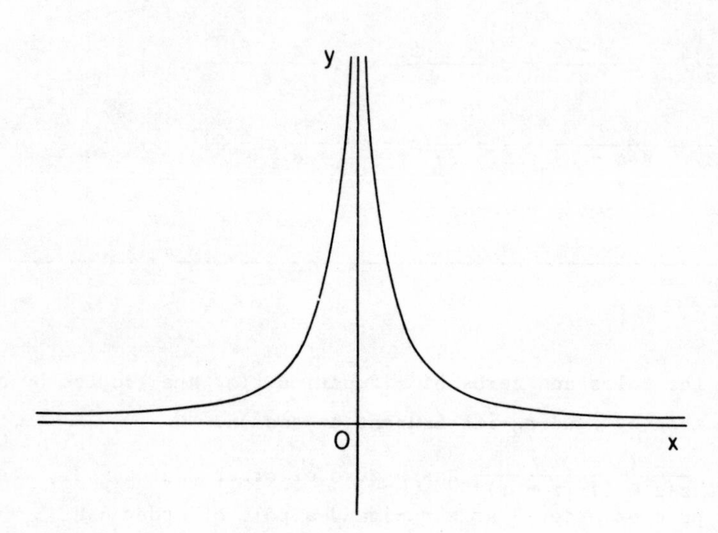

FRAME 50

In the present situation the function $w = |1/z|$ is sketched as $\frac{1}{z}$ has a simple pole at the origin.

However, the usual practice of drawing two diagrams, one in the z-plane and the other in the w-plane, is abandoned here and a three-dimensional picture drawn instead. The base plane is the usual z-plane and $|1/z|$ is shown on a vertical axis. The shape of the resulting figure can be obtained by rotating the curve in 49A about its y-axis. As is seen on the next page, a tapering pole (the ordinary sense of the word) is produced.

FRAME 50 continued

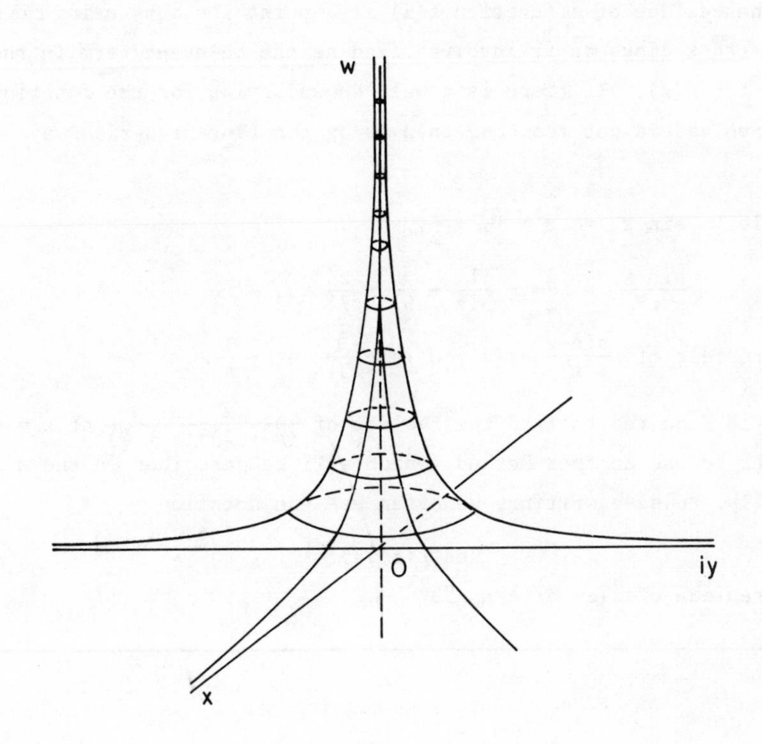

FRAME 51

Residues

The coefficient B_1 in the Laurent series (46.1) plays an important part in complex integration. It is called the RESIDUE of f(z) at a.

Thus, from (38.2), the residue of $\frac{1}{z^2(z - 2)}$ at the origin is $-\frac{1}{4}$, and, from (47A.1), its residue at 2 is $\frac{1}{4}$.

Note that the residue of f(z) at z = a can only be obtained from the Laurent series for f(z) if that series is expressed in powers of z - a.

FRAME 52

Finding the residue of a function f(z) at a point z = a by using the definition is very often tedious as it involves finding the relevant term in the Laurent expansion for f(z). If there is a well known series for the function or if the algebra involved is not too bad, then using the Laurent series may be quite easy.

For example, $\sin z = z - \frac{z^3}{3!} + \frac{z^5}{5!} - \ldots$

and so $\frac{\sin z}{z^4} = \frac{1}{z^3} - \frac{1}{3!z} + \frac{z}{5!} - \frac{z^3}{7!} \ldots$

Thus the residue of $\frac{\sin z}{z^4}$ at z = 0 is $-\frac{1}{3!} = -\frac{1}{6}$.

But if it is required to find the residue of $\frac{z - 3}{(z + 2)^3(z + 4)}$ at z = -4 it is much easier to use another method, which will be described in the next frame. Incidentally, to save writing, we often use the notation

$$\text{Res}\{f(z),a\}$$

for the "residue of f(z) at z = a".

FRAME 53

The simplest case arises when there is a simple pole at a point z = a. The Laurent series there for f(z) then takes the form

$$f(z) = A_0 + A_1(z - a) + A_2(z - a)^2 + A_3(z - a)^3 + \ldots + \frac{B_1}{z - a}$$

First multiply both sides of this equation by z - a to give

$$(z - a)f(z) = A_0(z - a) + A_1(z - a)^2 + A_2(z - a)^3 + A_3(z - a)^4 + \ldots + B_1 \quad (53.1)$$

As f(z) has a simple pole at z = a, it will contain a factor z - a in its denominator. Thus (z - a)f(z) will not contain a factor z - a in either its denominator or numerator.

Now let z → a. All terms on the R.H.S. of (53.1) will tend to zero, except B_1 and so we get

$$\lim_{z\to a}\{(z - a)f(z)\} = B_1$$

FRAME 53 continued

This gives an easy formula for the residue at a simple pole.

As an example, $$\text{Res}\left\{\frac{z-3}{(z+2)^3(z+4)}, -4\right\} = \lim_{z\to-4}\left\{(z+4)\frac{z-3}{(z+2)^3(z+4)}\right\}$$

$$= \lim_{z\to-4}\left\{\frac{z-3}{(z+2)^3}\right\}$$

$$= \frac{7}{8}$$

What is the residue of $\frac{z^2}{z^2+1}$ at (i) i, (ii) $-i$?

****

53A

$$\frac{z^2}{z^2+1} = \frac{z^2}{(z+i)(z-i)}$$

(i) $$\text{Res}\left\{\frac{z^2}{z^2+1}, i\right\} = \lim_{z\to i}\frac{z^2}{z+i}$$

$$= \frac{1}{2}i$$

(ii) $$\text{Res}\left\{\frac{z^2}{z^2+1}, -i\right\} = \lim_{z\to -i}\frac{z^2}{z-i}$$

$$= -\frac{1}{2}i$$

FRAME 54

Now suppose there is a pole of order 2 at $z = a$, then

$$f(z) = A_0 + A_1(z-a) + A_2(z-a)^2 + A_3(z-a)^3 + \ldots$$
$$+ \frac{B_1}{z-a} + \frac{B_2}{(z-a)^2}$$

This time we multiply both sides by $(z - a)^2$, thus

$$(z-a)^2 f(z) = A_0(z-a)^2 + A_1(z-a)^3 + A_2(z-a)^4 + A_3(z-a)^5 + \ldots$$
$$+ B_1(z-a) + B_2$$

Once again, the L.H.S. will not contain $z - a$ in either its numerator or denominator.

FRAME 54 continued

Now both sides of this equation are differentiated with respect to z.

$$\therefore \frac{d}{dz}\{(z-a)^2 f(z)\} = 2A_0(z-a) + 3A_1(z-a)^2 + 4A_2(z-a)^3 + 5A_3(z-a)^4 + \ldots + B_1$$

Finally, let $z \to a$ and

$$\lim_{z\to a}\left[\frac{d}{dz}\{(z-a)^2 f(z)\}\right] = B_1$$

The purpose of the differentiation is to make B_2 vanish and also ensure that the B_1 term does not disappear when $z \to a$.

As an example, the residue of $\dfrac{z}{(z+2)^2(z-i)}$ at $z = -2$ is given by

$$\lim_{z\to -2}\left[\frac{d}{dz}\left(\frac{z}{z-i}\right)\right]$$

Now complete the evaluation of this residue.

****

54A

$$\frac{d}{dz}\left(\frac{z}{z-i}\right) = \frac{-i}{(z-i)^2}$$

$$Res\left\{\frac{z}{(z+2)^2(z-i)}, -2\right\} = \frac{-i}{(-2-i)^2}$$

$$= \frac{-i}{3+4i}$$

$$= \frac{-4-3i}{25}$$

FRAME 55

Now see if you can work out the formula for the residue of f(z) at z = a if there is a pole of order 3 there.

****

55A

$$f(z) = A_0 + A_1(z - a) + A_2(z - a)^2 + \dots + \frac{B_1}{z - a} + \frac{B_2}{(z - a)^2} + \frac{B_3}{(z - a)^3}$$

$$(z - a)^3 f(z) = A_0(z - a)^3 + A_1(z - a)^4 + A_2(z - a)^5 + \dots + B_1(z - a)^2 + B_2(z - a) + B_3$$

This time differentiate twice.

$$\frac{d^2}{dz^2}\{(z - a)^3 f(z)\} = 3.2\,A_0(z - a) + 4.3\,A_1(z - a)^2 + 5.4\,A_2(z - a)^3 + \dots + 2.1\,B_1$$

$$\text{So } \lim_{z\to a}\left[\frac{d^2}{dz^2}\{(z - a)^3 f(z)\}\right] = 2B_1$$

$$B_1 = \frac{1}{2}\lim_{z\to a}\left[\frac{d^2}{dz^2}\{(z - a)^3 f(z)\}\right]$$

FRAME 56

Continuing in this way, we find that, for a pole of order n at a, the residue of f(z) is given by

$$\frac{1}{(n - 1)!}\lim_{z\to a}\left[\frac{d^{n-1}}{dz^{n-1}}\{(z - a)^n f(z)\}\right]$$

Now find the residue of $\frac{z - 1}{(z^2 - 4)(z + 1)^4}$ at each of its poles.

(HINT: When finding the residue at -1, you will find partial fractions useful.)

****

56A

$$\frac{z - 1}{(z^2 - 4)(z + 1)^4} = \frac{z - 1}{(z - 2)(z + 2)(z + 1)^4}$$

There are simple poles at 2 and -2 and a pole of order 4 at -1.

Residue at 2 is $\lim_{z\to 2}\frac{z - 1}{(z + 2)(z + 1)^4} = \frac{1}{324}$

Residue at -2 is $\lim_{z\to -2}\frac{z - 1}{(z - 2)(z + 1)^4} = \frac{3}{4}$

56A continued

Residue at -1 is $\dfrac{1}{3!}\lim_{z\to -1}\left[\dfrac{d^3}{dz^3}\left\{\dfrac{z-1}{(z-2)(z+2)}\right\}\right]$

$$= \frac{1}{3!}\lim_{z\to -1}\left[\frac{d^3}{dz^3}\left\{\frac{1}{4}\cdot\frac{1}{z-2}+\frac{3}{4}\cdot\frac{1}{z+2}\right\}\right]$$

$$= \frac{1}{3!}\lim_{z\to -1}\left\{\frac{1}{4}\cdot\frac{-6}{(z-2)^4}+\frac{3}{4}\cdot\frac{-6}{(z+2)^4}\right\}$$

$$= -\frac{61}{81}$$

FRAME 57

Cauchy's Residue Theorem

It has been seen (FRAMES 28-31) that integrals of the form $\displaystyle\int_C \frac{z-3}{(z+2)^3(z+4)}\,dz$ can be found by expressing the integrand in partial fractions. Cauchy's Residue Theorem enables us to find the value of such integrals without this preliminary algebra. As its name suggests, it makes use of the residues of the function at its poles.

Suppose it is desired to find $\displaystyle\int_C f(z)dz$ and that f(z) is such that its singularities within C are poles of finite order. For simplicity, we shall consider it to have just three poles within C. Let them be at P_1, P_2 and P_3 where $z = a_1$, $z = a_2$ and $z = a_3$ respectively. Then, as you already know,

$$\int_C f(z)dz = \int_{\gamma_1} f(z)dz + \int_{\gamma_2} f(z)dz + \int_{\gamma_3} f(z)dz \qquad (57.1)$$

where γ_1, γ_2 and γ_3 are small circles surrounding the singularities.

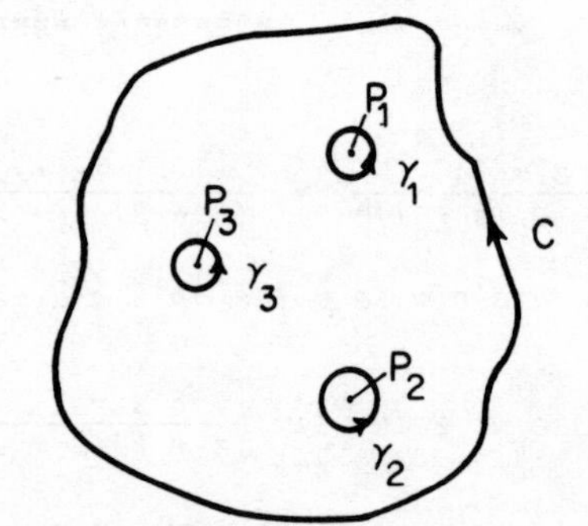

If the singularity at P_1 is a pole of order m, what will be the form of the Laurent expansion of f(z) at P_1?

****

57A

$$f(z) = A_0 + A_1(z - a_1) + A_2(z - a_1)^2 + A_3(z - a_1)^3 + \dots + \frac{B_1}{z - a_1} + \frac{B_2}{(z - a_1)^2} + \dots + \frac{B_m}{(z - a_1)^m}$$

FRAME 58

Thus

$$\int_{\gamma_1} f(z)dz = \int_{\gamma_1} A_0 dz + \int_{\gamma_1} A_1(z - a_1)dz + \int_{\gamma_1} A_2(z - a_1)^2 dz + \int_{\gamma_1} A_3(z - a_1)^3 dz + \dots$$

$$+ \int_{\gamma_1} \frac{B_1}{z - a_1} dz + \int_{\gamma_1} \frac{B_2}{(z - a_1)^2} dz + \dots + \int_{\gamma_1} \frac{B_m}{(z - a_1)^m} dz$$

What does the R.H.S. of this equation give?

****

58A

$2\pi i B_1$

All the other terms give zero. The A terms are all regular and so each of them gives zero by Cauchy's Theorem (FRAME 17). All the B terms apart from the first give zero as they are all examples of case (iv) in FRAME 27.

FRAME 59

Now B_1 is the residue of $f(z)$ at P_1. Thus

$$\int_{\gamma_1} f(z)dz = 2\pi i \times \text{Res}\{f(z), a_1\}$$

Similar results will hold for integration round γ_2 and γ_3.

So, incorporating these results into (57.1),

$$\int_C f(z)dz = 2\pi i \times \text{sum of residues of } f(z) \text{ at } P_1, P_2 \text{ and } P_3.$$

If there are more than 3 poles within C, then an extension of this gives

$$\int_C f(z)dz = 2\pi i \times \text{sum of residues of } f(z) \text{ at all poles within C.}$$

This is CAUCHY'S RESIDUE THEOREM.

FRAME 60

To demonstrate this result, let us find

$$\int_C \frac{z - 1}{(z^2 - 4)(z + 1)^4} \, dz$$

where C is (i) C_1, the circle $|z| = \frac{1}{2}$

(ii) C_2, the circle $\left|z + \frac{3}{2}\right| = 2$

(iii) C_3, the triangle whose vertices are at $-\frac{3}{2} + i$, $-\frac{3}{2} - i$ and 3.

First sketch the contours and insert the positions of the poles.

****

60A

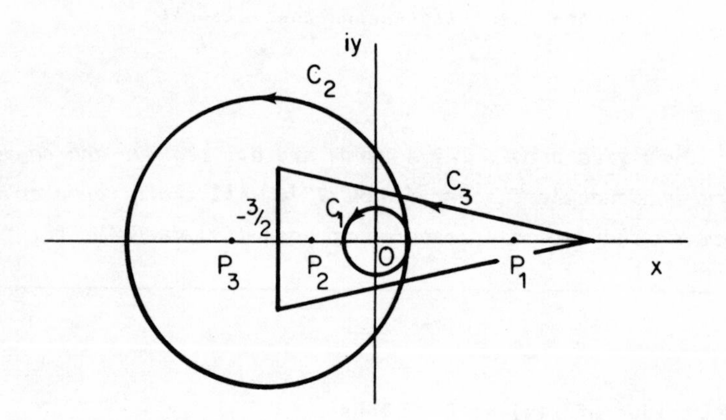

The poles are at $P_1(2)$, $P_2(-1)$ and $P_3(-2)$. That at P_2 is of order 4. The other two are simple.

FRAME 61

You have already found the residues at each of the poles (see 56A).

(i) C_1 encloses no pole and so

$$\int_{C_1} \frac{z - 1}{(z^2 - 4)(z + 1)^4} \, dz = 0$$

FRAME 61 continued

(ii) C_2 encloses P_2 and P_3

$$\therefore \int_{C_2} \frac{z - 1}{(z^2 - 4)(z + 1)^4}\, dz = 2\pi i\left(-\frac{61}{81} + \frac{3}{4}\right)$$
$$= -\frac{\pi i}{162}$$

Now write down the value of $\displaystyle\int_{C_3} \frac{z - 1}{(z^2 - 4)(z + 1)^4}\, dz$

****

61A

As C_3 encloses P_1 and P_2, the integral is

$$2\pi i\left(\frac{1}{324} - \frac{61}{81}\right) = -\frac{3\pi i}{2}$$

FRAME 62

Now try these examples, using this method of residues:

(i) $\displaystyle\int_C \frac{z^2}{z^2 + 4}\, dz$, where C is the square with vertices at ± 2, $\pm 2 + 4i$.

(ii) $\displaystyle\int_C \frac{\sinh z}{z^4}\, dz$, where C is $|z| = 1$ (HINT: Use the series for sinh z to obtain the residue, rather than the limit formula.)

****

62A

(i) $z^2 + 4 = (z + 2i)(z - 2i)$

$\dfrac{z^2}{z^2 + 4}$ *has simple poles at* $-2i$ *and* $2i$.

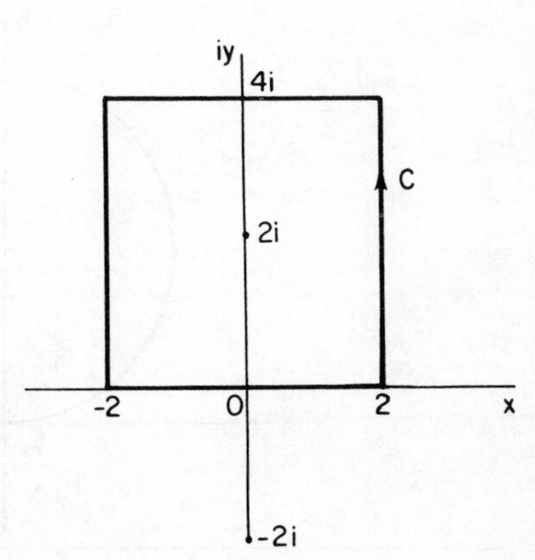

That at $2i$ is inside C

$$\text{Res}\left\{\frac{z^2}{z^2 + 4}, 2i\right\} = \lim_{z\to 2i}\left\{\frac{(z - 2i)z^2}{z^2 + 4}\right\}$$
$$= \lim_{z\to 2i}\left\{\frac{z^2}{z + 2i}\right\}$$
$$= i$$

$\therefore$ *Required integral* $= 2\pi i \times i$
$= -2\pi$

<u>62A</u> continued

(ii) $$\frac{\sinh z}{z^4} = \frac{z + \frac{1}{3!}z^3 + \frac{1}{5!}z^5 + \ldots}{z^4}$$

$$= \frac{1}{z^3} + \frac{1}{3!z} + \frac{z}{5!} + \ldots$$

There is a pole of order 3 at the origin and this is inside the contour.

$$Res\left\{\frac{\sinh z}{z^4}, 0\right\} = \frac{1}{3!}$$

Required integral $= \pi i/3$

Note that the order of the pole at the origin is 3 and not 4 as is suggested by the z^4 *term in the denominator. This is because sinh z has a zero of order 1 at the origin, and so, when expanded, has a factor z which cancels one of the z's in the denominator.*

<u>FRAME 63</u>

So far the examples that have been considered have all had denominators that had obvious factors. However, even if this is not the case, the method is still the same and the following example illustrates the procedure.

Find $\displaystyle\int_C \frac{1}{2z^2 + 2z + 3}\,dz$ where C is the circle $|z - 3i| = 3$.

Start by solving $2z^2 + 2z + 3 = 0$ as an ordinary quadratic to obtain the poles and showing these on a sketch with the contour.

****

<u>63A</u>

$$z = -\frac{1}{2} \pm i\,\frac{\sqrt{5}}{2}$$

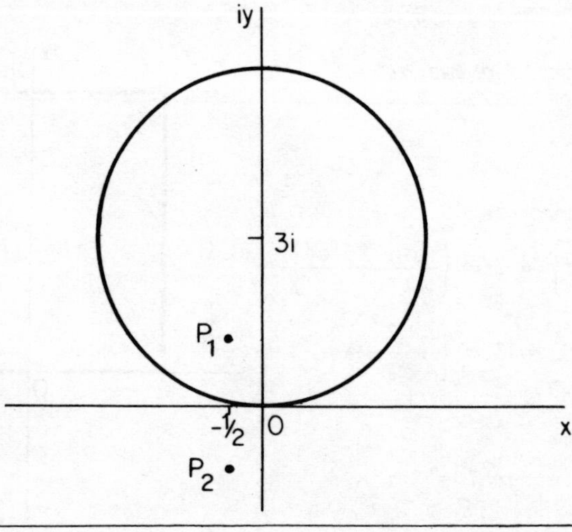

iy

3i

P_1

$-\frac{1}{2}$ O x

P_2

Poles at P_1 *and* P_2

FRAME 64

Let $-\frac{1}{2} + i\frac{\sqrt{5}}{2} = \alpha$ and $-\frac{1}{2} - i\frac{\sqrt{5}}{2} = \beta$.

We require the residue at α.

This is $\lim\limits_{z\to\alpha} \dfrac{z-\alpha}{2z^2 + 2z + 3} = \lim\limits_{z\to\alpha} \dfrac{z-\alpha}{2(z-\alpha)(z-\beta)}$

$$= \frac{1}{2(\alpha - \beta)}$$

$$= \frac{1}{2 \times i\sqrt{5}}$$

Required integral $= 2\pi i \times \dfrac{1}{2i\sqrt{5}}$

$$= \frac{\pi}{\sqrt{5}}$$

The use of α and β merely reduces the amount of writing.

FRAME 65

Application of Complex Integration to the Evaluation of Certain Real Integrals

Certain real definite integrals can be evaluated by either transforming them into complex integrals or by considering complex integrals that are similar in form to the real integral required. Many applications of contour integrals, such as those mentioned in FRAMES 11-15, do not require the technique and so it will not be discussed here. If you are interested in seeing how this technique works, you will find it applied to some examples in APPENDIX B at the end of this programme.

FRAME 66

Some Applications of Complex Integrals

We shall now return to some of the applications quoted in FRAMES 11-15 as we are now in a position to be able to evaluate the integrals given there.

First, a word about the contour mentioned in FRAME 11, i.e.

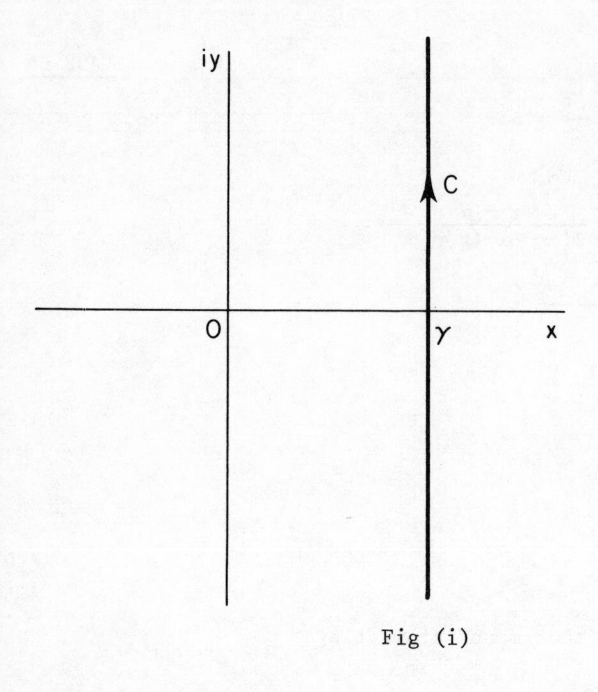

Fig (i)

FRAME 66 continued

As you now know, it is much better to have a closed contour, rather than an open one, so that Cauchy's Residue theorem can be used to write down the value of a complex integral. The closed contour taken in this case is as shown in the diagram below.

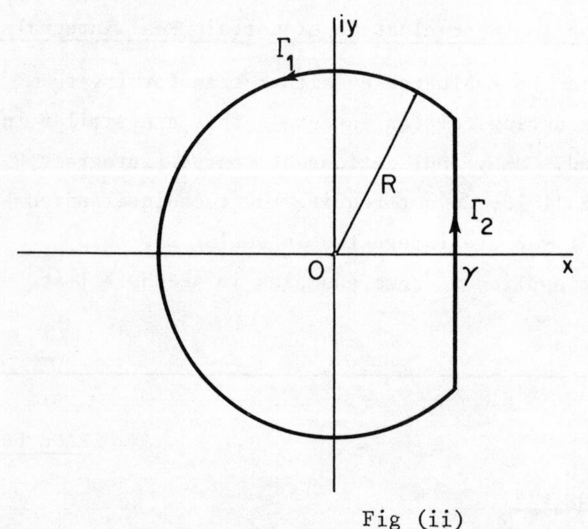

Fig (ii)

Then as $R \to \infty$, $\Gamma_2 \to C$. So, if the integrand $f(z)$ is such that $\int_{\Gamma_1} f(z)dz \to 0$, then $\int_C f(z)dz$ is given by $2\pi i \times$ sum of residues of $f(z)$ at all poles to the left of C. In FRAME 11, it was mentioned that γ is chosen sufficiently large so that all singularities do lie to the left of C.

FRAME 67

As an example on the use of complex integration to find an inverse Laplace transform, let us find $\mathcal{L}^{-1}\left\{\frac{s}{s^2 + \omega^2}\right\}$. This is well-known to be cos ωt, but taking a simple, well-known result will serve to illustrate the method. The formula quoted in FRAME 11 for the required inverse was

$$\frac{1}{2\pi i}\int_C e^{st}G(s)ds$$

C being the contour in the last frame. Remember that in this work s, like z, is of the form x + iy, and can therefore be treated in the same way. So, in this case,

$$\begin{aligned}\mathcal{L}^{-1}\left\{\frac{s}{s^2 + \omega^2}\right\} &= \frac{1}{2\pi i}\int_C e^{st}\frac{s}{s^2 + \omega^2}ds \\ &= \frac{1}{2\pi i}\left\{2\pi i \times \text{sum of residues of } e^{st}\frac{s}{s^2+\omega^2} \text{ at all poles to the left of C}\right\} \\ &= \text{sum of residues of } e^{st}\frac{s}{s^2+\omega^2} \text{ at these poles.}\end{aligned}$$

Where are the poles of $e^{st}\frac{s}{s^2+\omega^2}$ and what are its residues at these poles? (HINT: Regard t as a constant here.)

****

67A

Poles are at $\pm i\omega$.

Residue at $i\omega$ *is* $\frac{1}{2}e^{i\omega t}$

Residue at $-i\omega$ *is* $\frac{1}{2}e^{-i\omega t}$

FRAME 68

These poles both lie on the imaginary axis and so γ must be chosen to be > 0.

$$\begin{aligned}\text{Then } \mathcal{L}^{-1}\left\{\frac{s}{s^2 + \omega^2}\right\} &= \frac{1}{2}\left(e^{i\omega t} + e^{-i\omega t}\right) \\ &= \cos \omega t\end{aligned}$$

It can be shown that $\int_{\Gamma_1}$ does actually $\to 0$ as $R \to \infty$.

FRAME 69

Closely linked with the inversion of Laplace transforms is the question of stability in control theory, which is important in many branches of engineering. In this, just as in all our work on integration, the location of the poles of a function is of great importance. One way of describing a control system is to give what is called its transfer function. This concept will be treated more fully in Unit 13. Here, just the ideas involved will be illustrated, a simple electrical circuit being taken for this purpose.

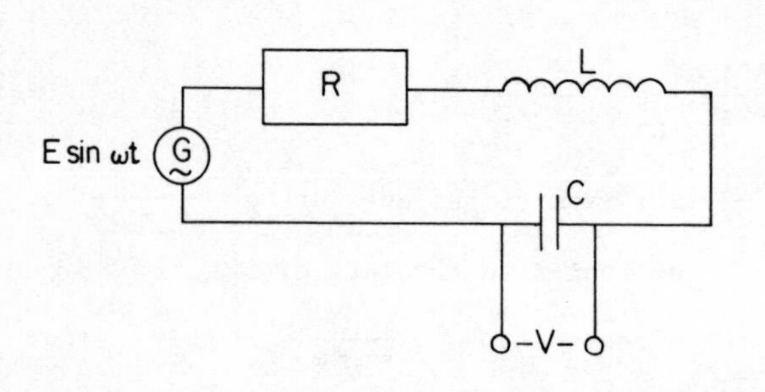

Suppose, in the circuit shown, it is desired to find the voltage V across C for a given input voltage E sin ωt, the charge on the capacitor and the current both being zero when t = 0, i.e. at the instant when the input voltage is applied to the circuit.

The equations $$E \sin \omega t = Ri + L\frac{di}{dt} + \frac{1}{C}\int_0^t i\,dt$$

and $$V = \frac{1}{C}\int_0^t i\,dt$$

will apply here and the transfer function is

$$\frac{\mathcal{L}\{V\}}{\mathcal{L}\{E \sin \omega t\}}, \text{ i.e. } \left\{\frac{E\omega}{C(s^2+\omega^2)(Ls^2+Rs+1/C)}\right\}\Bigg/\frac{E\omega}{s^2+\omega^2} = \frac{1}{LCs^2+RCs+1}$$

If you have read the programme on Laplace Transforms in Unit 5 of Vol I, you know that the inverse of $\dfrac{E\omega}{C(s^2+\omega^2)(Ls^2+Rs+1/C)}$ can be found by taking partial fractions and comparing these with a table of standard transforms. The partial fractions will be of the form

$$\frac{As+B}{s^2+\omega^2} + \frac{A's+B'}{Ls^2+Rs+1/C}$$

and it should be obvious that the former comes from the input voltage E sin ωt and the latter from the circuit components R, C and L. The inverse of $\dfrac{As+B}{s^2+\omega^2}$ involves cos ωt and sin ωt which oscillate finitely. These terms will give the steady state part of the solution - mathematically the particular integral.

FRAME 69 continued

The inverse of $\dfrac{A's + B'}{Ls^2 + Rs + 1/C}$ will give the complementary function, which, if it dies away, is the transient. The condition for it to do this is that all exponentials involved shall be of the form $e^{-\alpha t}$ where α is positive. Now if the method of FRAMES 67-8 is used to find the inverse, the residues at the poles of $e^{st}\dfrac{A's + B'}{Ls^2 + Rs + 1/C}$ will be necessary and these will give rise to terms involving exponentials of the form $e^{-\alpha t}$ if the real parts of the roots of $Ls^2 + Rs + 1/C$ are negative, i.e. if the real parts of the poles of the transfer function are negative. Thus, if these poles lie in the left-hand half of the complex plane, the system will be stable, but not if they lie in the right-hand half.

Find the poles of the transfer function $\dfrac{1}{LC(s^2 + \frac{R}{L}s + \frac{1}{LC})}$ in the two cases (i) $R^2 > 4L/C$ (ii) $R^2 < 4L/C$ and show their positions roughly on an Argand diagram. Will the system be stable in either or both (i) and (ii)?

****

69A

Poles are given by $s = \left(-\frac{R}{L} \pm \sqrt{\frac{R^2}{L^2} - \frac{4}{LC}}\right)\Big/2.$

(i) $\frac{R^2}{L^2} - \frac{4}{LC} > 0$ *and both roots are real, i.e. the poles lie on the real axis. Also as* $\sqrt{\frac{R^2}{L^2} - \frac{4}{LC}} < \frac{R}{L}$, *they both lie on the negative real axis.*

(ii) $\frac{R^2}{L^2} - \frac{4}{LC} < 0$ *and the poles are conjugate complexes with negative real parts.*

In both cases (i) and (ii) the system is stable.

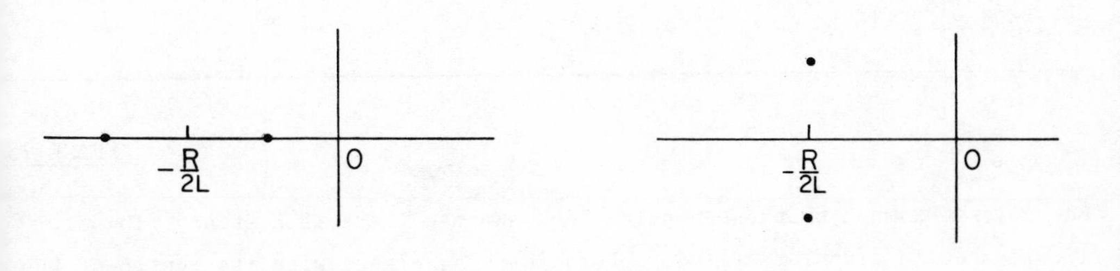

FRAME 70

In this example the poles of the transfer function were both simple. It can be shown that the system is still stable for multiple poles with negative real parts.

Poles on the imaginary axis have not been mentioned. It can be also shown that the system is stable if such poles are simple, but not otherwise.

FRAME 71

In FRAME 15, the formula given there for the pitching angle is

$\frac{9\cdot 56}{38\cdot 6} \times \frac{1}{2\pi i}\int_C \frac{e^{zt}}{z^3}\,dz$, with C as in Fig (i) in FRAME 66. This is equivalent to

$\frac{9\cdot 56}{38\cdot 6} \times \text{Res}\left\{\frac{e^{zt}}{z^3}, 0\right\}$ and here, once again, $\gamma > 0$, and t is treated as a constant.

What is $\text{Res}\left\{\frac{e^{zt}}{z^3}, 0\right\}$?

****

71A

$\frac{1}{2}t^2$

FRAME 72

Thus, in this particular case, for small t, the pitching angle is

$\frac{9\cdot 56}{38\cdot 6} \times \frac{1}{2}t^2 = 0\cdot 124t^2$. Initially, then, the angular displacement - time curve is parabolic.

FRAME 73

For a last example on these applications suppose there is a steady flow of liquid about a circular cylinder of radius < a, placed with its centre at the point ia, the x-axis being a rigid boundary.

FRAME 73 continued

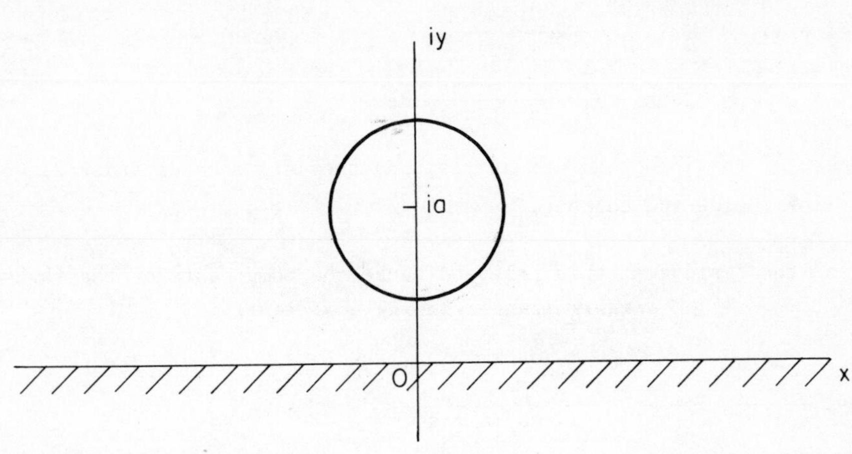

It was mentioned in FRAME 12 that there is a quantity w (a function of z) that is determined by the particular motion of the fluid. Incidentally w is such that $\frac{dw}{dz} = q_x - iq_y$, where q_x and q_y are the components of the velocity of the fluid parallel to the two axes. In this particular case, it can be shown that $w = ik \ln \frac{z - ia}{z + ia}$, where k is a constant. Sufficient information is now available to enable us to find, for example, the components of the thrust per unit length on the cylinder if, in addition, the fluid density is assumed to be ρ, a constant.

The formula for this is given in FRAME 12 as

$$X - iY = \frac{1}{2} i\rho \int_C \left(\frac{dw}{dz}\right)^2 dz$$

where X and Y are the components of the thrust and C is the boundary of the cylinder.

First, find $\frac{dw}{dz}$ here.

****

73A

$$w = ik\{\ln(z - ia) - \ln(z + ia)\}$$

$$\frac{dw}{dz} = ik\left(\frac{1}{z - ia} - \frac{1}{z + ia}\right)$$

$$= \frac{-2ak}{z^2 + a^2}$$

FRAME 74

Then
$$X - iY = \frac{1}{2} i\rho \int_C \frac{4a^2k^2}{(z^2 + a^2)^2} dz$$
$$= 2a^2k^2 i\rho \int_C \frac{1}{(z^2 + a^2)^2} dz$$

The poles of the integrand are at $\pm ia$, and they are each of order 2. Only that at ia lies inside the contour.

Now find the residue at this pole and hence the components of the thrust.

****

74A

$$Res\left\{\frac{1}{(z^2 + a^2)^2}, ia\right\} = \lim_{z \to ia} \frac{d}{dz} \frac{1}{(z + ai)^2}$$
$$= -\frac{i}{4a^3}$$
$$X - iY = 2a^2k^2 i\rho \times 2\pi i \times \left(-\frac{i}{4a^3}\right)$$
$$= \frac{\pi\rho k^2 i}{a}$$
$$\therefore \quad X = 0, \qquad Y = -\frac{\pi\rho k^2}{a}$$

FRAME 75

Finally, use the formula $-\frac{1}{2}\rho \, Re\int_C z\left(\frac{dw}{dz}\right)^2 dz$ to find the moment of this thrust about the origin.

****

75A

$$\text{Moment of thrust} = -\frac{1}{2}\rho \, Re\int_C \frac{4a^2k^2 z}{(z^2 + a^2)^2} dz$$
$$= -2a^2k^2\rho \, Re\int_C \frac{z}{(z^2 + a^2)^2} dz$$

C being once again the boundary of the cylinder.

$$Res\left\{\frac{z}{(z^2 + a^2)^2}, ia\right\} = \lim_{z \to ia} \frac{d}{dz} \frac{z}{(z + ia)^2}$$
$$= 0$$

$\therefore$ *Moment of thrust is zero also.*

FRAME 76

Miscellaneous Examples

In this frame a collection of miscellaneous examples is given for you to try. Answers are supplied in FRAME 77, together with such working as is considered helpful.

1. Evaluate $\int_C \frac{1}{(z+1)^3(z-1)(z-2)}\,dz$ where C is (i) $|z| = \frac{1}{2}$

   (ii) $|z+1| = 1$ (iii) the rectangle whose vertices are at $\pm i$, $3 \pm i$.

2. Evaluate $\int_C \frac{e^{2z}}{(z+1)^4}\,dz$ where C is the circle $|z| = 3$.

In questions 3 and 4 assume that $\int_{\Gamma_1} \to 0$ as $R \to \infty$, Γ_1 being the contour shown in Fig (ii) of FRAME 66.

3. The Laplace inverse of the transformed function F(s) is given by the contour integral

$$f(t) = \frac{1}{2\pi i}\int_C F(z)e^{zt}dz$$

   where C is a suitably chosen contour which encloses the poles of F(z). Use this result to show that the inverses of

$$(3s + 2)/(s^2 + 4) \quad \text{and} \quad 1/\{s(s + 1)^2\}$$

   are

$$(3\cos 2t + \sin 2t) \quad \text{and} \quad 1 - (t + 1)e^{-t}$$

   respectively. (C.E.I.)

4. A constant potential E_0 is applied to the circuit shown during the time interval $0 < t < T$.

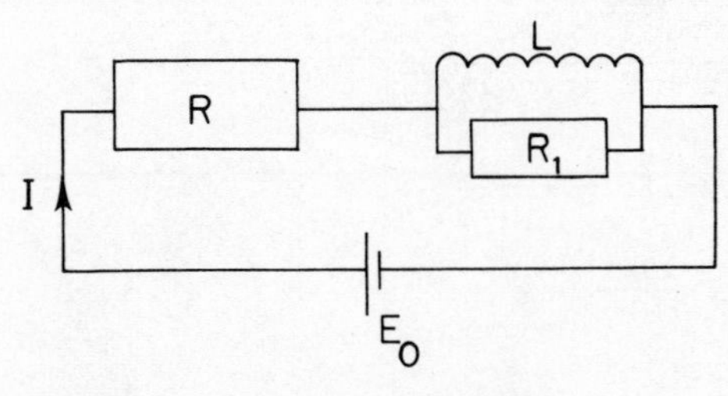

FRAME 76 continued

The current at any time $t > T$ is given by

$$I = \frac{E_0}{\alpha}\frac{1}{2\pi i}\int_C \frac{\{e^{zt} - e^{z(t-T)}\}(zL + R_1)dz}{z(z+\beta)}$$

where C is the contour in FRAME 11, $\alpha = L(R_1 + R)$, $\beta = \frac{RR_1}{L(R_1 + R)}$.

Find I when $E_0 = 100$, $T = 1$, $R = 90$, $R_1 = 10$, $L = 0\cdot2$.

5. $w = \frac{U(a^2 + z^2)}{z} + ik \ln \frac{z}{a}$ represents a certain two-dimensional fluid flow past a fixed cylinder whose cross-section is the circle $|z| = a$. Find the components of the thrust per unit length on this cylinder.

FRAME 77

Answers to Miscellaneous Examples

1. (i) No pole inside contour, so required integral is zero.

(ii) Triple pole at -1 inside contour.

$$\text{Res}\left\{\frac{1}{(z+1)^3(z-1)(z-2)}, -1\right\} = \frac{1}{2!}\lim_{z\to-1}\frac{d^2}{dz^2}\frac{1}{(z-1)(z-2)} = \frac{19}{216}$$

$$\text{Integral} = 2\pi i \times \frac{19}{216} = \frac{19\pi i}{108}$$

(iii) Simple poles at 1 and 2 are inside contour.

$$\text{Res}\left\{\frac{1}{(z+1)^3(z-1)(z-2)}, 1\right\} = -\frac{1}{8}$$

$$\text{Res}\left\{\frac{1}{(z+1)^3(z-1)(z-2)}, 2\right\} = \frac{1}{27}$$

$$\text{Integral} = 2\pi i\left(-\frac{1}{8} + \frac{1}{27}\right) = -\frac{19\pi i}{108}$$

FRAME 77 continued

2. Pole of order 4 at -1 is inside contour.

$$\text{Res}\left\{\frac{e^{2z}}{(z+1)^4}, -1\right\} = \frac{1}{3!}\lim_{z\to -1}\frac{d^3}{dz^3}e^{2z}$$

$$= \frac{4}{3}e^{-2}$$

Integral $= 8\pi i e^{-2}/3$

3. The poles of $\dfrac{(3z+2)e^{zt}}{z^2+4}$ are at $\pm 2i$

Res at $2i$ is $\dfrac{(6i+2)e^{2it}}{4i}$

Res at $-2i$ is $\dfrac{(-6i+2)e^{-2it}}{-4i}$

$$\text{Sum of residues} = \left(\frac{3}{2} - \frac{1}{2}i\right)e^{2it} + \left(\frac{3}{2} + \frac{1}{2}i\right)e^{-2it}$$

$$= \frac{3}{2}\left(e^{2it} + e^{-2it}\right) + \frac{1}{2i}\left(e^{2it} - e^{-2it}\right)$$

$$= 3\cos 2t + \sin 2t$$

The poles of $\dfrac{e^{zt}}{z(z+1)^2}$ are at 0, -1

Res at 0 is 1

Res at -1 is $\displaystyle\lim_{z\to -1}\frac{d}{dz}\frac{e^{zt}}{z} = -(t+1)e^{-t}$

Sum of residues $= 1 - (t+1)e^{-t}$

Note that, just as in real variable work, $\displaystyle\int_a^b F(x)dx = \int_a^b F(y)dy$, so, in complex variable work, $\displaystyle\int_C F(z)dz = \int_C F(s)ds$, i.e. the variable with respect to which integration is performed is immaterial when a contour is specified.

FRAME 77 continued

4. $\alpha = 20, \qquad \beta = 45,$

$$i = \frac{100}{20}\frac{1}{2\pi i}\int_C \frac{\{e^{zt} - e^{z(t-T)}\}(0\cdot 2z + 10)}{z(z + 45)}\,dz$$

$$= 5 \times \text{sum of residues of } \frac{\{e^{zt} - e^{z(t-T)}\}(0\cdot 2z + 10)}{z(z + 45)}$$

A simple pole occurs at -45. Note that there is no pole at the origin, as the Laurent expansion of $e^{zt} - e^{z(t-T)}$ starts with the term in z.

$$\text{Residue at } -45 = \frac{\{e^{-45t} - e^{-45(t-T)}\}(-9 + 10)}{-45}$$

$$i = \frac{1}{9}\{e^{-45(t-T)} - e^{-45t}\}$$

$$= \frac{1}{9}\,e^{-45t}(e^{45T} - 1)$$

5. $$\frac{dw}{dz} = -\frac{Ua^2}{z^2} + U + \frac{ik}{z}$$

$$X - iY = \frac{1}{2}\,i\rho\int_C\left(-\frac{Ua^2}{z^2} + U + \frac{ik}{z}\right)^2 dz \quad \text{where C is } |z| = a.$$

$$= \frac{1}{2}\,i\rho\int_C\left(\frac{2Uik}{z} + \text{terms involving other powers of } z\right)dz$$

$$= \frac{1}{2}\,i\rho \times 2\pi i \times 2Uik$$

$$= -2U\rho k\pi i$$

$\therefore X = 0, \qquad Y = 2U\rho k\pi$

APPENDIX A

Green's Theorem

If C is the curve enclosing the area S, as shown, then

$$\oint_C Pdx + Qdy = \iint_S \left(\frac{\partial Q}{\partial x} - \frac{\partial P}{\partial y}\right) dxdy,$$

where P and Q are functions of x and y.

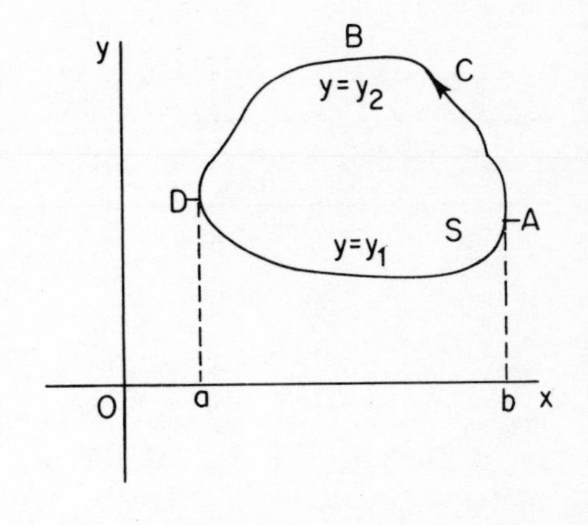

To show this, we start with $\iint_S \frac{\partial P}{\partial y}\, dxdy$, and insert limits so that the area S is covered by the integral.

Thus $\iint_S \frac{\partial P}{\partial y}\, dxdy$ becomes $\int_a^b dx \int_{y_1}^{y_2} \frac{\partial P}{\partial y}\, dy$

$$= \int_a^b \Big[P\Big]_{y_1}^{y_2} dx$$

As P is a function of x and y, $\Big[P\Big]_{y_1}^{y_2}$ means that all y's in P are first replaced by y_2, then by y_1 and the results subtracted, i.e. $\Big[P\Big]_{y_1}^{y_2} = P(x,y_2) - P(x,y_1)$.

Now $\int_a^b P(x,y_2)dx$ is effectively the line integral $\int_{DBA} P(x,y)dx$, as $y = y_2$ along the path DBA. Similarly, $\int_a^b P(x,y_1)dx$ is

$$\int_{DEA} P(x,y)dx \quad \text{as} \quad y = y_1 \text{ along DEA.}$$

$$\therefore \int_a^b \Big[P\Big]_{y_1}^{y_2} dx = \int_{DBA} P(x,y)dx - \int_{DEA} P(x,y)dx$$

$$= -\int_{ABD} P(x,y)dx - \int_{DEA} P(x,y)dx$$

$$= -\int_{ABDEA} P(x,y)dx$$

$$= -\oint_C Pdx$$

In a similar way, it can be shown that $\iint_S \frac{\partial Q}{\partial x}\,dxdy = \oint_C Qdy$

and so $\iint_S \left(\frac{\partial Q}{\partial x} - \frac{\partial P}{\partial y}\right) dxdy = \oint_C Qdy + Pdx$

or $\oint_C Pdx + Qdy = \iint_S \left(\frac{\partial Q}{\partial x} - \frac{\partial P}{\partial y}\right) dxdy.$

APPENDIX B

FRAME B1

The Evaluation of Certain Real Integrals

Algebraic Integrands

In this APPENDIX, some examples of real integrals which involve infinite limits will be discussed and it will be seen how complex integration can be adapted to help in such cases.

As a first example we shall consider $\int_{-\infty}^{\infty} \frac{1}{(x^2 + 4)^2}\,dx$. This can, of course, be quite easily done by ordinary integration, but it will illustrate the method of applying complex integration without giving rise to complicated working.

It is first necessary to write down a function of z which, when z is completely real, reduces to $\frac{1}{(x^2 + 4)^2}$. This is done by inspection. In this case, the required function is $\frac{1}{(z^2 + 4)^2}$. As $z = x + iy$, $z = x$ when $y = 0$, i.e. when z is real. Also, when $z = x$, $dz = dx$.

The next step is to consider $\int_C \frac{dz}{(z^2 + 4)^2}$, where C is a suitable contour. As the link between this and the required integral is the fact that $z = x$ if $y = 0$, i.e. z is on the real axis, part of the contour is chosen to lie along this axis. Further, it is arranged that the mid-point of this section of the contour is at the origin. Various paths can be chosen to close the contour, but a large semicircle in the upper half plane is found to be convenient. Let its radius be R.

C now consists of the two parts Γ_1 and Γ_2 shown, and so

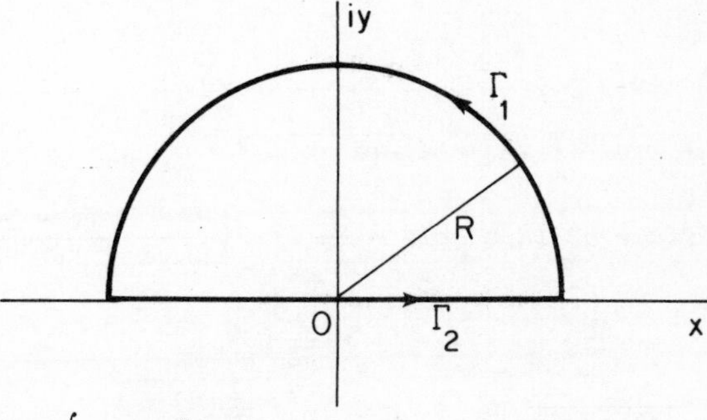

$$\int_C \frac{1}{(z^2 + 4)^2}\,dz = \int_{\Gamma_1} \frac{1}{(z^2 + 4)^2}\,dz + \int_{\Gamma_2} \frac{1}{(z^2 + 4)^2}\,dz$$

FRAME B1 continued

Now, seeing that along Γ_2, $z = x$, $\int_{\Gamma_2} \frac{1}{(z^2+4)^2}\,dz = \int_{-R}^{R} \frac{1}{(x^2+4)^2}\,dx$.

If we let $R \to \infty$, this will become the required integral.

The poles of $\frac{1}{(z^2+4)^2}$ are the next things to be considered. Where are the poles of this function?

****

B1A

$\pm 2i$

FRAME B2

In the last frame, it was simply stated that the semicircle was large without saying how large. It is necessary for it to be sufficiently large that it includes all singularities in the upper half plane. Then, when $R \to \infty$, Γ_1 will not have to cross any of these singularities. The new diagram shows the positions of the poles relative to C.

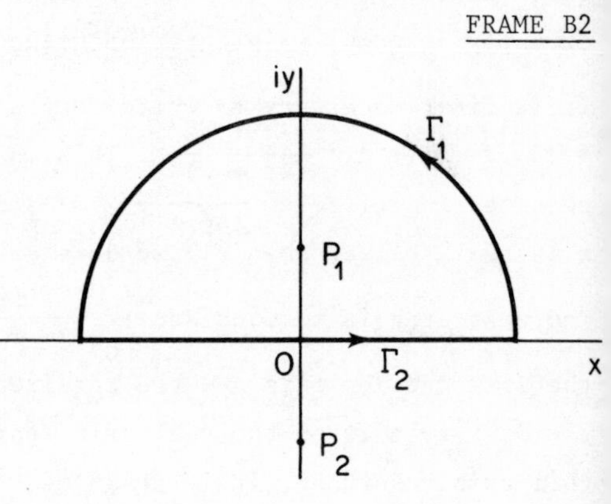

Now evaluate $\int_C \frac{1}{(z^2+4)^2}\,dz$.

****

B2A

The only pole inside the contour is P_1 and it is a pole of order 2 as

$$\frac{1}{(z^2+4)^2} = \frac{1}{(z+2i)^2(z-2i)^2}$$

$$\text{Residue at this pole} = \frac{1}{1!}\lim_{z\to 2i}\left\{\frac{d}{dz}\,\frac{(z-2i)^2}{(z+2i)^2(z-2i)^2}\right\}$$

$$= -\frac{1}{32}i$$

$$\int_C \frac{1}{(z^2+4)^2}\,dz = 2\pi i \times \left(-\frac{1}{32}i\right)$$

$$= \frac{\pi}{16}$$

FRAME B3

So now $\displaystyle\int_{\Gamma_1} \frac{1}{(z^2 + 4)^2}\,dz + \int_{\Gamma_2} \frac{1}{(z^2 + 4)^2}\,dz = \frac{\pi}{16}$

$$\therefore \quad \int_{\Gamma_1} \frac{1}{(z^2 + 4)^2}\,dz + \int_{-R}^{R} \frac{1}{(x^2 + 4)^2}\,dx = \frac{\pi}{16}$$

Now let $R \to \infty$, then

$$\lim_{R\to\infty} \int_{\Gamma_1} \frac{1}{(z^2 + 4)^2}\,dz + \int_{-\infty}^{\infty} \frac{1}{(x^2 + 4)^2}\,dx = \frac{\pi}{16}$$

The only problem remaining is the first term on the L.H.S. of this last equation.

The only integrals to which this method is applied are those for which this term is zero. We shall not go into details here, but it can be shown that this happens if the degree of the denominator of the integrand is at least 2 more than that of the numerator. In this case the degrees of the denominator and numerator are respectively 4 and 0. So, finally,

$$\int_{-\infty}^{\infty} \frac{1}{(x^2 + 4)^2}\,dx = \frac{\pi}{16}$$

FRAME B4

As a second example of this type, let us find $\displaystyle\int_0^{\infty} \frac{1}{x^4 + 16}\,dx$. The first thing to notice is that $\dfrac{1}{x^4 + 16}$ is an even function so that $\displaystyle\int_0^{\infty} \frac{1}{x^4 + 16}\,dx = \frac{1}{2}\int_{-\infty}^{\infty} \frac{1}{x^4 + 16}\,dx$. If this is not obvious, make a rough sketch of the function.

The complex integral in this case is $\displaystyle\int_C \frac{1}{z^4 + 16}\,dz$ where C is the same as that in FRAME B1.

Now find the poles of $\dfrac{1}{z^4 + 16}$, leaving them in the exponential form and draw a sketch to show them in relation to C.

****

B4A

$z^4 = -16 = 16e^{i(\pi+2k\pi)}$

$z = 2e^{i(\pi/4 + k\pi/2)}$ $(k = 0,1,2,3)$ *The poles are all simple.*

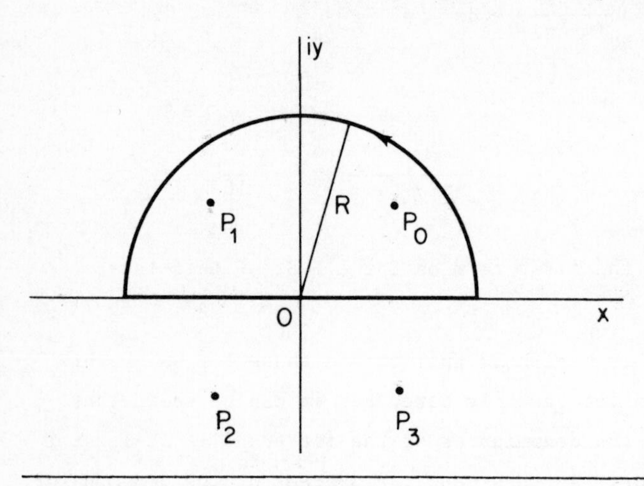

The suffix used for each pole corresponds to the value of k that gives the pole.
The semicircle must be large enough to include P_0 and P_1.

FRAME B5

If, for P_0, $z = \alpha_0$,

$$\text{Res}\left\{\frac{1}{z^4 + 16}, \alpha_0\right\} = \lim_{z\to\alpha_0} \frac{z - \alpha_0}{z^4 + 16}$$

Now, as $z = \alpha_0$ is a root of $z^4 + 16 = 0$ the denominator has a factor $z - \alpha_0$. Evaluation of this limit by cancelling this factor leads to cumbersome algebra. However, as both numerator and denominator of the limit are zero when $z = \alpha_0$, l'Hopital's rule can be applied.

$$\lim_{z\to\alpha_0} \frac{z - \alpha_0}{z^4 + 16} = \lim_{z\to\alpha_0} \frac{1}{4z^3}$$

$$= \lim_{z\to\alpha_0} \frac{z}{4z^4}$$

$$= \frac{2e^{i\pi/4}}{4 \times (-16)}$$

$$= -\frac{1}{32} e^{i\pi/4}$$

What will be the residue at P_1?

****

B5A

$$\lim_{z \to \alpha_1} \frac{z - \alpha_1}{z^4 + 16} \quad \text{where } z = \alpha_1 \text{ for } P_1$$

i.e. $$\lim_{z \to \alpha_1} \frac{z}{4z^4} = -\frac{1}{32} e^{3i\pi/4}$$

FRAME B6

Thus $$\int_C \frac{1}{z^4 + 16}\,dz = 2\pi i\left\{-\frac{1}{32}\left(e^{i\pi/4} + e^{3i\pi/4}\right)\right\}$$

$$= \frac{1}{16}\pi\sqrt{2}$$

$$\therefore \int_{\Gamma_1} \frac{1}{z^4 + 16}\,dz + \int_{\Gamma_2} \frac{1}{z^4 + 16}\,dz = \frac{1}{16}\pi\sqrt{2}$$

As $R \to \infty$, the first of these integrals $\to 0$ as $z^4 + 16$ is of degree 4 and the numerator is of degree zero. So

$$\int_{-\infty}^{\infty} \frac{1}{x^4 + 16}\,dx = \frac{1}{16}\pi\sqrt{2}$$

$$\therefore \int_0^{\infty} \frac{1}{x^4 + 16}\,dx = \frac{1}{32}\pi\sqrt{2}$$

FRAME B7

Algebraic-cum-Trigonometric Integrands

$\int_{-\infty}^{\infty} \frac{x \sin 2x}{(x^2 + 1)(x^2 + 4)}\,dx$ is an example of another type of integral which lends itself to treatment by a similar technique. It involves an algebraic fraction and, in the numerator, a term of the form $\cos mx$ or $\sin mx$.

The contour used is the same as in the previous few frames. This time, however, it is only necessary for the degree of the denominator to be one higher than that of the algebraic part of the numerator in order that the integral round the semicircular part tends to zero as R tends to infinity. In the example quoted at the beginning of this frame, these degrees are respectively 4 and 1, and so this condition is satisfied.

FRAME B7 continued

The function of z taken in this example is $\frac{ze^{2iz}}{(z^2 + 1)(z^2 + 4)}$. Part of this is obvious from our previous work. The reason for the e^{2iz} part will appear later.

Find the poles of this function and the residues at those poles which are inside the contour. Once again, R must be sufficiently large that C includes all the poles in the upper half plane.

****

B7A

Poles occur at $\pm i$, $\pm 2i$.
They are all simple.
Those at i and $2i$ are within the contour.

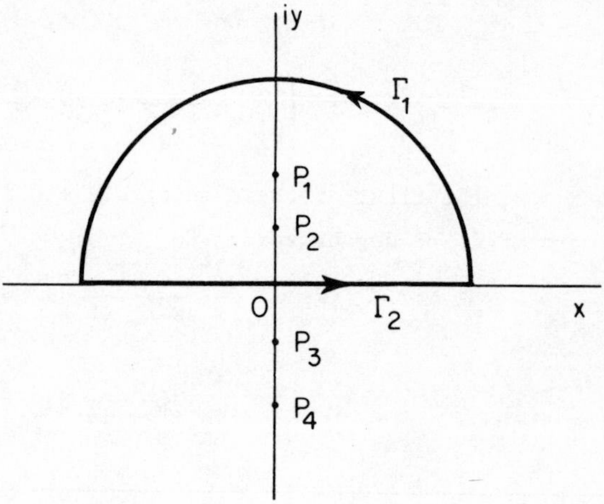

$$\text{Res}\left\{\frac{ze^{2iz}}{(z^2 + 1)(z^2 + 4)}, i\right\} = \lim_{z\to i} \frac{(z - i)ze^{2iz}}{(z^2 + 1)(z^2 + 4)}$$

$$= \lim_{z\to i} \frac{ze^{2iz}}{(z + i)(z^2 + 4)}$$

$$= \frac{ie^{-2}}{2i \times 3}$$

$$= \frac{1}{6} e^{-2}$$

$$\text{Res}\left\{\frac{ze^{2iz}}{(z^2 + 1)(z^2 + 4)}, 2i\right\} = \lim_{z\to 2i} \frac{(z - 2i)ze^{2iz}}{(z^2 + 1)(z^2 + 4)}$$

$$= -\frac{1}{6} e^{-4}$$

FRAME B8

Thus $\int_C \frac{ze^{2iz}}{(z^2+1)(z^2+4)}\,dz = 2\pi i\left(\frac{1}{6}e^{-2} - \frac{1}{6}e^{-4}\right)$

$= \frac{1}{3}\pi i\left(e^{-2} - e^{-4}\right)$

$\therefore \quad \int_{\Gamma_1} \frac{ze^{2iz}}{(z^2+1)(z^2+4)}\,dz + \int_{\Gamma_2} \frac{ze^{2iz}}{(z^2+1)(z^2+4)}\,dz = \frac{1}{3}\pi i\left(e^{-2} - e^{-4}\right)$

As $R \to \infty$, the first of these two integrals $\to 0$,

$\therefore \quad \int_{-\infty}^{\infty} \frac{xe^{2ix}}{(x^2+1)(x^2+4)}\,dx = \frac{1}{3}\pi i\left(e^{-2} - e^{-4}\right)$

Now $e^{2ix} = \cos 2x + i \sin 2x$, and so you should now be able to see why the e^{2iz} factor was inserted.

$$\therefore \quad \int_{-\infty}^{\infty} \frac{x(\cos 2x + i\sin 2x)}{(x^2+1)(x^2+4)}\,dx = \frac{1}{3}\pi i\left(e^{-2} - e^{-4}\right) \qquad \text{(B8.1)}$$

The value of the integral required is now simply obtained by equating the imaginary parts of this last equation. Thus

$$\int_{-\infty}^{\infty} \frac{x\sin 2x}{(x^2+1)(x^2+4)}\,dx = \frac{1}{3}\pi\left(e^{-2} - e^{-4}\right)$$

What happens if you equate the real parts of (B8.1)?

****

B8A

$$\int_{-\infty}^{\infty} \frac{x\cos 2x}{(x^2+1)(x^2+4)}\,dx = 0$$

Thus, although only one integral was asked for, this second one follows immediately.

FRAME B9

You may be wondering why it is necessary to take $\frac{ze^{2iz}}{(z^2+1)(z^2+4)}$ as the function of z in this example, rather than the more obvious $\frac{z\sin 2z}{(z^2+1)(z^2+4)}$. The reason is that $\int_{\Gamma_1} \frac{z\sin 2z}{(z^2+1)(z^2+4)}\,dz$ does not tend to zero as $R \to \infty$.

FRAME B9 continued

As the working of this example enabled us to find both $\int_{-\infty}^{\infty} \frac{x \sin 2x}{(x^2 + 1)(x^2 + 4)}\, dx$ and $\int_{-\infty}^{\infty} \frac{x \cos 2x}{(x^2 + 1)(x^2 + 4)}\, dx$, we see that if a term of the form $\cos mx$ or $\sin mx$ appears as a factor in the numerator of a real integral, then the corresponding complex integral contains a factor e^{imz}.

FRAME B10

Singularities on the Real Axis

You may have noticed that we have been very careful not to include any example where a singularity occurs on the contour. If it does, an additional complication arises. This problem will not be pursued here to any great extent, but just one example will be taken to indicate the general idea.

FRAME B11

$\int_{-\infty}^{\infty} \frac{\sin mx}{x(x^2 + a^2)}\, dx$ is an example which leads to a singularity on the contour.

What function of z do you suggest is taken to integrate here, and where are its poles?

****

B11A

$\frac{e^{imz}}{z(z^2 + a^2)}$. *Poles at 0, ±ia.*

FRAME B12

The contour that you have met earlier for integrals similar to this, but without a pole on the real axis, is the large semicircle of FRAME B1. As a pole exists here on that contour, a slight modification is made as shown on the next page.

FRAME B12 continued

Γ_3 is a semicircle whose centre is at O (i.e. at the pole on the real axis) and whose radius is ρ.

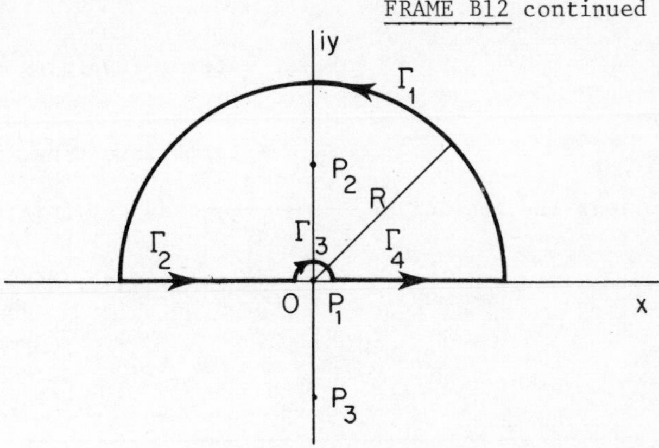

If C is the contour consisting of Γ_1, Γ_2, Γ_3 and Γ_4 what is the value of

$$\int_C \frac{e^{imz}}{z(z^2 + a^2)}\,dz?$$

****

B12A

$$Res\left\{\frac{e^{imz}}{z(z^2 + a^2)}, ia\right\} = -\frac{1}{2a^2}e^{-ma}$$

$$\int_C \frac{e^{imz}}{z(z^2 + a^2)}\,dz = -\frac{\pi i}{a^2}e^{-ma}$$

FRAME B13

Thus
$$\int_{\Gamma_1} + \int_{\Gamma_2} + \int_{\Gamma_3} + \int_{\Gamma_4} = -\frac{\pi i}{a^2}e^{-ma} \qquad \text{(B13.1)}$$

The next thing is to find out something about $\int_{\Gamma_3}$. To do this, $\frac{e^{imz}}{z(z^2 + a^2)}$ is expanded into a Laurent series about the origin. As there is a simple pole there, this series will be of the form $\frac{B_1}{z} + A_0 + A_1 z + A_2 z^2 + A_3 z^3 + \ldots\ldots$

and so
$$\int_{\Gamma_3} \frac{e^{imz}}{z(z^2 + a^2)}\,dz = \int_{\Gamma_3}\left(\frac{B_1}{z} + A_0 + A_1 z + A_2 z^2 + A_3 z^3 + \ldots\right)dz$$

Now as Γ_3 is a semicircle of radius ρ with centre at O, any point z on it can be represented by $\rho e^{i\theta}$ where $0 \leqslant \theta \leqslant \pi$.

$z = \rho e^{i\theta}$ is therefore substituted into the last integral, giving

$$\int_{\Gamma_3} \frac{e^{imz}}{z(z^2 + a^2)}\,dz = \int_{\pi}^{0}\left\{\frac{B_1}{\rho e^{i\theta}} + A_0 + A_1\rho e^{i\theta} + A_2\rho^2 e^{2i\theta} + A_3\rho^3 e^{3i\theta} + \ldots\right\}\rho i e^{i\theta}\,d\theta$$

FRAME B13 continued

$$= \int_{\pi}^{0} B_1 i d\theta + \text{terms involving } \rho \text{ and its powers}$$

$$= -B_1 i\pi + \text{terms involving } \rho \text{ and its powers.}$$

B_1 is the residue of $\dfrac{e^{imz}}{z(z^2 + a^2)}$ at the origin. What is its value?

****

B13A

$\dfrac{1}{a^2}$

FRAME B14

$$\therefore \int_{\Gamma_3} \frac{e^{imz}}{z(z^2 + a^2)} dz = -\frac{\pi i}{a^2} + \text{terms involving } \rho \text{ and its positive powers.}$$

We now take each term on the L.H.S. of (B13.1) and consider its value as $R \to \infty$ and $\rho \to 0$.

As the degree of the denominator of $\dfrac{1}{z(z^2 + a^2)}$ is 3 and that of the numerator is 0, $\int_{\Gamma_1} \to 0$ as $R \to \infty$.

$$\int_{\Gamma_2} = \int_{-R}^{-\rho} \frac{e^{imx}}{x(x^2 + a^2)} dx \quad \text{as on } \Gamma_2,\ z = x \text{ and so } dz = dx$$

$$\to \int_{-\infty}^{0} \frac{e^{imx}}{x(x^2 + a^2)} dx \quad \text{as } R \to \infty \text{ and } \rho \to 0.$$

$$\int_{\Gamma_3} \to -\frac{\pi i}{a^2} \quad \text{as } \rho \to 0.$$

$$\int_{\Gamma_4} = \int_{\rho}^{R} \frac{e^{imx}}{x(x^2 + a^2)} dx \quad \text{as on } \Gamma_4,\ z = x$$

$$\to \int_{0}^{\infty} \frac{e^{imx}}{x(x^2 + a^2)} dx \quad \text{as } \rho \to 0 \text{ and } R \to \infty.$$

$$\therefore \text{(B13.1) becomes } 0 + \int_{-\infty}^{0} \frac{e^{imx}}{x(x^2 + a^2)} dx - \frac{\pi i}{a^2} + \int_{0}^{\infty} \frac{e^{imx}}{x(x^2 + a^2)} dx = -\frac{\pi i}{a^2} e^{-m}$$

FRAME B14 continued

i.e. $$\int_{-\infty}^{\infty} \frac{e^{imx}}{x(x^2 + a^2)}\, dx = \frac{\pi i}{a^2}\left(1 - e^{-ma}\right)$$

Finally, $e^{imx} = \cos mx + i \sin mx$, and so equating imaginary parts gives

$$\int_{-\infty}^{\infty} \frac{\sin mx}{x(x^2 + a^2)}\, dx = \frac{\pi}{a^2}\left(1 - e^{-ma}\right)$$

FRAME B15

You will notice that the effect of the simple pole at the origin is to add the term $\frac{\pi i}{a^2}$ i.e. $\pi i \times$ residue at this pole to the value of the integral. This is a quite general result for any simple pole on the real axis, wherever such a pole may be on that axis, provided, of course, that the contour is such that part of it lies along the real axis.

FRAME B16

Miscellaneous Examples

In this frame a collection of miscellaneous examples is given for you to try. Answers are supplied in FRAME B17, together with such working as is considered helpful.

Evaluate the integrals numbers 1-3 by means of complex integration round a suitable contour.

1. $\displaystyle\int_{-\infty}^{\infty} \frac{1}{x^2 + x + 1}\, dx$

2. $\displaystyle\int_{-\infty}^{\infty} \frac{\sin 2x}{x^2 + x + 1}\, dx$

3. $\displaystyle\int_{0}^{\infty} \frac{\sin x}{x}\, dx$

FRAME B16 continued

4. In the calculation of a transition probability in quantum mechanics, $\int_{-\infty}^{\infty} \frac{2(1 - \cos \omega t)}{\omega^2} d\omega$ occurs. Show that the value of this integral is $2\pi t$.

FRAME B17

Answers to Miscellaneous Examples

1. Integrate $\frac{1}{z^2 + z + 1}$ round the contour shown and let $R \to \infty$.

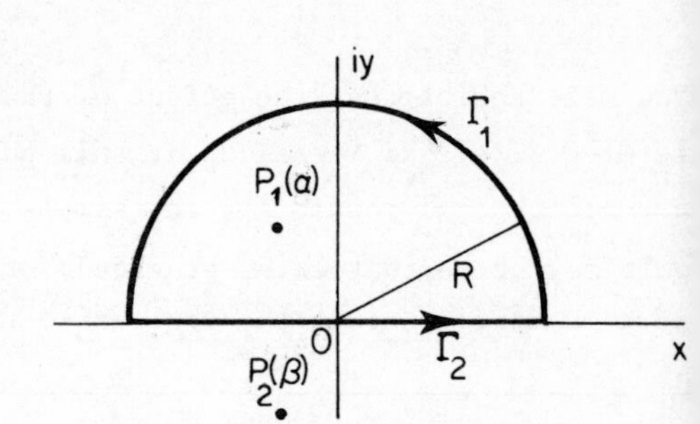

Simple poles occur at

$$z = \frac{-1 \pm \sqrt{1 - 4}}{2}$$

$$= -\frac{1}{2} \pm i\frac{\sqrt{3}}{2}$$

$$\text{Residue at } \alpha = \lim_{z\to\alpha} \frac{1}{z - \beta}$$

$$= \frac{1}{i\sqrt{3}}$$

$$\int_{\Gamma_1} \frac{1}{z^2 + z + 1} dz \to 0 \quad \text{as} \quad R \to \infty$$

$$\int_{-\infty}^{\infty} \frac{1}{x^2 + x + 1} dx = \frac{2\pi}{\sqrt{3}}$$

2. Integrate $\frac{e^{i2z}}{z^2 + z + 1}$ round the same contour as in Example 1.

The poles are at the same points.

$$\text{Residue at } \alpha = \lim_{z\to\alpha} \frac{e^{i2z}}{z - \beta}$$

$$= \frac{e^{i(-1+\sqrt{3}i)}}{i\sqrt{3}}$$

$$= \frac{e^{-\sqrt{3}-i}}{i\sqrt{3}}$$

FRAME B17 continued

$$\int_{\Gamma_1} \to 0 \quad \text{as} \quad R \to \infty$$

$$\int_{-\infty}^{\infty} \frac{e^{2ix}}{x^2 + x + 1}\,dx = \frac{2\pi}{\sqrt{3}} e^{-\sqrt{3}-i}$$

Equating imaginary parts,

$$\int_{-\infty}^{\infty} \frac{\sin 2x}{x^2 + x + 1}\,dx = -\frac{2\pi}{\sqrt{3}} e^{-\sqrt{3}} \sin 1$$

3. Integrate $\frac{e^{iz}}{z}$ round the same contour as in FRAME B2. There is a simple pole at the origin and the residue there is 1.

$$\therefore \int_{-\infty}^{\infty} \frac{e^{ix}}{x}\,dx = \pi i$$

Equating imaginary parts, $\int_{-\infty}^{\infty} \frac{\sin x}{x}\,dx = \pi$

$\therefore$, as the integrand is an even function, $\int_0^{\infty} \frac{\sin x}{x}\,dx = \frac{\pi}{2}$.

In the programme on integration in Vol I, you found $\int_0^{0\cdot 5} \frac{\sin u}{u}\,du$ by means of a series expansion because $\int \frac{\sin u}{u}\,du$ cannot be expressed as a simple function. As the example you have just done shows, it is possible to apply an analytical technique to find $\int_0^{\infty} \frac{\sin x}{x}\,dx$.

4. Integrate $\frac{2(1 - e^{izt})}{z^2}$ w.r.t. z where $z = \omega + iy$ round the same contour as used in Example 3.

There is a <u>simple</u> pole at the origin and the residue there is $-2it$.

$$\therefore \int_{-\infty}^{\infty} \frac{2(1 - e^{i\omega t})}{\omega^2}\,d\omega = \pi i(-2it) = 2\pi t$$

Equating real parts gives $\int_{-\infty}^{\infty} \frac{2(1 - \cos \omega t)}{\omega^2}\,d\omega = 2\pi t$

Note that $\int \frac{2(1 - e^{izt})}{z^2}\,dz$ must not be split up into two parts as the pole of $\frac{2}{z^2}$ is <u>not</u> simple.

CONFORMAL TRANSFORMATIONS

FRAME 1

Definition of Conformal Transformations

Some examples of complex mapping were considered in the first programme of this Unit. In the present programme the work on this topic will be extended and further applications of it to physical situations will be discussed.

You have seen that, given an equation of the form $w = f(z)$, e.g. $w = z^2$, $w = \dfrac{1}{z - 1}$, etc., there are definite points, curves and regions in the w-plane corresponding to specified points, curves and regions in the z-plane. A further geometrical correspondence which must now be considered is that of angles. As you will see later, what happens to angles under a mapping is very important in applications. In particular, right angles play a critical part. This is due to the fact that these occur in many practical situations. For example, in non-viscous incompressible fluid flow, equipotential lines and stream lines are perpendicular. Again, in electrostatics, equipotential lines and flux lines are perpendicular and in heat flow problems, isothermal lines and flux lines are perpendicular.

FRAME 2

To remind yourself of what is involved in complex mapping, try the following revision example:

Find the transformed curves corresponding to

(i) $|z| = 1$, (ii) $x = 0$, (iii) $y = 0$ under $w = \dfrac{z}{z - 1}$.

Illustrate by sketches, showing the three curves in the z-plane on one diagram and those in the w-plane on another.

****

2A

The two diagrams are shown below:

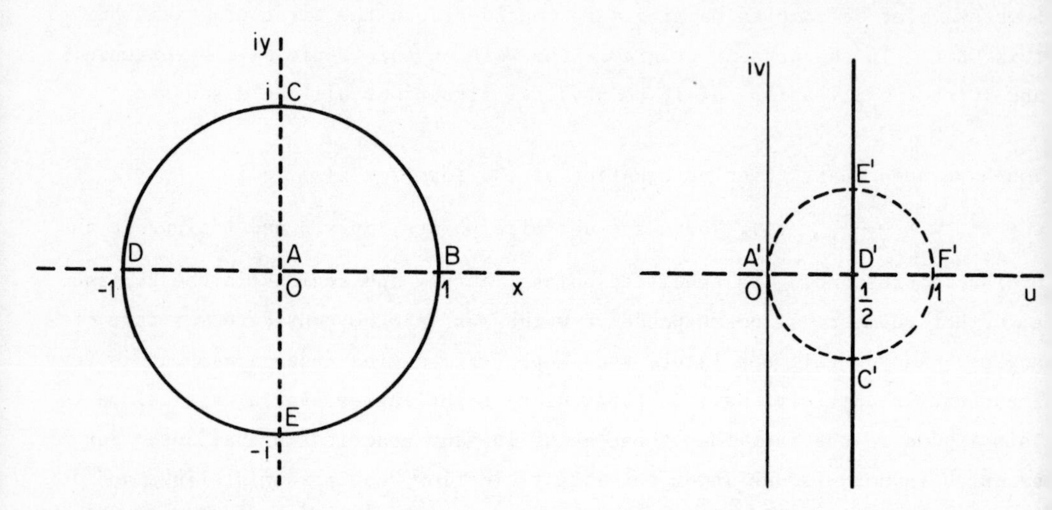

If you need any of the actual working, you will find it in FRAMES 62-65A on pages 12:45-12:47.

FRAME 3

Now write down the angles at which the intersecting curves meet at A and A', C and C', D and D', E and E'.

****

3A

All the curves at the given points in the two diagrams intersect at right angles.

FRAME 4

Thus, at all the points so far considered, if two curves intersect at an angle θ in the z-plane, the corresponding curves intersect at the same angle θ in the w-plane. The question now arises - is this coincidence or does it always happen?

FRAME 4 continued

To investigate the situation further, find out whether it happens at P and P', Q and Q', R and R' in the following diagrams, for which the transformation equation is $w = z^2$.

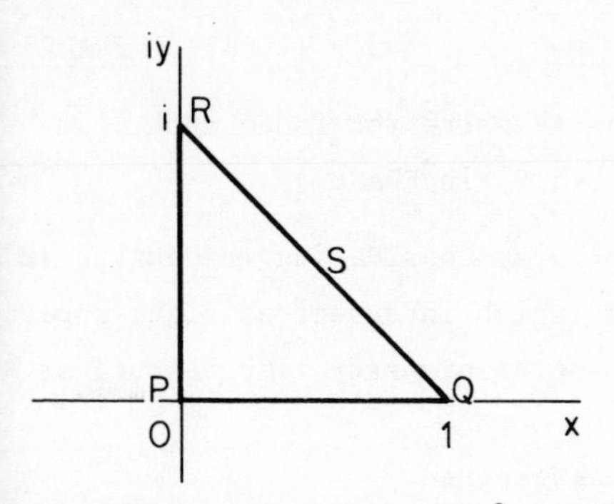

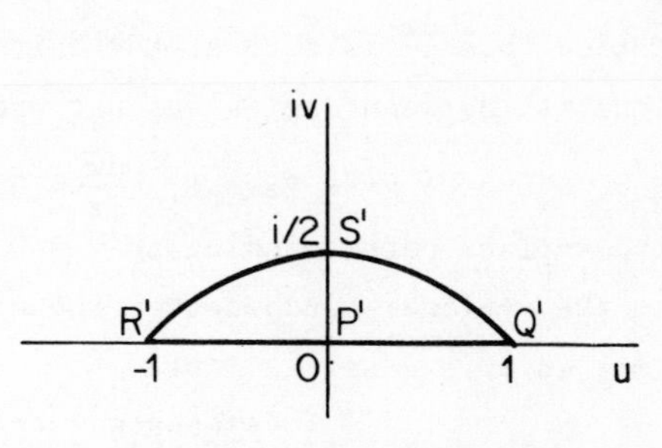

The equation of Q'S'R' is $u^2 = -2(v - \frac{1}{2})$. Apart from the use of different letters, this is the problem that was worked out in FRAMES 31-35A, pages 12:25-12:28.

****

4A

It does at Q and Q', R and R', but not at P and P'. In the first two cases the interior angles are all 45°, but at P it is 90°, whilst at P' it is 180°.

FRAME 5

So we now see that, although the angle between two intersecting curves in the w-plane is sometimes the same as that between the two corresponding curves in the z-plane, it is not always the same.

Now find the values of $\frac{dw}{dz}$ at z equal to

(i) 0, i, -1 and -i when $w = \frac{z}{z - 1}$,

(ii) 0, 1 and i when $w = z^2$.

(The points listed here are those in the z-plane at which you were considering the angles between intersecting curves in FRAMES 3 and 4.)

****

5A

(i) $-1,\ -\frac{1}{2}i,\ -\frac{1}{4},\ \frac{1}{2}i$ *(ii)* $0,\ 2,\ 2i$

FRAME 6

Notice that $\frac{dw}{dz}$ is zero at the only pair of points where you found that equality between angles was not preserved (P and P' in FRAME 4).

Now, when $w = z^2 + 2z - 1$, $\frac{dw}{dz} = 0$ if $z = -1$. By considering the curves in the w-plane corresponding to $y = 0$ and $x = -1$ (which intersect at right angles in the z-plane, find whether the equality of angles property is preserved or lost in this case.

****

6A

It is lost as both $y = 0$ and $x = -1$ become $v = 0$ in the w-plane.

FRAME 7

We shall now go on to show generally that if two curves intersect at an angle θ in the z-plane, then the two corresponding curves in the w-plane, under a transformation $w = f(z)$, intersect at the same angle θ provided that w is a regular function of z and $\frac{dw}{dz} \neq 0$.

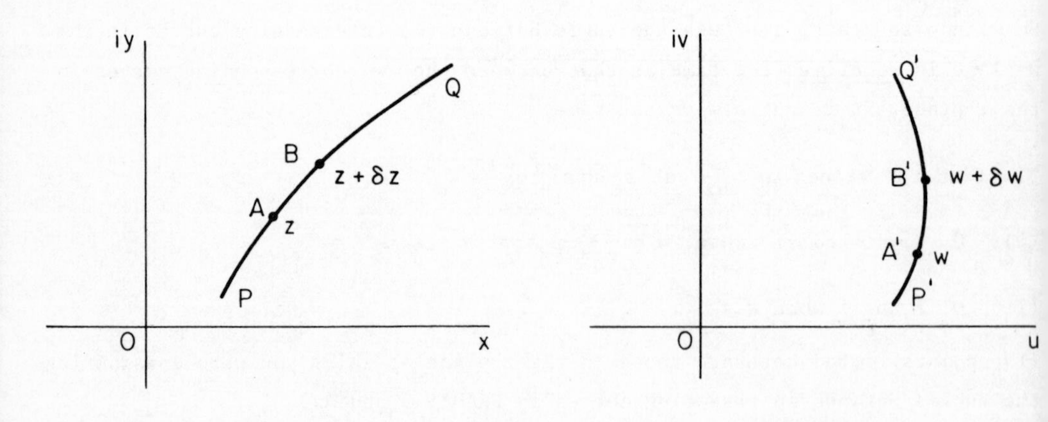

FRAME 7 continued

The two diagrams show a curve PQ in the z-plane, and the corresponding curve in the w-plane. A and A', B and B' are corresponding points on the two curves with complex representations as shown. Now the argument of δz is the angle the chord AB makes with Ox and consequently, as $\delta z \to 0$, this angle tends to that made by the tangent at A with Ox. Let this be θ. Similarly, as $\delta w \to 0$, the argument of δw tends to the angle made by the tangent at A' with Ou. Let this angle be ϕ.

Next let $\frac{dw}{dz}$ at A have the value $\rho e^{i\psi}$ $(\rho \neq 0)$ so that the argument of $\frac{dw}{dz}$ at A is ψ.

But also $\delta w \simeq \frac{dw}{dz}\delta z$ so that $\arg \delta w \simeq \arg \frac{dw}{dz} + \arg \delta z$.

Therefore, in the limit $\phi = \psi + \theta$.

Now suppose a second curve is drawn passing through A and that the angle between its tangent at A and Ox is ξ. If η is the angle between the tangent to the corresponding curve at A' and Ou, what will be the equation connecting ξ and η?

****

7A

$\eta = \psi + \xi$

FRAME 8

By subtraction, it follows immediately that

$$\eta - \phi = \xi - \theta$$

i.e. the angle between the two curves in the w-plane is the same as that between the two curves in the z-plane.

FRAME 9

In FRAME 7, use was made of $\delta w \simeq \frac{dw}{dz}\delta z$ to deduce information about angles. But also $|\delta w| \simeq \left|\frac{dw}{dz}\right||\delta z|$ or $|\delta w| \simeq \rho|\delta z|$, and so, in the diagrams in FRAME 7, $A'B' \simeq \rho AB$. Similarly if C and C', close to A and A', are corresponding points on the corresponding curves RS and R'S', then $A'C' \simeq \rho AC$.

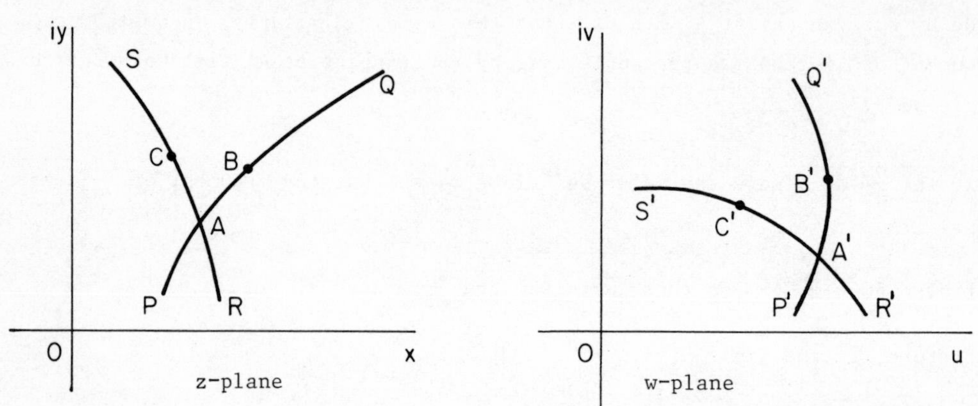

Thus $\frac{A'C'}{A'B'} \simeq \frac{AC}{AB}$ and so triangles ABC and A'B'C' are similar. Because very small areas in the z-plane transform into similar areas in the w-plane, the transformation is said to be CONFORMAL. Large areas are not, in general transformed into similar large areas because $\frac{dw}{dz}$ usually varies from point to point. That this is so will be immediately obvious if you compare the two figures in FRAME 4.

FRAME 10

The Transformation $w = z + \frac{a^2}{z}$, a being real and positive

There are two main transformations left for us to consider, of which the first is $w = z + \frac{a^2}{z}$. At what finite points will this not be conformal?

****

10A

0, ±a.

There is a singularity at z = 0 and $\frac{dw}{dz}$ does not exist there. Otherwise $\frac{dw}{dz} = 1 - \frac{a^2}{z^2}$ and this is zero at $z = \pm a$.

FRAME 11

Now find u and v when $w = z + \dfrac{a^2}{z}$.

****

11A

$$u = x\left(1 + \frac{a^2}{x^2 + y^2}\right) \qquad v = y\left(1 - \frac{a^2}{x^2 + y^2}\right)$$

FRAME 12

Now suppose we apply this transformation to the circle $|z| = k$. Can you find the equation connecting u and v in this case? What is the resulting curve?

****

12A

As $x^2 + y^2 = k^2$, $\quad u = x\left(1 + \dfrac{a^2}{k^2}\right)$, $\quad v = y\left(1 - \dfrac{a^2}{k^2}\right)$ (12A.1)

Hence $x = \dfrac{u}{1 + \dfrac{a^2}{k^2}}$ $\qquad y = \dfrac{v}{1 - \dfrac{a^2}{k^2}}$ *and so*

$$\frac{u^2}{\left(1 + \frac{a^2}{k^2}\right)^2} + \frac{v^2}{\left(1 - \frac{a^2}{k^2}\right)^2} = k^2 \qquad (12A.2)$$

or

$$\frac{u^2}{\left(\frac{k^2 + a^2}{k}\right)^2} + \frac{v^2}{\left(\frac{k^2 - a^2}{k}\right)^2} = 1$$

This is an ellipse, centre (0,0), semi-axes $\dfrac{k^2 + a^2}{k}$, $\dfrac{k^2 - a^2}{k}$.

(If $k < a$, *the semi-minor axis will be* $\dfrac{a^2 - k^2}{k}$.*)*

FRAME 13

What will happen to the ellipse in the following cases:

(i) $k >> a$ (ii) $k = a$ (iii) $k << a$?

(>> means 'is very much greater than'.)

****

13A

(i) $\frac{a^2}{k^2}$ *will be very small and (12A.2) will become approximately the circle* $u^2 + v^2 = k^2$.

(ii) $u = 2x$, $v = 0$, *from (12A.1). The ellipse will become that part of the real axis from* $u = -2k$ *to* $u = +2k$ *as* $\pm k$ *are the maximum and minimum values of* x *when* $x^2 + y^2 = k^2$.

(iii) *1 will be small in comparison with* $\frac{a^2}{k^2}$. *(12A.2) will become approximately the circle* $u^2 + v^2 = a^4/k^2$.

FRAME 14

To get a clearer indication of how the ellipse varies with k, suppose a is fixed at some specific value, say 2. The following table gives the magnitudes of the semi-axes of the ellipse for various values of k.

k	Semi-major axis	Semi-minor axis
100	100·04	99·96
10	10·4	9·6
4	5	3
3	4·67	1·67
2·1	4·00	0·20
2	4 Special case	0
1·9	4·00	0·20
1·5	4·17	1·17
1	5	3
0·5	8·5	7·5
0·1	40·1	39·9

You will notice that when k is very large or very small, the ellipse becomes approximately a circle and that when $k = 2$, i.e. when $k = a$, it degenerates into a straight line.

FRAME 15

The centre of $|z| = k$ is at the origin. If we had chosen a circle with centre elsewhere, the working would have been more complicated as there would have been no simple substitution for $x^2 + y^2$. However, a quite interesting shape results if the transformation $w = z + \frac{1}{z}$ is applied to a circle which passes through $z = 1$ and encloses $z = -1$. In this case $a = 1$ and you will remember that $z = a$ is one of the points at which $\frac{dw}{dz} = 0$ and the conformal property is lost.

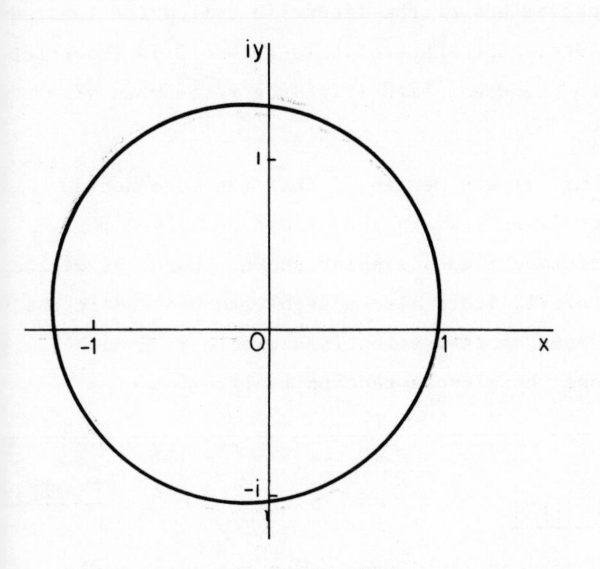

We shall not attempt to find the equation of the curve in the w-plane corresponding to the circle but it can be sketched quite simply by taking a number of points round the circle, then calculating and plotting the corresponding points in the w-plane.

If this is done, the result looks something like this:

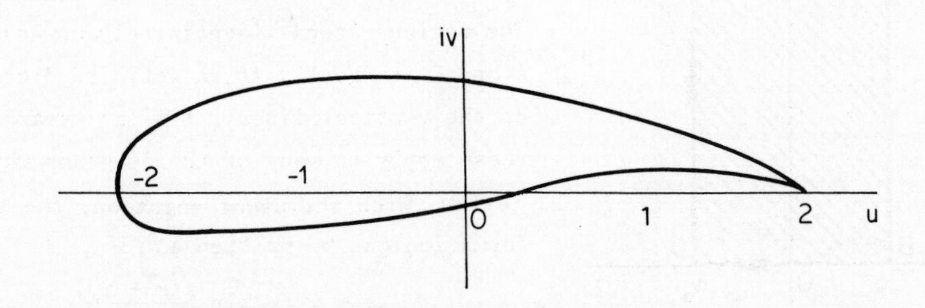

If you are interested, you can construct one of these for yourself.

FRAME 16

The actual shape obtained can be varied by changing the centre and radius of the circle. It resembles the cross-section of the wing of an aeroplane and is known as a Joukowski aerofoil. Consequently $w = z + \frac{1}{z}$ is called a JOUKOWSKI TRANSFORMATION. You will notice that, at the point $z = 1$ where the transformation is not conformal, the smooth curve in the z-plane produces a cusp in the w-plane (at $w = 2$). The rounded end of the figure is called the leading edge and the cusp the trailing edge. The shape of this aerofoil is important in aerodynamics as it is found to produce a high lift/drag ratio when viscosity is taken into account.

In the first programme in this Unit it was mentioned that the idea behind mapping is that some problems involving awkward shapes can be solved more easily if the shape can be transformed into a simpler shape. Here, as well as going from the circle to the aerofoil, it is also possible to proceed in the reverse direction. A cylinder whose cross-section is a circle is a much easier shape to deal with than one whose cross-section is like ⬭.

FRAME 17

The Schwarz-Christoffel Transformation

The other transformation left to be considered goes under the above title. Before saying what it actually does we shall take some examples of it and apply them to some particular figures.

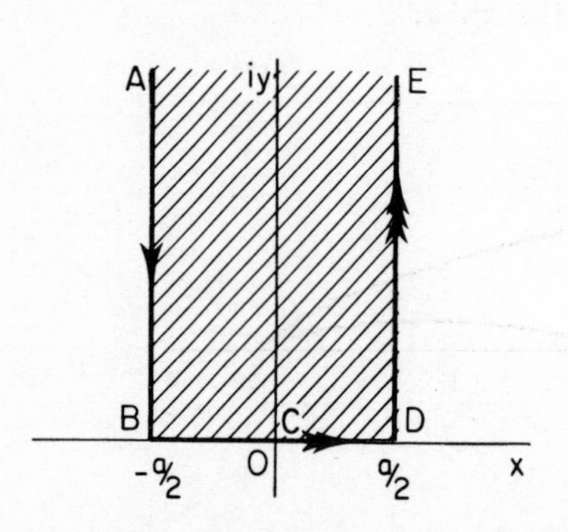

As a first example, let us apply $w = \sin\frac{\pi z}{a}$ to the region illustrated in the z-plane. The region extends indefinitely upwards. A and E are really at an infinite distance up the vertical lines. Similar remarks to these apply to many of the diagrams which follow. With the usual notation, the transformation can be written as

$$u + iv = \sin\frac{\pi(x + iy)}{a}$$

Start by finding u and v in terms of x and y, and also the w-points corresponding to B, C and D.

****

17A

$$u + iv = \sin\frac{\pi x}{a}\cos\frac{\pi iy}{a} + \cos\frac{\pi x}{a}\sin\frac{\pi iy}{a}$$

$$= \sin\frac{\pi x}{a}\cosh\frac{\pi y}{a} + i\cos\frac{\pi x}{a}\sinh\frac{\pi y}{a}$$

$$u = \sin\frac{\pi x}{a}\cosh\frac{\pi y}{a} \qquad v = \cos\frac{\pi x}{a}\sinh\frac{\pi y}{a}$$

When $z = -\frac{a}{2}$, 0 *and* $\frac{a}{2}$, $w = -1$, 0 *and* 1 *respectively.*

FRAME 18

Next it is necessary to find the curves in the w-plane corresponding to the three sides of the region in the z-plane.

Along AB, $x = -\frac{a}{2}$,

$$\therefore u = -\cosh\frac{\pi y}{a}, \qquad v = 0$$

So as z moves along the line of which AB is part, w moves along the real axis. However it doesn't move all the way along it, because, as y decreases from ∞ to 0, $\cosh\frac{\pi y}{a}$ decreases from ∞ to 1 and hence u changes from $-\infty$ to -1.

The commencement of the diagram in the w-plane is therefore

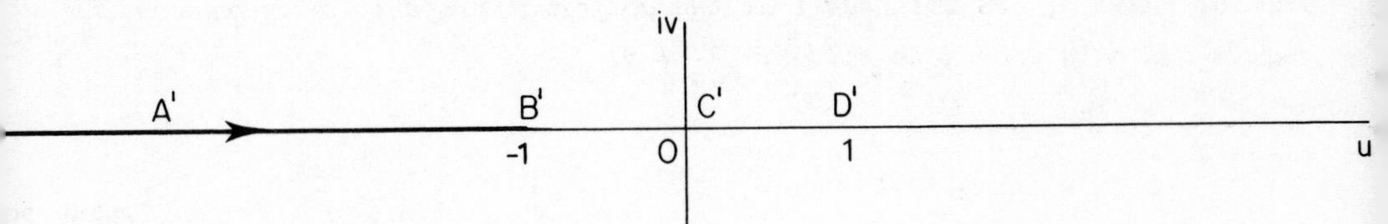

Now find the curves or parts of curves in the w-plane corresponding to the other two sides of the region in the z-plane.

****

18A

Along BCD, $y = 0$, $\therefore u = \sin\frac{\pi x}{a}$, $v = 0$.

From the diagram above, you can see that corresponding to BCD in the z-plane we have the part B'C'D' of the real w-axis. Alternatively, as x *increases from* $-\frac{a}{2}$ *through* 0 *to* $+\frac{a}{2}$, u *increases from* -1 *through* 0 *to* $+1$ *as* $u = \sin\frac{\pi x}{a}$.

Along DE, $x = \frac{a}{2}$, $\therefore u = \cosh\frac{\pi y}{a}$, $v = 0$

As y *increases from* 0 *to* ∞, u *will increase from* 1 *to* ∞.

FRAME 19

The complete diagram in the w-plane is therefore

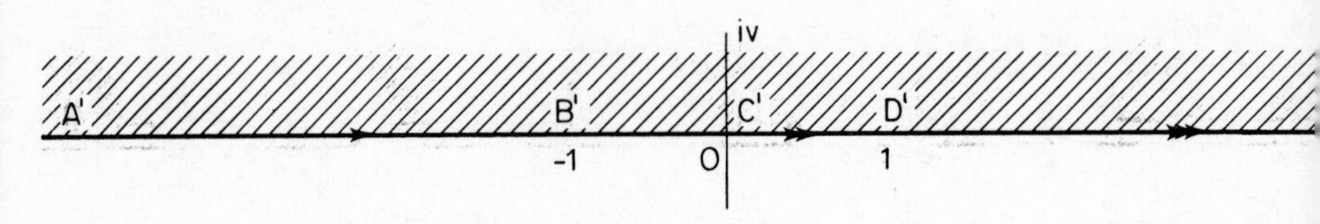

As the region shaded in the z-plane is on one's left as a journey is made round ABCDE, the corresponding region is on the left of the real axis in the w-plane. It extends to cover the whole of the region in the upper half of the w-plane.

What can you say about the values of $\frac{dw}{dz}$ at B and D and what do you notice about the angles at B and D and at B' and D'?

****

19A

$\frac{dw}{dz} = \frac{\pi}{a} \cos \frac{\pi z}{a}$, *which is zero when* $z = \mp \frac{a}{2}$.

The angles at B and D are $\frac{\pi}{2}$, *while those at B' and D' are* π. *If you look back at FRAME 7, you will see that the conformal property of corresponding curves was only proved to hold if* $\frac{dw}{dz} \neq 0$.

FRAME 20

Now see what happens if the transformation $w = e^{\pi z/a}$ is applied to the infinite strip in the z-plane shown below.

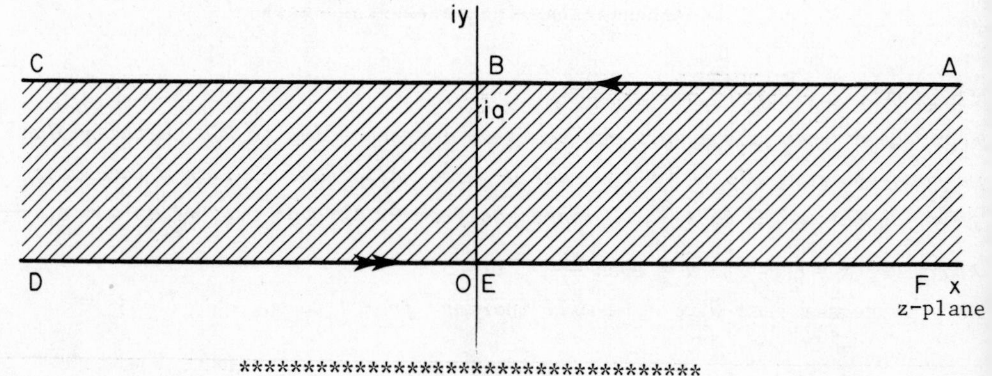

****

20A

$$u + iv = e^{\pi(x+iy)/a}$$

$$= e^{\pi x/a}\left(\cos \frac{\pi y}{a} + i \sin \frac{\pi y}{a}\right)$$

$$u = e^{\pi x/a} \cos \pi y/a \qquad v = e^{\pi x/a} \sin \pi y/a$$

When $y = a$, $\quad u = -e^{\pi x/a}$, $\quad v = 0$.

As x *decreases from* ∞ *through* 0 *to* $-\infty$,

u *increases from* $-\infty$ *to* 0.

When $y = 0$, $\quad u = e^{\pi x/a}$, $\quad v = 0$

As x *increases from* $-\infty$ *through* 0 *to* $+\infty$

u *increases from* 0 *to* ∞.

The points corresponding to $B(z = ia)$ *and* $E(z = 0)$ *are* $B'(w = -1)$ *and* $E'(w = 1)$.

The diagram shown below results.

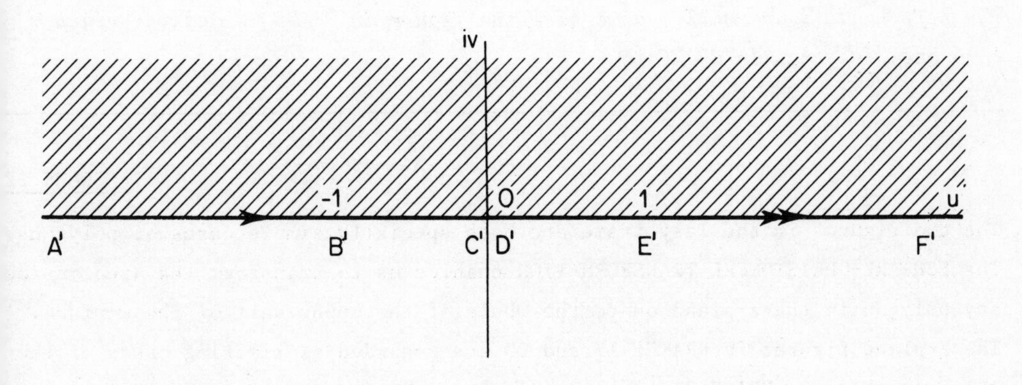

w-plane

FRAME 21

Now consider the following two figures:

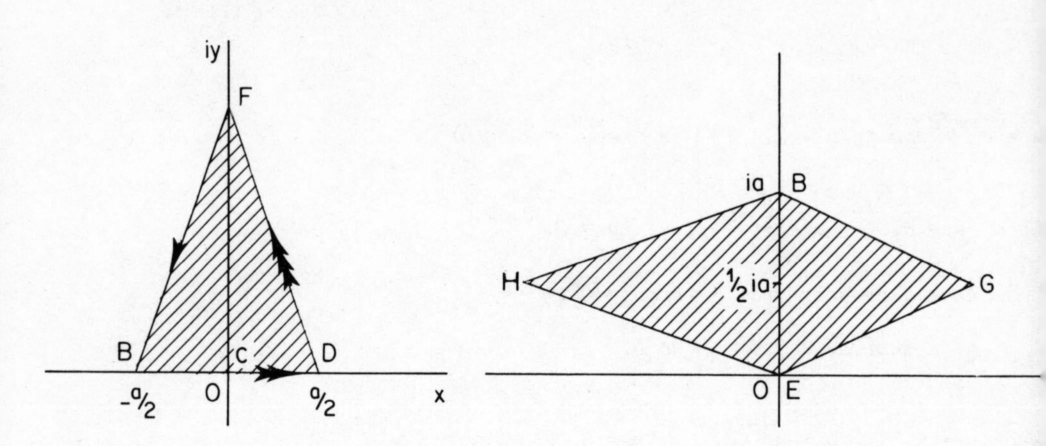

What will the shape of these two figures become over the sections shown if

(a) F moves an infinite distance vertically upwards, and

(b) H and G move infinite distances horizontally to the left and right respectively?

If it will help, imagine a window frame placed over each figure and think what you will see through the frame when the points move as indicated.

****

21A

The left hand figure will become like the figure in FRAME 17 and the right hand one like that in FRAME 20.

FRAME 22

The two figures in the last frame are both specially simple cases of polygons. The SCHWARZ-CHRISTOFFEL TRANSFORMATION enables us to transform the area inside any polygon in the z-plane on to the whole of the upper half of the w-plane. The z-plane figures in FRAMES 17 and 20 are regarded as limiting cases of the polygons shown in FRAME 21.

Now in each of those two examples the formula for w in terms of z was such that the upper half of the w-plane resulted. The question now arises: Given

FRAME 22 continued

any polygon in the z-plane, what f(z) should be used so that $w = f(z)$ has this effect?

The starting point in such a situation is to write down the expression for $\frac{dz}{dw}$. To understand the form that this takes, find $\frac{dz}{dw}$ for each of the transformation equations just used, i.e. $w = \sin\frac{\pi z}{a}$ and $w = e^{\pi z/a}$.

****

22A

$w = \sin\frac{\pi z}{a}$, $\quad \frac{dz}{dw} = \frac{a}{\pi\sqrt{1 - w^2}}$

$w = e^{\pi z/a}$, $\quad \frac{dz}{dw} = \frac{a}{\pi w}$

FRAME 23

The results obtained in 21A give a clue as to how to write down $\frac{dz}{dw}$ in any particular case. Taking the second of these results, the denominator vanishes when $w = 0$ and the power of the factor w in $\frac{dz}{dw}$ is -1.

Similarly, in the first result found in 22A for $\frac{dz}{dw}$, $w = 1$ makes the denominator vanish. The factor $w - 1$ appears in $\frac{dz}{dw}$, as $\frac{a}{\pi\sqrt{1 - w^2}}$ can be written as $\frac{a}{\pi\sqrt{w - 1}\sqrt{-1}\sqrt{w + 1}}$. Its power is $-\frac{1}{2}$.

What other value of w makes $\frac{dz}{dw}$ vanish in this case and to what power does the corresponding factor appear?

****

23A

-1. $w + 1$ appears to the power $-\frac{1}{2}$.

FRAME 24

As has been seen already, the points in the z-plane corresponding to $w = -1$ and 1 are B and D respectively in the figure in FRAME 17. The interior angles at these points are both $\pi/2$. If $\pi/2$ is divided by π and 1 is subtracted from the result, the number $-\frac{1}{2}$ (the power to which both $w + 1$ and $w - 1$ appeared) is obtained.

Is there any similar connection between the angle between BC and ED in FRAME 20 and the power of the factor w in the corresponding value of $\frac{dz}{dw}$, i.e. $\frac{a}{\pi w}$?

****

24A

Yes. The angle is 0 which gives -1 when divided by π and then has 1 subtracted from it.

FRAME 25

The general form for $\frac{dz}{dw}$ in a Schwarz-Christoffel transformation is given by

$$\frac{dz}{dw} = K(w - u_1)^{(\alpha_1/\pi)-1}(w - u_2)^{(\alpha_2/\pi)-1}(w - u_3)^{(\alpha_3/\pi)-1} \ldots$$

In this formula α_1, α_2, α_3 etc. are the interior angles of the polygon, taken in order, while u_1, u_2, u_3 ... are the points in the w-plane corresponding to the vertices of the polygon. K is a constant dependent upon the requirements of the problem. Thus, in the diagrams in FRAMES 17 and 19, the interior angle at B, which can be taken as α_1, is $\frac{1}{2}\pi$. The corresponding point B' is at $w = -1$ and so $u_1 = -1$. $\frac{dz}{dw}$ thus contains the factor $(w + 1)^{\frac{1}{2}}$. Similarly, taking the interior angle at C as α_2, $u_2 = 1$ and the factor $(w - 1)^{\frac{1}{2}}$ appears.

You will notice that there is no factor in the final form for $\frac{dz}{dw}$, i.e. $\frac{a}{\pi\sqrt{1 - w^2}}$, corresponding to the "vertex" A-E. Now in the w-plane the corresponding point to this is $u = \pm\infty$. It is a property of the Schwarz-Christoffel transformation that one of the vertices in the z-plane can always be made to give $u = \pm\infty$ in the w-plane and the corresponding factor in the expression for $\frac{dz}{dw}$ is then missing. It is also a property of this transformation that any two other vertices in the z-plane can be made to correspond to two preselected points on the u-axis in the w-plane.

FRAME 26

Now consider the two corresponding figures shown below.

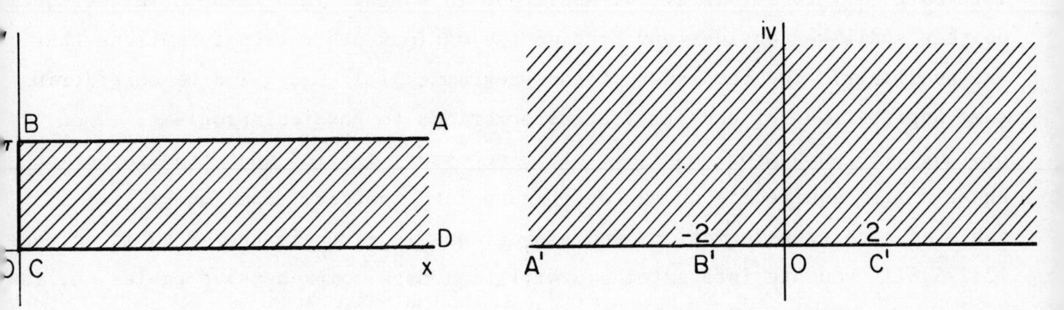

What factors would you expect in $\frac{dz}{dw}$ in this case and to what powers would you expect these factors to be raised? What, then, will be the form of $\frac{dz}{dw}$?

****

26A

$(w + 2)^{-\frac{1}{2}}$ *and* $(w - 2)^{-\frac{1}{2}}$ *would be expected.*

$$\frac{dz}{dw} = \frac{K}{\sqrt{w^2 - 4}}$$

FRAME 27

Integration gives

$$z = K \cosh^{-1} \frac{w}{2} + K_1$$

Now when $z = 0$, $w = 2$, $\therefore 0 = K \cosh^{-1} 1 + K_1$, i.e. $K_1 = 0$.

Use the corresponding points B and B' to show that $\cosh \frac{i\pi}{K} = -1$ and hence find K.

****

27A

$K = 1$

Thus $w = 2 \cosh z$

FRAME 28

As you will realise, it is not always easy (and may not even be possible) to integrate $\frac{dz}{dw}$ to obtain z. We shall not go further into these cases here, neither shall we consider any more of the various other transformations that are available. The remainder of the programme will be devoted to considering some applications of conformal transformations to physical problems. When dealing with such problems it is often necessary to find what transformation equation is required to change one figure into another. Some books give a table of such transformations and a short one is given in the APPENDIX on page 12:225. If you are interested you will find more comprehensive tables in, for example, the following two books:

Churchill, Ruel V., Complex Variables and Applications, 2nd ed., McGraw-Hill, 1960.

Spiegel, Murray R., Theory and Problems of Complex Variables, Schaum Outline Series, McGraw-Hill, 1964.

FRAME 29

Applications of Conformal Transformations

You will remember that some applications of complex mapping to electrical problems were considered in the first programme in this Unit. There are many other practical situations where complex mapping is useful and some of these will now be illustrated by means of further examples. It is not, however, proposed to give here an exhaustive treatment by going into details of all the applications of the subject.

As a first example, we shall have a look at the following problem:

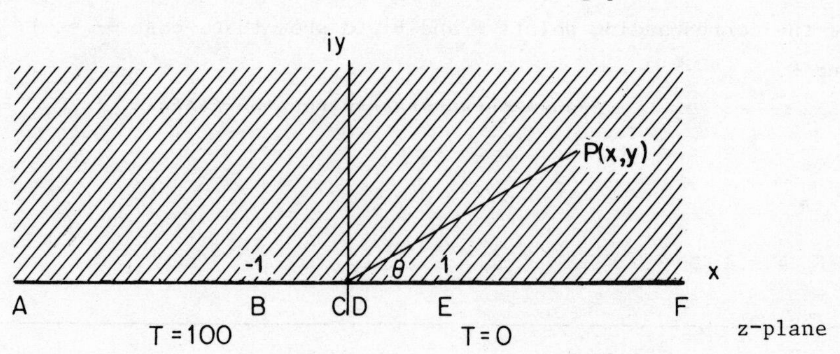

FRAME 29 continued

The diagram represents the cross-section of a slab of material which, theoretically, extends to infinity in all the directions indicated and is also infinitely thick. In practice it is assumed that the slab is sufficiently thick that, if the face of which ABCDEF is the cross-section is heated in a particular way, then no heat flows into or out of the section of the slab shown. In order to get a complete picture of the temperature distribution throughout the slab, it is therefore only necessary to obtain the temperature distribution throughout the section shown in the diagram. The problem can thus be regarded as one in two dimensions.

Now suppose that ABC is maintained at temperature 100°C, while DEF is maintained at temperature 0°C. It is required to find the temperature at any point P(x,y) in the section after a sufficient time has elapsed so that the temperature at P no longer varies with time.

FRAME 30

Now you may know the solution to this problem as it is relatively simple. If you do it doesn't matter. The purpose of this example is to illustrate, without introducing complicated algebra, how to proceed in more difficult cases.

If you turn back to FRAME 20 and 20A you will see that the figure in the w-plane there is the same as the one in the z-plane that we have now. The figure in our present problem can therefore be transformed into a figure in the w-plane that looks like the one in FRAME 20 in the z-plane. However, to do this, it is necessary to interchange z and w in the transformation equation $w = e^{\pi z/a}$. At the same time a can be put equal to 1 for simplicity, so we arrive at $z = e^{\pi w}$. Thus, the present problem is converted into the corresponding one in the w-plane as shown.

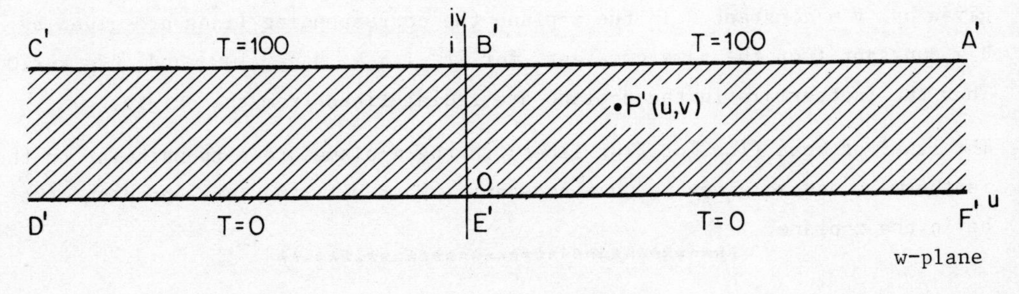

FRAME 30 continued

This problem has a very simple solution. Can you write down the temperature T at any point P'(u,v) if A'B'C' is maintained at T = 100 while D'E'F' is kept at T = 0? (The temperature falls cff linearly from 100^oC on the face A'B'C' to 0^oC on D'E'F'.)

****

30A

$T = 100v$

FRAME 31

To find the temperature at P in the x-y plane, it is necessary to find v in terms of x and y.

Now as $z = e^{\pi w}$, $w = \frac{1}{\pi}\ln z$. Find v from this equation, taking the principal value of the logarithm.

****

31A

$u + iv = \frac{1}{\pi}\{ln|z| + i\ arg\ z\}$

$v = \frac{1}{\pi}\tan^{-1}\frac{y}{x}$

FRAME 32

$\therefore T = \frac{100}{\pi}\tan^{-1}\frac{y}{x}$

For any point such as P, $\tan^{-1}\frac{y}{x}$ is the angle θ shown in the diagram in FRAME 29 and so must lie between 0 and π.

The lines of constant temperature (i.e. the isothermals) in the w-plane are given by v = constant. In the z-plane the corresponding lines are given by θ = constant (not the same constant, for if T = k, v = k/100 and θ = kπ/100). Thus the isothermals in the z-plane are radial lines from the origin.

The lines of heat flow (or flux lines) in the w-plane are perpendicular to the isothermals and so given by u = constant. What will the corresponding lines be in the z-plane?

****

32A

As $u = \frac{1}{\pi} \ln \sqrt{x^2 + y^2}$,

$u = c$ *becomes* $x^2 + y^2 = C$ *where* $C = e^{2c\pi}$.

The corresponding lines are therefore semicircles, each with its centre at the origin.

You will notice that these 'lines' of flux are again perpendicular to the isothermals. This links up with what was mentioned in FRAME 1 about the angles between these curves. It is only because the angles between curves are preserved under conformal transformations that this process is so useful in practical situations.

FRAME 33

There are two further points in connection with this problem, which link it up with FRAME 26 (page 12:83) in the Theory of Functions programme. In that frame it was mentioned that T satisfies the equation $\frac{\partial^2 T}{\partial x^2} + \frac{\partial^2 T}{\partial y^2} = 0$, i.e. T is harmonic. This is, of course, when T is a function of x and y. When T is a function of u and v (as in 30A where T = 100v) then the equation satisfied by T is $\frac{\partial^2 T}{\partial u^2} + \frac{\partial^2 T}{\partial v^2} = 0$. T = 100v obviously satisfies this equation and it can also be verified, without too much difficulty, that $\frac{\partial^2 T}{\partial x^2} + \frac{\partial^2 T}{\partial y^2} = 0$ when $T = \frac{100}{\pi} \tan^{-1} \frac{y}{x}$.

A second point mentioned in FRAME 26, page 12:83, was that the lines of heat flow are given by S = constant, where T and S are conjugate harmonic functions. As you know, in the z-plane, T and S are conjugate functions if $\frac{\partial T}{\partial x} = \frac{\partial S}{\partial y}$ and $\frac{\partial T}{\partial y} = -\frac{\partial S}{\partial x}$. In a similar way, in the w-plane, $\frac{\partial T}{\partial u} = \frac{\partial S}{\partial v}$ and $\frac{\partial T}{\partial v} = -\frac{\partial S}{\partial u}$. Find S in the two cases,

(i) T = 100v, (ii) $T = \frac{100}{\pi} \tan^{-1} \frac{y}{x}$.

****

33A

(i) $S = -100u$, *apart from a constant.*

(ii) $S = -\frac{50}{\pi} \ln (x^2 + y^2)$, *again apart from a constant.*

S = *constant then reduces to* u = *constant and* $x^2 + y^2$ = *constant as also given in FRAME 32 and 32A for the flux lines. The expression* $T + iS$ *is called the complex temperature.*

FRAME 34

A similar problem is the following:

Find T at any point P(x,y) in the section of the slab shown, the faces being maintained at the stated temperatures. (The same conditions about the thickness of the slab etc. will be assumed to hold as in the last example.)

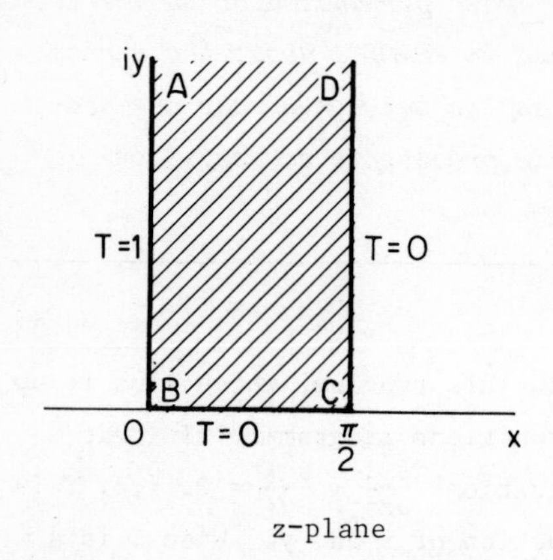

z-plane

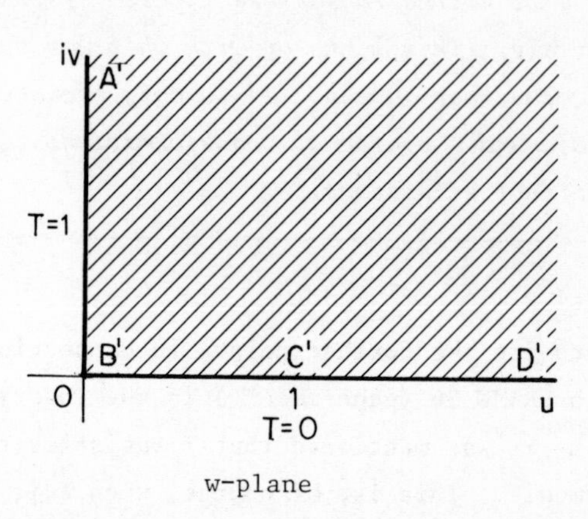

w-plane

The diagram in the z-plane can be converted into that shown in the w-plane by the transformation $w = \sin z$. Verify that, when $w = \sin z$ is applied to the z-figure, the stated w-figure is obtained.

****

34A

$$u + iv = \sin(x + iy)$$

$$= \sin x \cosh y + i \cos x \sinh y$$

$$u = \sin x \cosh y \qquad v = \cos x \sinh y \qquad (34A.1)$$

Along AB, $x = 0$, $\therefore u = 0$, $v = \sinh y$

As y *decreases from* ∞ *to* 0, *so also does* v *(but not at the same rate).*

Along BC, $y = 0$, $\therefore u = \sin x$, $v = 0$

As x *increases from* 0 *to* $\frac{1}{2}\pi$, u *increases from* 0 *to* 1.

Along CD, $x = \frac{1}{2}\pi$, $\therefore u = \cosh y$, $v = 0$

As y *increases from* 0 *to* ∞, u *increases from* 1 *to* ∞.

FRAME 35

By comparing the problem in the w-plane with that in the z-plane in the previous example, you may now be able to write down the solution for the present problem in the w-plane immediately. If you can write this down correctly, you may then proceed directly to FRAME 38. If you can't, carry straight on.

****

35A

$T = \frac{2}{\pi} \tan^{-1} \frac{v}{u}$ *The inverse tangent to be taken in the range 0 to ½π as u and v are both positive.*

FRAME 36

What will the problem in the w-plane become if the transformation $W = w^2$ is applied to the area in the w-plane.

****

36A

As $W = w^2$, $|W| = |w|^2$ *and* $\arg W = 2 \arg w$.

Along A'B', $\arg w = \frac{1}{2}\pi$, $\therefore \arg W = \pi$.

Also $|w|$ *decreases from* ∞ *to 0,* $\therefore |W|$ *decreases from* ∞ *to 0.*

Along B'C'D', $\arg w = 0$, $\therefore \arg W = 0$.

As $|w|$ *increases from 0 to* ∞, $|W|$ *increases from 0 to* ∞.

We therefore get

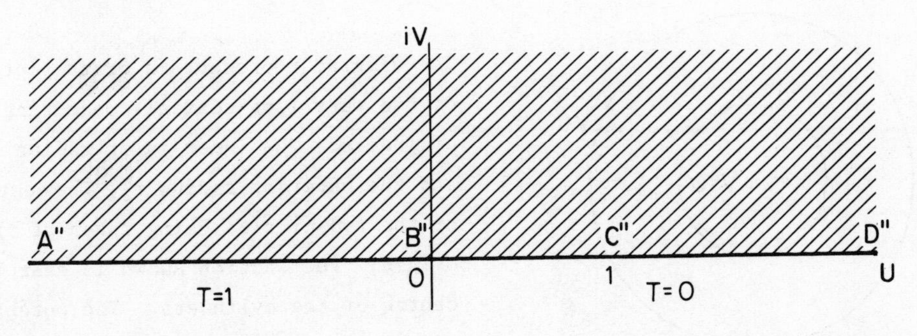

FRAME 37

Apart from the fact that $T = 1$ instead of 100 along A"B" this is now virtually the same problem as in FRAME 29 and so $T = \frac{1}{\pi}\tan^{-1}\frac{V}{U}$, $(0 \leqslant \tan^{-1}\frac{V}{U} \leqslant \pi)$ which can also be written as $T = \frac{1}{\pi}\arg W$. Then, as $\arg W = 2\arg w$ it follows that $T = \frac{2}{\pi}\arg w = \frac{2}{\pi}\tan^{-1}\frac{v}{u}$, $\left(0 \leqslant \tan^{-1}\frac{v}{u} \leqslant \frac{\pi}{2}\right)$.

FRAME 38

Using the equations (33A.1) then gives

$$T = \frac{2}{\pi}\tan^{-1}\frac{\cos x \sinh y}{\sin x \cosh y}$$

$$= \frac{2}{\pi}\tan^{-1}(\cot x \tanh y)$$

The isothermals are given by $T =$ constant, i.e. by

$\cot x \tanh y =$ constant (not the same one).

The lines of flux are given by $u^2 + v^2 =$ constant, i.e. by

$\sin^2 x \cosh^2 y + \cos^2 x \sinh^2 y =$ constant

or $\cosh^2 y - \cos^2 x =$ constant

or $\cosh 2y - \cos 2x =$ constant

as $\cosh^2 y = \frac{1}{2}(1 + \cosh 2y)$ and $\cos^2 x = \frac{1}{2}(1 + \cos 2x)$.

FRAME 39

The following problem may appear to be of an entirely different nature, but mathematically it is very similar to those just considered.

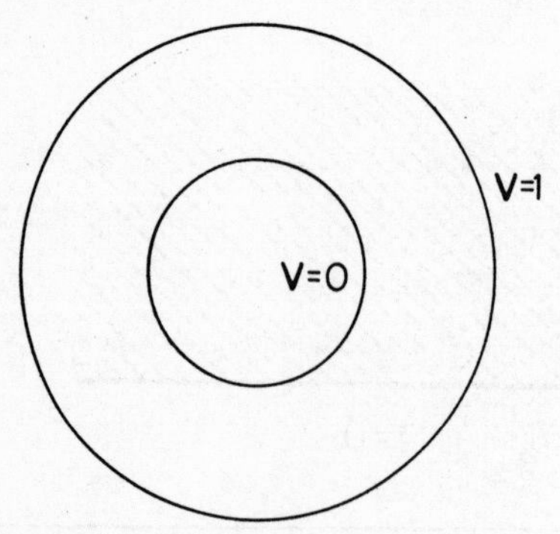

The diagram represents a cross-section of two very long coaxial conducting cylinders. The inner is of radius 1 and the outer of radius r_0, the centres of the two circles being taken at the origin. The section shown is near the centre of the cylinders. The potential on the inner cylinder is kept at $V = 0$

FRAME 39 continued

while that on the outer is maintained at V = 1. The problem is to find the electrostatic potential at any point in the annulus between the two circles. (Electrostatic is often abbreviated to e.s.) By taking the section to be somewhere near to the centre of the cylinders, any flow of electricity into or out of the section may be ignored. Start by finding the region in the w-plane corresponding to that shown below in the z-plane under the transformation $w = \ln z$, taking the principal value of the logarithm.

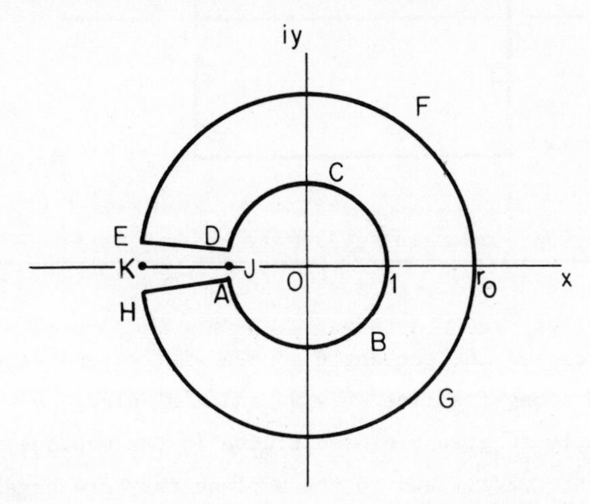

At the moment there appears to be no reason for the cut. Don't worry about this for the time being but assume that everywhere along DE, $\arg z = \pi - \delta$ and along HA it is $-\pi + \varepsilon$.

****

39A

$w = \ln z$

$\therefore\ u + iv = \ln |z| + i \arg z$

or $u = \ln|z|$, $v = \arg z$.

Around ABCD, $|z| = 1$, $\therefore\ u = 0$.

Also $\arg z$ *increases from* $-\pi + \varepsilon$ *to* $\pi - \delta$, $\therefore$ *so also does* v.

Along DE, $\arg z = \pi - \delta$, $\therefore\ v = \pi - \delta$

Also $|z|$ *increases from 1 to* r_0, $\therefore\ u$ *increases from 0 to* $\ln r_0$.

Around EFGH, $|z| = r_0$, $\therefore\ u = \ln r_0$.

Also $\arg z$ *decreases from* $\pi - \delta$ *to* $-\pi + \varepsilon$, $\therefore$ *so also does* v.

39A continued

Along HA, $\arg z = -\pi + \varepsilon$, $\therefore v = -\pi + \varepsilon$

Also $|z|$ *decreases from* r_0 *to 1,* $\therefore$ *u decreases from* $\ln r_0$ *to 0.*

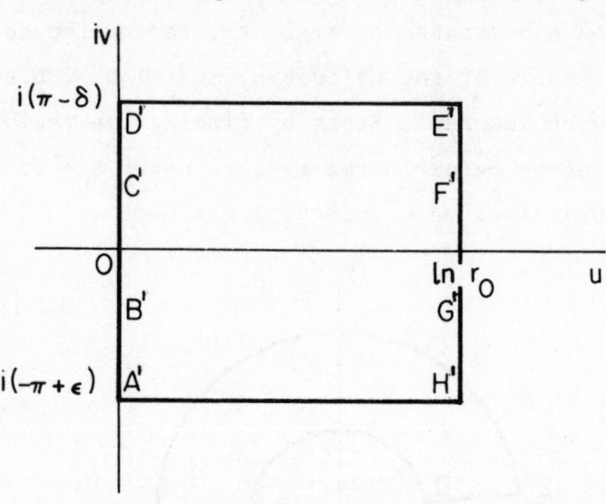

FRAME 40

If the cut is now closed the rectangle in 39A will extend vertically so that eventually D'E' is along the line $v = \pi$ and A'H' along $v = -\pi$. The purpose of the cut was simply to give a closed figure in the w-plane. In the z-plane the lines of flux are radial and in the w-plane they are parallel to Ou. The equipotential lines in the z-plane are concentric circles and in the w-plane they are perpendicular to Ou. In the w-plane the problem is as shown.

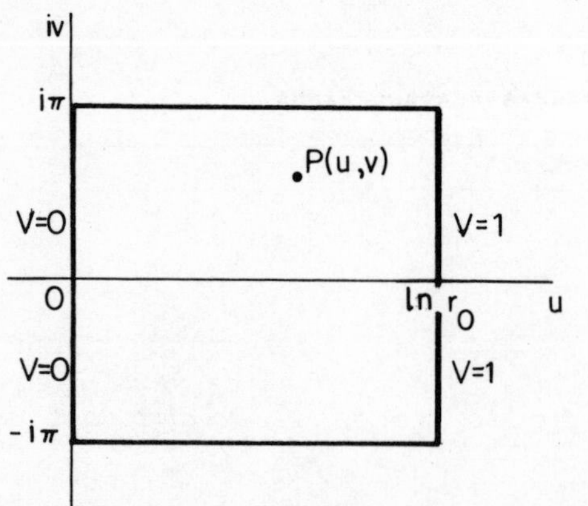

As the flux lines are parallel to Ou, the e.s. potential at any point P is given by $V = \dfrac{u}{\ln r_0}$.

(The value of V at any point is given by a similar formula to that for T had the two faces perpendicular to Ou been at temperatures T = 0 and T = 1.)

What will now be the formula for V in terms of x and y remembering that $w = \ln z$?

****

40A

$$u = \ln |z| = \ln \sqrt{x^2 + y^2} = \frac{1}{2} \ln(x^2 + y^2)$$

$$V = \frac{\ln(x^2 + y^2)}{2 \ln r_0}$$

FRAME 41

The forms of the flux and equipotential lines have already been mentioned. These were not obtained mathematically but were stated from the physical properties of the problem. However, they can now be linked mathematically as follows. With regard to the flux lines, these are given by v = constant in the w-plane. But as $v = \arg z$, v = constant becomes $\arg z$ = constant in the z-plane, i.e. radial lines. Similarly with regard to equipotential lines these are given by u = constant in the w-plane which becomes $\ln |z|$ = constant in the z-plane. But $\ln |z|$ = constant means that $|z|$ is constant and so the lines are concentric circles. A further link up is provided by the result in 40A, for the equipotential lines are given by V = constant, from which $x^2 + y^2$ = constant, corroborating our previous statements.

Now check that V, as found in 40A, is harmonic and that V and ψ are conjugate functions where $\psi = \frac{1}{\ln r_0} \tan^{-1} \frac{y}{x}$.

****

41A

$$\frac{\partial V}{\partial x} = \frac{1}{\ln r_0} \frac{x}{x^2 + y^2} \qquad \frac{\partial V}{\partial y} = \frac{1}{\ln r_0} \frac{y}{x^2 + y^2}$$

$$\frac{\partial^2 V}{\partial x^2} = \frac{1}{\ln r_0} \frac{y^2 - x^2}{(x^2 + y^2)^2} \qquad \frac{\partial^2 V}{\partial y^2} = \frac{1}{\ln r_0} \frac{x^2 - y^2}{(x^2 + y^2)^2}$$

$$\frac{\partial \psi}{\partial x} = \frac{1}{\ln r_0} \frac{-y}{x^2 + y^2} \qquad \frac{\partial \psi}{\partial y} = \frac{1}{\ln r_0} \frac{x}{x^2 + y^2}$$

Results follow.

FRAME 42

The lines given by ψ = constant reduce to $\tan^{-1}\frac{y}{x}$ = constant, i.e. arg z = constant, and these have already been seen to be the flux lines. If you compare the results that have been obtained here with those in the second temperature problem (see FRAMES 34-38) you will see that all the quantities occurring in a heat flow problem have their counterparts in an electrostatics problem. The analogy is completed by defining the complex e.s. potential to be $V + i\psi$. By virtue of the analogy which exists between the two types of problem, once the solution to a heat problem of this type has been obtained, the solution of the corresponding e.s. problem can be written down immediately and vice versa.

FRAME 43

Now try the following problems. In the last two, assume conditions such that the problems can be treated by the methods of the last few frames.

1. Find the region in the w-plane corresponding to that enclosed by the upper half of the circle $|z| = 1$ and its diameter under the transformation $z = \frac{i - w}{i + w}$.

2. Find the e.s. potential V in the space bounded by the upper half of the cylinder $x^2 + y^2 = 1$ and the plane $y = 0$ if $V = 0$ on the cylindrical surface and $V = 1$ on the plane surface.

3. Find the temperature T at any point inside a semicircular cylinder of material if the plane face is maintained at $T = 1$ and the cylindrical surface at $T = 0$. The cross-section of the semicircle in the x-y plane is given by $x^2 + y^2 = 1$, $y \gtrless 0$.

****

43A

1. *As the transformation is bilinear, the transformed curves will be parts of circles or straight lines.*

As $z = \dfrac{i - w}{i + w}$, $\quad w = \dfrac{-zi + i}{z + 1}$

Corresponding points: $A(z = 1) \to A'(w = 0)$

$B(z = i) \to B'(w = 1)$

$C(z = -1) \to C'(w = \infty)$

$D(z = 0) \to D'(w = i)$

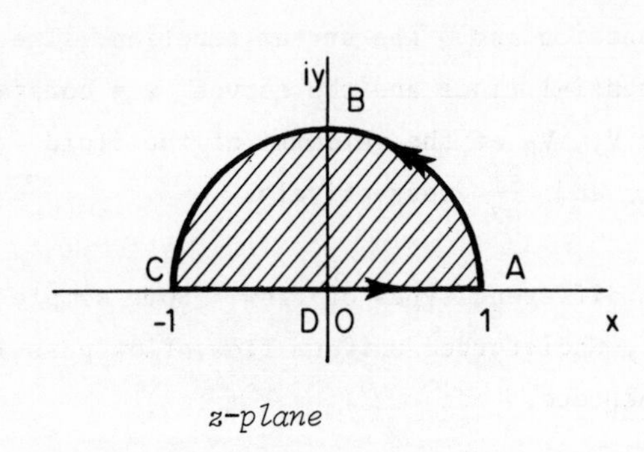

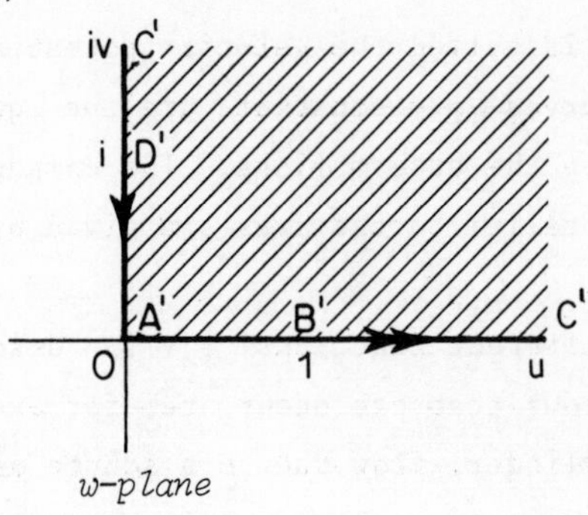

2. *Problem corresponds to* $V = 1$ *along* $C'D'A'$ *and* $V = 0$ *along* $A'B'C'$ *in the second diagram in the answer to Question 1. The solution to this is*

$V = \dfrac{2}{\pi} \tan^{-1} \dfrac{v}{u}$, $\quad \left(0 \leq \tan^{-1} \dfrac{v}{u} \leq \dfrac{\pi}{2}\right)$, $\quad$ *(see FRAMES 34-37).*

From $w = \dfrac{-zi + i}{z + 1}$,

$$u + iv = \frac{2y + i(1 - x^2 - y^2)}{(x + 1)^2 + y^2}$$

$$u = \frac{2y}{(x + 1)^2 + y^2}, \qquad v = \frac{1 - x^2 - y^2}{(x + 1)^2 + y^2}$$

$$V = \frac{2}{\pi} \tan^{-1} \frac{1 - x^2 - y^2}{2y}$$

3. $T = \dfrac{2}{\pi} \tan^{-1} \dfrac{1 - x^2 - y^2}{2y}$

FRAME 44

Another physical situation in which complex variable theory is used is that of two-dimensional fluid flow. By two-dimensional flow we mean that the flow pattern is the same in all parallel planes, and so it is only necessary to consider one of these planes. The plane considered is taken as the z-plane. Other assumptions that will be made are that the flow is steady (i.e. independent of time) and that the fluid is incompressible.

Under these conditions a function $\phi + i\psi$ exists, called the complex potential. ϕ is called the velocity potential function and ψ the stream function. The curves ϕ = constant are the equipotential lines and the curves ψ = constant are the stream lines. The components V_1, V_2 of the velocity of the fluid parallel to the axes are given by $\frac{\partial\phi}{\partial x}$ and $\frac{\partial\phi}{\partial y}$ respectively.

Different functions $\phi + i\psi$ describe different types of flow. Some simple flows that can occur are, for example, undisturbed uniform flow, flow past a cylinder, flow due to a source or sink, etc.

FRAME 45

If a source is placed at the origin and no other motion is involved, this effectively means that fluid is being injected into the system at that point and will flow out equally in all directions from it. The stream lines will be radial from the origin and the velocity at any point will be along the stream line there. Under these conditions it can be shown that $\phi + i\psi = k \ln z$ where k is a real constant and is a measure of how much fluid is being injected into the system at the origin.

For this particular formula for $\phi + i\psi$, find ϕ and ψ in terms of x and y. Verify that the stream lines are radial from the origin and find the velocity components parallel to the axes.

****

45A

$k\ ln\ z = k\{ln\ |z| + i\ arg\ z\}$

$\quad = k\{\tfrac{1}{2}ln(x^2 + y^2) + i\ tan^{-1}\frac{y}{x}\}$

$\therefore \quad \phi = \tfrac{1}{2}k\ ln(x^2 + y^2), \qquad \psi = k\ tan^{-1}\frac{y}{x}$

The stream lines are given by ψ = *constant, i.e.* $\frac{y}{x}$ = *constant, and these are radial lines.*

The velocity components are $V_1 = \frac{\partial\phi}{\partial x} = \frac{kx}{x^2 + y^2}$ *and* $V_2 = \frac{\partial\phi}{\partial y} = \frac{ky}{x^2 + y^2}$.

The actual velocity at any point is $\sqrt{V_1^2 + V_2^2} = k/\sqrt{x^2 + y^2}$.
This is the same for all points on a circle centre the origin and decreases as one moves out from the origin. This, of course, is exactly what one would expect from the physical implications of the situation.

FRAME 46

Fluid flowing with constant velocity V_0 parallel to the x-axis is disturbed by placing in it at the origin a cylinder of radius a, the axis of the cylinder being perpendicular to the x-y plane. Under these circumstances it can be shown that $\phi + i\psi = V_0\left(z + \frac{a^2}{z}\right)$. The flow is as shown in the diagram.

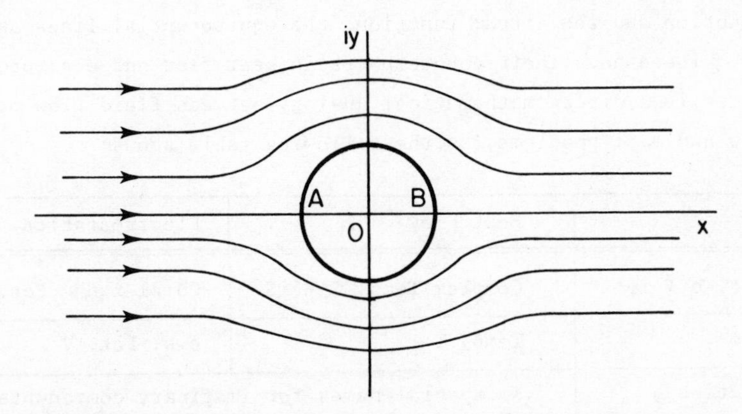

Find the equations of the equipotential and stream lines and the velocity components at any point. Show that the cylindrical surface is part of one of the stream lines and that there are stagnation points (i.e. points where the velocity is zero) at A and B.

****

46A

$\phi = V_0 x\left(1 + \dfrac{a^2}{x^2 + y^2}\right)$ $\qquad\qquad \psi = V_0 y\left(1 - \dfrac{a^2}{x^2 + y^2}\right)$

Equipotential lines are $x\left(1 + \dfrac{a^2}{x^2 + y^2}\right) = \text{constant}$

Stream lines are $y\left(1 - \dfrac{a^2}{x^2 + y^2}\right) = \text{constant}$ (46A.1)

$V_1 = V_0\left\{1 + a^2 \dfrac{y^2 - x^2}{(x^2 + y^2)^2}\right\}$ $\qquad\qquad V_2 = -2a^2 V_0 \dfrac{xy}{(x^2 + y^2)^2}$

$x^2 + y^2 = a^2$ *makes the L.H.S. of (46A.1) zero and this is the stream line obtained when the constant on the R.H.S. is put equal to zero. The other part of this stream line is* $y = 0$, *and this is also given by* $y\left[1 - \dfrac{a^2}{x^2 + y^2}\right] = 0$.

At A, $x = -a$, $y = 0$ *and at B,* $x = a$, $y = 0$. *Both of these sets of values make* $V_1 = 0$ *and* $V_2 = 0$, *so that A and B are stagnation points.*

FRAME 47

There are some points worth noticing about two-dimensional fluid flow that emerge from what has been done in the last few frames.

The concept of a complex potential is exactly the same mathematically as that of complex temperature and complex e.s. potential. It gives the velocity potential function and the stream function, the equipotential lines and the stream lines. These have their counterparts in heat flow and e.s. problems. So again, there is a direct mathematical analogy between fluid flow problems and heat flow and e.s. problems, as the following table shows:

Fluid Flow	Heat Flow	Electrostatics
Complex Pot. $\phi + i\psi$	Complex Temp. $T + iS$	Complex e.s. Pot. $V + i\psi$
Velocity Pot. ϕ	Temp. T	e.s. Pot. V
Stream Function ψ	No special names for imaginary components	
Equipotential Lines ϕ = constant	Isothermals T = constant	Equipotential Lines V = constant
Stream Lines ψ = constant	Flux Lines S = constant	Flux Lines ψ = constant

FRAME 48

The complex potential taken in the last problem, i.e. $V_0\left(z + \frac{a^2}{z}\right)$ is the same, apart from the multiplying constant V_0, as the Joukowski transformation considered in FRAMES 10-14. You will further see that the boundary of the circle in FRAME 46 corresponds to the special case in the table in FRAME 14. If the transformation $w = z + \frac{a^2}{z}$ is applied to the problem in FRAME 46, the complex potential becomes $\phi + i\psi = V_0 w$ and the cylinder becomes a flat plate perpendicular to the w-plane and meeting that plane along the real axis from -2a to +2a. Taking $w = u + iv$, and consequently working in terms of u and v instead of x and y, what expressions will be obtained from $\phi + i\psi = V_0 w$ for

(a) ϕ, (b) ψ, (c) the equipotential lines, (d) the stream lines,

(e) the velocity components parallel to the u and v axes?

****

48A

(a) $\phi = V_0 u$ *(b)* $\psi = V_0 v$

(c) $u = constant$ *(d)* $v = constant$

(e) V_0 *and 0.*

FRAME 49

This therefore represents the uniform flow past a very thin plate A'B' as shown.

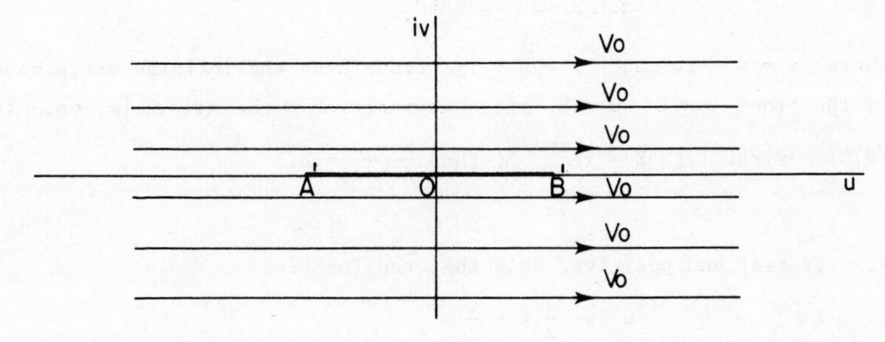

FRAME 49 continued

As the stream lines are all parallel to Ou and the plate is along Ou it has no effect and can be ignored. The flow is now uniform (in that the velocity is everywhere constant) and parallel to Ou. $\phi + i\psi = V_0 w$ is thus the complex potential for this particular flow.

The work done in this last section should be sufficient to give you some idea as to the uses of conformal transformations. You should now have sufficient background to enable you to proceed further in the study of any particular topic in your field of interest which uses these techniques.

FRAME 50

Miscellaneous Examples

In this frame a collection of miscellaneous examples is given for you to try. Answers are supplied in FRAME 51, together with such working as is considered helpful.

1. If $w = \cosh(\pi z/a)$, where a is real and positive, find the region in the w-plane which corresponds to the rectangle bounded by the lines $x = 0$, $y = 0$, $x = N > 0$, $y = a$. (L.U.)

2. Show that, if a and b are real, the transformation

$$\frac{z - a}{z + a} = ie^{iw\pi/b},$$

where $w = u + iv$ and $z = x + iy$, transforms the infinite strip bounded by the lines $u = 0$, $u = b$, into the interior of the circle $|z| = a$. (L.U.)

(HINT: First let $Z = ie^{iw\pi/b}$, then $\frac{z - a}{z + a} = Z$.)

3. If a is real and positive, show that the function

$$w = \frac{a}{2}\left(z + \frac{1}{z}\right)$$

transforms the circles of centre $z = 0$ in the z-plane into ellipses of a confocal system in the w-plane in such a manner that each ellipse corres-

ponds in general to two cricles.
Show that the unit circle $|z| = 1$ transforms into the part of the real axis between the foci of the ellipses.
Find a transformation which maps the inside of the ellipse $\frac{u^2}{25} + \frac{v^2}{16} = 1$, cut between the foci, on to the annulus $1 \leqslant |z| \leqslant 3$. (L.U.)
[HINT: For the ellipse $\frac{x^2}{A^2} + \frac{y^2}{B^2} = 1$, the eccentricity e is given by $B^2 = A^2(1 - e^2)$, and the foci are at the points $(\pm Ae, 0)$.]

4. The figure represents a cross-section of a cylinder of radius $\frac{1}{2}$ lying on a plane. The cylinder is maintained at e.s. potential V = 100 and the plane at V = 50. (Suitable insulation is inserted at the point of contact for this to be possible.) By the use of the mapping function $w = 1/z$, find the potential at any point exterior to the circle and above the plane.

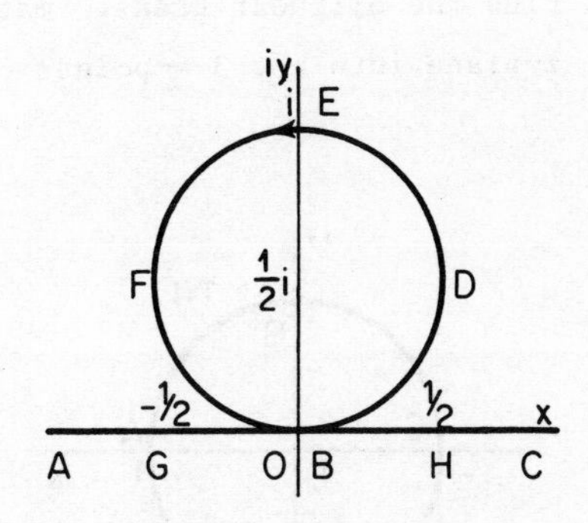

5. 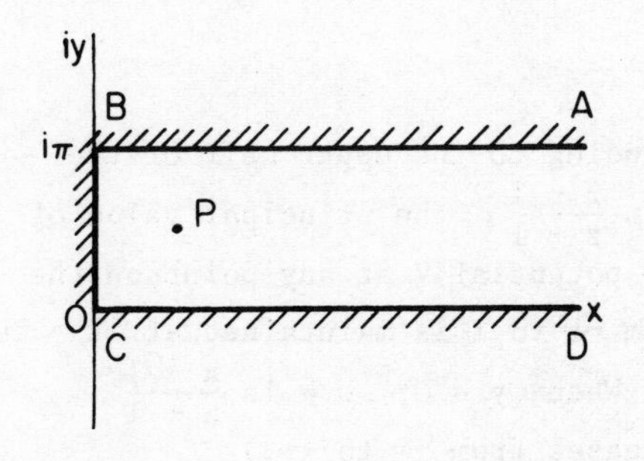

Two-dimensional fluid motion is taking place in the channel shown, there being a source at the point $P[z = (1+i)\pi/2]$. Find the region in the w-plane corresponding to the channel under the transformation $w = 2 \cosh z$ and the point P' corresponding to P.

The corresponding situation in the w-plane is that of two-dimensional fluid motion in the region with a source at P'. If the boundary be removed

FRAME 50 continued

and an equal source placed at the point Q', the image of P' in the u-axis, the flow in the original w-region is unaltered.

Under these conditions it can be shown that, in the w-plane,

$\phi + i\psi = k\{\ln(w - 2i \sinh \pi/2) + \ln(w + 2i \sinh \pi/2)\}$,

i.e. $\phi + i\psi = k \ln\{w^2 + 2(\cosh \pi - 1)\}$ where k is a real constant.

Convert this to a function of z and so find the equations of the stream lines in the original channel. Verify that the boundaries of the channel and the line $y = \pi/2$ are all stream lines, as you would expect.

6. Find the bilinear transformation that carries the points 1, -1, i in the z-plane into the 3 w-points 0, ∞, 1 respectively.

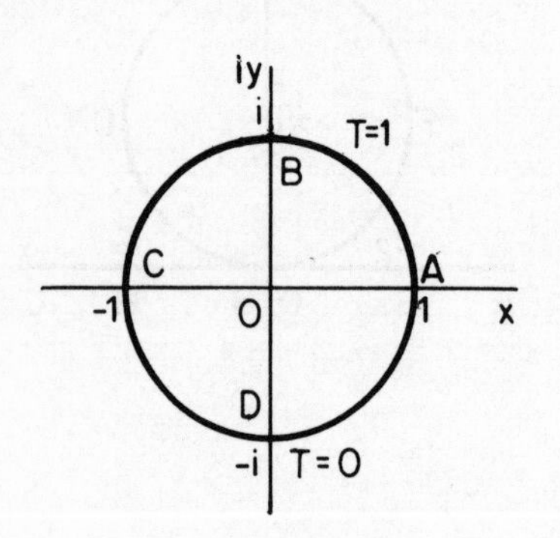

The diagram represents the cross-section of a cylindrical slab of material. The half of the surface for which y is positive is maintained at temperature T = 1, the other half at T = 0. By means of the transformation obtained in the first part of this question, find the temperature T at any other point in the cross-section. (You may be able to write down the solution in the w-plane straight away. Alternatively, you may find it helpful to refer to the table of transformations in the APPENDIX.)

7. Find the region in the w-plane corresponding to the upper half of the z-plane under the transformation $w = \ln \dfrac{z - 1}{z + 1}$, the principal value of the log being inferred. Hence find the potential V at any point on the y-axis if the section of the x-axis from -1 to 1 is maintained at V = 100, the remainder being at V = 0. (HINT: When y = 0, $w = \ln \dfrac{x - 1}{x + 1}$. Investigate what happens to w as x increases from $-\infty$ to $+\infty$.)

FRAME 51

Answers to Miscellaneous Examples

1. $w = \cosh \frac{\pi z}{a}$ gives

$u = \cosh \frac{\pi x}{a} \cos \frac{\pi y}{a}$ $\qquad$ $v = \sinh \frac{\pi x}{a} \sin \frac{\pi y}{a}$

Along AB, $y = 0$,

$\therefore u = \cosh(\pi x/a)$, $v = 0$

u increases from 1 to $\cosh(\pi N/a)$.

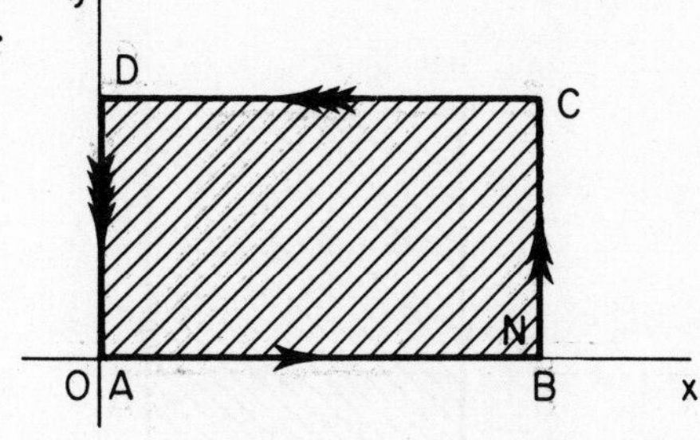

Along BC, $x = N$

$\therefore u = \cosh \frac{\pi N}{a} \cos \frac{\pi y}{a}$

$v = \sinh \frac{\pi N}{a} \sin \frac{\pi y}{a}$

Eliminating y,

$$\frac{u^2}{\cosh^2(\pi N/a)} + \frac{v^2}{\sinh^2(\pi N/a)} = 1$$

which is the equation of an ellipse.

The points corresponding to B and C are $B'\{\cosh(\pi N/a),0\}$ and $C'\{-\cosh(\pi N/a),0\}$ and as $0 \leqslant y \leqslant a$, v is always positive. The required path is thus the upper half of the ellipse.

Along CD, $y = a$, $\therefore u = -\cosh(\pi x/a)$, $v = 0$,

u increases from $-\cosh(\pi N/a)$ to -1.

Along DA, $x = 0$, $\therefore u = \cos(\pi y/a)$, $v = 0$,

u increases from -1 to 1.

The required region is as shown.

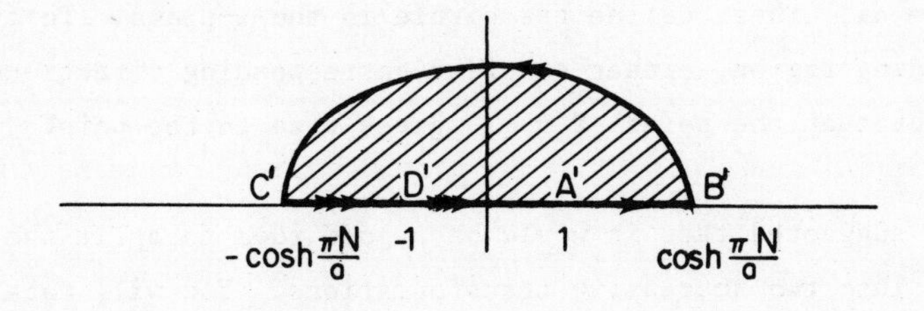

2. If $Z = ie^{iw\pi/b} = X + iY$,

$X = -e^{-\pi v/b} \sin(\pi u/b)$, $\quad Y = e^{-\pi v/b} \cos(\pi u/b)$

$u = 0$, $\quad X = 0$, $\quad Y = e^{-\pi v/b}$

$u = b$, $\quad X = 0$, $\quad Y = -e^{-\pi v/b}$

The corresponding figures in the w- and Z-planes are

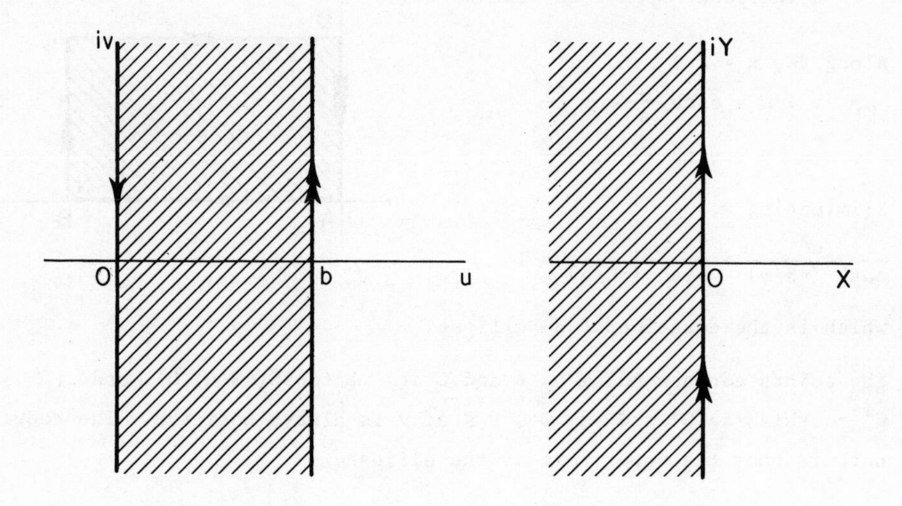

$\dfrac{z - a}{z + a} = Z$ is a bilinear transformation and therefore the Y-axis in the Z-plane becomes a circle (or straight line) in the z-plane.

From $\dfrac{z - a}{z + a} = Z$, $\quad z = a\,\dfrac{1 + Z}{1 - Z}$

Some corresponding points are: $Z = 0$, $z = a$; $Z = \infty$, $z = -a$; $Z = i$, $z = ai$. These define the circle in the z-plane. To find the corresponding region, either consider corresponding directions of travel or the fact that the point $Z = -1$ gives rise to the point $z = 0$.

The hint suggested that it would be a good idea to split the transformation into two successive transformations. You will recall that a similar approach was adopted in the first programme in this Unit when the bilinear transformation was being considered.

FRAME 51 continued

3. Similar working to that in FRAMES 8-10A leads to the equations

$$u = \frac{ax}{2}\left(1 + \frac{1}{k^2}\right), \qquad v = \frac{ay}{2}\left(1 - \frac{1}{k^2}\right),$$

$$\frac{u^2}{\left\{\frac{a}{2}\left(\frac{k^2 + 1}{k}\right)\right\}^2} + \frac{v^2}{\left\{\frac{a}{2}\left(\frac{k^2 - 1}{k}\right)\right\}^2} = 1 \qquad (51.1)$$

Taking, say, the semi-major axis, if this is given a fixed value C, say, then $\frac{a}{2}\frac{k^2 + 1}{k} = C$ which is a quadratic in k and thus, in general, for a given C, there are two values of k. This means that for a given ellipse, there are, in general, two circles.

Also, from (51.1), using $B^2 = A^2(1 - e^2)$, we get $A^2e^2 = A^2 - B^2 = a^2$, i.e. the foci of all the ellipses are at $(\pm a, 0)$.

When $k = 1$, $v = 0$, $u = ax$ and as x then varies between -1 and 1, u varies between -a and a.

The circle $|z| = 1$ corresponds to the cut between the foci. When $|z| = 3$, i.e. $k = 3$, the ellipse becomes

$$\frac{u^2}{\frac{25a^2}{9}} + \frac{v^2}{\frac{16a^2}{9}} = 1$$

For this to be the given ellipse $a^2 = 9$ and transformation is $w = \frac{3}{2}\left(z + \frac{1}{z}\right)$. (Due to the fact that for each w there are two z's, this transformation will also map the interior of the ellipse on to the annulus $\frac{1}{3} \leqslant |z| \leqslant 1$.)

4. As $w = 1/z$, circles in the z-plane become circles in the w-plane, the term circle in each case including that of infinite radius, i.e. a straight line.

The region exterior to the circle and above ABC becomes, in the w-plane, as shown:

FRAME 51 continued

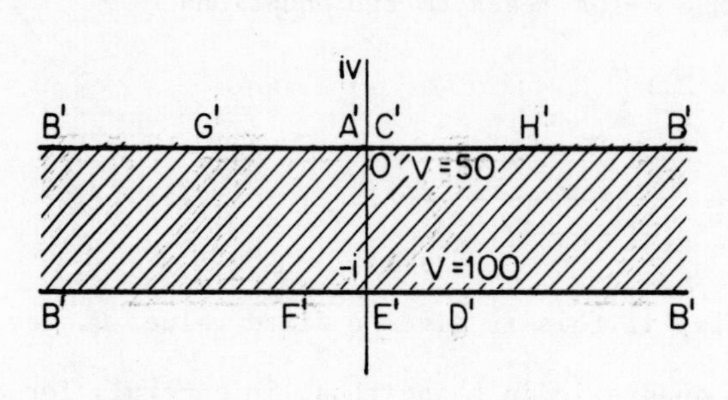

The solution in the w-plane is

$V = 50(1 - v)$

$w = \dfrac{1}{z}$,

$\therefore\ u + iv = \dfrac{x - iy}{x^2 + y^2}$

$v = \dfrac{-y}{x^2 + y^2}$

$V = 50\left(1 + \dfrac{y}{x^2 + y^2}\right)$

5. The region corresponding to the channel is as shown:

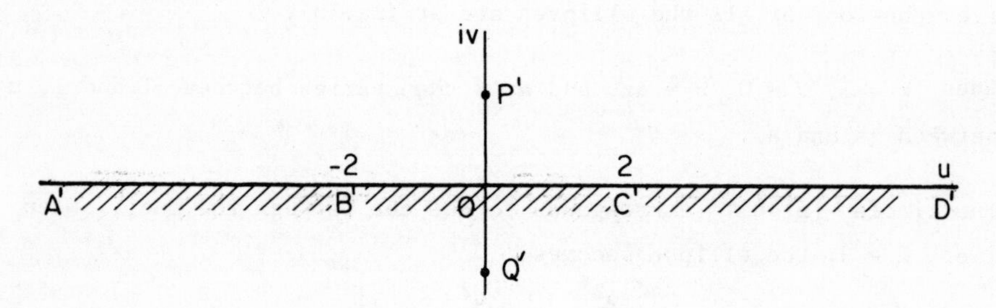

At P', $u = 0$, $v = 2 \sinh \tfrac{1}{2}\pi$

As a function of z,

$$\begin{aligned}\phi + i\psi &= k \ln\{4 \cosh^2 z + 2(\cosh \pi - 1)\} \\ &= k \ln 2(\cosh 2z + \cosh \pi) \\ &= k \ln 2(\cosh 2x \cos 2y + i \sinh 2x \sin 2y + \cosh \pi)\end{aligned}$$

ψ = constant implies that the argument of the contents of the brackets is constant.

$$\text{i.e.} \quad \tan^{-1} \frac{\sinh 2x \sin 2y}{\cosh 2x \cos 2y + \cosh \pi} = \text{const}$$

$$\text{i.e.} \quad \frac{\sinh 2x \sin 2y}{\cosh 2x \cos 2y + \cosh \pi} = \text{const}$$

The lines $x = 0$, $y = 0$, $y = \pi$ and $y = \tfrac{1}{2}\pi$ all make the L.H.S. of this last equation zero and so are all stream lines.

FRAME 51 continued

6. Taking $w = \dfrac{az + b}{cz + d}$ and substituting corresponding points into this equation leads to $w = \dfrac{1 - z}{1 + z}i$.

The interior of the circle is transformed into the upper half of the w-plane, as shown.

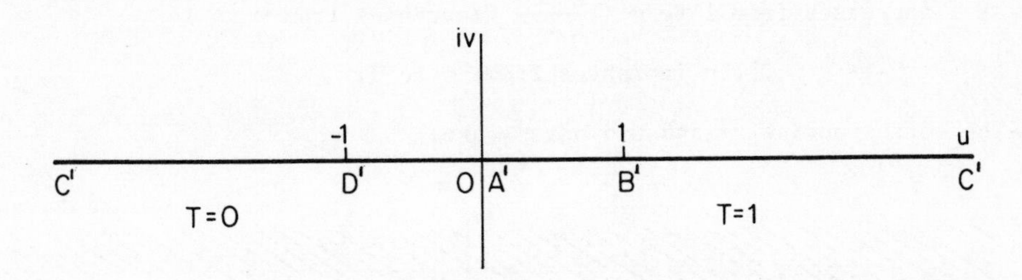

The solution in this plane is $T = 1 - \frac{1}{\pi}\tan^{-1}\frac{v}{u}$, $(0 \leqslant \tan^{-1}\frac{v}{u} \leqslant \pi)$.

(Compare with FRAMES 29-32 if in difficulty.)

$w = \dfrac{1 - z}{1 + z}i$ leads to $u + iv = \dfrac{2y + i(1 - x^2 - y^2)}{(1 + x)^2 + y^2}$

from which $\dfrac{v}{u} = \dfrac{1 - x^2 - y^2}{2y}$

$\therefore T = 1 - \frac{1}{\pi}\tan^{-1}\dfrac{1 - x^2 - y^2}{2y}$

7. Along $y = 0$, $w = \ln\dfrac{x - 1}{x + 1}$

When $x < -1$, $\dfrac{x - 1}{x + 1} > 0$, $\therefore v = 0$

As x increases from $-\infty$ to -1, $\dfrac{x - 1}{x + 1}$ increases from 1 to ∞

$\therefore$ u increases from 0 to ∞.

When $-1 < x < 0$, $\dfrac{x - 1}{x + 1} < 0$, $\therefore v = \pi$

As x increases from -1 to 0, $\dfrac{x - 1}{x + 1}$ increases from $-\infty$ to -1

$\therefore$ u decreases from ∞ to 0.

FRAME 51 continued

When $0 < x < 1$, $\dfrac{x-1}{x+1} < 0$, $\therefore\ v = \pi$

As x increases from 0 to 1, $\dfrac{x-1}{x+1}$ increases from -1 to 0

$\therefore$ u decreases from 0 to $-\infty$

When $x > 1$, $\dfrac{x-1}{x+1} > 0$, $\therefore\ v = 0$

As x increases from 1 to ∞, $\dfrac{x-1}{x+1}$ increases from 0 to 1

$\therefore$ u increases from $-\infty$ to 0.

The corresponding z- and w-diagrams are:

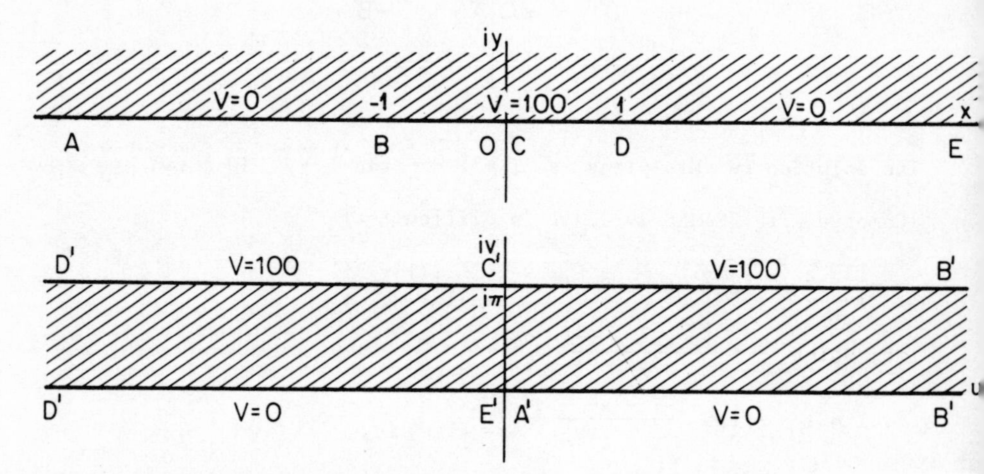

The solution in the w-plane is $V = \dfrac{100}{\pi}v$

$w = \ln\dfrac{z-1}{z+1}$ leads to

$$u + iv = \ln\frac{(x^2 + y^2 - 1) + 2iy}{(x+1)^2 + y^2}$$

from which $v = \tan^{-1}\dfrac{2y}{x^2 + y^2 - 1}$ $\left(0 \leqslant \tan^{-1}\dfrac{2y}{x^2 + y^2 - 1} \leqslant \pi\right)$

$$\therefore\ V = \frac{100}{\pi}\tan^{-1}\frac{2y}{x^2 + y^2 - 1}$$

When $x = 0$, $V = \dfrac{100}{\pi}\tan^{-1}\dfrac{2y}{y^2 - 1}$

This gives the solution for $y > 0$. From symmetry the value of V when $y = -k$ $(k > 0)$ will be the same as that when $y = k$.

APPENDIX

Table of Transformations

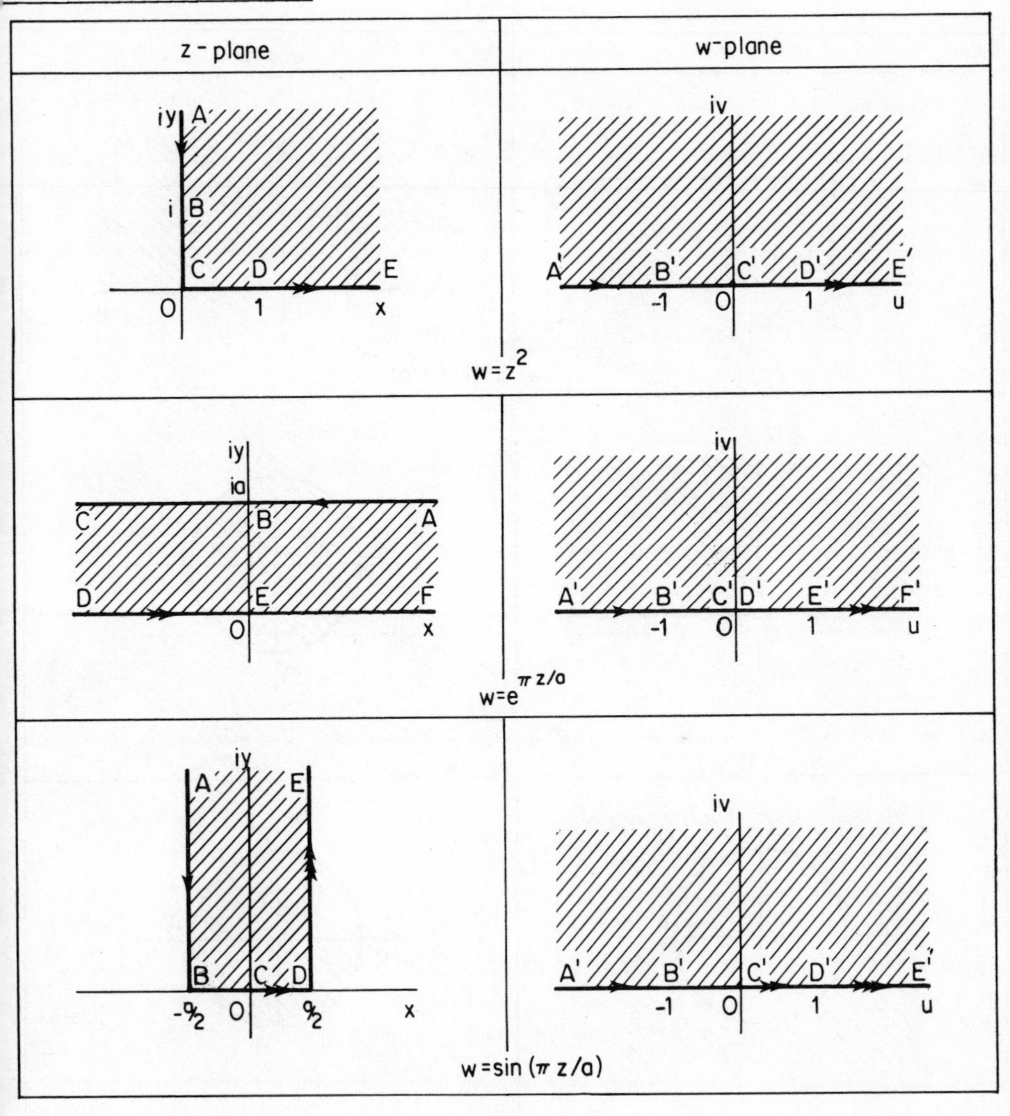

APPENDIX

Table of Transformations - continued

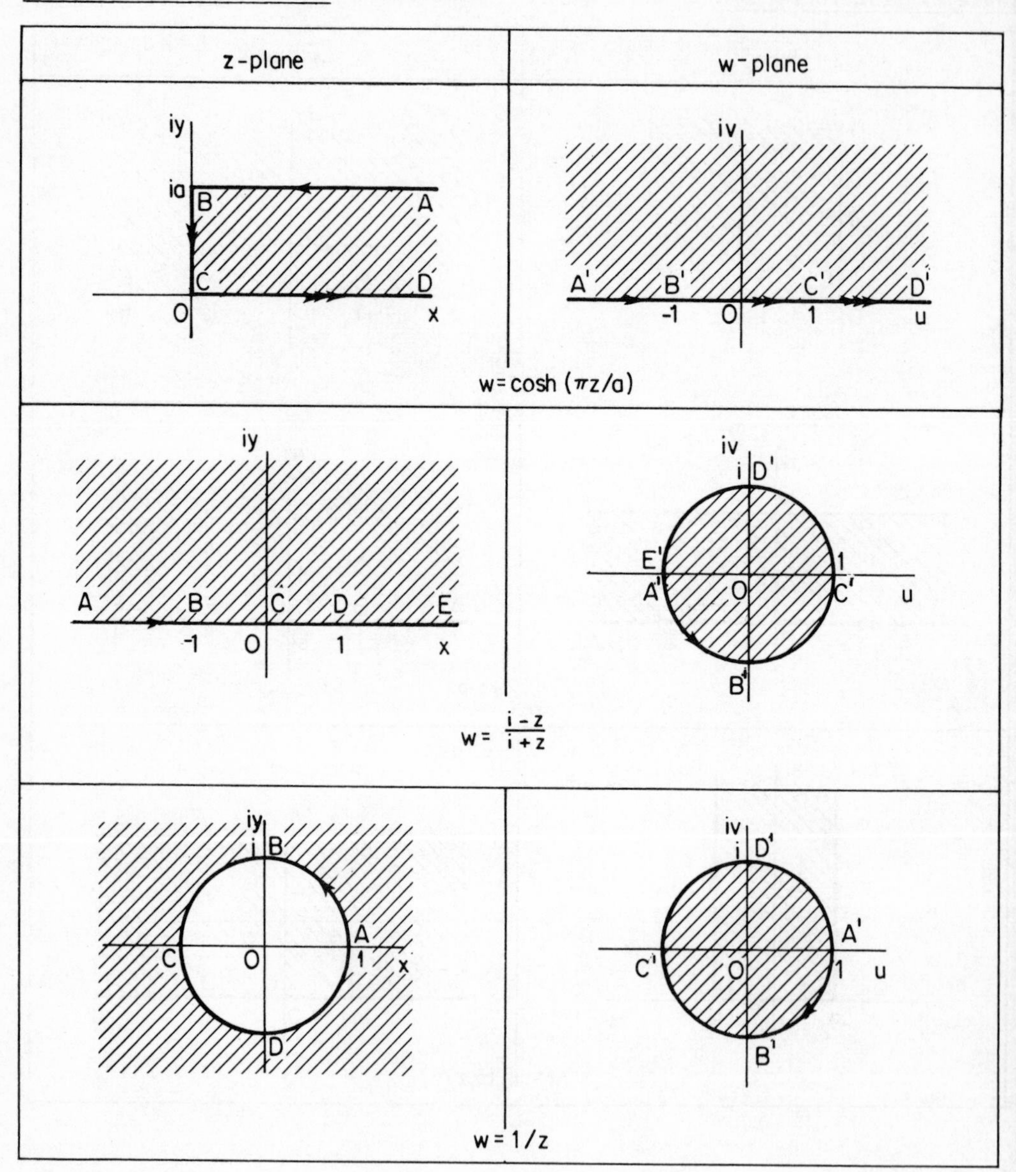

APPENDIX

Table of Transformations - continued

z-plane	w-plane
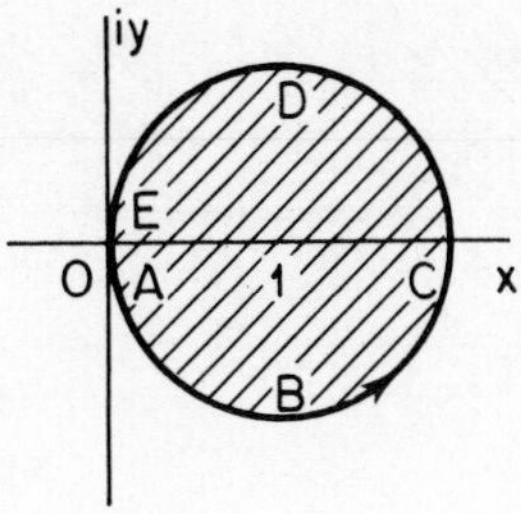	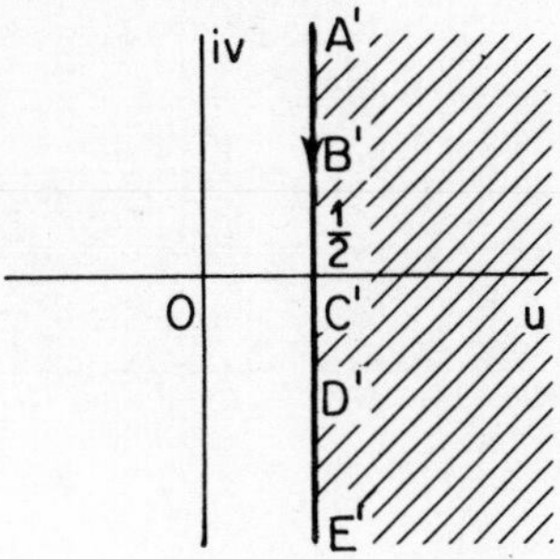

$w = 1/z$

UNIT 13

FURTHER LAPLACE TRANSFORMS

A.C. Bajpai
I.M. Calus
J.A. Fairley

Loughborough University of Technology

INSTRUCTIONS

This programme is divided up into a number of FRAMES which are to be worked *in the order given*. You will be required to participate in many of these frames and in such cases the answers are provided in ANSWER FRAMES, designated by the letter A following the frame number. Steps in the working are given where this is considered helpful. The answer frame is separated from the main frame by a line of asterisks: *********. Keep the answers covered until you have written your own response. If your answer is wrong, go back and try to see why. Do not proceed to the next frame until you have corrected any mistakes in your attempt and are satisfied that you understand the contents up to this point.

Before reading this programme, it is necessary that you are familiar with the following

Prerequisites

The contents of the programme on the Solution of Differential Equations by Laplace Transform Methods in Unit 5 of Vol I.

The contents of FRAMES 20-30 of the Simultaneous Differential Equations programme in Unit 5.

Differentiation under the Integral Sign as covered in FRAMES 14-20 of Unit 7.

CONTENTS

FRAME 1

Introduction

In Unit 5 of Vol I you were introduced to the Laplace transform and shown its use in the solution of certain differential equations. In this programme further transform pairs will be obtained so that a wider range of d.e.'s can be solved. You will also see that information about a system (such as an electrical network or control system) may sometimes be more conveniently obtained by working in terms of Laplace transforms rather than in terms of the corresponding time functions.

You will find a table of Laplace transform pairs in the APPENDIX (see pages 13:58-13:63) where s is used as the parameter as was the case in Unit 5, i.e.

$$F(s) = \mathcal{L}\{f(t)\} = \int_0^\infty e^{-st} f(t)dt$$

You will notice that at the top of the table it states that the initial conditions are those just prior to $t = 0$. This is specified by the notation 0-. Sometimes initial conditions are used which are those obtaining just after $t = 0$ and the notation for this is 0+. Usually this distinction is unimportant and you can interpret initial conditions as being those at $t = 0$ instead of either $t = 0-$ or $t = 0+$. Where the distinction _is_ necessary it will be mentioned in the text.

FRAME 2

The Transform of an Integral

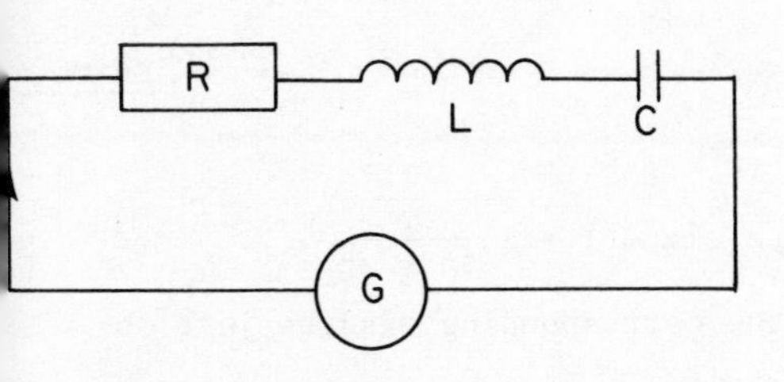

If a generator supplies, from time $t = 0$, a voltage E, which may be direct or alternating, to the circuit shown, then the differential equation giving i, the current, is

$$L\frac{di}{dt} + Ri + \frac{1}{C}\int_0^t i\,dt = E \qquad (2.1)$$

assuming that the charge on the capacitor is solely due to the passage of the current i for time t.

FRAME 2 continued

One way of finding the solution of this equation is to use the fact that $i = \frac{dq}{dt}$ where q is the charge. Then $q = \int_0^t idt$, and so the d.e. in i can be replaced by one in q, i.e.

$$L\frac{d^2q}{dt^2} + R\frac{dq}{dt} + \frac{1}{C}q = E$$

This can now be solved for q and then i can be found immediately.

An alternative way of finding i would be possible if an expression can be obtained for $\mathcal{L}\{\int_0^t idt\}$. The equation (2.1) could then be solved directly for i without the necessity of introducing q.

FRAME 3

Before seeing if we can find a formula for $\mathcal{L}\{\int_0^t idt\}$ let us see what happens in particular cases. Taking first $i = e^{-3t}$, find $\int_0^t idt$ and then use the table to obtain $\mathcal{L}\{\int_0^t idt\}$. Then repeat for $i = a \sin \omega t$.

****

3A

When $i = e^{-3t}$: $\int_0^t idt = \frac{1}{3}(1 - e^{-3t})$ $\mathcal{L}\{\int_0^t idt\} = \frac{1}{s(s+3)}$

When $i = a \sin \omega t$: $\int_0^t idt = \frac{a}{\omega}(1 - \cos \omega t)$ $\mathcal{L}\{\int_0^t idt\} = a\frac{\omega}{s(s^2+\omega^2)}$

FRAME 4

Also from the table

$$\mathcal{L}\{e^{-3t}\} = \frac{1}{s+3} \quad \text{and} \quad \mathcal{L}\{a \sin \omega t\} = a\frac{\omega}{s^2+\omega^2}$$

Comparing these results for $\mathcal{L}\{i\}$ with the corresponding results just obtained for $\mathcal{L}\{\int_0^t idt\}$, do you notice any connection between them?

****

4A

In each case $\mathcal{L}\{\int_0^t i dt\} = \frac{1}{s}\mathcal{L}\{i\}$.

FRAME 5

The question now arises: Is this coincidence, or is it always true? To investigate, we proceed as follows.

Start by letting $f(t) = \int_0^t i dt$

What will be the value of (i) f(0) and (ii) f'(t)? (If you have any difficulty in stating either of these results, the simple functions used in the previous two frames will give you a clue.)

****

5A

(i) $f(0) = 0$ *(ii)* $f'(t) = i$

[*For example, if* $i = e^{-3t}$, $f(t) = \frac{1}{3}(1 - e^{-3t})$ *and from this it immediately follows that* $f(0) = 0$ *and* $f'(t) = e^{-3t} = i$.]

FRAME 6

Now, using standard notations, let $\mathcal{L}\{f(t)\} = F(s)$ and $\mathcal{L}(i) = \bar{i}$. Then, as $i = f'(t)$, $\bar{i} = \mathcal{L}\{f'(t)\}$. But, as you saw in Unit 5,

$$\mathcal{L}\{f'(t)\} = sF(s) - f(0)$$

$$\therefore \quad \bar{i} = sF(s)$$

$$\therefore F(s) = \frac{1}{s}\bar{i}$$

$$\text{i.e.} \quad \mathcal{L}\{\int_0^t i dt\} = \frac{1}{s}\bar{i}$$

So it wasn't just coincidence after all.

Returning to the circuit in FRAME 2, you should now have no difficulty in solving directly equation (2.1) for i if $L = 1\,H$, $R = 250\,\Omega$, $C = 10^{-4}F$, $E = 10\,V$ and the current is initially zero.

****

6A

The transformed equation is

$$s\bar{i} + 250\bar{i} + 10^4 \frac{\bar{i}}{s} = \frac{10}{s}$$

$$\bar{i} = \frac{10}{s^2 + 250s + 10\,000}$$

$$= \frac{10}{(s + 50)(s + 200)}$$

$$= \frac{1}{15}\left(\frac{1}{s + 50} - \frac{1}{s + 200}\right)$$

$$i = \frac{1}{15}\left(e^{-50t} - e^{-200t}\right)$$

FRAME 7

The result obtained for the transform of an integral can also be used in an example such as the following:

From entry 16 in the table, find $\mathcal{L}^{-1}\left\{\frac{\omega^2}{s^2(s^2 + \omega^2)}\right\}$

We have $\mathcal{L}\{1 - \cos \omega t\} = \frac{\omega^2}{s(s^2 + \omega^2)}$

$$\therefore \quad \mathcal{L}\left\{\int_0^t (1 - \cos \omega t)dt\right\} = \frac{1}{s}\,\frac{\omega^2}{s(s^2 + \omega^2)}$$

From this, find the required inverse.

****

7A

$$\mathcal{L}^{-1}\left\{\frac{\omega^2}{s^2(s^2 + \omega^2)}\right\} = \left[t - \frac{1}{\omega}\sin \omega t\right]_0^t$$

$$= t - \frac{1}{\omega}\sin \omega t$$

This is an alternative method, in some cases, to that of the use of partial fractions.

FRAME 8

Use of Differentiation under the Integral Sign

The process of differentiation under the integral sign was discussed in Unit 7 and one of the examples taken there (see FRAME 19A, page 7:12) gave a hint that the technique can be of use in finding Laplace transforms.

You were reminded in FRAME 1 that $F(s) = \int_0^\infty e^{-st}f(t)dt$. Differentiate both sides of this equation w.r.t. s. Then look at the R.H.S. of your result and say what transform this represents.

****

8A

$$\frac{d}{ds}F(s) = \int_0^\infty (-t)e^{-st}f(t)dt \qquad (8A.1)$$

The R.H.S. can be written as $\int_0^\infty e^{-st}\{-tf(t)\}dt$ *which, by definition, is* $\mathcal{L}\{-tf(t)\}$. *This is obviously the same as* $-\mathcal{L}\{tf(t)\}$.

FRAME 9

Thus $\mathcal{L}\{tf(t)\} = -\frac{d}{ds}F(s)$.

If, for example, this result is applied to entry 5 in the table,

i.e. $\mathcal{L}\{e^{-\alpha t}\} = \dfrac{1}{s+\alpha}$, we shall get

$$\mathcal{L}\{te^{-\alpha t}\} = -\frac{-1}{(s+\alpha)^2}$$

$$= \frac{1}{(s+\alpha)^2}$$

which is entry 7.

Now differentiate (8A.1) w.r.t. s to obtain $\frac{d^2}{ds^2}F(s)$ and then express the result as a Laplace transform.

****

9A

$$\frac{d^2}{ds^2}F(s) = \int_0^\infty t^2 e^{-st} f(t)dt$$

$$= \int_0^\infty e^{-st}\{t^2 f(t)\}dt$$

$$= \mathcal{L}\{t^2 f(t)\}$$

FRAME 10

You will notice that every time e^{-st} is differentiated w.r.t. s, a factor -1 is introduced.

Thus $\mathcal{L}\{tf(t)\} = (-1)\frac{d}{ds}F(s)$

$\mathcal{L}\{t^2 f(t)\} = (-1)^2 \frac{d^2}{ds^2}F(s)$

The extension to $\mathcal{L}\{t^n f(t)\}$, where n is a +ve integer, should now be obvious to you. Can you suggest what the formula for this will be?

****

10A

$$(-1)^n \frac{d^n}{ds^n}F(s)$$

FRAME 11

Thus $\mathcal{L}\{t^n f(t)\} = (-1)^n \frac{d^n}{ds^n}F(s)$

Now use this result, with $f(t) = e^{-\alpha t}$, to obtain $\mathcal{L}\{t^n e^{-\alpha t}\}$.

Then check your answer by using the exponential multiplier theorem

i.e. $\mathcal{L}\{e^{-\alpha t}f(t)\} = F(s + \alpha)$

with $f(t) = t^n$.

****

11A

$$\frac{n!}{(s + \alpha)^{n+1}}$$

FRAME 12

For further practice, use this method to find $\mathcal{L}\{t \cosh \omega t\}$ and $\mathcal{L}\{t^2 \sin \omega t\}$ by choosing suitable entries in the table.

****

12A

From entry 22, $$\mathcal{L}\{t \cosh \omega t\} = -\frac{d}{ds}\frac{s}{s^2 - \omega^2} = \frac{s^2 + \omega^2}{(s^2 - \omega^2)^2}$$

From entry 12, $$\mathcal{L}\{t^2 \sin \omega t\} = -\frac{d}{ds}\frac{2\omega s}{(s^2 + \omega^2)^2} = \frac{2\omega(3s^2 - \omega^2)}{(s^2 + \omega^2)^3}$$

FRAME 13

The Delayed Function

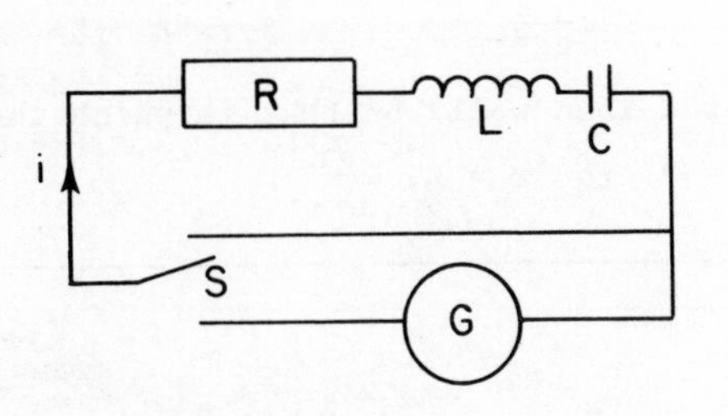

The diagram represents a slightly modified version of the circuit shown in FRAME 2. S is a two-way switch which allows the generator either to be included in the circuit or by-passed. Suppose S is switched so that the generator is included in the circuit for one second, from $t = 0$ to $t = 1$. As before, suppose the charge on the capacitor is due only to the passage of the current i for time t. For the first second, the same conditions will apply as in FRAME 2, and so we shall have

$$L\frac{di}{dt} + Ri + \frac{1}{C}\int_0^t i\,dt = E$$

i and q both being zero when $t = 0$. However, at the instant when the generator is by-passed, there will be a certain charge on the capacitor, given by $\int_0^1 i\,dt$, and due to this charge, a current will continue to flow. The d.e. will now be

$$L\frac{di}{dt} + Ri + \frac{1}{C}\int_0^t i\,dt = 0$$

FRAME 13 continued

but the conditions which will now have to be applied to find the constants in the solution are those obtaining when $t = 1$. If the figures quoted in FRAME 6 are used, then when $t = 1$, i is to all intents and purposes zero but

$$q = \int_0^1 \frac{1}{15}\left(e^{-50t} - e^{-200t}\right)dt = 0{\cdot}001.$$

FRAME 14

Other physical situations can arise where a single d.e. is not sufficient to describe them completely. For example, suppose a light beam of length ℓ is clamped at one end, freely supported at the other and loaded as shown. The differential equation for y, the deflection, is then given by

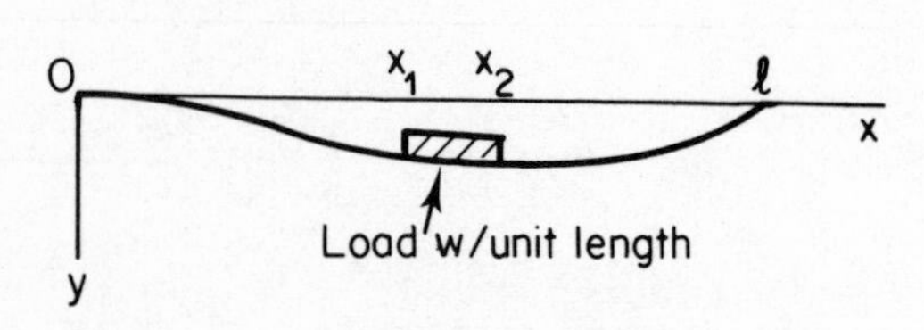

$$EI\frac{d^4y}{dx^4} = 0 \qquad 0 < x < x_1$$

$$EI\frac{d^4y}{dx^4} = w \qquad x_1 < x < x_2$$

$$EI\frac{d^4y}{dx^4} = 0 \qquad x_2 < x < \ell$$

This situation is obviously more complicated than would be that in which the load w/unit length extends right from $x = 0$ to $x = \ell$.

FRAME 15

The method of Laplace transforms can quite easily be extended to cope with situations such as those described in the last two frames. This is accomplished by considering functions such as those shown in the following graphs:

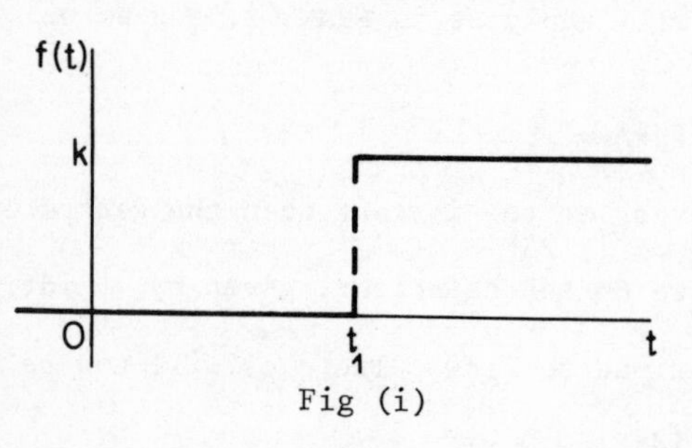

Fig (i)

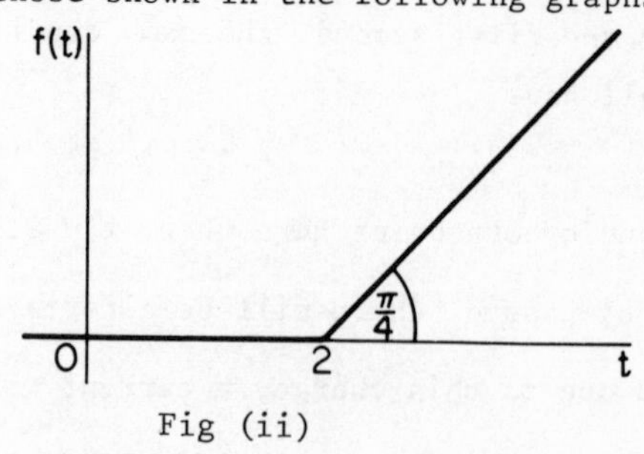

Fig (ii)

FRAME 15 continued

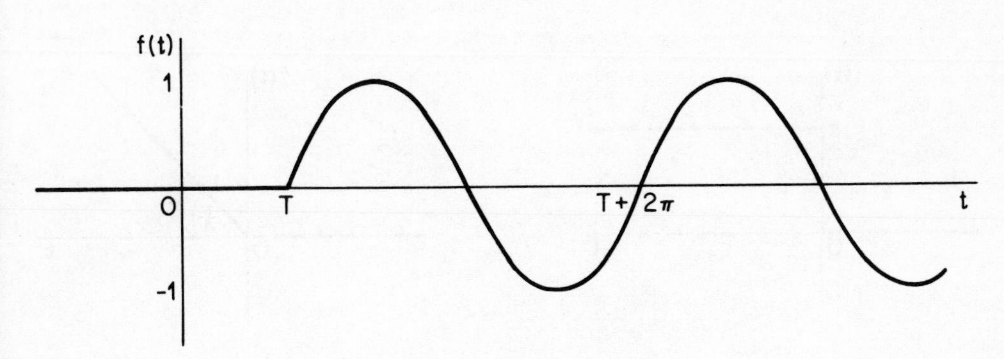

Fig (iii)

The section of Fig (iii) to the right of $t = T$ is a sine wave.

Can you write down sets of equations which will define the graphs shown?

****

15A

For Fig (i) $\begin{cases} f(t) = 0 & \text{for } t < t_1 \\ f(t) = k & \text{for } t > t_1 \end{cases}$

For Fig (ii) $\begin{cases} f(t) = 0 & \text{for } t < 2 \\ f(t) = t - 2 & \text{for } t > 2 \end{cases}$

For Fig (iii) $\begin{cases} f(t) = 0 & \text{for } t < T \\ f(t) = \sin(t - T) & \text{for } t > T \end{cases}$

FRAME 16

When you find the Laplace transform of a function you integrate between 0 and ∞, e.g. $\mathcal{L}\{k\} = \int_0^\infty ke^{-st}dt$, $\mathcal{L}\{t\} = \int_0^\infty te^{-st}dt$, $\mathcal{L}\{\sin t\} = \int_0^\infty e^{-st}\sin t\, dt$.
The fact that the integral does not extend to cover negative values of t can be interpreted as putting the function equal to zero for such values of t. Thus the three functions whose transforms have just been quoted can be interpreted as the three functions shown in the following diagrams.

FRAME 16 continued

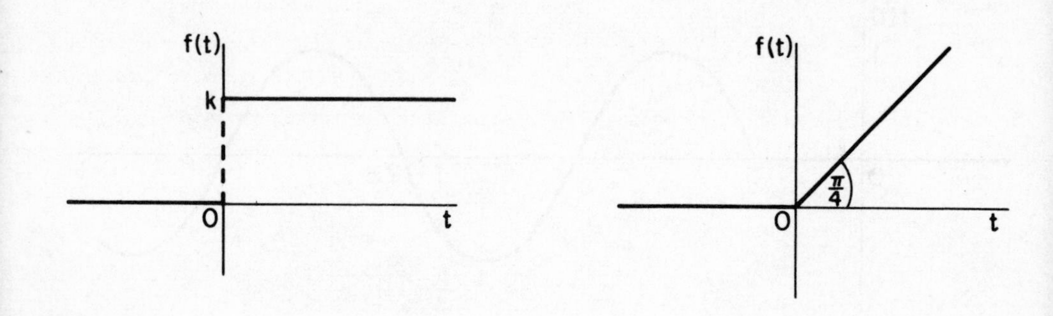

Fig (1) Fig (2)

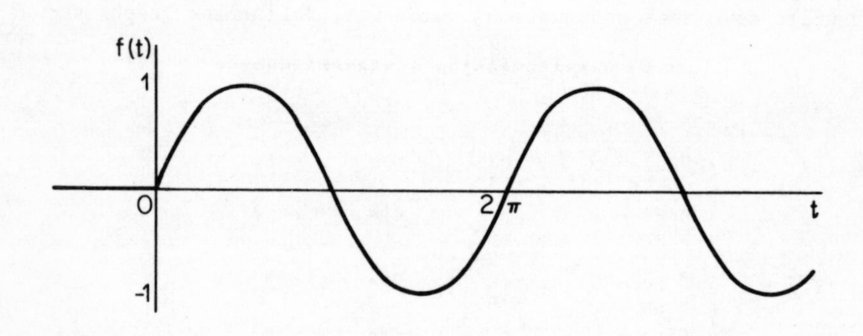

Fig (3)

The equations of these three curves are, in each case,

$$f(t) = 0 \quad \text{for} \quad t < 0$$

and $f(t) = k$, $f(t) = t$, $f(t) = \sin t$ respectively for $t > 0$.

The three curves shown in FRAME 15 are the same shapes as those above and can be obtained from them simply by displacing them distances t_1, 2 and T respectively to the right, i.e. by delaying the start of the non-zero parts by these amounts. Such functions as those shown in FRAME 15 are called DELAYED FUNCTIONS.

FRAME 17

The Heaviside Unit Step Function

Instead of having to define functions such as those shown in FRAME 15 by a series of equations and inequalities, it is useful to have a notation which has the same meaning but is shorter. To achieve this, we introduce the HEAVISIDE UNIT STEP FUNCTION. It is denoted by H(t) or u(t) and is defined as

$$\begin{cases} u(t) = 0 & \text{for} \quad t < 0 \\ u(t) = 1 & \text{for} \quad t > 0 \end{cases}$$

Its graph is

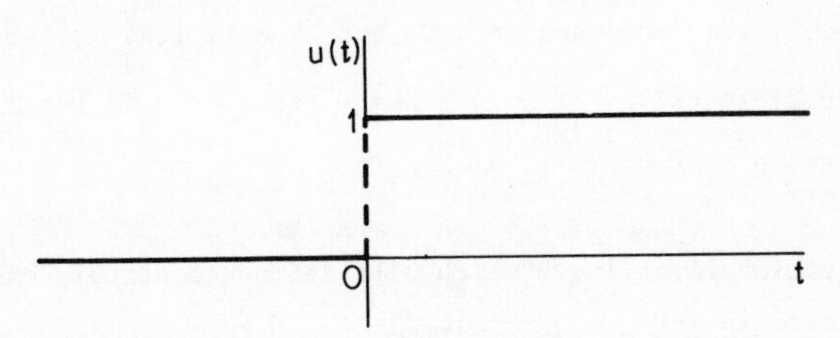

Using the definition of the Laplace transform find $\mathcal{L}\{u(t)\}$.

****

17A

$$\int_0^\infty e^{-st} 1 \, dt = \frac{1}{s}$$

Over the range 0 to ∞, f(t) = u(t) and f(t) = 1 coincide and so their Laplace transforms are the same.

FRAME 18

Linked with the function just defined is one similar but displaced a distance a to the right, i.e.

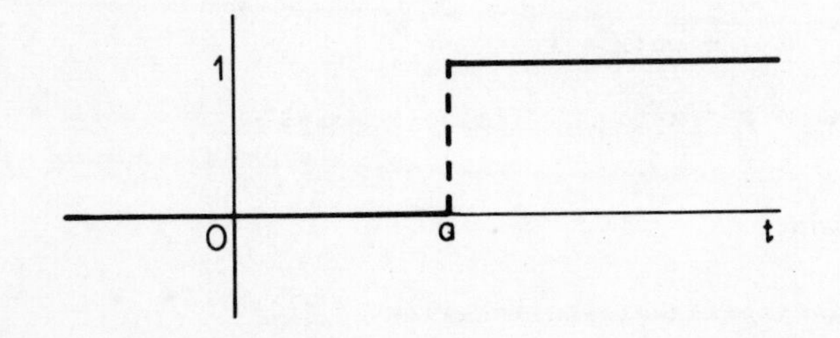

This is denoted by H(t - a) or u(t - a) and is defined by the equations

$$\begin{cases} u(t - a) = 0 & \text{for} \quad t < a \\ u(t - a) = 1 & \text{for} \quad t > a. \end{cases}$$

FRAME 19

This function is now used in conjunction with the equation defining the right hand section of a curve such as any of those in the graphs in FRAMES 15 or 16.

Thus the function shown in Fig (1), FRAME 16, is given by the equation

$$f(t) = ku(t)$$

and that shown in Fig (ii), FRAME 15, is given by

$$f(t) = (t - 2)u(t - 2)$$

$ku(t)$ is $k \times 0$, i.e. 0, when $t < 0$ but $k \times 1$, i.e. k, when $t > 0$.

Similarly $(t - 2)u(t - 2) = (t - 2) \times 0 = 0$ when $t < 2$, but is $(t - 2) \times 1 = t - 2$ when $t > 2$.

Now give an equation involving u which will define completely each of the functions depicted in

(a) Fig (2), FRAME 16 (b) Fig (3), FRAME 16

(c) Fig (i), FRAME 15 (d) Fig (iii), FRAME 15.

****

19A

(a) $f(t) = tu(t)$ *(b)* $f(t) = \sin t\ u(t)$

(c) $f(t) = ku(t - t_1)$ *(d)* $f(t) = \sin(t - T)\ u(t - T)$

FRAME 20

What will be the value of

$$wu(x - x_1) - wu(x - x_2)$$

when (i) $x < x_1$ (ii) $x_1 < x < x_2$ (iii) $x > x_2$?

Sketch the graph of this function.

****

20A

(i) $w \times 0 - w \times 0 = 0$ *(ii)* $w \times 1 - w \times 0 = w$

(iii) $w \times 1 - w \times 1 = 0$

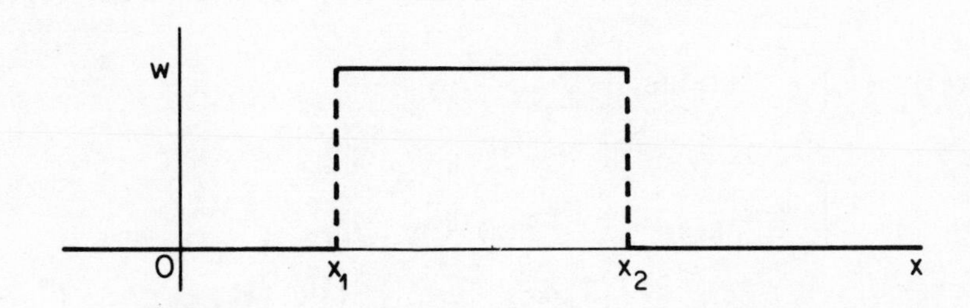

FRAME 21

If you look back to FRAME 14 you will see that these are the values of $EI\,\dfrac{d^4y}{dx^4}$ used there. Thus the equations given in FRAME 14 can be combined into the single equation

$$EI\,\frac{d^4y}{dx^4} = wu(x - x_1) - wu(x - x_2)$$

Can you now give a single expression involving u which will define completely the functions shown:

(i) (ii)

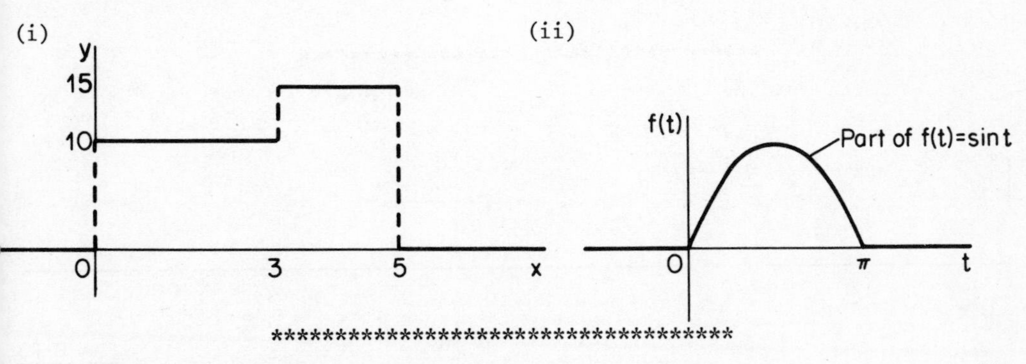

****

21A

(i) $y = 10u(x) + 5u(x - 3) - 15u(x - 5)$

(ii) $f(t) = \sin t\ u(t) + \sin(t - \pi)\ u(t - \pi)$

or $f(t) = \sin t\ u(t) - \sin t\ u(t - \pi)$

FRAME 22

In order to make further use of the ideas just introduced, it is necessary to consider the Laplace transforms of functions such as those depicted in FRAME 15. Taking first of all the function shown in Fig (i), its transform is given by

$$\mathcal{L}\{f(t)\} = \int_0^\infty e^{-st} f(t)dt$$

$$= \int_{t_1}^\infty e^{-st} k\,dt \quad \text{as} \quad f(t) = 0 \quad \text{for} \quad t < t_1 \quad \text{and}$$

$$\int_0^\infty e^{-st} f(t)dt = \int_0^{t_1} e^{-st} f(t)dt + \int_{t_1}^\infty e^{-st} f(t)dt$$

$$= \left[-\frac{k}{s} e^{-st} \right]_{t_1}^\infty$$

$$= \frac{k}{s} e^{-st_1}$$

It therefore follows that $\mathcal{L}\{ku(t - t_1)\} = \frac{k}{s} e^{-st_1}$ and hence that $\mathcal{L}\{u(t - t_1)\} = \frac{1}{s} e^{-st_1}$.

Now find the transform of the function shown in Fig (ii) of FRAME 15.

****

22A

$$\int_2^\infty (t - 2)e^{-st} dt = \frac{1}{s^2} e^{-2s}$$

FRAME 23

Now, as was found in Unit 5, the transforms of the functions shown in Figs (1) and (2) of FRAME 16 are $\frac{k}{s}$ and $\frac{1}{s^2}$. From the results just obtained, it would appear that, if a function is delayed by an amount a, then its Laplace transform can be obtained simply by multiplying the transform of the original function by e^{-sa}. We shall now go on to show that this is true for any function.

FRAME 23 continued

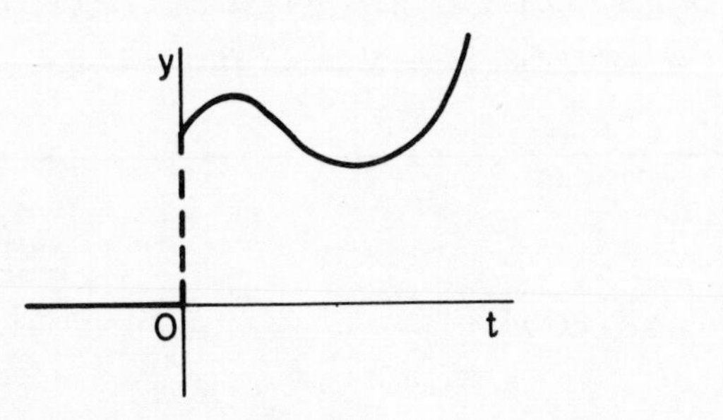

Fig (i)

Fig (ii)

Fig (i) shows the graph of the function $y = f(t)u(t)$

Fig (ii) shows the graph obtained when the curve in Fig (i) is displaced a distance a to the right. This graph therefore represents $y = f(t - a)u(t - a)$.

Now let $\mathcal{L}\{f(t)u(t)\} = F(s)$.

With any luck we shall be able to show that the transform of the function depicted in Fig (ii) is $e^{-as}F(s)$.

From the definition of the transform

$$\mathcal{L}\{f(t - a)u(t - a)\} = \int_a^\infty e^{-st}f(t - a)dt \quad \text{as the function is zero for } t < a$$

The substitution $t - a = \theta$, say, is now made in the integral.

What will you get if you do this?

****

23A

$$\int_0^\infty e^{-s(\theta+a)}f(\theta)d\theta$$

FRAME 24

$$\text{Thus } \mathcal{L}\{f(t - a)u(t - a)\} = \int_0^\infty e^{-s\theta}e^{-sa}f(\theta)d\theta$$

$$= e^{-sa}\int_0^\infty e^{-s\theta}f(\theta)d\theta$$

What can you say about $\int_0^\infty e^{-s\theta}f(\theta)d\theta$ and $\int_0^\infty e^{-st}f(t)dt$?

****

24A

They are equal as the variable of integration does not affect the result of a definite integral, provided it is the same function that is integrated.

FRAME 25

$$\therefore \quad \mathcal{L}\{f(t-a)u(t-a)\} = e^{-sa}\int_0^\infty e^{-st}f(t)dt$$

But $$\int_0^\infty e^{-st}f(t)dt = \int_0^\infty e^{-st}f(t)u(t)dt$$

$$= \mathcal{L}\{f(t)u(t)\}, \quad \text{i.e. } F(s)$$

$$\therefore \quad \mathcal{L}\{f(t-a)u(t-a)\} = e^{-sa}F(s) \qquad (25.1)$$

Now use this theorem in conjunction with the relevant entries in the table to write down the transforms of

(i) $\{1 - e^{-2(t-3)}\}u(t-3)$ (ii) $wu(x - x_1) - wu(x - x_2)$

****

25A

(i) $\dfrac{2e^{-3s}}{s(s+2)}$ *(ii)* $\dfrac{w}{s}e^{-sx_1} - \dfrac{w}{s}e^{-sx_2}$

FRAME 26

When solving d.e.'s with R.H.S.'s which involve the u function it will be necessary to find the inverse transforms of functions of s involving exponentials. These are not given, except in a certain few cases, directly in the table. However, you can find them by making use of the theorem that has just been proved. A couple of examples should make the process clear.

As a first example, we shall find $\mathcal{L}^{-1}\left\{\dfrac{se^{-st_1}}{s^2+\omega^2}\right\}$.

$\dfrac{se^{-st_1}}{s^2+\omega^2}$ is a particular case of the R.H.S. of (25.1) in which $a = t_1$ and $F(s) = \dfrac{s}{s^2+\omega^2}$.

FRAME 26 continued

From the table, $\mathcal{L}^{-1}\left\{\frac{s}{s^2 + \omega^2}\right\} = \cos \omega t$ where here, $\cos \omega t\, u(t)$ is really meant, i.e. $f(t) = \cos \omega t\, u(t)$.

Then $f(t - t_1)u(t - t_1) = \cos \omega(t - t_1)\, u(t - t_1)$ and is the curve $\cos \omega t\, u(t)$ shifted a distance t_1 to the right.

$$\therefore \mathcal{L}^{-1}\left\{\frac{se^{-st_1}}{s^2 + \omega^2}\right\} = \cos \omega(t - t_1)\, u(t - t_1)$$

Its graph looks like

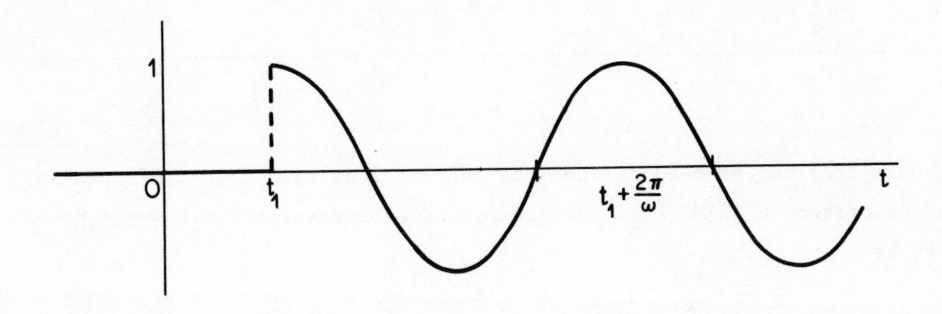

FRAME 27

As a second example, let us find $\mathcal{L}^{-1}\left\{\frac{e^{-4s}}{(s + 1)(s + 2)(s + 3)}\right\}$

First it is necessary to find $\mathcal{L}^{-1}\left\{\frac{1}{(s + 1)(s + 2)(s + 3)}\right\}$. What will this be?

****

27A

There is no entry in the table giving this inverse directly and so partial fractions must first be used.

$$\frac{1}{(s + 1)(s + 2)(s + 3)} = \frac{1}{2}\frac{1}{s + 1} - \frac{1}{s + 2} + \frac{1}{2}\frac{1}{s + 3}$$

the inverse of which is $\frac{1}{2}e^{-t} - e^{-2t} + \frac{1}{2}e^{-3t}$ *by which is really meant*

$(\frac{1}{2}e^{-t} - e^{-2t} + \frac{1}{2}e^{-3t})u(t)$

FRAME 28

The required inverse is therefore

$$\left\{\frac{1}{2}e^{-(t-4)} - e^{-2(t-4)} + \frac{1}{2}e^{-3(t-4)}\right\}u(t-4)$$

Now find (i) $\mathcal{L}^{-1}\left\{\frac{e^{-3s}}{(s+4)^2}\right\}$ (ii) $\mathcal{L}^{-1}\left\{\frac{\omega e^{-sT}}{s^2+\omega^2}\right\}$

****

28A

(i) $(t-3)e^{-4(t-3)}u(t-3)$ *(ii)* $\sin\omega(t-T)\,u(t-T)$

FRAME 29

The unit step can be used to find the current i at time t in a circuit such as that described in FRAME 13. The d.e.'s describing the circuit were there given as

$$L\frac{di}{dt} + Ri + \frac{1}{C}\int_0^t i\,dt = E \qquad 0 < t < 1$$

$$L\frac{di}{dt} + Ri + \frac{1}{C}\int_0^t i\,dt = 0 \qquad t > 1$$

Using the unit step function these can be combined in the one equation

$$L\frac{di}{dt} + Ri + \frac{1}{C}\int_0^t i\,dt = Eu(t) - Eu(t-1)$$

For the values of R,L,C and E given in FRAME 6 this becomes

$$\frac{di}{dt} + 250i + 10^4\int_0^t i\,dt = 10u(t) - 10u(t-1)$$

What will be the transformed equation for this d.e., the current being initially zero?

****

29A

$$s\bar{i} + 250\bar{i} + 10^4\frac{\bar{i}}{s} = \frac{10}{s} - \frac{10}{s}e^{-s}$$

FRAME 30

From this equation, $\bar{i} = \dfrac{10}{s^2 + 250s + 10\,000} - \dfrac{10e^{-s}}{s^2 + 250s + 10\,000}$

The inverse of the first term on the R.H.S. has already been found (in 6A) to be $\frac{1}{15}\left(e^{-50t} - e^{-200t}\right)$. Can you now write down the complete expression for i?

****

30A

$$i = \frac{1}{15}\left(e^{-50t} - e^{-200t}\right)u(t) - \{e^{-50(t-1)} - e^{-200(t-1)}\}u(t-1)$$

You will probably have realised by now that in earlier work, an expression such as $\frac{1}{15}\left(e^{-50t} - e^{-200t}\right)$ *has been used when* $\frac{1}{15}\left(e^{-50t} - e^{-200t}\right)u(t)$ *has really been meant. Where there is no possibility of ambiguity, the former notation can, of course, still be used.*

FRAME 31

Turning next to the beam described in FRAME 14, you will now see that to obtain the deflection y in terms of x we have to solve

$$EI\frac{d^4y}{dx^4} = wu(x - x_1) - wu(x - x_2) \qquad (31.1)$$

subject to the conditions that when $x = 0$, $y = 0$ and $\frac{dy}{dx} = 0$, and when $x = \ell$, $y = 0$ and $\frac{d^2y}{dx^2} = 0$.

You have already written down (in 25A) the transform of the R.H.S. and so, for the complete equation,

$$EI(s^4\bar{y} - s^3y_0 - s^2y_1 - sy_2 - y_3) = \frac{w}{s}e^{-sx_1} - \frac{w}{s}e^{-sx_2}$$

where y_0 is the value of y when $x = 0$, y_1 that of $\left(\frac{dy}{dx}\right)_{x=0}$ and so on.

Now $y_0 = y_1 = 0$ but y_2 and y_3 are unknown

$$\therefore EI\,s^4\bar{y} = \frac{w}{s}e^{-sx_1} - \frac{w}{s}e^{-sx_2} + EI(sy_2 + y_3)$$

or $$\bar{y} = \frac{w}{EIs^5}e^{-sx_1} - \frac{w}{EIs^5}e^{-sx_2} + \frac{y_2}{s^3} + \frac{y_3}{s^4}$$

What will be the inverse of this equation?

****

31A

$$y = \frac{w}{24EI}(x - x_1)^4 u(x - x_1) - \frac{w}{24EI}(x - x_2)^4 u(x - x_2) + \frac{1}{2} y_2 x^2 + \frac{1}{6} y_3 x^3 \quad (31A.1)$$

FRAME 32

When $x = \ell$, $y = 0$. What will (31A.1) become when these values are substituted into it?

****

32A

$$0 = \frac{w}{24EI}(\ell - x_1)^4 - \frac{w}{24EI}(\ell - x_2)^4 + \frac{1}{2} y_2 \ell^2 + \frac{1}{6} y_3 \ell^3 \qquad (32A.1)$$

When $x = \ell$, both $u(\ell - x_1)$ and $u(\ell - x_2)$ are, of course, unity.

FRAME 33

To use the remaining condition, i.e. $\frac{d^2y}{dx^2} = 0$ when $x = \ell$, it is necessary to differentiate (31A.1) twice. So far, an expression involving a u function has not been differentiated. But give it a moment's thought and see if you can decide what will happen. Then try writing down first $\frac{dy}{dx}$ and then $\frac{d^2y}{dx^2}$ from (31A.1).

****

33A

As the u function is a constant, it will have the same effect in differentiation as any other constant, even though its value changes abruptly at one point.

Thus, for example,

$$\frac{d}{dt} tu(t) = u(t)$$

$$\frac{d}{dt} \sin(t - T)\, u(t - T) = \cos(t - T)\, u(t - T)$$

The first of these equations is simply stating that the slope of the function shown in Fig (2), FRAME 16, is zero when $t < 0$ and 1 when $t > 0$.

From (31A.1)

$$\frac{dy}{dx} = \frac{w}{6EI}(x - x_1)^3 u(x - x_1) - \frac{w}{6EI}(x - x_2)^3 u(x - x_2) + y_2 x + \frac{1}{2} y_3 x^2$$

$$\frac{d^2y}{dx^2} = \frac{w}{2EI}(x - x_1)^2 u(x - x_1) - \frac{w}{2EI}(x - x_2)^2 u(x - x_2) + y_2 + y_3 x$$

FRAME 34

Then, as $\frac{d^2y}{dx^2} = 0$ when $x = \ell$,

$$0 = \frac{w}{2EI}(\ell - x_1)^2 - \frac{w}{2EI}(\ell - x_2)^2 + y_2 + y_3\ell \qquad (34.1)$$

The two equations (32A.1) and (34.1) are now sufficient to determine y_2 and y_3. As this is just pure algebra, the solution will not lead to any greater understanding of unit functions and Laplace transforms and so will not be pursued here.

FRAME 35

The Impulse Function

The total load on the beam in the problem just worked out is $w(x_2 - x_1)$. Let this be W. Now suppose we let x_2 decrease (but still keeping $x_2 > x_1$) and at the same time increase w so that the total load remains the same, i.e. W, but is spread over a shorter length of the beam. Obviously (31A.1) will still apply. Continuing this process, as $x_2 - x_1$ gets very small, w will have to increase greatly in order to keep the product $w(x_2 - x_1)$ the same. Theoretically now let $x_2 \to x_1$ and $w \to \infty$ in such a way that

$$\lim_{\substack{x_2 \to x_1 \\ w \to \infty}} w(x_2 - x_1) = W$$

Practically the result is a loading of very high intensity over a very short distance and is regarded as a concentrated load W at the point x_1. The question then arises: What effect will this have on (31.1) and (31A.1)? You might think that if $x_2 \to x_1$ then the R.H.S. of (31.1) will become zero and that of (31A.1) will become $\frac{1}{2}y_2x^2 + \frac{1}{6}y_3x^3$. But if you do this, it is equivalent to saying that the load W will have no effect whatsoever, which obviously just isn't so.

FRAME 36

Before the method for dealing with this problem is considered, two other cases will be mentioned where, although the branches of science in which they occur are different, the mathematical representation is effectively the same.

FRAME 36 continued

In the first case, returning to the L,R,C circuit of FRAME 13, suppose the generator applies an extremely large voltage for a very short time, i.e. suppose an electrical impulse is applied to the circuit. If this voltage is E and it is applied from $t = 0$ to $t = \varepsilon$, the circuit equation becomes

$$L\frac{di}{dt} + Ri + \frac{1}{C}\int_0^t i\,dt = Eu(t) - Eu(t - \varepsilon)$$

The question that has to be answered this time is: What happens when $\varepsilon \to 0$ and $E \to \infty$?

Secondly, taking a mechanical situation, what happens to a mass m if an impulsive force P is applied to it, the mass being at rest at $x = 0$ before the impulse is applied? If you have studied dynamics you probably know the answer to this, but, as will be seen later, a method involving Laplace transforms can also be used for its solution.

FRAME 37

In order to deal with a circuit to which a voltage is applied over a finite time or a beam where a load is applied over a finite distance, the unit step function was introduced. Furthermore, you saw how the Laplace transform of a function was modified if the graph of the function was displaced a distance to the right. In order to cope with impulses, another new function is introduced

A start is made with the function shown below:

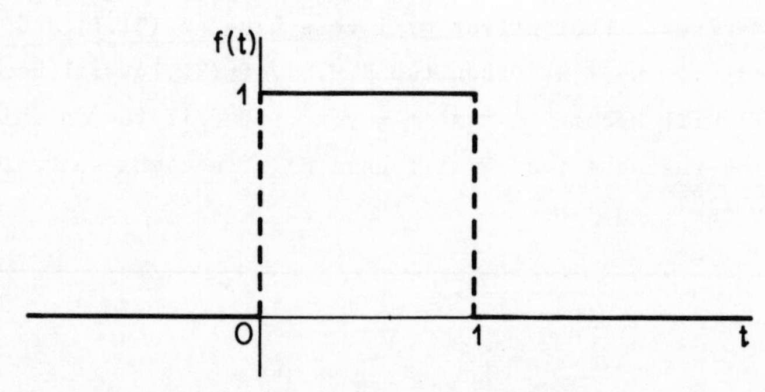

What will be the formula giving f(t) for this function and what will be its Laplace transform?

****

37A

$f(t) = u(t) - u(t - 1);$ $\frac{1}{s} - \frac{1}{s}e^{-s}$

FRAME 38

You will notice that the area enclosed between the curve and the t-axis in the figure in the last frame is 1. Following the idea suggested in FRAME 35 this square is now deformed into a rectangle which gradually gets narrower and narrower but still retains area 1. In order for this to happen, its height must, of course, increase. The following diagrams show some such rectangles and the corresponding f(t) functions.

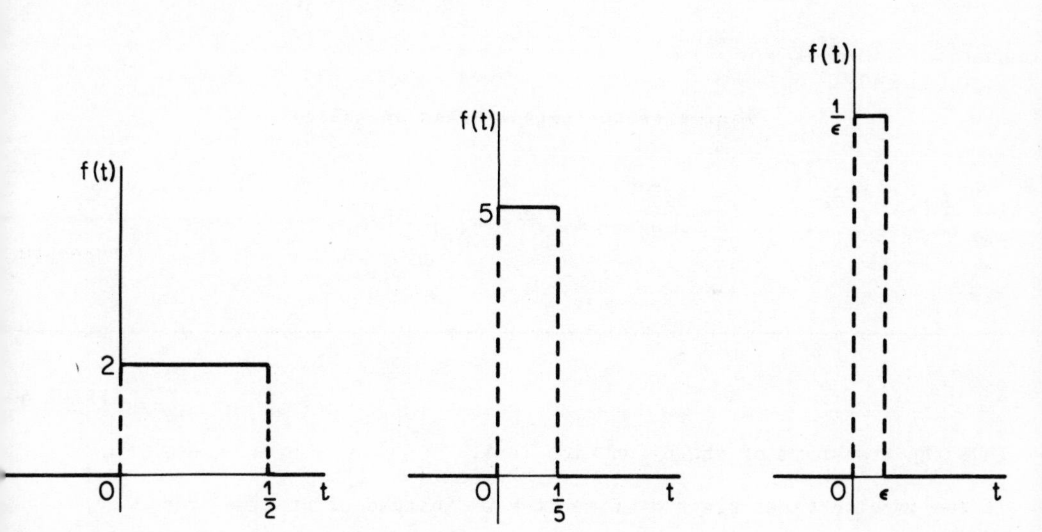

Write down the formula for f(t) in the last of these three figures and state its Laplace transform.

****

38A

$f(t) = \frac{1}{\varepsilon}\{u(t) - u(t - \varepsilon)\};$ $\frac{1}{\varepsilon s}\left(1 - e^{-s\varepsilon}\right)$

FRAME 39

Now let ε get very small. $\frac{1}{\varepsilon}$ will then get very large. We define the DIRAC δ FUNCTION or UNIT IMPULSE FUNCTION as

$$\delta(t) = \begin{cases} 0 & \text{for} \quad t < 0 \\ \frac{1}{\varepsilon} & \text{for} \quad 0 < t < \varepsilon \\ 0 & \text{for} \quad t > \varepsilon \end{cases}$$

As the area of the rectangle is still 1, $\int_{-\infty}^{\infty} \delta(t)dt = 1$.

Theoretically we now let $\varepsilon \to 0$ and consequently $\frac{1}{\varepsilon} \to \infty$. In practice ε simply becomes extremely small as, for example, the time taken by the blow of a hammer.

What is $\lim\limits_{\varepsilon \to 0} \frac{1}{\varepsilon s}\left(1 - e^{-s\varepsilon}\right)$?

****

39A

$$\lim_{\varepsilon \to 0} \frac{1}{\varepsilon s}\left(1 - e^{-s\varepsilon}\right) = \lim_{\varepsilon \to 0} \frac{se^{-s\varepsilon}}{s} \quad \textit{by l'Hopital's rule}$$

$$= 1$$

FRAME 40

Thus the transform of the δ function is 1.

If the impulse takes place at time $t = T$ instead of at time $t = 0$ (practically from $t = T$ to $t = T + \varepsilon$) the corresponding δ function is then denoted by $\delta(t - T)$ and is defined by

$$\delta(t - T) = \begin{cases} 0 & \text{for} \quad t < T \\ \frac{1}{\varepsilon} & \text{for} \quad T < t < T + \varepsilon \\ 0 & \text{for} \quad t > T + \varepsilon \end{cases}$$

where again, theoretically, $\varepsilon \to 0$.

What will be the transform of this function?

****

40A

e^{-sT}, *as the displacement of a curve through a distance T to the right multiplies its transform by* e^{-sT}.

FRAME 41

The situation described in FRAME 35 was commenced by putting $w(x_2 - x_1) = W$ and then letting $x_2 \to x_1$ and $w \to \infty$ in such a way that $\lim_{\substack{x_2 \to x_1 \\ w \to \infty}} w(x_2 - x_1) = W$.

The area under the w-curve from $-\infty$ to $+\infty$ is therefore to be W, i.e. $\int_{-\infty}^{\infty} w\,dx = W$. The δ function (the area under whose graph, you will remember, is 1) is then used in conjunction with W to denote a concentrated load W at the point x_1 by writing

$$W\delta(x - x_1)$$

Similarly, an impulsive voltage E_o applied to a circuit at time $t = 0$ is denoted by $E_o\delta(t)$ and an impulsive force P applied to a mechanical system at time $t = 0$ by $P\delta(t)$.

What will be the transforms of (i) $W\delta(x - x_1)$, (ii) $E_o\delta(t)$, (iii) $P\delta(t)$?

****

41A

(i) We^{-sx_1} *(ii)* E_o *(iii)* P

FRAME 42

We are now in a position to deal with the situations described in FRAMES 35 and 36.

Taking first the mechanical problem in FRAME 36, the equation of motion can be written as

$$m\ddot{x} = P\delta(t)$$

assuming that no force other than the impulse acts on the mass m. As the particle is at rest at $x = 0$ before the impulse is applied, the transformed

FRAME 42 continued

equation is

$$ms^2\bar{x} = P$$

from which

$$\bar{x} = \frac{P}{m}\frac{1}{s^2}$$

and so

$$x = \frac{P}{m}t, \quad \text{or to be more precise} \quad x = \frac{P}{m}tu(t)$$

as $x = 0$ for $t < 0$.

If you have previously studied impulsive motion in dynamics, you will remember that if an impulse P is applied to a mass m, it imparts to it a velocity P/m. If no other force acts, the distance thus travelled in time t is $(P/m)t$ as the velocity remains constant. This problem can therefore be solved quite easily without the use of the δ function or Laplace transforms, but it was used here to give a simple illustration of the method.

You will remember that it was mentioned earlier that in the majority of cases the distinction between $t = 0$, $t = 0+$ and $t = 0-$ does not affect the solution of a problem. This is one instance where, from the way in which the problem has been tackled, it is important as the initial value of the velocity must be taken as zero, i.e. its value at $t = 0-$, not P/m which is its value at $t = 0+$.

FRAME 43

In the electrical problem mentioned in FRAME 36, if an impulsive voltage E_o is applied at time $t = 0$ to an R,C,L series circuit, the d.e. is then

$$L\frac{di}{dt} + Ri + \frac{1}{C}\int_0^t i\,dt = E_o\delta(t)$$

Taking, as before, $L = 1$, $R = 250$, $C = 10^{-4}$ and putting $E_o = 1000$, the transformed equation is

$$s\bar{i} + 250\bar{i} + 10^4\frac{\bar{i}}{s} = 1000$$

assuming, as before, no current or charge before the impulse is applied.

Solve this equation for $\bar{i}$ and hence find i.

****

43A

$$\bar{i} = \frac{1000s}{s^2 + 250s + 10\,000} = \frac{4000}{3}\frac{1}{s + 200} - \frac{1000}{3}\frac{1}{s + 50}$$

$$i = \frac{4000}{3}e^{-200t} - \frac{1000}{3}e^{-50t},$$ *or to be more precise,*

$$i = \frac{4000}{3}e^{-200t}u(t) - \frac{1000}{3}e^{-50t}u(t) = \frac{1000}{3}(4e^{-200t} - e^{-50t})u(t)$$

FRAME 44

Finally, the problem of finding the deflection of the beam in FRAME 35 will now be solved. The equation for a beam loaded as shown is

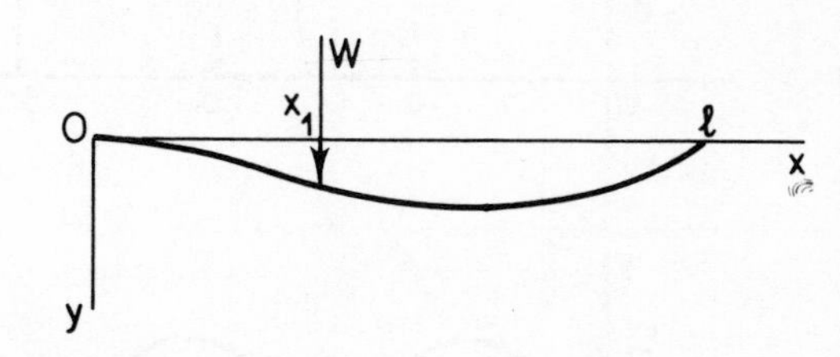

$$EI\frac{d^4y}{dx^4} = W\delta(x - x_1)$$

Assuming as before that the beam is clamped horizontally at $x = 0$ and is freely supported at $x = \ell$, write down the transformed equation. Then obtain $\bar{y}$ and hence y. Your equation for y will still involve y_2 and y_3. Using the fact that $y = 0$ and $\frac{d^2y}{dx^2} = 0$ when $x = \ell$, what will be the two equations for y_2 and y_3?

****

44A

$$EI(s^4\bar{y} - sy_2 - y_3) = We^{-sx_1}$$

$$\bar{y} = \frac{y_2}{s^3} + \frac{y_3}{s^4} + \frac{We^{-sx_1}}{EIs^4}$$

$$y = \frac{1}{2}y_2x^2 + \frac{1}{6}y_3x^3 + \frac{1}{6}\frac{W}{EI}(x - x_1)^3u(x - x_1)$$

$$y'' = y_2 + y_3x + \frac{W}{EI}(x - x_1)u(x - x_1)$$

$$0 = 3y_2\ell^2 + y_3\ell^3 + \frac{W}{EI}(\ell - x_1)^3$$

$$0 = y_2 + y_3\ell + \frac{W}{EI}(\ell - x_1)$$

Once again, no further insight into the operation of the method is gained by the algebraic solution of these equations for y_2 and y_3.

FRAME 45

Periodic Functions

You have already met some periodic functions when dealing with Laplace transforms, e.g. $\sin \omega t$ and $\cos \omega t$. There are, of course, other periodic functions, some being illustrated below.

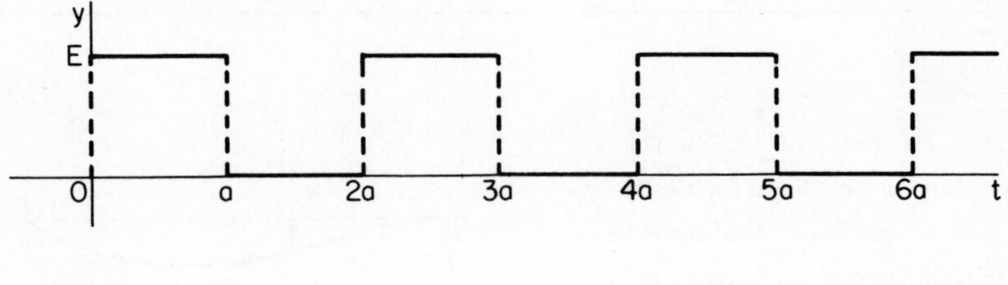

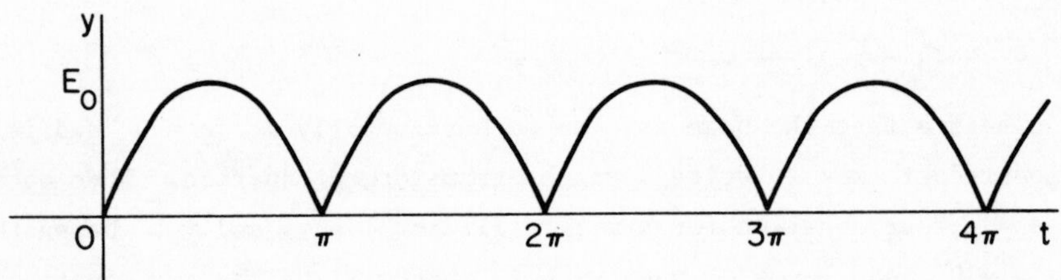

The first of these is a square wave function. The second is a rectified sine wave and its equation is $y = E_0|\sin t|$.

The square wave function is a very simple periodic function to deal with as regards finding its transform. We shall first of all find its transform directly, and then consider the general periodic function before going on to the rectified sine wave.

Write down a single equation which will describe this square wave function, and then deduce its Laplace transform.

****

45A

$$y = E\left\{u(t) - u(t-a) + u(t-2a) - u(t-3a) + \ldots\right\}$$

$$\bar{y} = \frac{E}{s}\left\{1 - e^{-sa} + e^{-2sa} - e^{-3sa} + \ldots\ldots\right\} \qquad (45A.1)$$

FRAME 46

The R.H.S. of (45A.1) is an infinite geometric progression whose first term is E/s and whose common ratio is $-e^{-sa}$. This series can therefore be summed to give

$$\bar{y} = \frac{E}{s}\frac{1}{1 + e^{-sa}}$$

Now, looking at a general periodic function, this can be represented by a graph of the form

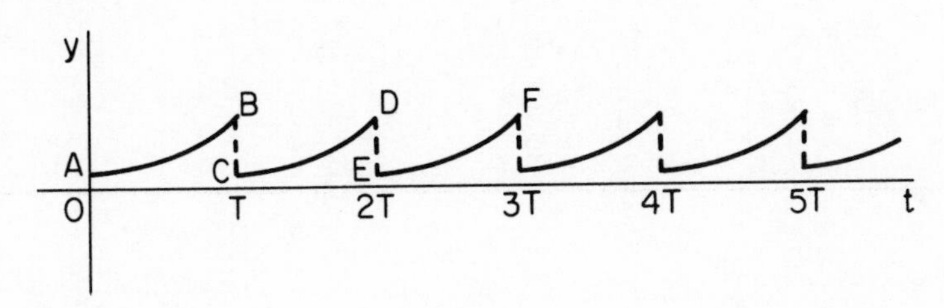

Let the equation of AB be $y = f(t) \quad 0 < t < T$. Can you write down the equations for the sections CD and EF?

****

46A

For CD: $y = f(t - T) \quad T < t < 2T$

For EF: $y = f(t - 2T) \quad 2T < t < 3T$

The equations of the other sections follow a similar pattern.

FRAME 47

In this case, in order to find $\bar{y}$, we shall make use of the original definition of the Laplace transform,

i.e. $$\bar{y} = \int_0^\infty e^{-st} y \, dt$$

As the equation giving y changes at T, 2T, 3T etc., this integral is split up into the ranges 0 - T, T - 2T, 2T - 3T, etc. Then

$$\bar{y} = \int_0^T e^{-st} f(t)dt + \int_T^{2T} e^{-st} f(t - T)dt + \int_{2T}^{3T} e^{-st} f(t - 2T)dt + \ldots \qquad (46.1)$$

As before, there is still an infinite series on the R.H.S. To put this into a

FRAME 47 continued

form which can be summed, each integral after the first is expressed in terms of the first one. This can be done quite simply by means of a substitution.

Now what does $\int_T^{2T} e^{-st} f(t - T)dt$ become if you make the substitution $r = t - T$?

****

47A

$$\int_0^T e^{-s(r+T)} f(r)dr$$

FRAME 48

This integral can be rewritten as $e^{-sT}\int_0^T e^{-sr} f(r)dr$ as e^{-sT} is a constant.

As it is a definite integral,

$$\int_0^T e^{-sr} f(r)dr = \int_0^T e^{-st} f(t)dt$$

Taking next the third integral on the R.H.S. of (47.1), find what this will become when expressed in terms of $\int_0^T e^{-st} f(t)dt$

****

48A

Putting $r = t - 2T$ *leads to* $e^{-2sT}\int_0^T e^{-st} f(t)dt$

FRAME 49

Proceeding similarly with the other integrals on the R.H.S. of (47.1) gives

$$\bar{y} = \left(1 + e^{-sT} + e^{-2sT} + e^{-3sT} + \ldots\right)\int_0^T e^{-st} f(t)dt$$

$$= \frac{1}{1 - e^{-sT}}\int_0^T e^{-st} f(t)dt$$

FRAME 49 continued

Verify that this formula gives the same result as that obtained in FRAME 46 for the square wave and then obtain the transform of the rectified sine wave in FRAME 45.

****

49A

For the square wave, $T = 2a$ *and thus*

$$\int_0^T e^{-st} f(t)dt = \int_0^a e^{-st} E dt + \int_a^{2a} e^{-st} 0 dt$$

$$= \frac{E}{s}\left(1 - e^{-sa}\right)$$

$$\bar{y} = \frac{1}{1 - e^{-2sa}} \frac{E}{s}\left(1 - e^{-sa}\right)$$

$$= \frac{E}{s} \frac{1}{1 + e^{-sa}}$$

For the rectified sine wave, the equation of the function from 0 to T is $y = E_o \sin t$ *(*$T = \pi$ *here).*

$$\int_0^T e^{-st} f(t)dt = \int_0^\pi e^{-st} E_o \sin t \, dt$$

$$= E_o \frac{1}{1 + s^2}\left(1 + e^{-s\pi}\right)$$

$$\bar{y} = \frac{1}{1 - e^{-s\pi}} E_o \frac{1}{1 + s^2}\left(1 + e^{-s\pi}\right)$$

This can be further simplified to $\dfrac{E_o}{1 + s^2} \dfrac{e^{\frac{1}{2}s\pi} + e^{-\frac{1}{2}s\pi}}{e^{\frac{1}{2}s\pi} - e^{-\frac{1}{2}s\pi}}$

$$= \frac{E_o}{1 + s^2} \coth \frac{1}{2} s\pi$$

Examples

Now try these examples to make sure that you have understood the programme so far. Then turn to FRAME 51 to check your answers.

1. A voltage $E \sin \omega t$ is applied from time $t = 0$ to a circuit consisting of an inductance L and a capacitance C in series. At $t = 0$ both the charge on the capacitor and the current are zero. The current i satisfies the equation

$$L \frac{di}{dt} + \frac{1}{C}\int_0^t i\,dt = E \sin \omega t$$

Solve this equation for i, if $\omega^2 = 1/LC$.

2. Use entry 13 in the table to obtain entry 15.

3. A pulse y(t) is given by

$$y(t) = E \sin \omega t \qquad 0 < t < \pi/\omega$$
$$y(t) = 0 \qquad t > \pi/\omega$$

Use the unit step function to obtain $\mathcal{L}\{y(t)\}$.
(The function y is similar to that in (ii) in FRAME 21 and there are two possible ways of representing it. Choose the one which is more appropriate when finding the Laplace transform.)

4. A waveform is given by

$$f(t) = k(t - nT) \qquad nT < t < (n + 1)T, \text{ when } n = 0,1,2,\ldots$$

Sketch the waveform and find its Laplace transform.

5. For a certain beam with specified loading the bending moment M is given by

$$M = -2Wx/3 \qquad 0 < x < \ell/3$$
$$M = W(x - \ell/3) - 2Wx/3 \qquad x > \ell/3$$

Find the Laplace transform of M.

FRAME 50 continued

6. A voltage $E_o\delta(t)$ is applied to a circuit consisting of an inductance L and a capacitance C in series. Before the voltage is applied, both the charge on the capacitor and the current are zero. The current i satisfies the equation

$$L\frac{di}{dt} + \frac{1}{C}\int_0^t i\,dt = E_o\delta(t)$$

Solve this equation for i.

FRAME 51

Answers to Examples

1. $i = (E/2L)t \sin \omega t$

2. Use $\mathcal{L}\{tf(t)\} = -\frac{d}{ds}F(s)$ with $f(t) = \cos \omega t$ and $F(s) = \frac{s}{s^2 + \omega^2}$.

3. $y(t) = E \sin \omega t + E \sin \omega(t - \pi/\omega)\, u(t - \pi/\omega)$

$$\mathcal{L}\{y(t)\} = \frac{E\omega}{s^2 + \omega^2}\left(1 + e^{-\pi s/\omega}\right)$$

4. $\mathcal{L}\{f(t)\} = \frac{1}{1 - e^{-sT}}\int_0^T e^{-st}kt\,dt$

$$= \frac{k}{s}\left(\frac{1}{s} - \frac{Te^{-sT}}{1 - e^{-sT}}\right)$$

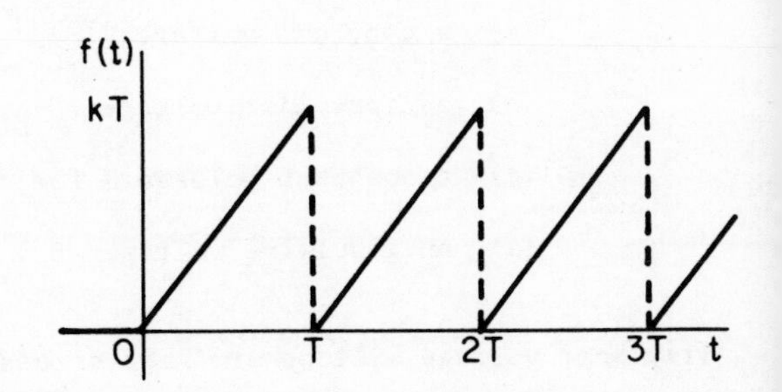

5. $M = W(x - \ell/3)u(x - \ell/3) - 2Wx/3$

$$\mathcal{L}\{M\} = \frac{We^{-\ell s/3}}{s^2} - \frac{2W}{3}\frac{1}{s^2}$$

FRAME 51 continued

6. $i = \frac{E_o}{L} \cos \frac{t}{\sqrt{LC}}$

This is another example where, when considering initial conditions, the distinction between $t = 0-$ and $t = 0+$ is important. Here we have an electrical impulse just as in FRAME 42 there was a mechanical impulse.

FRAME 52

Transfer Functions

If you have read the programme on Complex Integration in Unit 12, you will have already met the idea of a transfer function. However, in case Complex Integration is not part of your course, we shall here start from scratch as far as this topic is concerned.

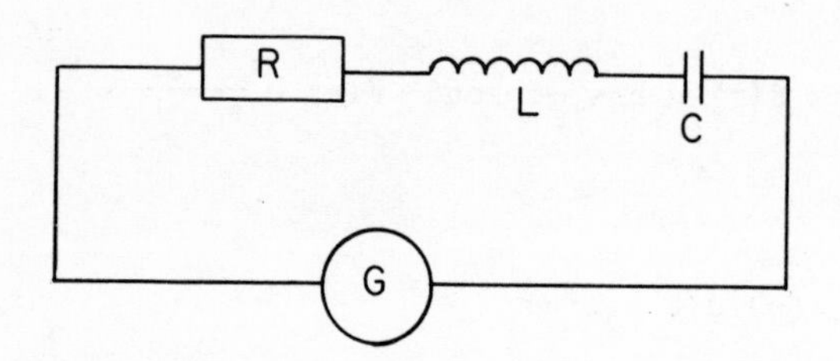

Taking the circuit shown, there are various questions that might be asked about it, for example:

(i) What will be the current i flowing in the circuit if the initial current and charge are zero and, at time $t = 0$, the generator supplies

(a) a constant voltage E,

(b) a sinusoidal voltage $E \sin \omega t$,

(c) a constant voltage E for one second only,

(d) an impulsive voltage E_o?

(ii) What voltage will be produced across the capacitor in each of the four cases in (i) if the same initial conditions apply?

FRAME 53

ooking first at question (i) the circuit equation in each of the four cases ill be

$$\text{(a)}\quad L\frac{di}{dt} + Ri + \frac{1}{C}\int_0^t idt = Eu(t)$$

$$\text{(b)}\quad L\frac{di}{dt} + Ri + \frac{1}{C}\int_0^t idt = E\sin\omega t\,u(t)$$

$$\text{(c)}\quad L\frac{di}{dt} + Ri + \frac{1}{C}\int_0^t idt = E\{u(t) - u(t-1)\}$$

$$\text{(d)}\quad L\frac{di}{dt} + Ri + \frac{1}{C}\int_0^t idt = E_o\delta(t)$$

he work that has been done so far should enable you to write down the trans-orm of each of the equations (a) to (d).

****

53A

a) $\left(Ls + R + \frac{1}{Cs}\right)\bar{i} = \frac{E}{s}$ (b) $\left(Ls + R + \frac{1}{Cs}\right)\bar{i} = \frac{E\omega}{s^2 + \omega^2}$

c) $\left(Ls + R + \frac{1}{Cs}\right)\bar{i} = \frac{E}{s}\left(1 - e^{-s}\right)$ (d) $\left(Ls + R + \frac{1}{Cs}\right)\bar{i} = E_o$

FRAME 54

hese then give, respectively,

$= \frac{s}{Ls^2 + Rs + 1/C}\cdot\frac{E}{s}$ $\bar{i} = \frac{s}{Ls^2 + Rs + 1/C}\cdot\frac{E\omega}{s^2 + \omega^2}$

$= \frac{s}{Ls^2 + Rs + 1/C}\cdot\frac{E}{s}\left(1 - e^{-s}\right)$ $\bar{i} = \frac{s}{Ls^2 + Rs + 1/C}\cdot E_o$

can therefore be regarded as made up of a product of two parts, $\frac{s}{Ls^2 + Rs + 1/C}$ nd the transform of the applied voltage. The first of these occurs in each ase and is not affected by the applied voltage. It would, however, change if he components of the circuit were altered. In view of this, it can be egarded as an expression which describes the actual circuit itself. For this

FRAME 54 continue

particular circuit, then, it is possible to write

$$\bar{i} = \frac{s}{Ls^2 + Rs + 1/C} \cdot \bar{E}_1$$

or

$$\frac{\bar{i}}{\bar{E}_1} = \frac{s}{Ls^2 + Rs + 1/C}$$

where E_1 is the input voltage, whatever that may be. The expression $\frac{s}{Ls^2 + Rs + 1/C}$ is an example of what is called a TRANSFER FUNCTION. In this case, i can be regarded as the response or output of the circuit to the applied voltage or input. In general, the transfer function of a system is defined as the ratio

$$\frac{\text{Transform of Output}}{\text{Transform of Input}}$$

FRAME 5

Turning now to question (ii) posed in FRAME 52, the voltage across the capacitor is given by $\frac{1}{C}\int_0^t idt = e_1$, say. If this is now regarded as the output, the transfer function for an input E_1 will be $\bar{e}_1/\bar{E}_1$. What expression will give this function now?

****

55

$$\bar{e}_1 = \frac{1}{Cs}\bar{i} = \frac{1}{Cs}\frac{s}{Ls^2 + Rs + 1/C}\bar{E}_1$$

Transfer function is $\frac{1}{LCs^2 + RCs + 1}$

FRAME 5

You will see, therefore, that once the transfer function for a system is know the output transform can immediately be written down for any given input, provided the input transform is known or can be found.

A similar idea also occurs in mechanical systems. For example, suppose a mas m is attached to one end of a horizontal spring of stiffness k and, from time

FRAME 56 continued

t = 0, a force F is applied to the mass as shown. The other end of the spring is fixed. Initially the mass is at rest and the spring is unstretched. If x is the displacement of the mass at time t, then

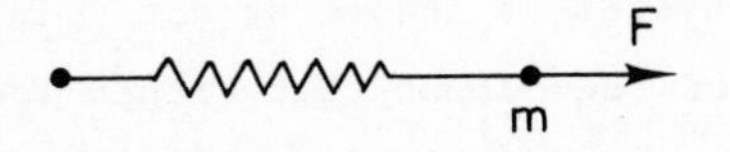

$$m\ddot{x} = F - kx$$

or $$m\ddot{x} + kx = F.$$

In this case F is regarded as the input and x as the output. What will be

(i) the transformed equation,

(ii) the transfer function for this system?

****

56A

(i) $ms^2\bar{x} + k\bar{x} = \bar{F}$ (ii) $\dfrac{1}{ms^2 + k}$

FRAME 57

What will $\bar{x}$ be if (i) $F = F_o$, a constant

(ii) $F = F_o \sin \omega t$

(iii) $F = F_o\delta(t)$?

****

57A

(i) $\dfrac{1}{ms^2 + k} \cdot \dfrac{F_o}{s}$ (ii) $\dfrac{1}{ms^2 + k} \cdot \dfrac{F_o\omega}{s^2 + \omega^2}$ (iii) $\dfrac{1}{ms^2 + k} \cdot F_o$

FRAME 58

The idea of a transfer function has been introduced by considering two very simple systems, one electrical and the other mechanical. The concept of a system is a much wider one, and those which occur in present day technology are more complicated as, for instance, in the automatic control of chemical and other processes.

FRAME 59

s as a Complex Number. Stability

As was mentioned in the Complex Integration programme in Unit 12, it is often necessary in control theory to know if a system is stable or not. It is unstable if terms exist in the solution for the output that either increase without limit or oscillate with ever increasing amplitude. Thus terms of the form e^t, t^2, $t \sin nt$ in the solution would lead to an unstable system, but those of the form e^{-t}, $\sin nt$, $te^{-2t} \cos 3t$ would indicate a stable system.

Now, as you will realise if you look at the table of transforms, there is, to some extent, a pattern in the form of the transforms of various functions. Thus, for example, if the transform of a function involving a sine or cosine is taken, the denominator of the transform always contains the sum of two squares. This suggests that it might be possible to determine the stability or otherwise of a function by considering the form of its output transform.

FRAME 60

To start with, look down the table of transforms and write down the values of s which make the transform infinite, i.e. the denominator of the transform zero, in the cases of transforms numbers 3, 5, 6, 7, 9, 10, 12, 16, 18, 20, 21, 22. Include all values, real, imaginary and complex and indicate those where the same value occurs more than once.

****

60A

(3) 0 *twice*	*(5)* $-\alpha$	*(6)* $0, -\alpha$	*(7)* $-\alpha$ *twice*
(9) $-\alpha, -\beta$	*(10)* $\pm j\omega$	*(12)* $\pm j\omega$ *twice*	*(16)* $0, \pm j\omega$
(18) $-\alpha \pm j\omega$	*(20)* $-\alpha \pm j\omega$	*(21)* $\pm\omega$	*(22)* $\pm\omega$

j has been used instead of i as this work is mainly concerned with electrical circuits.

FRAME 61

It was mentioned in Unit 5 that s can be real or complex. So far, except for the work that was done on Laplace transforms in Unit 12, it has not been found necessary to bother about this distinction. However, the results in 60A suggest that sometimes it may be necessary to be aware that s can be complex. When this is the case, we usually write

$$s = \sigma + j\omega$$

(In Unit 12, so as not to cause confusion with the work being done there, the notation $s = x + iy$ was used.)

Now, a d.e. for a circuit can either be solved by trial/D-operator methods or by Laplace transform methods. If either of the former is used, the working is done entirely in terms of the variable t, the time. If the latter is used, the actual working is done in terms of s. When the working is in terms of t, we say that it is being done in the TIME DOMAIN. But when s is used, we are then said to be in the COMPLEX FREQUENCY DOMAIN. If you go back to the answers in 60A which contain $j\omega$ you will see, on reference to the transform table, that the corresponding inverse functions all contain $\sin \omega t$ or $\cos \omega t$ and ω is, of course, $2\pi \times$ frequency.

FRAME 62

A further link is seen when the impedance of a circuit is written down. Taking the example in FRAME 52 (i), the complex impedance of the circuit is $R + j\omega L + \frac{1}{j\omega C}$. You will notice that, in 53A, the coefficient of $\bar{i}$ is $R + Ls + \frac{1}{Cs}$ which is effectively the impedance with $j\omega$ replaced by s.

Returning now to the question of stability, take the transform pairs listed in FRAME 60 and state whether the time function in each case is stable or unstable, assuming α, β and ω to be positive.

****

62A

5, 6, 7, 9, 10, 16, 18, 20 are stable.

3, 12, 21, 22 are unstable.

FRAME 63

It is possible to plot, on an Argand diagram, the points in the s-plane where the denominator of a transform is zero. If you have read Unit 12, you will know that these are called the POLES of the function. The location of these poles gives information about the stability of the system. This information can be summarised as follows.

If all poles occur in the left hand half of the Argand diagram, the system is stable.

If any pole occurs in the right hand half of the Argand diagram, the system is unstable.

If a pole occurs on the imaginary axis of the Argand diagram, the system is unstable if the same pole occurs more than once, but is otherwise stable.

Thus, for example, taking the transform $\dfrac{\omega \cos \phi + s \sin \phi}{s^2 + \omega^2}$, the denominator is zero when $s = \pm j\omega$. These poles are both on the imaginary axis, and each occurs only once, and so this transform relates to a stable system.

Now look back at the values in 60A and check that the values given there, taken in conjunction with the statements just quoted, agree with the results you obtained in 62A. Then examine the transformed functions in entries 8, 14, 15, 17, 19, and say which of these will relate to stable systems.

****

63A

8, 14, 19

FRAME 64

There is just one further point to be mentioned when looking at transformed functions for system stability. You will remember that

Transform of Output = Transfer Function × Transform of Input

As the input to the system can be expected to behave itself, i.e. not increase to infinity, it is really only necessary to consider the transfer function when looking at stability.

FRAME 65

The Convolution Theorem

The last property of Laplace transforms that will be looked at is what is known as convolution. Having found the transform of the output, it is necessary to obtain its inverse in order to find the output itself. This means that

$$\mathcal{L}^{-1}\{\text{Transfer Function} \times \text{Transform of Input}\} \qquad (65.1)$$

is required. The question that the convolution theorem seeks to answer is this:

> Given two functions of s, $F(s)$ and $G(s)$, whose inverses are known, is it possible, from these inverses, to find the inverse of the product $F(s)G(s)$? That is, can $\mathcal{L}^{-1}\{F(s)G(s)\}$ be found if $\mathcal{L}^{-1}\{F(s)\}$ and $\mathcal{L}^{-1}\{G(s)\}$ are known?

As (65.1) is the inverse of a product, you will immediately see the relevance of this question.

The answer is in the affirmative and the CONVOLUTION THEOREM states:

If $\mathcal{L}^{-1}\{F(s)\} = f(t)$ and $\mathcal{L}^{-1}\{G(s)\} = g(t)$, then

$$\mathcal{L}^{-1}\{F(s)G(s)\} = \int_0^t f(\tau)g(t-\tau)\,d\tau \qquad (65.2)$$

The integral on the R.H.S. of this equation is called the CONVOLUTION of $f(t)$ and $g(t)$ and is denoted by $f*g$.

FRAME 66

Before going on to the proof of the theorem, we shall consider a simple example of it, to see how it actually works. Taking $\frac{1}{s(s+\alpha)}$, this can be regarded as the product of $\frac{1}{s}$ and $\frac{1}{s+\alpha}$, so let $F(s) = \frac{1}{s}$ and $G(s) = \frac{1}{s+\alpha}$. Then, from entries 2 and 5 in the table $f(t) = 1$ and $g(t) = e^{-\alpha t}$. (The actual entry 2 in the table gives $\mathcal{L}^{-1}\{\frac{1}{s}\} = u(t)$, but as the convolution integral is only taken between 0 and t, f(t) can be taken as 1 as $u(t) = 1$ between these values.)

What will be $g(t - \tau)$ and hence the integral corresponding to the R.H.S. of (65.2) in this case?

****

66A

$e^{-\alpha(t-\tau)}$ $\int_0^t e^{-\alpha(t-\tau)}\,d\tau$

FRAME 67

$$\text{Thus,}\quad \mathcal{L}^{-1}\left\{\frac{1}{s(s+\alpha)}\right\} = \int_0^t e^{-\alpha(t-\tau)}\,d\tau$$

$$= e^{-\alpha t}\int_0^t e^{\alpha\tau}\,d\tau$$

$$= e^{-\alpha t}\,\frac{e^{\alpha t} - 1}{\alpha}$$

$$= \frac{1}{\alpha}\left(1 - e^{-\alpha t}\right)$$

If you look at entry 6 in the table, you will see that this result agrees with that given there. Obviously it could have been obtained from the table without the use of this theorem, but it served very well as a simple illustration.

Note that, even although t is one of the limits in the integral, $e^{-\alpha t}$ can be taken outside of the integral (though this is not essential) as it is to be treated as a constant seeing that the integration is w.r.t. τ.

FRAME 67 continued

The proof of the theorem will now be given. It involves a knowledge of double integrals, so you will not be able to follow it if you have not read Unit 10. In that case, proceed directly to FRAME 71.

FRAME 68

An alternative way of stating (65.2) is

$$F(s)G(s) = \mathcal{L}\left\{\int_0^t f(\tau)g(t - \tau)d\tau\right\}$$

and this is what will actually be proved.

$$\text{Now}\quad G(s) = \mathcal{L}\{g(t)\}$$

$$= \int_0^\infty e^{-st}g(t)dt \quad \text{by definition}$$

which, as the integral is a definite one, can be written as

$$\int_0^\infty e^{-sv}g(v)dv$$

As you know, it is possible to make a substitution in an integral, and, in this case, the substitution $v = t - \tau$ is made where t is regarded as the new variable and at this stage τ is regarded as a constant.

What will $\int_0^\infty e^{-sv}g(v)dv$ become if this substitution is made?

****

68A

$$\int_\tau^\infty e^{-s(t-\tau)}g(t - \tau)dt$$

FRAME 69

$$\text{Next,}\quad F(s) = \mathcal{L}\{f(t)\}$$

$$= \int_0^\infty e^{-st}f(t)dt$$

which can be written as $\int_0^\infty e^{-s\tau}f(\tau)d\tau$

FRAME 69 continued

$$\therefore\ F(s)G(s) = \int_0^\infty e^{-s\tau}f(\tau)d\tau \int_\tau^\infty e^{-s(t-\tau)}g(t - \tau)dt$$

As the second integral is w.r.t. t, $e^{-s\tau}f(\tau)$ can be taken inside it, as this is independent of t,

$$\therefore\ F(s)G(s) = \int_0^\infty d\tau \int_\tau^\infty e^{-s\tau}f(\tau)e^{-s(t-\tau)}g(t - \tau)dt$$

$$= \int_0^\infty d\tau \int_\tau^\infty e^{-st}f(\tau)g(t - \tau)dt$$

This is now regarded as a double integral and consequently the order of integration can be changed. What do you get when you do this?

****

69A

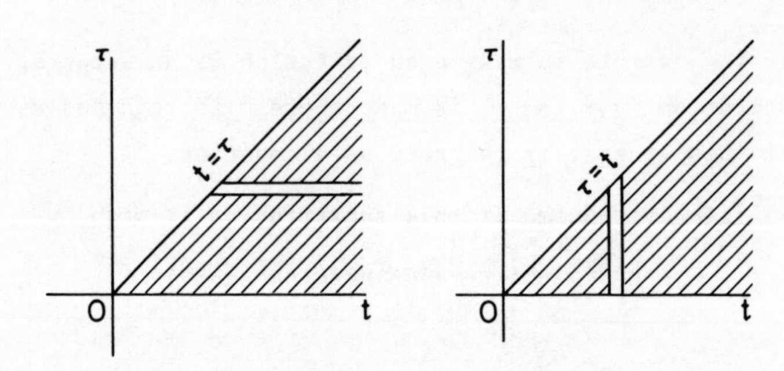

Assuming that axes are taken as shown, the shaded region is that which is covered by the integral.

$$\int_0^\infty dt \int_0^t e^{-st}f(\tau)g(t - \tau)d\tau$$

FRAME 70

As e^{-st} is constant as far as integration w.r.t. τ is concerned this result can be written as

$$\int_0^\infty e^{-st}dt \int_0^t f(\tau)g(t-\tau)d\tau = \int_0^\infty e^{-st}\left\{\int_0^t f(\tau)g(t-\tau)d\tau\right\}dt$$

$$= \mathcal{L}\left\{\int_0^t f(\tau)g(t-\tau)d\tau\right\} \quad \text{by definition}$$

$$\therefore \quad F(s)G(s) = \mathcal{L}\left\{\int_0^t f(\tau)g(t-\tau)d\tau\right\}$$

$$\mathcal{L}^{-1}\{F(s)G(s)\} = \int_0^t f(\tau)g(t-\tau)d\tau$$

Remembering that the R.H.S. of this equation is often denoted by $f*g$, can you see any relationship between $f*g$ and $g*f$?

****

70A

They are equal, as $F(s)G(s) = G(s)F(s)$ *and hence*

$\mathcal{L}^{-1}\{F(s)G(s)\} = \mathcal{L}^{-1}\{G(s)F(s)\}$

FRAME 71

As you know, $\mathcal{L}^{-1}\left\{\dfrac{1}{s^2(s+4)^2}\right\}$ can be obtained using partial fractions.

An alternative method is to make use of the Convolution Theorem.

Putting $F(s) = \dfrac{1}{(s+4)^2}$ and $G(s) = \dfrac{1}{s^2}$

$f(t) = te^{-4t}$ and $g(t) = t$

$$\mathcal{L}^{-1}\{F(s)G(s)\} = \int_0^t \tau e^{-4\tau}(t-\tau)d\tau$$

This can then be evaluated using ordinary integration techniques to give

$$\frac{1}{32}\{(2t+1)e^{-4t} + 2t - 1\}$$

Note that it is better to choose F(s) and G(s) in such a way that g(t) is the

FRAME 71 continued

simpler of the two functions f(t) and g(t).

Now use this method to find $\mathcal{L}^{-1}\left\{\dfrac{1}{s(s+4)^2}\right\}$.

****

71A

$$F(s) = \frac{1}{(s+4)^2} \qquad G(s) = \frac{1}{s}$$

$$f(t) = te^{-4t} \qquad g(t) = u(t)$$

$$\mathcal{L}^{-1}\{F(s)G(s)\} = \int_0^t \tau e^{-4\tau} u(t-\tau)\,d\tau$$

$$= \int_0^t \tau e^{-4\tau}\,d\tau \quad \textit{as} \quad u(t-\tau) = 1 \textit{ over the range of integration}$$

$$= \frac{1}{16}\left\{1 - (4t+1)e^{-4t}\right\}$$

Alternatively g(t) can be taken as 1 and then $g(t-\tau) = 1$.

FRAME 72

If we now take the transfer function that occurred in FRAME 54, i.e. $\dfrac{s}{Ls^2 + Rs + 1/C}$ and use the values of L, R and C that were taken in FRAME 6, i.e. $L = 1$, $R = 250$, $C = 10^{-4}$, this transfer function becomes $\dfrac{s}{s^2 + 250s + 10\,000}$, the inverse of which is $\frac{1}{3}\left(4e^{-200t} - e^{-50t}\right)$. What integral will give the output if the voltage

(i) $E \sin \omega t$, (ii) $E\{u(t) - u(t-1)\}$,

is applied to the circuit?

****

72A

(i) $\int_0^t \frac{1}{3}(4e^{-200\tau} - e^{-50\tau})E \sin \omega(t - \tau)d\tau$

or $\int_0^t E \sin \omega\tau\left\{\frac{4}{3} e^{-200(t-\tau)} - \frac{1}{3} e^{-50(t-\tau)}\right\}d\tau$

(ii) $\int_0^t \frac{1}{3}(4e^{-200\tau} - e^{-50\tau})E\{u(t - \tau) - u(t - \tau - 1)\}d\tau$

or $\int_0^t E\{u(\tau) - u(\tau - 1)\}\{\frac{4}{3} e^{-200(t-\tau)} - \frac{1}{3} e^{-50(t-\tau)}\}d\tau$

FRAME 73

In case (i) the integration is straightforward whichever form is used, but in (ii) care is necessary in the interpretation of the u functions in the integrals. To understand this, let us take a simpler, but similar, example, i.e.

$$\int_0^t e^{-2\tau}\{u(t - \tau) - u(t - \tau - 1)\}d\tau \qquad (73.1)$$

and

$$\int_0^t \{u(\tau) - u(\tau - 1)\}e^{-2(t-\tau)}d\tau \qquad (73.2)$$

Effectively this means that, for illustration purposes, we are putting $E = 1$ and using the function e^{-2t} instead of $\frac{1}{3}(4e^{-200t} - e^{-50t})$.

Taking (73.2) first, as that is the easier to interpret, the diagram shows an $O\tau$ axis, with t at some point on it. Now, for the integral $\int_0^t u(\tau)e^{-2(t-\tau)}d\tau$, $\tau > 0$ over the range of integration and thus $u(\tau) = 1$.

0 t τ

The integral can therefore be written as $\int_0^t e^{-2(t-\tau)}d\tau$, i.e. $e^{-2t}\int_0^t e^{2\tau}d\tau$.

FRAME 73 continued

If $t < 1$, $\tau < 1$ over the range of integration, and so $u(\tau - 1) = 0$.

(73.2) therefore just becomes $\int_0^t u(\tau)e^{-2(t-\tau)}d\tau$, i.e. $e^{-2t}\int_0^t e^{2\tau}d\tau$

$$= e^{-2t}\left[\frac{1}{2}e^{2\tau}\right]_0^t$$

$$= \frac{1}{2}\left(1 - e^{-2t}\right)$$

Can you now say what (73.2) will become if $t > 1$?

****

73A

$\int_0^1 e^{-2(t-\tau)}d\tau$ as the integrand is $e^{-2(t-\tau)}$ when $0 < \tau < 1$ but zero when $\tau > 1$. On integration this gives $\frac{1}{2}e^{-2t}(e^2 - 1)$.

FRAME 74

(73.2) is easier to interpret than (73.1) so we suggest that if you meet an integral which can be put into either form you should use that corresponding to (73.2). If you are interested in seeing how to interpret (73.1) read straight on but otherwise you can proceed directly to FRAME 76.

Turning now to (73.1), $u(t - \tau) = 0$ if $t - \tau < 0$, i.e. $\tau > t$

$u(t - \tau) = 1$ if $t - \tau > 0$, i.e. $\tau < t$

For what values of τ will $u(t - \tau - 1)$ be

(a) 0, (b) 1?

****

74A

(a) $\tau > t - 1$ (b) $\tau < t - 1$ As $\tau > 0$, (b) will not occur if $t < 1$.

FRAME 75

Thus, if $t < 1$, the only part of (73.1) which exists is

$$\int_0^t e^{-2\tau}d\tau = \frac{1}{2}\left(1 - e^{-2t}\right).$$

But when $t > 1$, $u(t - \tau) - u(t - \tau - 1) = 1$ only when $t - 1 < \tau < t$ and so (73.1) then becomes $\int_{t-1}^{t} e^{-2\tau}d\tau = -\frac{1}{2}\left\{e^{-2t} - e^{-2(t-1)}\right\} = \frac{1}{2}e^{-2t}\left(e^2 - 1\right).$

(73.1) and (73.2) both give the same results, as indeed they must seeing that $f*g = g*f$.

FRAME 76

If your course includes circuit analysis you may be familiar with the term impulse response. If the impulse response is used to describe a circuit, instead of the transfer function, then the relation between input and output is

$$\text{Output} = \text{Input} * \text{Impulse Response}$$

i.e. if the input is $f(t)$ and the impulse response is $g(t)$, then

$$\text{Output} = \int_0^t f(\tau)g(t - \tau)d\tau$$

So far, it has been assumed that it is possible to express the input as a function of t whose equation is known. This may not, however, always be the case and the convolution integral provides a method whereby the output can then be found at any time. This is because (65.2) can be considered as giving the output in terms of a time integral. This can be represented as the area under a graph if the integrand is known. In order to use the method, it is necessary for some means of measuring the input to be available. Just one very simple example will be taken to illustrate what is involved.

FRAME 77

For this example, suppose that a voltage, shown in the graph at the top of the next page, is applied for one second, from $t = 0$ to $t = 1$, to a circuit whose transfer function is $\frac{1}{s + 2}$. Let the voltage be denoted by $V(t)$.

FRAME 77 continued

Now the inverse of $\dfrac{1}{s + 2}$ is e^{-2t} and the output at time t is therefore given by

$$\int_0^t V(\tau)e^{-2(t-\tau)}d\tau.$$

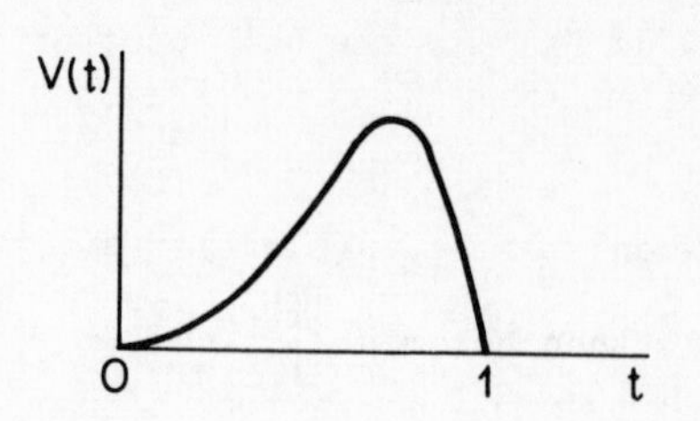

A series of graphs is now drawn, representing the following functions plotted against τ:

(i) $V(\tau)$ (ii) $e^{-2\tau}$ (iii) $e^{-2(-\tau)}$ (iv) $e^{-2(t-\tau)}$

(v) $V(\tau)e^{-2(t-\tau)}$

Start by sketching (i) and (ii).

****

77A

(i)

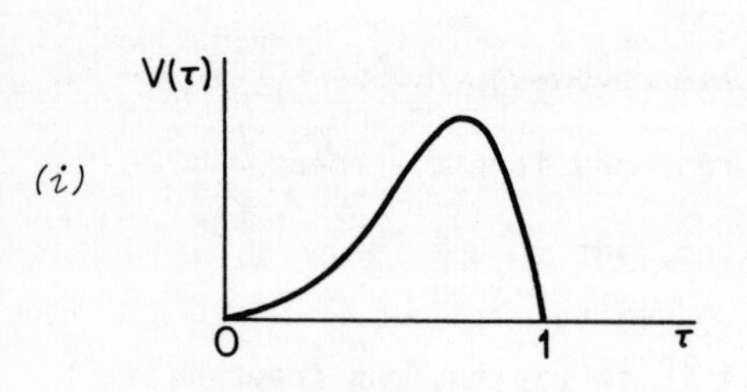

(ii)

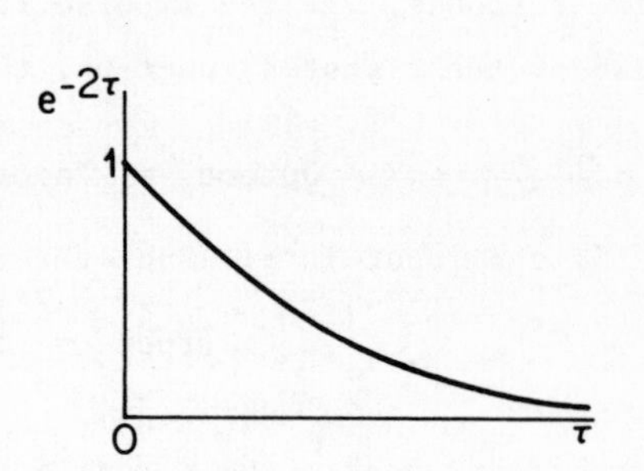

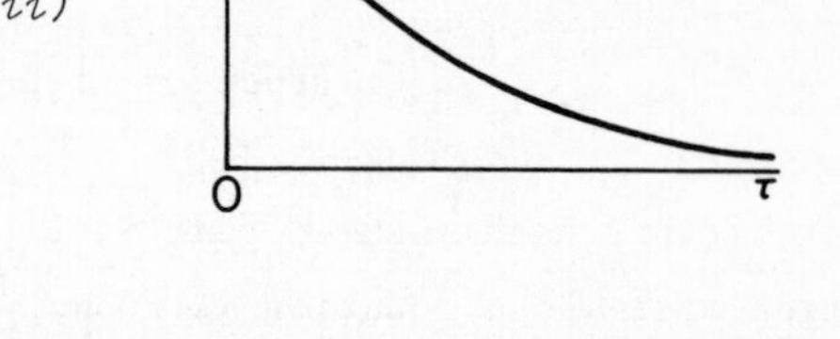

FRAME 78

(iii) is obtained from (ii) by taking its mirror image in the vertical axis. This gives

(iii)

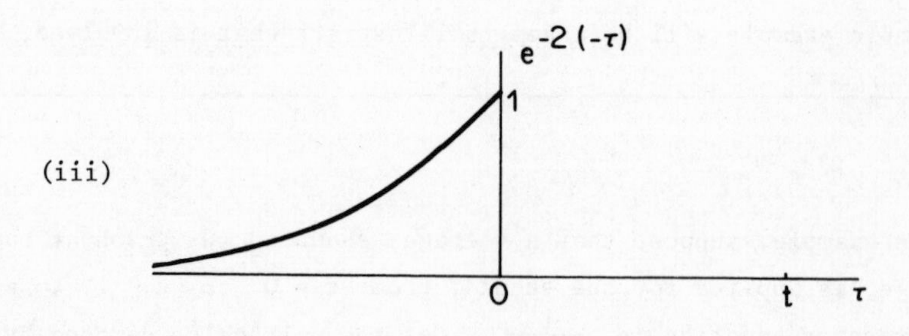

Only that section is drawn which corresponds to that in (ii).

FRAME 78 continued

Now, taking t at the point on the τ scale indicated in this diagram, sketch the graph of $e^{-2(t-\tau)}$, going to the right as far as $\tau = t$.

****

78A

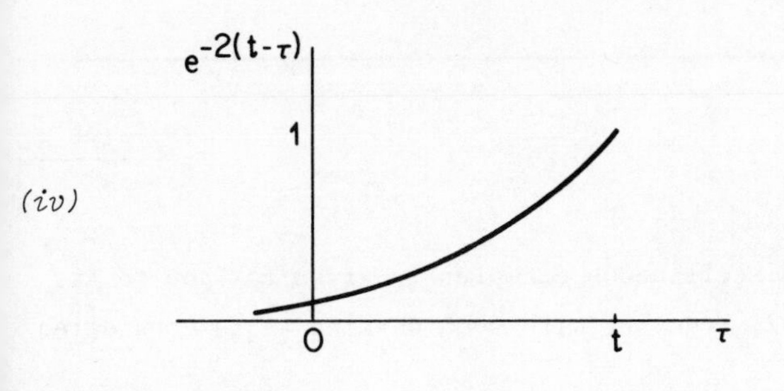

This is the curve in (iii) shifted a distance t to the right.

FRAME 79

(v) is obtained by multiplying together corresponding ordinates of (i) and (iv). This gives

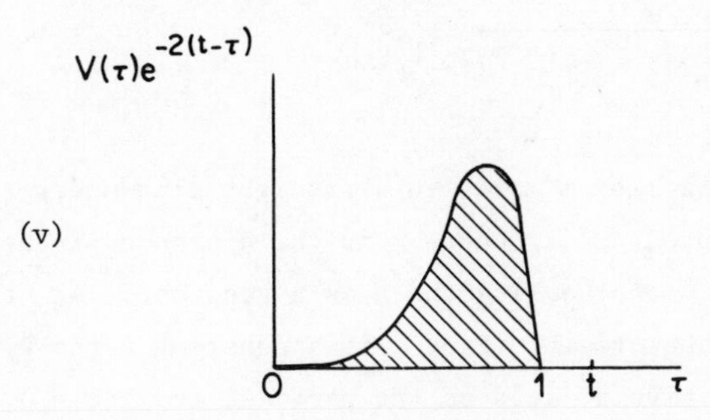

The output at time t is then the shaded area in this diagram.

FRAME 80

During the course of this Unit and that programme of Unit 5 on Laplace transforms, you have encountered various ways of obtaining inverses. In Unit 12 a further method was mentioned, involving Complex Integration, in which f(t) is given by the formula $\frac{1}{2\pi i}\int_C e^{st}F(s)ds$. This is a very powerful method and

FRAME 80 continued

enables the inverses of some of the more complicated functions to be found, such as those which would occur if you had to solve d.e.'s involving the transforms of some of the more awkward periodic functions. However, this is somewhat beyond the scope of this book and so will not be investigated here.

FRAME 81

Miscellaneous Examples

In this frame a collection of miscellaneous examples is given for you to try. Answers are supplied in FRAME 82, together with such working as is considered helpful.

1. Show that the transform of the half-wave rectification of $\sin \omega t$ given over one period by the function

$$g(t) = \begin{cases} \sin \omega t & 0 \leqslant t < \pi/\omega \\ 0 & \pi/\omega < t \leqslant 2\pi/\omega \end{cases}$$

is $\dfrac{\omega}{(s^2 + \omega^2)(1 - e^{-\pi s/\omega})}$. (C.E.I.)

2. A particle of mass m moves in a straight line under the action of a restoring force mn^2x, where x is the distance of the particle from a fixed point O on the line and n is a constant. At $t = 0$, the particle is at rest at the origin. Find x if a constant force P, whose direction is that of positive x, acts on the particle from $t = 0$ to $t = T$.

3. A light beam of length ℓ rests on supports at each end, the supports being at the same level. A load W is placed at $x = \ell/3$, x being measured from one end of the beam. The deflection y measured vertically downwards is given by the equation

$$EI\,\frac{d^2y}{dx^2} = W\left(x - \frac{\ell}{3}\right)u\left(x - \frac{\ell}{3}\right) - \frac{2W}{3}\,x$$

Solve this equation for y in terms of x.

FRAME 81 continued

4. Given that $\mathcal{L}\{\sin \omega t\} = \dfrac{\omega}{s^2 + \omega^2}$, use the Convolution Theorem to deduce that $\mathcal{L}^{-1}\left\{\dfrac{1}{(s^2 + \omega^2)^2}\right\} = \dfrac{1}{2\omega^3}\left(\sin \omega t - \omega t \cos \omega t\right)$.

   Hence determine $\mathcal{L}^{-1}\left\{\dfrac{1}{(s^2 + 2s + 5)^2}\right\}$.

5. The input θ_i and output θ_o of a servomechanism are related by the equation

$$\frac{d^2\theta_o}{dt^2} + 8\frac{d\theta_o}{dt} + 16\theta_o = \theta_i \qquad t \geqslant 0$$

   and initially $\theta_o = 0 = \dfrac{d\theta_o}{dt}$.

   If $\theta_i = f(t)$ where $f(t) = 1 - t \quad 0 < t < 1$

   $f(t) = 0 \quad t > 1$

   show that $\mathcal{L}\{f(t)\} = \dfrac{p - 1}{p^2} + \dfrac{1}{p^2}e^{-p}$ and hence obtain an expression for θ_o in terms of t. (L.U.)

   NOTE: Here $\mathcal{L}\{f(t)\}$ is defined as $\displaystyle\int_0^\infty e^{-pt}f(t)dt$, i.e. p is used instead of s.

   (You will find the inverses worked out in FRAME 71 and 71A useful.)

6. A function f(t) is periodic, of period $2\pi/b$, and is defined as

$$\begin{aligned} f(t) &= 0, & 0 < t < c \\ &= \sin b(t - c), & c < t < c + \pi/b \\ &= 0, & c + \pi/b < t < c + 2\pi/b \end{aligned}$$

   Obtain the Laplace transform of f(t). (L.U.)

7. Find whether the system whose transfer function is $\dfrac{2s + 1}{s^3 + 4s^2 + 7s + 6}$ is stable or otherwise.

FRAME 81 continued

8. A constant voltage E is applied from time $t = 0$ to an L, C, R series circuit which is initially quiescent. In addition, an impulsive voltage 2E is applied at time $t = T$. Find the current i at time t, assuming that $R^2 < 4L/C$.

9. The equation representing the forced oscillations of a damped harmonic oscillator is

$$\frac{d^2x}{dt^2} + 2k\frac{dx}{dt} + n^2x = f(t)$$

subject to $x = x_0$ and $\frac{dx}{dt} = v_0$ at $t = 0$. (x_0, v_0, n and k are constants with $k < n$.) Show that $x(t) = x_1(t) + x_2(t)$ where

$$x_1(t) = x_0e^{-kt}\cos\omega t + \frac{1}{\omega}(kx_0 + v_0)e^{-kt}\sin\omega t$$

$$x_2(t) = \frac{1}{\omega}\int_0^t f(y)e^{-k(t-y)}\sin\omega(t-y)dy$$

and $\qquad \omega^2 = n^2 - k^2$ (C.E.I.)

10. State the Convolution Theorem for Laplace transforms. Hence show that the solution of the system of differential equations

$$\dot{x} - 2\dot{y} = e^{-t^2}$$

$$\ddot{x} - \ddot{y} + y = 0$$

subject to the conditions $x(0) = \dot{x}(0) = \dot{y}(0) = y(0) = 0$ with the usual notation, may be written as

$$x(t) = A - 2C\cos t - 2S\sin t$$

$$y(t) = -C\cos t - S\sin t$$

where $A = \int_0^t e^{-\tau^2}d\tau$, $\quad C = \int_0^t e^{-\tau^2}\cos\tau\, d\tau \quad$ and $\quad S = \int_0^t e^{-\tau^2}\sin\tau\, d$

NOTE: In the solution of this problem, you will not need to evaluate the Laplace transform of e^{-t^2}. (C.E.I.)

FRAME 81 continued

11. The equation $y(t) = t^2 + \int_0^t y(\tau)\sin(t - \tau)d\tau$ is an example of what is called an integral equation. By taking Laplace transforms solve this equation for $y(t)$.

Attempt the next question only if you have read the programme on Complex Integration in Unit 12.

12. The Laplace inverse of the transformed function $F(s)$ is given by the contour integral

$$f(t) = \frac{1}{2\pi i}\oint_C F(z)e^{zt}dz$$

where C is a suitably chosen closed contour which encloses the poles of $F(z)$. Use this result to show that the inverses of

$(3s + 2)/(s^2 + 4)$ and $1/\{s(s + 1)^2\}$

are $(3\cos 2t + \sin 2t)$ and $1 - (t + 1)e^{-t}$ respectively. (C.E.I.)

FRAME 82

Answers to Miscellaneous Examples

1. $$\mathcal{L}\{g(t)\} = \frac{1}{1 - e^{-2\pi s/\omega}}\int_0^{\pi/\omega} e^{-st}\sin \omega t\, dt = \text{answer}$$

2. $$m\ddot{x} = -mn^2x + P\{1 - u(t - T)\}$$

$$\bar{x} = \frac{P}{m}\left\{\frac{1}{s(s^2 + n^2)} - \frac{1}{s(s^2 + n^2)}e^{-Ts}\right\}$$

$$x = \frac{P}{mn^2}\left[1 - \cos nt - \{1 - \cos n(t - T)\}u(t - T)\right]$$

3. $$y = \frac{5}{81}W\ell^2x + \frac{W}{6EI}\left(x - \frac{\ell}{3}\right)^3 u\left(x - \frac{\ell}{3}\right) - \frac{W}{9EI}x^3$$

FRAME 82 continued

4. $$\mathcal{L}^{-1}\left\{\frac{1}{s^2+\omega^2}\cdot\frac{1}{s^2+\omega^2}\right\} = \int_0^t \frac{1}{\omega}\sin\omega\tau\,\frac{1}{\omega}\sin\omega(t-\tau)d\tau$$
$$= \frac{1}{2\omega^2}\int_0^t \{\cos\omega(t-2\tau) - \cos\omega t\}d\tau$$

and integration then gives result.

Write $\frac{1}{(s^2+2s+5)^2}$ as $\frac{1}{\{(s+1)^2+2^2\}^2}$. Then use exponential multiplier theorem in conjunction with result just obtained to arrive at required inverse, which is $\frac{e^{-t}}{16}(\sin 2t - 2t\cos 2t)$.

5. $$\frac{1}{32}\left[3 - 2t - (3+10t)e^{-4t} + \{2t - 3 + (2t-1)e^{-4(t-1)}\}u(t-1)\right]$$

6. $$\mathcal{L}\{f(t)\} = \frac{1}{1-e^{-2\pi s/b}}\int_c^{c+\pi/b} e^{-st}\sin b(t-c)dt$$
$$= \frac{be^{-cs}}{\left(1-e^{-\pi s/b}\right)(s^2+b^2)}$$

7. $s^3 + 4s^2 + 7s + 6 = (s+2)(s^2+2s+3)$

Poles of the transfer function are at -2, $-1 \pm j\sqrt{2}$. System is stable.

8. $$Ls\bar{i} + R\bar{i} + \frac{1}{Cs}\bar{i} = E\left(\frac{1}{s} + 2e^{-sT}\right)$$
$$\bar{i} = \frac{E}{L}\,\frac{1+2se^{-sT}}{(s+R/2L)^2+\omega^2} \quad \text{where } \omega^2 = \frac{1}{CL} - \frac{R^2}{4L^2}$$
$$i = \frac{E}{L}\left[\frac{1}{\omega}e^{-Rt/2L}\sin\omega t + 2e^{-R(t-T)/2L}\{\cos\omega(t-T) - \frac{R}{2L\omega}\sin\omega(t-T)\}u(t-T)\right]$$

9. $$\bar{x} = \frac{x_0(s+k)}{(s+k)^2+\omega^2} + \frac{kx_0+v_0}{(s+k)^2+\omega^2} + \frac{F(s)}{(s+k)^2+\omega^2} \quad \text{where } F(s) = \mathcal{L}\{f(t)\}$$

On inversion, the first two terms on the R.H.S. give $x_1(t)$, and the third term $x_2(t)$.

FRAME 82 continued

10. $s\bar{x} - 2s\bar{y} = \mathcal{L}\{e^{-t^2}\} = F(s)$ say

$$s^2\bar{x} - s^2\bar{y} + \bar{y} = 0$$

$$\bar{y} = -\frac{sF(s)}{s^2+1}$$

$$y = -\int_0^t e^{-\tau^2}\cos(t-\tau)d\tau$$

$$= -C\cos t - S\sin t$$

$$\bar{x} = \bar{y} + \frac{F(s)}{s(s^2+1)}$$

$$x = y + \int_0^t e^{-\tau^2}\{1-\cos(t-\tau)\}d\tau$$

$$= A - 2C\cos t - 2S\sin t$$

11. $\bar{y} = \frac{2}{s^3} + \frac{\bar{y}}{s^2+1}$ using the Convolution Theorem

$$\bar{y} = \frac{2}{s^3} + \frac{2}{s^5}$$

$$y = t^2 + t^4/12$$

12. Poles of $\frac{(3z+2)e^{zt}}{z^2+4}$ occur at $z = \pm 2i$

Res. at 2i is $\frac{1}{2}(3-i)e^{2it}$

Res. at -2i is $\frac{1}{2}(3+i)e^{-2it}$

$$\mathcal{L}^{-1}\left\{\frac{3s+2}{s^2+4}\right\} = \frac{3}{2}\left(e^{2it}+e^{-2it}\right) - \frac{1}{2}i\left(e^{2it}-e^{-2it}\right)$$

$$= 3\cos 2t + \sin 2t$$

Poles of $\frac{e^{zt}}{z(z+1)^2}$ occur at 0, -1

Res. at 0 is 1

Res. at -1 is $\lim_{z\to -1}\frac{d}{dz}\frac{e^{zt}}{z} = -te^{-t} - e^{-t}$

$$\mathcal{L}^{-1}\left\{\frac{1}{s(s+1)^2}\right\} = 1 - (t+1)e^{-t}$$

APPENDIX

TABLE OF LAPLACE TRANSFORMS

This table is adapted from that used in the Part 2 Examination of the Council of Engineering Institutions.

$\mathcal{L}\{f(t)\}$ is defined by $\int_0^\infty f(t)\exp(-st)dt$ and is written as $F(s)$.

Initial conditions are those just prior to $t = 0$.

	Time Function	Laplace Transform
Sum	$af_1(t) + bf_2(t)$	$aF_1(s) + bF_2(s)$
First derivative	$\frac{d}{dt} f(t)$	$sF(s) - f(0-)$
n^{th} derivative	$\frac{d^n}{dt^n} f(t)$	$s^nF(s) - s^{n-1}f(0-) - s^{n-2}f^{(1)}(0-)\ldots-f^{(n-1)}(0-)$
Definite integral	$\int_{0-}^{t} f(t)dt$	$\frac{1}{s} F(s)$
Exponential multiplier	$\exp(-\alpha t)f(t)$	$F(s + \alpha)$
Shift	$f(t - T)$	$\exp(-sT)F(s)$
	$t^n f(t)$	$(-1)^n \frac{d^n}{ds^n} F(s)$
Periodic function, period T	$f(t)$	$\frac{1}{\{1 - \exp(-sT)\}} \int_{0-}^{T} \exp(-st)f(t)dt$
1. (0, t) Unit impulse	$\delta(t)$	1

APPENDIX continued

	Time Function	Laplace Transform
2.  Unit step	$u(t)$	$\frac{1}{s}$
3. 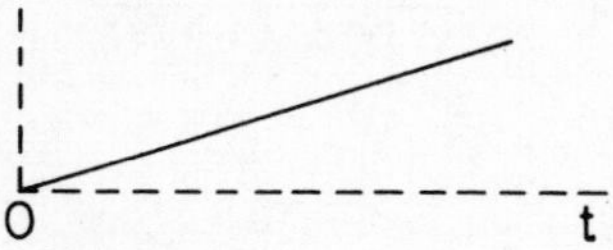 Ramp	t	$\frac{1}{s^2}$
4. 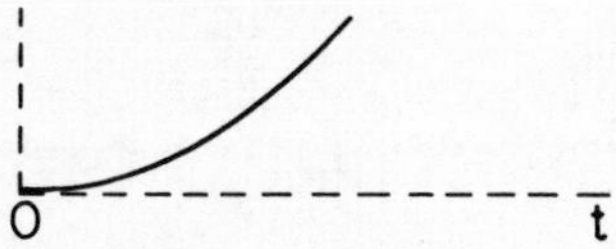 n^{th} order ramp	t^n	$\frac{n!}{s^{n+1}}$
5. 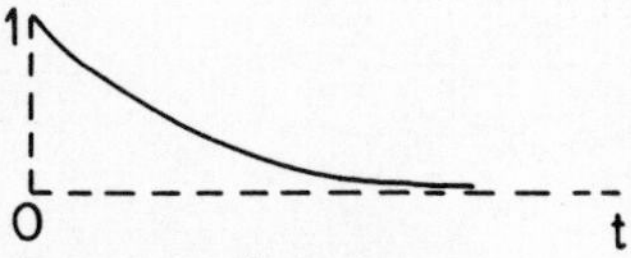 Exponential decay	$\exp(-\alpha t)$	$\frac{1}{s + \alpha}$
6. 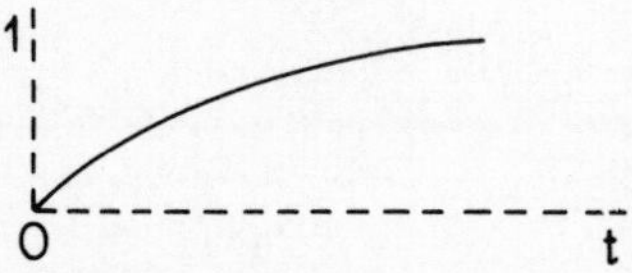	$1 - \exp(-\alpha t)$	$\frac{\alpha}{s(s + \alpha)}$
7. 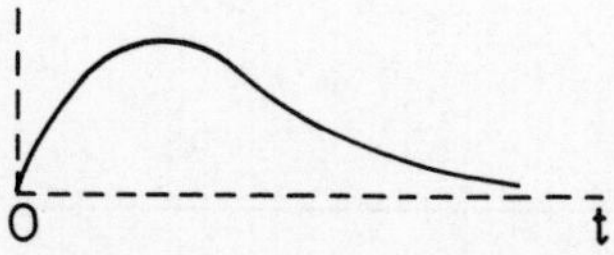	$t \exp(-\alpha t)$	$\frac{1}{(s + \alpha)^2}$

APPENDIX continued

	Time Function	Laplace Transform
8. 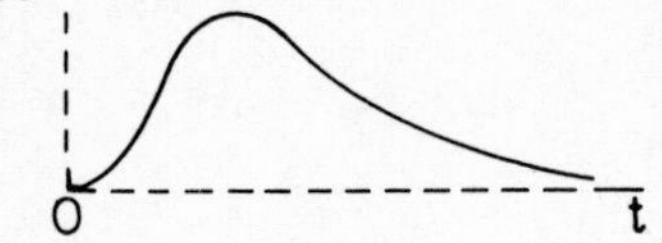 O t	$t^n \exp(-\alpha t)$	$\frac{n!}{(s+\alpha)^{n+1}}$
9.	$\exp(-\alpha t) - \exp(-\beta t)$	$\frac{\beta - \alpha}{(s+\alpha)(s+\beta)}$
10. 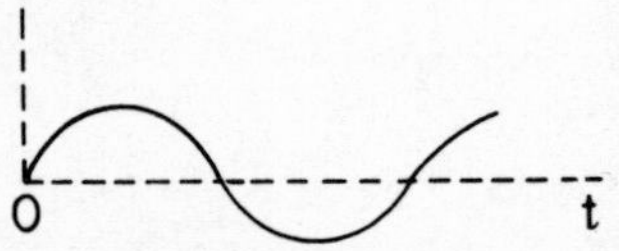 O t Sine wave	$\sin \omega t$	$\frac{\omega}{s^2 + \omega^2}$
11. 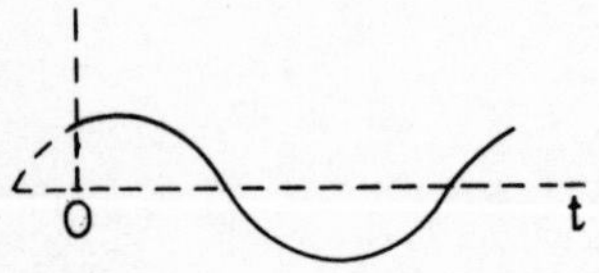 O t	$\sin(\omega t + \phi)$	$\frac{\omega \cos \phi + s \sin \phi}{s^2 + \omega^2}$
12. 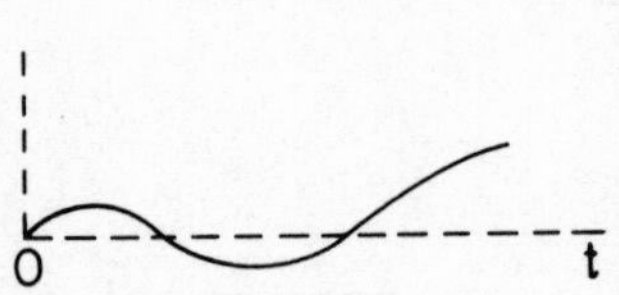 O t	$t \sin \omega t$	$\frac{2\omega s}{(s^2 + \omega^2)^2}$
13. 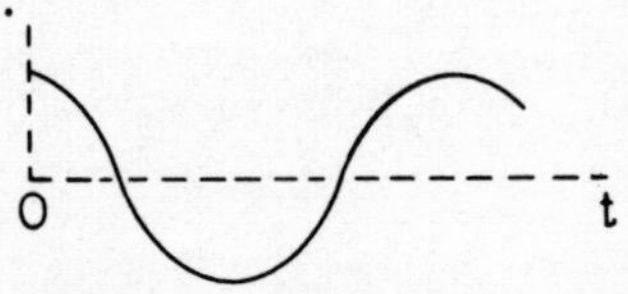 O t Cosine wave	$\cos \omega t$	$\frac{s}{s^2 + \omega^2}$

APPENDIX continued

	Time Function	Laplace Transform
14. 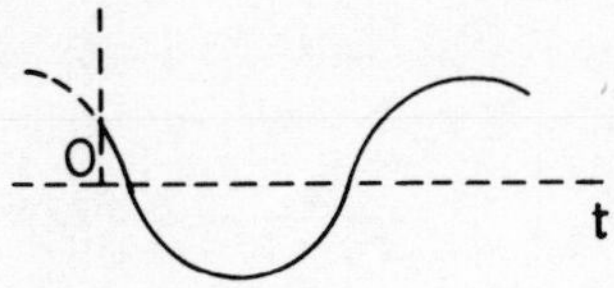	$\cos(\omega t + \phi)$	$\frac{s \cos \phi - \omega \sin \phi}{s^2 + \omega^2}$
15. 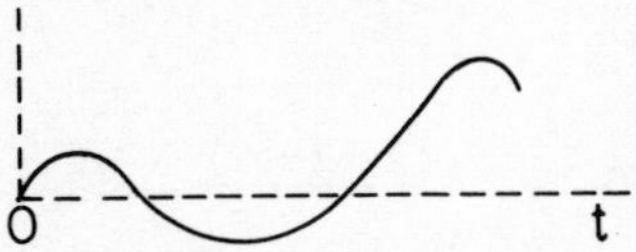	$t \cos \omega t$	$\frac{s^2 - \omega^2}{(s^2 + \omega^2)^2}$
16. 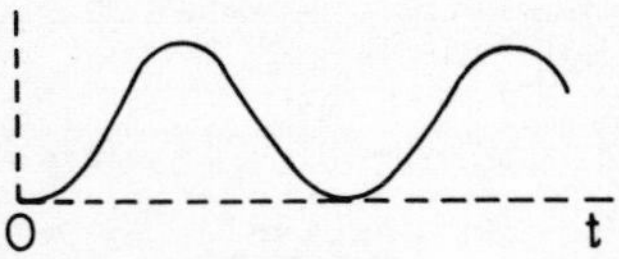	$1 - \cos \omega t$	$\frac{\omega^2}{s(s^2 + \omega^2)}$
17.	$\sin \omega t - \omega t \cos \omega t$	$\frac{2\omega^3}{(s^2 + \omega^2)^2}$
18. 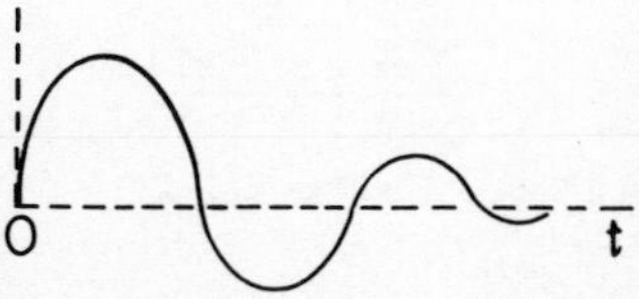	$\exp(-\alpha t) \sin \omega t$	$\frac{\omega}{(s + \alpha)^2 + \omega^2}$
19. 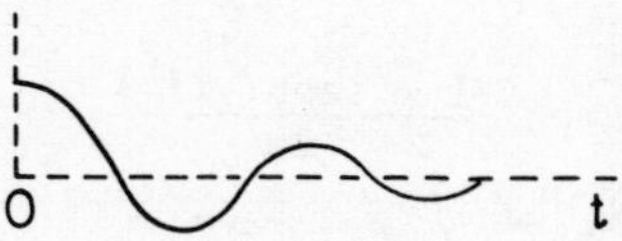	$\exp(-\alpha t) \cos \omega t$	$\frac{(s + \alpha)}{(s + \alpha)^2 + \omega^2}$
20.	$\exp(-\alpha t)(\sin \omega t - \omega t \cos \omega t)$	$\frac{2\omega^3}{\{(s + \alpha)^2 + \omega^2\}^2}$

APPENDIX continued

	Time Function	Laplace Transform
21.	$\sinh \omega t$	$\frac{\omega}{s^2 - \omega^2}$
22.	$\cosh \omega t$	$\frac{s}{s^2 - \omega^2}$
23. 1, O, T, t — Delayed step	$u(t - T)$	$\frac{\exp(-sT)}{s}$
24. O, T, t — Delayed ramp	$(t - T)u(t - T)$	$\frac{\exp(-sT)}{s^2}$
25. 1, O, T, t — Rectangular pulse	$1 - u(t - T)$	$\frac{1 - \exp(-sT)}{s}$
26. 1, O, T/2, T, 2T, t — Rectangular periodic wave, period T		$\frac{1 + \tanh(sT/4)}{2s}$

APPENDIX continued

	Time Function	Laplace Transform
27. 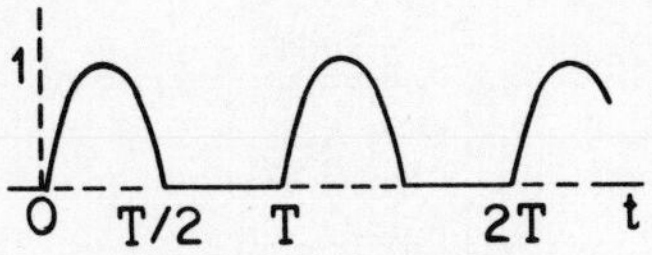 Half wave rectified sine, period $T = \frac{2\pi}{\omega}$		$\dfrac{\omega \exp(\frac{1}{2}sT) \operatorname{cosech}(\frac{1}{2}sT)}{2(s^2 + \omega^2)}$
28. 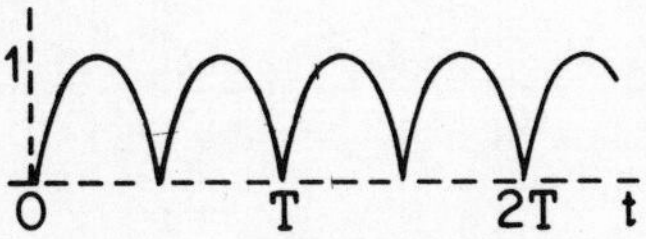 Full wave rectified sine, period $T = \frac{2\pi}{\omega}$		$\dfrac{\omega \coth(\frac{1}{2}sT)}{s^2 + \omega^2}$